ANTIBIOTIC RESISTANCE

TRANSPOSITION AND OTHER MECHANISMS

ANTIBIOTIC RESISTANCE

TRANSPOSITION AND OTHER MECHANISMS

Fourth International Symposium on
ANTIBIOTIC RESISTANCE
Castle of Smolenice, Czechoslovakia, 1979

Editors

S. MITSUHASHI, L. ROSIVAL, V. KRČMÉRY

Springer-Verlag Berlin Heidelberg GmbH

1980

ISBN 978-3-540-10322-6   ISBN 978-3-642-67790-8 (eBook)
DOI 10.1007/978-3-642-67790-8

Originally published by AVUCENUM, Czechoslovak Medical Press, Prague in 1980.

Sofcover reprint of the hardcover 1st edition 1980

FOURTH INTERNATIONAL SYMPOSIUM on ANTIBIOTIC RESISTANCE
ANTIBIOTIC RESISTANCE:
TRANSPOSITION AND OTHER MECHANISMS

Under the Auspices of ISC-ICC
and Professor G. Čatár, M.D., DrSc, Dean of the Medical Faculty,
Commenius University, Bratislava

Castle of Smolenice, Czechoslovakia
4th to 8th June, 1979

Organizers:

SLOVAK MEDICAL SOCIETY (on behalf of Czechoslovak Medical Society
    J. E. Purkyně)
    President: Professor R. T. Niederland, M.D.,
    Vice-President: Professor J. Štefanovič, M.D.
    Secretary General: J. Mariányi, M.D.

CZECHOSLOVAK SOCIETY OF MICROBIOLOGY (Czechoslovak Academy of
    Sciences)
    President: Professor T. Martinec, PhD.
    Vice-President: Professor J. Štefanovič, M.D.
    Secretary: V. Havel, PhD.

SLOVAK SOCIETY OF HYGIENE
    President: Professor J. Kukura, M.D.

Chairman of the Symposium:
    Professor Susumu Mitsuhashi, M.D., Japan

Secretary General of the Symposium:
    Professor L. Rosival, M.D., Czechoslovakia

International Organizing Committee:
    Professor B. Holloway (Australia)
    Professor S. M. Navašin (U.S.S.R.)
    Professor J. Štefanovič (Czechoslovakia)
    Professor P. Kontomichalou (Greece)
    Professor R. Goméz - Lus (Spain)
    Professor W. Goebel (F.R.G.)

Organising Secretariat:
    J. Kalač, PhD.
    V. Krčméry, PhD.
    F. Výmola, M.D.

Publishers of the Proceedings:
    AVICENUM Czechoslovak Medical Press, Prague
    Manager: Dr. M. Knejfl
    SPRINGER VERLAG Berlin-Heidelberg-New York
    Manager: H. Grossmann

PREFACE

Proceedings from the International Symposium on Antibiotic Resistance 1979, the fourth volume in the series of material edited from the Smolenice symposium, is a continuation in the tradition of communicating new scientific results on a currently important and interesting problem. It concerns not only human and veterinary medicine and practice, but also such important fields of contemporary biology as molecular biology and genetics.

Indeed, in both these disciplines, we have recently noted a strong and dynamic development of scientific knowledge directed toward such outgrowths of the study of plasmids as gene manipulations, transposons, and transposition, as well as the evolution of plasmids in general.

Because of a very real global development of resistance to the newest antibiotics, and the active participation of experts from medical science and practice, it was reasonable to separate the medical part of the program from the theoretical one. Consequently, priorities of the fourth symposium were transposons and the transposition of resistance of to antibiotics, with their serious impact on epidemiology, the resistance of *Pseudomonas aeruginosa*, that is becoming a prominent nosocomial pathogen, the ecology and epidemiology of R plasmids, and, last though not least, the computer-assisted surveillance of resistance to antibiotics.

Transposons which code for resistance to several antibiotics have been identified as a major mechanism by which this resistance is presently spread to new species and genera from its natural reservoir of bacteria. This has been found true for resistance to ampicillin, which appeared in *Neisseria* or *Hemophilus*, and additional transposons, e.g., for resistance to trimethoprim and even to gentamicin, have also been identified.

Gentamicin resistance is found to be an increasing problem in dealing with staphylococci, *Enterobacteriaceae*, and *Pseudomonas aeruginosa*. Enzymatic mechanisms for its inactivation are intensively studied so that we can now even address the question of the epidemiology of drug—inactivating enzymes in bacteria.

To guide antibiotic policy, computer-assisted systems for surveillance and monitoring of antibiotic resistance in selected species of "problem bacteria" have been developed in several countries and on an international basis. Comparison of the present and developing situation on this highest level, e.g., between selected hospitals and wards, seems to be highly productive and useful.

When setting forth this volume, many thanks of all participants should be expressed to AVICENUM, Czechoslovak Medical Press, and to SPRINGER-Verlag, for their efforts and help in editing these proceedings.

L. ROSIVAL

CONTENTS

I.D. Genetics and Molecular Biology

Chairwoman: P. Kontomichalou

I.E. Plasmids in the Nature

Chairman: W. Goebel

II. MEDICAL PART

II.A. New Drugs Against Resistant Bacteria

Chairman: S. M. Navashin

II.E. Computer Surveillance of Antibiotic Resistance

Chairman: B. Wiedemann

I. THEORETICAL PART
A) INTRODUCTORY LECTURES

NONCONJUGATIVE DRUG RESISTANCE PLASMIDS

S. MITSUHASHI

Department of Microbiology and Laboratory of Bacterial Resistance
School of Medicine Gunma University
Maebashi, Japan

It is a real pleasure for me to give the opening address at the 4th International Symposium on Antibiotic Resistance at the Castle of Smolenice in Czechoslovakia.

We are very glad once again to have met many representatives from all over the world who are studying plasmids and trying to solve the problems caused by new biohazards resulting from drug resistance plasmids. First of all, I want to express our sincere thanks to the Minister of Health of the Slovak Socialist Republic, the Minister of Agriculture of the Slovak Socialist Republic, and the Slovak Academy of Sciensces, and also to the Organizers of this symposium. Special thanks are due to the Secretary General, Organizing and Program Comittee, and to Publishers of the Proceedings.

TABLE I.

Isolation frequency of R plasmids from resistant strains

Drug	Isolation frequency of R plasmids from strains resistant to (%)
Tc	65.2
Cm	58.0
Sm	64.3
Su	55.8
Km	70.2
Ap	69.3
Gm	79.3

The results are based on surveys of 19,984 clinical isolates. Abbreviation of drugs: tetracycline (Tc), chloramphenicol (Cm), streptomycin (Sm), sulfanilamide (Su), kanamycin (Km), ampicillin (Ap), and gentamicin (Gm).

The introduction of antibacterial agents has contributed greatly to improved treatment of infectious diseases and to the progress of practical medicine. Only half a century after the real start of chemotherapy, however, we are now faced with the problems of bacterial resistance, and the prevalence of resistant bacteria has caused many problems in medicine, stock farming and fish breeding.

Transmissible drug resistance(R) plasmids were discovered in 1960, and since then the extensive studies of R plasmids have disclosed the epidemiology, genetics, and molecular biology of R plasmids. R plasmids are conjugally transferable and have a wide host range so that resistance spreads quickly from bacteria to bacteria, compounding the infectious spread by multiplying resistant organisms. It was further found that the drug resistance determinants on plasmids were easily transferred from plasmid to bacterial chromosomes, bacteriophages and to other plasmids, or *vice versa*, the transferable unit is called "transposon" (Cohen, 1976, Mitsuhashi, 1977, Mitsuhashi et al., 1977). Easy transferability of resistance determinants has recently been explained by the presence of an insertion sequence(IS) (Starlinger and Saedler, 1976).

TABLE II.

Isolation frequency of R plasmids from strains resistant to Tc, Cm, Sm, and Su

Pattern of [a] resistance	Isolation frequency of R plasmids from strains resistant to (%)
Quadruple	67.9
Triple	55.2
Double	25.1
Single	4.3

The results are based on surveys of 14,530 strains.
[a] Resistance to Tc, Cm, Sm, and Su.

TABLE III.

Resistance patterns of R plasmids carrying Km or Ap resistance

R plasmids carrying	Resistance patterns	No. of R plasmids (%)
Ap resistance	Tc.Cm.Sm.Su.Km.Ap	87 (11.9)
	Quintuple resistance	252 (34.6)
	Quadruple resistance	219 (30.0)
	Triple resistance	75 (10.3)
	Double resistance	50 (6.9)
	Single Ap resistance	46 (6.3)
Total		729
Km resistance	Tc.Cm.Sm.Su.Ap.Km	87 (18.5)
	Quintuple resistance	78 (16.6)
	Quadruple resistance	91 (19.3)
	Triple resistance	137 (29.1)
	Double resistance	59 (12.5)
	Single Km resistance	19 (4.0)
Total		471

The R plasmids were isolated from 15,884 strains of gram-negative bacteria.

Surveys of clinical isolates disclosed that R plasmids were demonstrated at high frequencies, ranging from 50 to 80 percent, from drug-resistant bacteria (Table I.). Among drug-resistant strains to Tc, Cm, Sm and Su, R plasmids were most frequently seen in quadruply resistant strains, followed by triply, doubly and singly resistant strains, in that order (Table II.).

Resistance patterns of R plasmids carrying kanamycin (Km) and ampicillin (Ap) resistance are shown in Table III. R plasmids carrying Km- or Ap-resistance were isolated at a high frequency of 70% from Km- or Ap-resistant strains of gram-negative bacteria. Of those carrying Ap resistance, R plasmids with quintuple resistance were isolated most frequently, followed by those encoding quadruple, sexiduple, triple, double,

4

and single resistance, in that order, this distribution indicates the prevalence of Ap plasmids encoding multiple resistance. Of those with Km resistance, R plasmids carrying triple resistance were the most common, but the various patterns of multiple resistance were seen with very similar frequencies. These results are accountable for by the following factors:

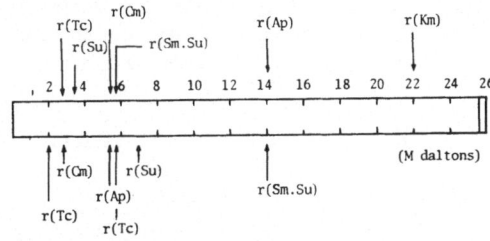

Fig. 1. The molecular size of nonconjugative resistance(r) plasmid DNAs.

(1) selection of multiple resistance plasmids by various drugs used,
(2) translocation of new resistance determinant on R plasmids, and
(3) spread of R plasmids by conjugal transmission.

Studies of multiple resistance in *Shigella* strains opened a way to the discovery of R plasmids (see Rev. Mitsuhashi, 1977, 1979). Strains of *S. aureus* are frequently isolated from clinical sources and play an important role in pathological lesions. This laboratory examined the reasons for the prevalence of drug-resistant strains and for the acquisition

TABLE IV.

Isolation of nontransferable (r) plasmids
from bacteria carrying nontransmissible resistance

Bacteria	Isolation frequency of r plasmids (%)
S. aureus	85.0
S. pyogenes	95.0
H. influenzae	60.0
E. coli	96.0
Shigella	98.0
Salmonella	95.0
P. mirabilis	67.0
S. marcescens	63.0

The presence of r plasmids was examined using singly or doubly resistant strains.

of multiple resistance in staphylococci. The strains triple-and quadruple-resistant to tetracycline (Tc), streptomycin (Tc), penicillin (Pc) and sulfanilamide (Su) accounted for more than half of the strains showing multiple resistance. The isolation frequency of strains resistant to macrolide antibiotics (Mac), chloramphenicol (Cm), synthetic penicillins (Dmp and Mci), and kanamycin (Km) was highest among the triple- and quadruple-resistant strains. These results indicate that staphylococci easily develop multiple resistance when new drugs are introduced. Soon after the discovery of conjugative R plasmids in gram-negative bacteria, we examined the presence of plasmids in staphylococci. In spite of the nontransmissibility of drug resistance in staphylococci, we found in 1963 that cross-resistance to Mac antibiotics was irreversibly eliminated by treatment with acriflavine (Mitsuhashi, Morimura, Kono, and Oshima, 1963). It was subsequently found that Pc resistance in staphylococcal strains was eliminated by treatment with acriflavine. Loss of the capacity to produce penicillinase explained the reversion from Pc resistance to Pc sensitivity (Hashimoto, Kono, and Mitsuhashi, 1964). These results indicate that drug resistance determinants in staphylococci are located on nonconjugative resistance (r) plasmids. We now have various methods to confirm

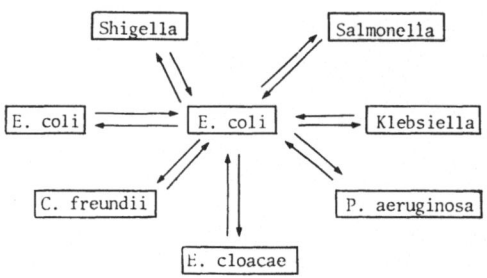

Fig. 2. Host range in the transformation of r plasmid DNAs.

the presence of nonconjugative r plasmids in bacteria: (1) artificial and spontaneous elimination of drug resistance, (2) demonstration of a satellite band of plasmid DNA by the density gradient centrifugation, (3) transformation of drug resistance by means of isolated plasmid DNA to a *rec⁻* recipient, and (4) transduction of drug ressistance to a *rec⁻* recipient. Using these techniques, we have isolated many r plasmids from staphylococcal strains, about 85% of the strains examined were found to carry r plasmids, indicating that drug resistance determinants in staphylococci are located primarily on these plasmids (Mitsuhashi et al., 1973, Mitsuhashi et al. 1976).

Further studies of drug resistance have disclosed that most single or double resistance in gram-positive or gram-negative bacteria is due to the presence of r plasmids that are present as multiple copies. Epidemiology of the r plasmid distribution in singly or double resistant strains is shown in Table IV. The r plasmids were demonstrated at a high frequency from singly and double resistant strains carrying nontransferable resistance. Drug resistance plasmids with molecular weight of less than 20×10^6 daltons are nonconjugative and lack the transfer region that is on the R plasmid (Fig. 1). The physical properties of r plasmid DNAs isolated from gram-positive and gram-negative bacteria are shown in Table V. Contour length of most r plasmids ranged from 1 to 3 μm. It is characteristic that r plasmids are present as multiple copies in a cell and most of them encode single resistance, except for r (Mac. Pc) and r (Sm.Su). Therefore, multiple resistance in staphylococci is due to the presence of various r plasmids in a cell, which are present as multiple copies.

The r plasmid DNAs in gram-negative bacteria were found to be easily transmitted to various species of bacteria through transformation, resulting in the stable existence and the expression of resistance in a new host cell. It is interesting to note that the r plasmid DNAs from *P. aeruginosa* strains are transmitted through transformation to *E. coli* strains, although most R plasmids from *P. aeruginosa* are transmissible only between *P. aeruginosa* strains and not to *E. coli*. One of the representative results is shown in Fig. 2 (Inoue and Mitsuhashi, unpublished observation).

6

It is well known that almost all of *S. aureus* strains are lysogenic and plasmid resistance in staphylococci is transduced with phage lysates obtained from multiply resistant strains. In the transduction of Tc resistance by means of phage lysates obtained from *S. aureus* strains, every strain was found to be competent when tested in various combinations of donor and recipient. The percentage of competent recipient strains in the various phage groups was: group I including 80/81 (89%), group II (22%), group III (0%),

TABLE V.

*Summary of the physical properties of nonconjugative (r) plasmid
DNAs isolated from gram-negative and gramm-positive bacteria*

DNA sources	Resistance pattern	Contour length (μm)	No. of copies per chromosome
S. aureus	Tc	1.4	22–25
	Cm	1.4	5–7
S. pyogenes	Tc	1.0	NT
H. influenzae	Ap	12.5	NT
	Su	1.8	22–29
E. coli	Tc	3.1	3–6
	Sm.Su	2.5	15–19
Shigella	Tc	2.8	NT
	Su	1.8	22–23
	Sm.Su	2.5	17–19
Salmonella	Su	1.9	14–23
	Sm.Su	2.5	17–19
Proteus	Su	1.7	16–18
	Sm.Su	2.5	16–18
P. aeruginosa	Sm.Su	2.4	NT
A. hydrophilla	Tc. Su. Ap	1.0	NT
		2.8	NT

NT = not tested.

TABLE VI.

Mobilization of r plasmid with conjugative plasmid

Donor carrying	Selective drug	Resistance patterns (%) of transconjugants
RP4 (Km.Tc.CBPC) + rMS21 (Sm.Su)	Sm	rMS21 (3.3) rMS21+RP4 (96.7)
	Km	RP4 (25.1) RP4 + rMS21 (74.9)
R9-5 (Cm) + rMS76 (Ap)	Ap	rMS76 (7.0) rMS76 + R9-5 (93.0)
	Cm	R-5 (27.3) R9-5 + rMS76 (72.7)

Donor, *E. coli* X2207 Nal^r; recipient, *E. coli* X1037 Rif^r. After 3 hr of incubation at 37C, the mixed culture of donor and recipient was spread on selective plates for either r plasmid or R plasmid marker.

and group IV (40%). The results coincide with the distributions of multiple resistance in staphylococcal phage groups (Mitsuhashi, Oshima, Kawaharada, and Hashimoto, 1965). Similarly, most *P. aeruginosa* strains are also lysogenic, and the r plasmids in these strains are transduced with phage lysates obtained from multiply resistant strains of *P. aeruginosa*.

The r plasmids were found to be transmitted by means of mobilization with conjugative plasmid. One of the results is shown in Table VI. indicating that more than 96 percent of the transconjugants carried both R and r plasmids when selected for r plasmid marker. Even when selected for R plasmid marker, about 70 percent of the transconjugants carried both R and r plasmid. In the epidemiologic studies of R plasmids in naturally

TABLE VII.

Types of resistance mediated by plasmid

Bacteria	Resistance mediated by
Gram-positive	r $r_1 + r_2 + r_3 + ...$
Gram-negative	R $R_1 + R_2$ R + r $R + r_1 + r_2 + r_3...$

occurring resistant strains, we must be careful of the difference in resistance patterns of transconjugants after selection with various drugs, resulting from the mobilization of r plasmid with conjugative R plasmid.

According to the surveys of drug resistance in more than fifty thousands strains of bacteria isolated from clinical materials, livestock and cultured fish, we can conclude that most resistance in these strains is due to the presence of drug resistance plasmids. Gram-positive bacteria usually carry r plasmid and multiple resistance is due to the concomitant presence of various r plasmids in a cell. There are many types of resistance mediated by plasmids in gram-negative bacteria due to the presence of R, (R_1+R_2), $(R+r)$, $(R+r_1+r_2+...)$, etc. (Table VII.). The studies of r plasmids have disclosed that the situations of resistance mediated by plasmid are rather complicated in naturally occurring resistant strains.

REFERENCES

COHEN, S. N. (1976): Transposable genetic elements and plasmid evolution. Nature 263, 731–738.
HASHIMOTO, H., M. KONO and S. MITUHASHI (1964): Elimination of penicillin resistance of *Staphylococcus aureus* by treatment with acriflavine. J. Bacteriol. 88, 261–262.
MITUHASHI, S., M. MORIMURA, M. KONO and H. OSHIMA (1963): Elimination of drug resistance of *Staphylococcus aureus* by treatment with acriflavine. J. Bacteriol. 86, 162–164.
MITSUHASHI, S., H. OSHIMA, U. KAWAHARADA and H. HASHIMOTO (1965): Drug resistance of staphylococci. 1. Transduction of tetracycline resistance with phage lysates obtained from multiply resistant staphylococci. J. Bacteriol. 89, 967–976.
MITSUHASHI, S., M. INOUE, H. OSHIMA, H. KAWABE and T. OKUBO (1973): Genetics and biochemical studies of drug resistance in staphylococci. *In* Staphylococci and Staphylococcal Infections (J. Jelijaszewicz, edt), pp. 144–165, Karger, Basel.
MITSUHASHI, S., M. INOUE, H. OSHIMA, T. OKUBO and T. SAITO (1976): Epidemiologic and genetic

studies of drug resistance in *Staphylococcus aureus*. *In* Staphylococci and Staphylococcal Infections (J. Jelijaszewicz, edt), pp. 255–274, Gustav Fischer Verlag, Stuttgart, New York.

MITSUHASHI, S. (1977): Translocable drug resistance determinants. *In* R factor (S. Mitsuhashi, edt), pp. 73–88, University of Tokyo Press, Tokyo.

MITSUHASHI, S., H. HASHIMOTO, S. IYOBE and M. INOUE (1977): Formation of conjugative drug resistance (R) plasmids. *In* DNA insertion elements, plasmids, and episomes (A. I. Bukhari, J. A. Shapiro and S. L. Adhya, edt), pp. 139–146, Cold Spring Harbor Laboratory, New York.

MITSUHASHI, S. (1979): Drug Resistance Plasmids. Molecular and Cellular Biochemistry, in press.

STARLINGER, P. and H. SAEDLER (1976): IS-elements in microorganisms. Curr. Top. Microbiol. Immunol. 75, 111—152.

S. M., Dept. of Microbiology, School of Medicine,
Gunma Univ. Maebashi, Japan

GENETIC STUDIES OF NONCONJUGATIVE SULFANILAMIDE RESISTANCE PLASMIDS IN GRAM-NEGATIVE BACTERIA

M. INOUE, K. INOUE, T. NAGATE AND S. MITSUHASHI

Laboratory of Microbial Resistance and Department of Microbiology
School of Medicine, Gunma University, Maebahsi, Japan.

INTRODUCTION

A large number of papers have been published in recent years about the conjugal transferable plasmids in the field of epidemiology, biochemical mechanisms, or molecular biology (Schlessinger, 1978). Epidemiological surveys have disclosed that the conjugal transferable (R) plasmids isolated most frequently were about 68% quadruply resistant strains to Tc, Cm, Sm and Su, and followed by triply, doubly and singly resistant strains in that order (Mitsuhashi, 1977). It should be noted that the isolation frequency of R plasmid from singly Su-resistant strains was less than 5%. However, the studies of nonconjugative resistance(r) plasmids lagged far behind those of conjugative R plasmids in the epidemiology, genetics and biochemistry because of the technical difficulty of conjugal transmissibility. Along with the technical advancement, we could demonstrate the nonconjugative plasmids were isolated from strains carrying nontransferable resistance at much higher frequencies than that of R plasmids, and the genetics and biochemical studies of r plasmids is now in rapid progress as were those of R plasmids. The frequency of conjugative Su plasmids is rather low in strains singly resistant to Su. Accordingly, we focused attention on the anlysis of nonconjugative resistance to Su in gram-negative bacteria from the standpoint to study the genetic of plasmid.

RESULTS AND DISCUSSION

According to the results described in our previous paper (Nagate, Inoue, Inoue and Mitsuhashi, 1978), we concluded that there were two types of biochemical mechanism of Su resistance: Type 1 Su resistance due to the formation of Su-resistant dihydropteroate synthetase (DHPS) and type 2 Su resistance due to decrease in the permeability of the drug into cells. Most of the conjugative R (Tc.Cm.Sm.Su), R (Cm.Sm.Su) and R(Tc.Sm.Su) and some of the R(Sm.Su) plasmids mediated type 2 Su resistance and they are most often demonstrated from clinical isolates. It is interesting to note that most of the nonconjugative (Sm.Su) and conjugative R(Su) and some of R(Sm.Su) plasmids mediate the formation of Su-resistant DHPS, i.e., the type 1 Su resistance.

We investigated the biochemical mechanisms of nonconjugative Su resistance by examining the rate of incorporation of ^{14}C-labeled paraaminobenzoic acid (PABA) into dihydropteroate synthetase fraction with extracts from clinical isolates encoding single Su resistance. It was found that there are two types of DHPS activities, i.e., a normal Su-sensitive enzyme and an altered enzyme. The inhibitory concentration of Su against these enzymes is 1000 times higher than that of Su sensitive enzyme. To confirm the presence of Su-resistant DHPS activity in addition to the normal Su-sensitive

11

TABLE I.

Properties of nonconjugative Su resistance plasmids

Plasmid	MIC of Su[a] (µg/ml)	50% Inhibitory conc. of Su DHPS (10^{-3}M)	Contour length (µm)	Restriction[b] site with EcoR1	Molecular weight from agarose gels	Origin
pMS31	6400	8.5	1.88	1	3.78	E. coli
pMS33	3200	10.2	1.87	1	3.74	E. coli
pMS36	6400	4.8	1.91	1	3.79	Shigella
pMS57	6400	6.9	1.80, 3.27	1	3.70, 7.42	E. coli
pMS67	6400	6.0	1.80	1	3.75	Salmonella
pMS71	6400	7.5	2.01	1	3.50	Proteus
pMS75	1600	5×10^{-3}	2.07	1	3.99	E. coli

The determination of specific DHPS activities and of 50% inhibitory concentration of Su against DHPS was carried out by the method described in the previous paper (Nagate et al., 1978). [a]MIC values of Su against Su-resistant transformants of *E. coli* K12. [b]Purified plasmid DNA was digested with a restriction endonuclease and subjected to electrophoresis in 0.8% agarose gels.

enzyme, we examined the heat sensitivity of DHPS activities in singly Su-and (Tc.Su)-resistant strains, i.e., a Su-resistant and heat-resistant DHPS. Accordingly, all of the singly Su resistance strains were found to be due to the formation of Su-resistant DHPS, i.e., the type 1 Su resistance.

Next we randomly selected from our stock cultures 24 *E. coli*, 4 *Shigella*, 3 *Salmonella*, and 5 *Proteus* strains possessing nonconjugative singly Su-and (Tc.Su)-resistance and examined the presence of plasmids encoding Su resistance by transformation analysis. All of the nonconjugative Su resistance determinants were transformed to both *E. coli* K12 and *E. coli* K12 *recA* strains at the same frequency of 10^4 to 10^5 per µg of covalently closed circular (CCC) DNA. Then we examined the plasmid DNA from the Su-resistant transformants of *E. coli* K12 by dye-buoyant density gradient centrifugation. DNA from all of the strains contained visible amounts of CCC DNA from 20 ml cultured cells under the ultraviolet (UV) lamp. The CCC DNA was prepared for electron micros-copy, and electron micrographs of open circular molecules were enlarged, traced and measured. Su plasmids were measured by electronmicroscopy and their size ranged from 1.8 to 2.0 µm.

Su resistance was transformed to *E. coli* K12 r+m+ strain with Su plasmid DNA isolated from *E. coli* C carrying nonconjugative Su plasmid. It was found that no differen-ce was found in transformation frequency from *E. coli* C to *E. coli* K12 r+m+ compared with that between r−m− strains. This result is shown in Table I. The results indicated that Su plasmid from gram-negative bacteria, such as *E. coli*, *Shigella*, *Salmonella*, and *Proteus* are not restricted in *E. coli* as a host.

Incompatibility is most often manifested by segregation of one of the plasmids from the cell or by recombination between two plasmids to form a single unit of replication. Plasmids which are mutually incompatible with each other form a compatibility group, and plasmids of the same compatibility group have a significant proportion of poly-nucleotide sequences in common (Crosa, Brenner and Falkow, 1973) but are only margi-nally related to plasmids of other compatibility groups.

It is interesting to determine whether nonconjugative Su plasmids belong to any of the known compatibility groups or are a representative of a unique plasmid group. From our results, Su plasmids did not belong to any of the compatibility groups among standard compatibility plasmids (Datta, 1973). Then we isolated a series of Su-sensitive

TABLE II.

Incompatibility test of nonconjugative Su resistance plasmid

Plasmid in donor	Plasmid in recipient strain						
	pMS21(Sm.Su)	pMS76(Ap)	pMS79(Ap)	pMS80(Ap)	pMS81(Ap)	pCR1	pSC204
pMS31	N	c	i	i	i	c	c
pMS33	N	c	i		i	c	c
pMS36	N	c	i	i	i	c	c
pMS57	N	c	i	i	i	c	c
pMS71	N	c	i	i	i	c	c
pMS67	N	c	i	i	i	c	c
pMS75	N	c	i	i	i	c	c
pMS76	c	N	N	N	N	c	N
pMS79	c	N	N	N	N	c	N
pMS21	N	c	c	c	c	c	c
pCR1	c	c	c	c	c	N	c
pSC204	c	c	c	c	c	c	N

Incompatibility experiments were carried out by the method described by Datta (1974). pMS79(Ap), pMS80(Ap) and pMS81(Ap) plasmids were obtained by the loss of Su resistance from the mutant plasmids, respectively, which were obtained by insertion of TnA to pMS31(Su), pMS71(Su) and pMS75(Su) plasmids. N = not tested, c = compatible, i = incompatible.

but Ap-resistant plasmid from several Su resistance plasmids that had contained a TnA (Ampicillin) from R1 plasmid (Kopecko and Cohen, 1975). The TnA was stably integrated into Su resistance plasmids and Su-resistant revertants were not obtained at a frequency of 10^{-8} or less than 10^{-8}. The spontaneous loss of either Su or Ap resistance from *E. coli* carrying any of pMS(Su) plasmids or any of Su::TnA plasmids was not observed among 1000 colonies examined. These Su::TnA plasmids were named pMS79, pMS80, pMS81 and pMS82. All pMS(Su) plasmids were used to test incompatibility between Su::TnA plasmids or between nonconjugative plasmids carrying resistance other than Ap or Su, i.e., pMS21, pMS76, PCR1 and pSC204. As shown in TableII., all of the pMS(Su) plasmids tested can efficiently replace the TnA inserted plasmids, i.e., pMS79, pMS80, pMS81 and pMS82, indicating that they belong to the same compatibility group. But all pMS(Su) plasmids coexist stably with pMS21, pMS76, pCR1, and pSC204, indicating that they are compatible.

Restriction endonuclease digestion patterns have revealed striking similarities between pMS(Su) plasmids isolated from *E. coli*, *Shigella*, *Salmonella* but differences between those and pMS(Su) plasmids from *Proteus*, which correspond to the homology between the plasmid DNAs. We used pMS201–68 DNA as a molecular weight marker. The two restriction endonucleases, SmaI and Hind III, have no cleavage sites on all the of Su resistance plasmids. When Su resistance plasmid DNAs were subjected to complete digestion with EcoR1 and BamH1, specific fragments were visualized with only one cleavage site on all of the pMS(Su) plasmids, converting the supercoiled molecules to the linear form. We determined the molecular weight of Su plasmids by using the relative mobilities in agarose gels of the linear of the plasmid (produced by EcoR1) together with pMS201–68. The value of plasmids obtained from *E. coli*, *Shigella*, *Salmonella*, was 3.8 Mdaltons and that of plasmids from *Proteus* was 3.55 Mdaltons. We also detected one cleavage site on all of the Su plasmid DNAs after digestion with BamH1. On the other hand, Su plasmid mutants inserted with TnA were not digested with

13

EcoR1 and BamH1. Therefore, the restriction site of these enzymes is predicted to be located between the resistance genes of Su.

An analysis of Su resistance plasmid was undertaken to determine whether a plasmid function affected its maintenance in the mutants for DNA replication of *E. coli* host chromosome. Su resistance plasmids, pMS(Su), isolated from *E. coli*, *Shigella* and *Salmonella* were found to be unstable in *E. coli* K12 *polA* $_{214}$ host but a plasmid from *Protesu* stably existed after incubation at 42°C. Su plasmids were transformed to *E. coli* K12 *polA*$_{214}$ and Su-resistant transformants were selected at 30, 37 or 42°C. Transformation frequency of Su resistance plasmids isolated from *E. coli*, *Shigella* and *Salmo-*

TABLE III.

Transformation of Su resistance plasmids to polA strain

Plasmid	polA$^+$ recipient selected at (C)		polA$^-$ recipient selected at (C)	
	37	42	37	42
pMS31	1.0	0.51	1.0	<0.002
pMS33	1.0	0.31	1.0	<0.002
pMS36	1.0	1.0	1.0	<0.002
pMS57	1.0	1.1	1.0	<0.008
pMS71	1.0	0.52	1.0	0.97
pMS72	1.0	0.94	1.0	0.73
pMS75	1.0	1.03	1.0	<0.002
pCR1	1.0	0.79	1.0	<0.002

After the transformation, transformants were selected at 30 or 42 C. Number indicates the relative ratio of transformants.

nella was enormously decreased when selected at 42°C compared with the selection at 30 and 37°C. On the other hand, Su plasmids from *Proteus mirabilis* were transformed almost at the same frequency when selected at 30, 37 and 42°C. pCR1 plasmid whith one of derivatives from ColE1 plasmid also showed the decreased frequency of transformation when selected at 42°C compared with the selection at 30 and 37°C in *E. coli* K12 *polA*$_{214}$. This result indicated that: (1) Su plasmids isolated from *E. coli*, *Shigella* and *Salmonella* require DNA polymerase 1 for their plasmid replication, (2) Su plasmids isolated from *Proteus* do not require DNA polymerase 1 for their replication. ColE1 plasmid was reported to be the best known example of the requirement of DNA polymerase 1 products for its plasmid replication (Kingsbury and Helinski, 1973) (Table III.).

We tested the nature of the replication of Su plasmids in the presence of chloramphenicol. It was really interesting to know that plasmid which require DNA polymerase 1 for its replication can amplify in the presence of chloramphenicol (Clewell, 1972. Crosa, Luttropp and Falkow, 1975. Veltkamp, Barendsen and Nijkamp, 1974). When protein synthesis is inhibited, cells can complete the current round of chromosome replication but cannot initiate a new round of replication. The rate of Su resistance plasmid replication under this condition can be examined by the following increase in the ^{14}C/^{3}H ratio of plasmid DNA after a ^{3}H-tymidine-labelled culture is shifted to a medium containing both ^{14}C-thymine and chloramphenicol. DNA synthesis of Su plasmids from *E. coli*, *Shigella* and *Salmonella* continues in a linear fashion for at least several hours after addition of chloramphenicol.

To follow plasmid replication for longer periods of time, chloramphenicol was added to a log culture (5 ml) growing in M9 Casamino Acids medium containing ^{14}C-thymine, and subsequently samples were removed at 12 hr. Lysates were prepared by the sarkosyl methods and centrifuged to equilibrium density gradient in CsCl-ethidium bromide. The gradients were photographed with an ultraviolet (UV) lamp to promote ethidium bromide-mediated fluorescence of the DNA band. The gradients were then fractionated, and radioactivity was determined in a sample of the fraction. In this case, the CCC DNA

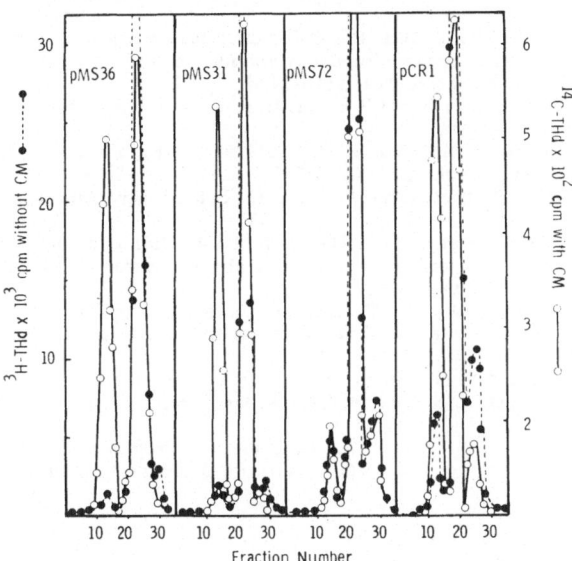

Fig. 1. A culture of *E. coli* ML4907 *thyA* strain carrying plasmid was grown in the M9 Casamino Acid medium containing ^{3}H-thymidine at 37 C for several generations in log phase. The cells were shifted to the medium containing ^{14}C-thymine in place of ^{3}H-thymidine. Chloramphenicol was quickly added, and incubated for 12 hr at 37 C. Samples were lysed with sarkosyl and subjected to dye-CsCl density gradient centrifugation. A portion of each fraction was counted.
○, ^{3}H-labelled cell; ●, ^{14}C-labeled cells.

represented about 65 to 82% level of the chromosome DNA. Assuming the size of a chromosome genome equivalent is 2.5×10^{9} daltons, we can estimate that these are plasmid DNAs from 1.6 to 2.1×10^{9} equivalent. Su plasmid DNA, which was isolated from *E. coli*, *Salmonella*, and *Shigella* has a molecular weight of 3.75 to 3.78×10^{6} daltons. This corresponds to about 470 to 535 copies per chromosome equivalent. Accordingly, it can be concluded that plasmids which require DNA polymerase 1 for DNA replication can amplify after treatment with chloramphenicol (Fig. 1). In the case of Su plasmid isolated from *Proteus* can little replicate in the presence of chloramphenicol. The amount of CCC DNA represented about only a 3 or 4 times increase after addition of chloramphenicol compared with the CCC DNA in the absence of chloramphenicol. Timmis et al. have reported the ColE1-pSC101 constructed hybrid plasmid exists as a stable replicon in *E. coli*, where it can utilize the replication functions specified by both of its parent plasmid molecules (Timmis, Cabello and Cohen, 1974). However, it is not presently known whether Su plasmids, which were isolated from *Proteus*, have more than one set of replication functions.

REFERENCES

CROSA, J. H., L. K. LUTTROPP and S. FALKOW (1975): Nature of R-factor replication in the presence of chloramphenicol. Proc. Nat. Acad. Sci. USA 72, 654–658.

CLEWELL, D. (1972): Nature of ColE1 plasmid replication in *E. coli* in the presence of chloramphenicol. J. Bacteriol. 110, 667–676.

DATTA, N. and R. W. HEDGES (1971): Compatibility groups among fi⁻ R factors. Nature (London), 234, 222–223.

DATTA, N. and R. W. HEDGES (1973): R factors of compatibility group A. J. Gen. Microbiol. 74, 335–336.

KINGSBURY, D. T. and D. R. HELINSKI (1973): Temperature sensitive mutants for the replication of plasmids in *E. coli*: Requirement for deoxyribonucleic acid polymerase 1 in the replication of the plasmid ColE1. J. Bacteriol. 114, 1116–1124.

MITSUHASHI, S. (1977): Epidemiology of R factor. In R factor (ed. S. Mitsuhashi) 26–45, University of Tokyo Press.

NAGATE, T., M. INOUE, K. INOUE and S. MITSUHASHI (1978): Plasmid mediated sulfanilamide resistance. Microbiol. Immunol. 122, 367–375.

SCHLESINGER, D. (1978): Microbiology-1978 (ed. D. Schlesinger) American Society for Microbiology.

VELTKAMP, E., W. BARENDSEN and H. J. NIJKAMP (1974): Influence of protein and ribonucleic acid syntheiss of the replication of the bacteriocinogenic factor ColDF13 in *E. coli* cells and minicells. J. Bacteriol. 118, 165–174.

KOPECKO, D. J. and S. N. COHEN (1975): Site-specific recA-independent recombination between bacterial plasmids: involvement of palindoromes at the recombination loci. Proc. Natl. Acad. Sci. U.S.A. 72, 1373–1377.

TIMMIS, K., F. CABELLO and S. N. COHEN (1974): Utilization of two distinct modes of replication by a hybrid plasmid constructed in vitro from separate replicon. Proc. Nat. Acad. Sci. U.S.A. 71, 4556–4560.

M. I., Dept. of Microbiology, School of Medicine,
Gunma Univ. Maebashi, Japan

I. THEORETICAL PART
B) R PLASMIDS IN *P. AERUGINOSA*
Chairman: B. W. HOLLOWAY

R PLASMIDS AND BACTERIAL CHROMOSOME TRANSFER

B. W. HOLLOWAY, C. CROWTHER, P. ROYLE and M. NAYUDU

Department of Genetics, Monash University,
Clayton, Victoria, Australia

INTRODUCTION

As our knowledge of R plasmids expands, the range of phenotypic properties which they have been shown to contribute to the bacterial phenotype becomes more extensive. One such property, the ability to promote the transfer of bacterial host chromosome, is of special interest to microbial geneticists. The plasmid F in *Escherichia coli* has been intensively studied for this function and the genetic and physical basis of its abilities in this respect are well known. However, F is by no means unique in this respect and various plasmids have the ability to transfer bacterial chromosomes for a variety of genera. Some, but not all R plasmids of the incompatibility group P1 (IncP-1) have been shown to have chromosome mobilizing ability (Cma) for various bacterial genera including *Acinetobacter calcoaceticus*, *Proteus mirabilis*, *Rhizobium leguminosarum* and some species of *Pseudomonas*.

For those bacteria where native IncP-1 plasmids can only mobilize the chromosome inefficiently, various attempts have been made to change the plasmid to improve its performance in this respect. Dénarié et al. (1977) constructed a hybrid between RP4 and the bacteriophage mu and showed that this had enhanced chromosome mobilization in mu lysogens of *E. coli* and *Klebsiella pneumoniae*. Barth (1979) constructed RP4 primes by *in vitro* insertion of *E. coli* chromosome fragments at the HindIII and EcoRI cutting sites of RP4 and found that such hybrid plasmids can promote chromosome transfer from a variety of sites in *E. coli* K12. Similarly Watson and Scaife (1978) found that the hybrid RP4 λ*att* can promote low frequency transfer by *int*-promoted integration into the host λ attachment site *att*λ.

The isolation of chromosome mobilizing plasmids has been intensively studied in various *Pseudomonas* species (Holloway, 1979, Dean et al., 1979, Mylroie et al., 1977) and these investigations have been prompted by the need to have an adequate chromosome map for genetic studies in those organisms. We have previously described the isolation of R68.45, a derivative of the wide host range IncP1 plasmid R68, which can promote chromosome transfer in *Pseudomonas aeruginosa*, *P. putida*, *P. glycinea*, *Rhizobium leguminosarum*, *R. meliloti*, *Rhodopseudomonas sphaeroides*, *E. coli*, *Azospirillum braziliensis* and *Agrobacterium tumefaciens* (see Holloway [1979] for review). In this paper, the isolation and characterization of a variety of R plasmids with Cma will be described and the importance of these studies for bacterial genetics and for further understanding of antibiotic resistance will be discussed. The special properties of a new Cma plasmid in *P. aeruginosa*, FP110, will be discussed, in particular its relationship to antibiotic resistance plasmids of that organism.

IS R68.45 A SPECIAL CASE?

R68.45 is a derivative of the IncP-1 plasmid R68 and was found amongst the progeny of a cross between a *P. aeruginosa* PAO multi-auxotrophic recipient and a PAO donor strain carrying R68, selection being for *arg*B⁺ (located at 20 min on the PAO map, see Fig. 1). One of these rare (10^{-8}/donor parent) *arg*B⁺ recombinants carried a plasmid which was subsequently characterized as R68.45. We have now shown that selection for other regions of the PAO chromosome in similar crosses can result in the isolation of other plasmids of the R68.45 type and these have been categorized as ECM plasmids (Enhanced Chromosome Mobilization). The other regions identified involve selection for *cys-5605* and *his*I (*ca* 12 min), *ilv-218* (*ca* 28 min), *arg*G, or *arg*F (55 min) and *leu-10*, *met-9011* (*ca* 60 min) (Holloway et al., 1979, Holloway, 1979) (see Fig. 1). The frequency with which such plasmids are found in the progeny of these crosses is similar to that associated with the original isolation of R68.45.

ECM plasmids have now been isolated by similar techniques from IncP-1 plasmids other than R68, including R18 (Holloway and Richmond, 1973) and R906 (Hedges et al., 1974). In general the properties of ECM plasmids derived from these other sources are similar to those found in R68.45.

A plasmid with properties very similar to R68.45, pMO60 (also named Rm16b, Jacoby and Matthews, 1979) was isolated from a collection of *P. aeruginosa* isolates obtained from Japan. The identification of this plasmid as an ECM-like plasmid leads to the conclusion that ECM plasmids occur in nature, are not merely laboratory curiosities and that R68.45 is not unique. Different ECM plasmids show variations in plasmid properties, for example, not all ECM plasmids show the same pattern of chromosome transfer in *P. putida*, there being variability in the number of origins and the amount of chromosome transferred (Dean and Morgan, personal communication).

It has been shown that R68.45 differs from R68 in the addition of an extra segment of DNA, inserted near the kanamycin resistance determinant (van Montagu and Schell, personal communication, Burkhardt et al., 1978). This insertion, termed ISP, is about 1800 base pairs long and from restriction endonuclease cutting data has been shown to occur in different ECM plasmids (Reiss et al., 1979).

Much remains to be determined as to the nature of ISP. For example, in view of its potential for interaction with sites on a wide range of bacterial chromosomes, does it originate solely from the *P. aeruginosa* genome? We have attempted to look for ECM plasmids in *E. coli* K12 amongst the progeny of crosses of the type *E. coli* F⁻ × *E. coli* R68. By selection for *arg*A⁺, some of the rare (10^{-8}/donor parent) recombinants were found to carry a plasmid which can promote chromosome transfer in *E. coli* at frequencies of $1-3 \times 10^{-6}$, a property not found in R68 or other IncP-1 plasmids. It will be of interest to see if such plasmids have properties comparable with R68.45 and whether there is an addition to the R68 genome similar to IPS as found in R68.45. The molecular differences between IncP-1 plasmids with and without Cma are discussed in the accompanying paper by Reiss et al. (1979).

PLASMID-CHROMOSOME HYBRIDS

In addition to promoting chromosome transfer in a variety of bacterial hosts, ECM plasmids have the property of forming hybrid plasmids in which a section of baciaterl chromosome is inserted into the plasmid genome (R prime, or R′ plasmids). The isolation and characterization of one such plasmid, R′PA1 derived from R68.45 in *P. aeruginosa*

PAO, has been described (Holloway, 1978). It has been found that other R' derivatives of R68.45 can be isolated carrying various regions of the *P. aeruginosa* PAO chromosome (Royle, Hill, Crowther and Holloway, unpublished data). The technique used in these isolations was to cross a donor carrying R68.45 to a *rec*A deficient recipient, selection being made for recombinants which carry the normal donor allele of an auxotrophic marker in the recipient. Such recombinants are unlikely to arise by integration into the recipient chromosome of the donor allele due to the RecA⁻ phenotype of the recipient. A similar technique has been used to isolate R prime plasmids of *E. coli*. The *E. coli* K12 strain 1230 (*pro, met*) carrying R68.45 was patch mated on nutrient agar to DG1210 (*trp, rec*A, *uvr*A). The overnight growth was washed off, resuspended in saline and plated on minimal media containing kanamycin. Any colonies arising on the minimal agar were tested for the presence of R68.45 plasmid markers and the ability to transfer *trp*⁺ genes to BH1, a rifampicin resistant derivative of DG1210. One such clone was studied in detail and shown to carry a derivative of R68.45 which has a segment of the *E. coli* chromosome carrying the entire *trp* operon and the *pyr*F and *cys*B genes. This plasmid has been called R'EC1.

The evidence to support this view is as follows:
1. All detectable markers of R68.45 are present (CBʳ, KMʳ, TCʳ, ARʳ, Tra⁺ and Cma).
2. When transferred to *P. aeruginosa* PAO, all six *trp* loci (*trp*A-F) are complemented by R'EC1. The resulting hybrids grow as well as wild type PAO and still retain their *P. aeruginosa trp* allele as shown by transduction tests using F116L.
3. R'EC1 is stable for plasmid and *trp* markers in *E. coli* DG1210, *P. aeruginosa* and *P. putida*. The plasmid can be transferred from *P. aeruginosa* back to *E. coli* with high coinheritance of *trp* markers and plasmid markers.

It will be interesting to examine the regulation of the tryptophan biosynthetic enzymes of *E. coli* in *P. aeruginosa* by means of this plasmid. The tryptophan enzymes are very suitable for such a study because, while in *E. coli* the five enzymes are all repressible by tryptophan, in *P. aeruginosa*, the enzymes of *trp*A and *trp*B are inducible by tryptophan, the enzymes of *trp*C, D and E are repressible by tryptophan and the enzyme of *trp*F is constitutive (Crawford, 1975). In a similar study, Nagari et al. (1977) have constructed by *in vitro* genetic engineering techniques an RP4 derivative with part of the *E. coli trp* operon. Regulation of the *trp*A and *trp*B gene products of this plasmid when transferred to *P. aeruginosa* was that characteristic of *P. aeruginosa* rather than that of *E. coli*.

MUTATOR FUNCTION OF R68.45

On the available evidence, the best explanation for the origin of ISP, the 1800 base pair addition to R68 by which it is changed into R68.45, is that it comes from the PAO chromosome. What then is the function of ISP when present in the PAO chromosome and does ISP take part in any other genetic phenomena? We have evidence that R68.45 may have a mutator role. The *P. aeruginosa* strain PAO25 (R68.45), (*leu-10, arg*F) has been found to have a spontaneous rate of mutation to high level (500 μg/ml) aminoglycoside resistance 50—100 times higher than that of other PAO strains. Many of these resistant mutants have entirely lost or have a much reduced Cma activity. The aminoglycoside resistance mutant is chromosomal, not plasmid borne, and for several independently isolated mutants has been mapped in the region of *leu-10* and *arg*F (55–60 min) on the *P. aeruginosa* chromosome. Strains cured of R68.45 do not show the mutator effect, but the effect reappears if such cured strains are reinfected with R68.45. It does not reappear if such a cured strain is reinfected with R68. The mutator effect

21

seems to be strain specific, because other PAO strains carrying R68.45 do not show the mutator effect. It is possible that strain PAO25 may have a chromosomal mutation which is a necessary prerequisite for the mutator effect. It is worthwhile noting that R68 and R68.45 do not carry any genes for streptomycin resistance, that the aminoglycoside resistance mutants are highly resistant (M.I.C. >600 µg/ml) to streptomycin, and that the chromosomal site of the marker for high level streptomycin resistance (strA) is at 33 min on the PAO chromosome (Holloway et al., 1979, Crowther and Holloway, in preparation). While this mutator effect of R68.45 seems to be rather limited by its specificity, in view of similar reports with other plasmids (Pearce and Meynell, 1968), it is possible that this may be one source of aminoglycoside resistant mutants in nature. It would be worthwhile to see if any variation occurs in the map site of streptomycin resistant mutations in *P. aeruginosa* in strains isolated from hospitals by crossing them with genetically marked PAO derivatives. This could be carried out by a Cma plasmid like R68.45 or a transducing phage like E79*tv*1 (Morgan, 1979) which can transduce markers in strains other than PAO, strA being cotransducible with argC.

CMA PLASMIDS OF *P. aeruginosa*

Several plasmids have been used as sex factors for mapping and conjugation experiments in *P. aeruginosa*, notably FP2 (Holloway, 1969), FP5 (Matsumoto and Tazaki, 1973). It has been shown that such Cma plasmids (FP plasmids) occur in 15–30% of hospital isolates of *P. aeruginosa*, the number depending on the source of the strains. A technique has been developed (Dean et al., 1979) for the identification of FP plasmids directly in hospital strains, without the need to isolate auxotrophic mutants of the hospital isolates as had been necessary in previous surveys of this type (Pemberton and Holloway, 1973). FP plasmids may carry other determinants such as resistance to inorganic mercury, e.g. FP2 or FP5, or more rarely, a leucine biosynthetic determinant (e.g. FP39) but in most instances they do not possess any phenotypic characteristic which can be used for selection in plasmid transfer. Most FP plasmids examined have been shown to have an origin for bacterial chromosome transfer at the same site as that for FP2 (fig. 1) (for mapping details, see Holloway et al., 1979). One new plasmid, FP110, has been shown

TABLE I.

Comparison of the Cma plasmids F (in E. coli K12) and FP2 and FP110 (in P. aeruginosa PAO)

Property	F	FP2	FP110
Stability	high	very high	very high
Host RecA function for chromosome transfer	low frequency host chromosome transfer in RecA⁻ donor	low frequency host chromosome transfer in RecA⁻ donor	no host chromosome transfer in RecA⁻ donor
Stable integration into host chromosome	yea	no	no
Plasmid chromosome hybrids (F' or FP') found	yes	not yet demonstrated	rarely
Oriented transfer of chromosome	yes	yes	yes (however, see text)

22

to have a quite different origin and also to transfer chromosome in a direction opposite to that of FP2.

Evidence for the major origin of transfer for FP110 being at the 28 min site (FP2 is 0 min) comes from recombination frequencies obtained for various markers and coinheritance data obtained in plate matings. It is not possible to use the interrupted mating technique to obtain time of entry data from FP110 matings because nalidixic acid does not inhibit chromosome transfer in such matings as it has been found to do with R68.45 or FP2 matings (Haas et al., 1977).

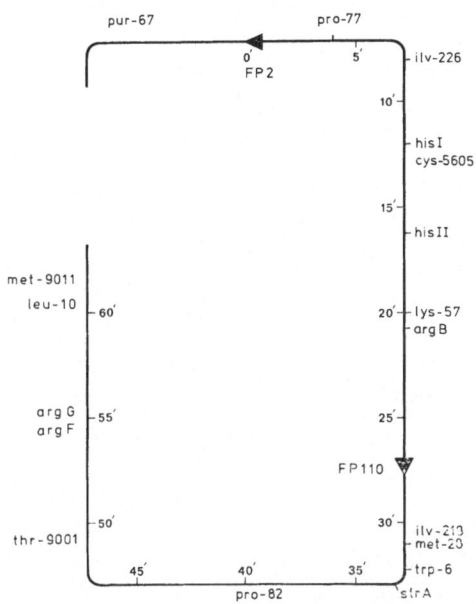

Fig. 1. Chromosome map of *P. aeruginosa* PAO showing markers mentioned in text. This map is derived from that previously published in Holloway et al. (1979) and the same system of symbols is used. Markers whose location is indicated by a bar joining the locus designation to the map were located by interrupted matings using FP2 donors. The following abbreviations have been used in this figure and in the text: *arg*, arginine requirement; *cys*, cysteine requirement; *chl*, chloramphenicol resistance; *his*, histidine requirement; *ilv*, isoleucine plus valine requirement; *leu*, leucine requirement; *lys*, lysine requirement; *met*, methionine requirement; *pro*, proline requirement; *pur*, adenine requirement; *str*, streptomycin resistance; *thr*, threonine requirement; *trp*, tryptophan requirement. In the text, other markers are referred to which are closely linked (allelic) to some of those in Fig. 1. The relevant allelic markers are: *leu-10*, *leu-13* and *leu-9001*; *lys-57* and *lys-12*; *his*II and *his-4*.

FP110 has a variety of characteristics which distinguish it from FP2 and F, both well characterized Cma plasmids in *P. aeruginosa* and *E. coli* respectively.

There is abundant evidence to support the view that in *E. coli*, F can promote chromosome transfer from a variety of sites in contrast to the situation with FP2 in *P. aeruginosa* where transfer is primarily from one site, with some evidence for low frequency transfer from other minor origins. FP110 shows features of transfer different from either of these patterns. The length of the chromosome fragment transferred by FP110 is dependent upon both the marker for which selection is made and the contraselective marker of the donor strain. If we compare crosses with FP2 and FP110 using the same contraselective

23

TABLE II.

A comparison of the relative recovery of markers in plate matings for the crosses PAO227 × PAO315 (FP2) and PAO222 × PAO315 (FP110). PAO227 has the genotype ilv-226, his-4, lys-12, met-28, trp-6, pro-82, leu-13. PAO315 has the genotype argG9, chl-2. Positions of these auxotrophic markers on the PAO chromosome map are shown in Fig. 1.

Selected marker and map location (min. from FP2 origin)	Recombination frequency/donor parent	
	FP2 donor	FP110 donor
ilv-226^+ (8)	7.8×10^{-5}	3.0×10^{-4}
his-4^+ (16)	6.8×10^{-5}	4.4×10^{-4}
lys-12^+ (20)	2.0×10^{-5}	5.1×10^{-4}
met-28^+ (32)	3.4×10^{-6}	4.4×10^{-7}
trp-6^+ (35)	8.0×10^{-7}	8.0×10^{-8}
pro-82^+ (40)	1.3×10^{-6}	4.0×10^{-7}
leu-13^+ (ca 65)	5.0×10^{-7}	1.2×10^{-6}

TABLE III.

Recombination frequencies for selected markers and segregation of markers in recombinants in the plate mating of PAO325 (FP110) × PAO134. PAO134 has the genotype hisI, pur-67, met-9011, leu-9001, argF23, thr-9001, strA66 and PAO325 has the genotype argB18, lys-57. The positions of these markers on the PAO chromosome map are shown in Fig. 1.

Marker selected and map location (min. from FP2 origin)	Recombination frequency /donor parent	Coinheritance (%) of other markers					
		$hisI^+$	pur-67^+	met-9011^+	$argF23^+$	thr-9001^+	$strA66$
$hisI^+$ (12)	3.2×10^{-5}	—	13	0	0	0	0
pur-67^+ (>65)	2.5×10^{-5}	82	—	82	51	82	66
met-9011^+ (ca 65)	7.7×10^{-7}	1	3	—	30	10	0
leu-9001^+ (ca 65)	1.1×10^{-6}	16	15	91	27	24	6
$argF23^+$ (55)	1.5×10^{-6}	24	22	36	—	49	13
thr-9001^+ (48)	2.5×10^{-6}	82	86	86	72	—	78

marker $argG$ (ca 55 min), the recombination frequencies of various markers are shown in Table II.

Analysis of the coinheritance of the markers in each cross is consistent with the view that there is a major origin at 0 min for FP2 and a major origin for FP110 between 25 and 28 min (see Fig. 1). However, with FP110, if contraselection of the donor is made using markers in the 20 min region, quite a different pattern of coinheritance is obtained (Table III). No such effect has ever been found with FP2 crosses.

It is seen that selection for $hisI$ results in low linkage of this marker to pur-67^+ gives a recombinant frequency much the same as that for $hisI^+$, but involves coinheritance of a much longer segment of chromosome in most recombinants. For example, most recombinants carry in addition to pur-67^+ the markers $hisI^+$, met-9011^+ and thr-9001^+. By contrast, selection for met-9011^+ or leu-9001^+ recombinants results in the inheritance of a shorter segment of chromosome (i.e. reduced coinheritance of other markers) and much lower recovery of the selected markers. In addition linkage of markers is non-reciprocal depending upon the marker selected. For example, when thr-9001^+ is selected it shows 86% coinheritance with met-9011^+. When met-9011^+ is selected, coinheritance of thr-9001^+ is found in only 10% of the recombinants.

From this and other crosses (data not given) it can be concluded that the pattern of chromosome transfer by FP110 is different from that of either FP2 in *P. aeruginosa* or F in *E. coli*. By forcing crossovers in certain chromosomal regions by selection for particular donor or recipient markers, the length of the segment of chromosome inherited by the recipient parent may vary substantially. One possible explanation is that crossovers do not occur with equal probability for all regions of the chromosome. These results with FP110 crosses and the resulting uncertainty as to the reason for these patterns of re-combinant recovery mean that FP110 data cannot be used for location of genes on the PAO chromosome or for establishing genetic circularity of the PAO chromosome. For the present, additional mapping of the 'late' region (i.e. more than 50 min from the FP2 origin) will have to be carried out by the analysis of short chromosomal sections with R68.45 crosses. In PAO, R68.45 usually transfers 10 to 15 min segments of chromosome. For the region later than 50 mins from the FP2 origin, FP2 matings give recombination frequencies which are too low for accurate mapping.

RELATIONSHIP OF FP110 TO OTHER R PLASMIDS

The high frequency with which FP plasmids are found in *P. aeruginosa* strains isolated from hospitals has raised the question as to the relationship between these FP plasmids, which invariably do not carry genetic determinants for antibiotic resistance, and R plasmids carrying resistance determinants. There is evidence to suggest that FP plasmids could be a source of some R plasmids.

Firstly, FP plasmids may acquire antibiotic resistance determinants by transposition. By mating a PAO (R18 *tra*) (FP110) strain with a PAO recipient it was possible to get a derivative of FP110 which had acquired resistance to carbenicillin presumably by acquisition of Tn1 from R18, without losing any FP110 plasmid properties. This plasmid is referred to as FP110::Tn1. Similarly, by mating a PAO (RP4::Tn7 *tra*) (FP110) strain with a PAO recipient it was possible to transpose Tn7 to FP110, the frequency of transfer being 7×10^{-5} per donor parent.

Using these derivatives of FP110, it was possible to test for entry exclusion between FP110 and two other plasmids, RP1-1, isolated in England (Ingram et al., 1972) and R56Be isolated in France (Michel-Briand et al., 1977). It had already been shown that RP-1 and R56Be were closely related (Michel-Briand et al., 1977). The entry exclusion data concerning these three plasmids is shown in Table 4. R18-1 is very likely identical with RP1-1 (Holloway and Richmond, 1973). From the data in Table 4 it can be con-cluded that all these plasmids show mutual entry exclusion, which is one criterion of plasmid relationship.

The derivatives of FP110 carrying the transposon were used to test the incompatibility relationships of this plasmid with R18-1 and R56Be. In the first instance FP110::Tn7 was used as the resident plasmid and FP110::Tn1 as the incoming plasmid. Here no incompatibility is found and both plasmids can coexist stably in the same cell. When FP110::Tn1 is the resident plasmid and FP110::Tn7 is the incoming plasmid, there is 100% incompatibility between these two derivatives of FP110, as would be expected. If R56Be or R18-1 is the resident plasmid, then there is 100% incompatibility with the incoming FP110::Tn7 and this result supports the close relationship of these three plasmids suggested by the entry exclusion data given above. We have no explanation at present for the apparent lack of incompatibility between FP110::Tn1 and FP110::Tn7 when FP110::Tn7 is the resident plasmid.

25

TABLE IV.

Entry exclusion relationships between the P. aeruginosa plasmids RP1-1, R18-1, R56Be and FP110

Resident Plasmid	Incoming Plasmid	Transfer frequency of incoming plasmid/donor parent
none	R56Be	3.0×10^{-1}*
RP1-1	R56Be	4.0×10^{-3}*
none	FP110::Tn1	1.0×10^{-1}
FP110	FP110::Tn1	1.6×10^{-4}
FP110::Tn7	R18-1	1.3×10^{-4}
R18-1	FP110::Tn7	1.4×10^{-4}
FP110::Tn7	R56Be	2.0×10^{-3}
R56Be	FP110::Tn7	1.0×10^{-4}

*data of Michel-Briand et al., 1977. *R18-1* and *RP1-1* are probably identical plasmids (Holloway and Richmond, 1973).

DISCUSSION

The study of Cma in R plasmids is important for a variety of reasons. Firstly, it demonstrates the importance of R plasmids as a source of genetic variation in bacteria through their ability to mobilize the bacterial chromosome. The identification of R prime plasmids is of particular significance in view of the wide host range of IncP-1 plasmids and as a means of exchanging bacterial chromosomal material between unrelated genera.

The identification of IncP-1 plasmids as Cma plasmids has opened up the possibility of genetic analysis for a wide variety of bacteria for which genetic analysis did not previously exist. Further, R prime plasmids provide a means for *in vivo* genetic engineering of microorganisms with desired genetic properties. Through such plasmids it will be possible to compare the expression of bacterial genes when in their normal genome and when transferred to other bacteria.

The extensive study of FP plasmids with Cma in *P. aeruginosa* has shown that this type of plasmid can acquire transposons carrying antibiotic resistance and in effect such recombinant plasmids are newly synthesized R plasmids. As a particular example, one such plasmid, FP110, has a close relationship with two other R plasmids, one isolated in England, the other in France, demonstrating the potential role of FP plasmids in the formation of plasmids resistant to antibiotics in *P. aeruginosa*.

ACKNOWLEDGEMENTS

Work in the authors' laboratory is supported by the Australian Research Grants Committee and the National Health and Medical Research Council.

REFERENCES

BARTH, P. T. (1979): Plasmid RP4, with *Escherichia coli* DNA inserted *in vitro*, mediates chromosomal transfer. Plasmid 2, 130–136.

BURKHARDT, H., G. REISS and A. PÜHLER (1978): Molecular relationship between R plasmids RP1, RP4, RP8, RK2, R68 and R68.45 revealed by electron microscopical techniques. Hoppe-Seyler's. Z. Physiol. Chem. 359, 1068.

CRAWFORD, I. P. (1975): Gene arrangements in the evolution of the tryptophan synthetic pathway. Bacteriol. Rev. 39, 87–120.

DEAN, H. F., P. ROYLE and A. F. MORGAN (1979): Detection of FP plasmids in hospital isolates of *Pseudomonas aeruginosa*. J. Bact. (in press).

DÉNARIÉ, J., C. ROSENBERG, B. BERGERON, C. BOUCHER, M. MICHEL and M. BARATE de BERTALMIO (1977): "Potential of RP4: Mu Plasmids for *in vivo* Genetic Engineering of Gram-Negative Bacteria". *In*: DNA Insertion Elements, Plasmids, and Episomes (Bukhari, Shapiro and Adhya, eds.), pp. 507–520, Cold Spring Harbor Laboratory.

HAAS, D., B. W. HOLLOWAY, A. SCHAMBÖCK and T. LEISINGER (1977): The genetic organization of arginine biosynthesis in *Pseudomonas aeruginosa*. Molec. gen. Genet. 154, 7–22.

HEDGES, R. W., A. E. JACOB and J. T. SMITH (1974): Properties of an R factor from *Bordetella bronchiseptica*. J. gen. Microbiol. 84, 199–204.

HOLLOWAY, B. W. (1969): Genetics of Pseudomonas. Bacteriol. Rev. 33, 419 – 443.

HOLLOWAY, B. W. (1978): Isolation and characterization of an R' plasmid in *Pseudomonas aeruginosa*. J. Bact. 133, 1078–1082.

HOLLOWAY, B. W. (1979): Plasmids that mobilise bacterial chromosome. Plasmid 2, 1–19.

HOLLOWAY, B. W. and M. H. RICHMOND (1973): R-factors used for genetic studies in strains of *Pseudomonas aeruginosa* and their origin. Genet. Res. 21, 103–105.

HOLLOWAY, B. W., V. KRISHNAPILLAI and A. F. MORGAN (1979): Chromosomal Genetics of *Pseudomonas*. Microbiol. Rev. (in press).

INGRAM, L. C., R. B. SYKES, J. GRINSTEAD, J. R. SAUNDERS and M. H. RICHMOND (1972): A transmissible resistance element from a strain of *Pseudomonas aeruginosa* containing no detectable extrachromosomal DNA. J. gen. Microbiol. 72, 269–279.

JACOBY, G. A. and M. MATTHEWS (1979): The distribution of β-lactamase genes on plasmids found in *Pseudomonas*. Plasmid 2, 41–47.

MATSUMOTO, H. and T. TAZAKI (1973): FP5 factor, an undescribed sex factor of *Pseudomonas aeruginosa*. Jap. J. Microbiol. 17, 409–417.

MICHEL-BRIAND, Y., V. STANISICH and M. JOUVENOT (1977): *Pseudomonas aeruginosa* strain isolated in France that carries a plasmid determining carbenicillin resistance. Antimicrob. Ag. Chemother. 11, 589–593.

MORGAN, A. F. (1979): Transduction of *Pseudomonas aeruginosa* with a mutant of E79. J. Bact. (in press).

MYLROIE, J. R., D. A. FRIELLO, T. V. SIEMENS and A. M. CHAKRABARTY (1977): Mapping of *Pseudomonas putida* chromosomal genes with a recombinant sex factor plasmid. Molec. gen. Genet. 157, 231–237.

NAGAHARI, K., Y. SANO and K. SAKAGUCHI (1977): Derepression of *E. coli trp* operon on interfamilial transfer. Nature (London) 266, 745–746.

PEARCE, L. E. and E. MEYNELL (1968): Mutation to high level streptomycin resistance in R+ bacteria. J. gen. Microbiol. 50, 173–176.

PEMBERTON, J. M. and B. W. HOLLOWAY (1973): A new sex factor of *Pseudomonas aeruginosa*. Genet. Res. 21, 263–272.

REISS, G., A. PÜHLER and B. W. HOLLOWAY (1979): An insertion element of *Pseudomonas aeruginosa* is responsible for chromosomal mobilisation ability of IncP-1 plasmids. *In*: Proceedings of the 4th International Symposium on Antibiotic Resistance, Smolenice June 1979 (in press).

WATSON, M. D. and J. G. SCAIFE (1978): Chromosomal transfer promoted by the promiscuous plasmid RP4. Plasmid 1, 226–237.

R68.45, A PLASMID WITH CHROMOSOME MOBILIZING ABILITY DIFFERS FROM R68 BY A DUPLICATED DNA REGION

G. RIESS, H. J. BURKARDT, A. PÜHLER and B. W. HOLLOWAY

Institute of Microbiology and Biochemistry, University of Erlangen, Erlangen, F.R.G.

INTRODUCTION

Recently a derivative of plasmid R68, designated R68.45, with chromosome mobilizing ability (Cma) was isolated in *Pseudomonas aeruginosa* strain PAO. In this strain R68.45 is able to mobilize the chromosome from a number of different origins (1). As R68.45 belongs to incompatibility class Inc P1 it can be transferred to a variety of gram⁻ bacteria. In some of these strains, such as *Rhizobium meliloti*, *Rhizobium leguminosarum*, *Rhizobium trifolii* and *Rhodopseudomonas sphaeroides*, it has been shown that R68.45 can also be used for mobilization of chromosomal genes, (2, 3, 4, 5). We report now that R68.45 is able to mobilize the *Escherichia coli* chromosome. In addition we show that R68.45 differs from its parent R68 by a duplicated DNA region.

RESULTS

The E. coli chromosome is mobilized by R68.45

Both R68 and R68.45 confer resistance to ampicillin, kanamycin and tetracycline. The only difference between these two plasmids is the Cma property of R68.45. In order to distinguish between these two plasmids we tested their chromosome mobilizing ability

TABLE I.

Mobilization of E. coli chromosomal genes by R68.45

donor	recipient	R-plasmid transfer frequency	chromosomal gene transfer selected marker and frequency	
E.coli C(R68.45) wt	E.coli C his nal^r	1.2×10^{-1}	3.3×10^{-6}	His$^+$
E.coli C(R68)	"	0.3×10^{-1}	$> 2 \times 10^{-8}$	His$^+$
CSH51(R68.45) ara Δ(lac pro) strA thi (φ80d lac$^+$)	CSH65 leu lac nalA strA thi	0.7×10^{-1}	1.5×10^{-6}	Leu$^+$
CSH51(R68)	"	1.5×10^{-1}	$> 2 \times 10^{-8}$	Leu$^+$
CSH56(R68.45) ara Δ(lac pro) nalA thi	CSH59 pyrC trp strA thi	1.2×10^{-1}	4.0×10^{-6}	Trp$^+$
CSH56(R68)	"	1.3×10^{-1}	$> 2 \times 10^{-8}$	Trp$^+$

in *E. coli*. For this reason R68 and R68.45 were transferred from *P. aeruginosa* to *E. coli* C and *E. coli* K12. *E. coli* exconjugants were then used as donors in *E. coli* matings which were performed as follows: 0.1 ml of a mixture of a logarithmic donor and a stationary recipient were spread on a membrane filter attached to Penassay agar. Matings were carried out overnight at 37°C. In order to harvest the mating mixture the filters were vortexed in saline buffer. Appropriate dilutions of this mating suspension were plated on selective agar. Antibiotic containing agar was used to select for R-plasmid

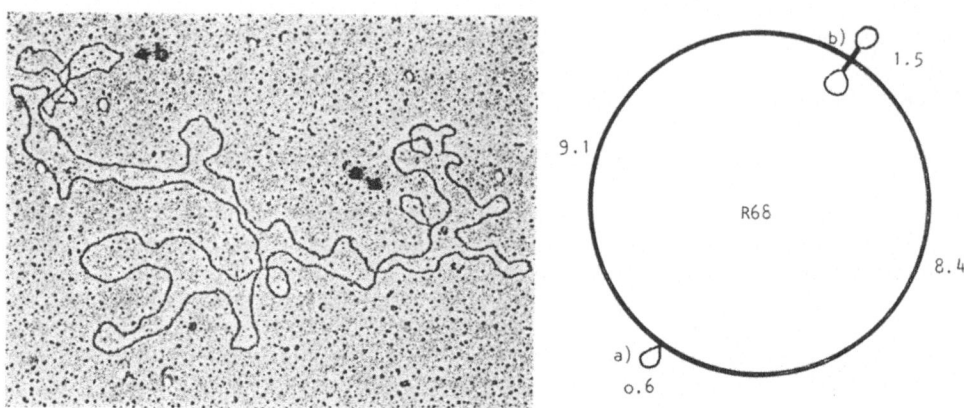

Fig. 1. Heteroduplex molecule between R68.45 and R68.
The electron micrograph of a typical heteroduplex molecule between R68.45 and R68 shows two special structures:
a) insertion loop   b) the outlooped DNA structure of Transposon Tn1
Lengths are given in μm and were calculated from 5 different molecules.

transfer, whereas minimal agar with special supplements was applied to recover proto-trophic recombinants. In Table I. the frequencies for R-plasmid transfer (R68 and R68.45) and for chromosomal gene transfer (His, Leu, Trp) are listed. As can be seen from the different crosses, chromosome mobilizing ability with R68.45 is at least 100-fold more efficient than with R68. The mobilization frequency is also similar for all tested markers. This is in agreement with other bacterial systems where R68.45 was used for chromosomal gene transfer.

R68.45 carries an additional DNA segment

We have shown that R68.45 in contrast to R68 mobilizes the *E. coli* chromosome. We now tried to decide whether this Cma property is connected with a special molecular structure of R68.45. We therefore isolated DNA of the plasmids and measured their contour lengths in the electron microscope. R68 shows an average length of 19.1 ± 0.3 μm, whereas R68.45 is slightly longer (19.7 ± 0.4 μm). This difference in contour length could be demonstrated in an heteroduplex experiment. A typical heteroduplex molecule between R68 and R68.45 is shown in Figure 1. This molecule carries two special DNA structures. One of them (b) is due to the special renaturation behaviour of Transposon Tn1[6]. The other (a) located on the opposite to Tn1 represents a single stranded DNA loop of 0.6 μm length. This experiment indicates that R68.45 differs from R68 only by an additional DNA segment which is integrated into R68, 9.1 μm or 8.4 μm away from the location. We propose that this additional DNA segment is responsible for the Cma property of R68.45. This assumption could be confirmed by experiments reported in the next section.

R68.45 carries additional SmaI and PstI restriction rites

R68.45 and R68 DNA was digested by the restriction endonucleases SmaI and PstI. The generated fragments separated by agarose gel electrophoresis are shown in Figure 2. In the SmaI digest of R68.45 an additional 2120 bp fragment can be detected, whereas

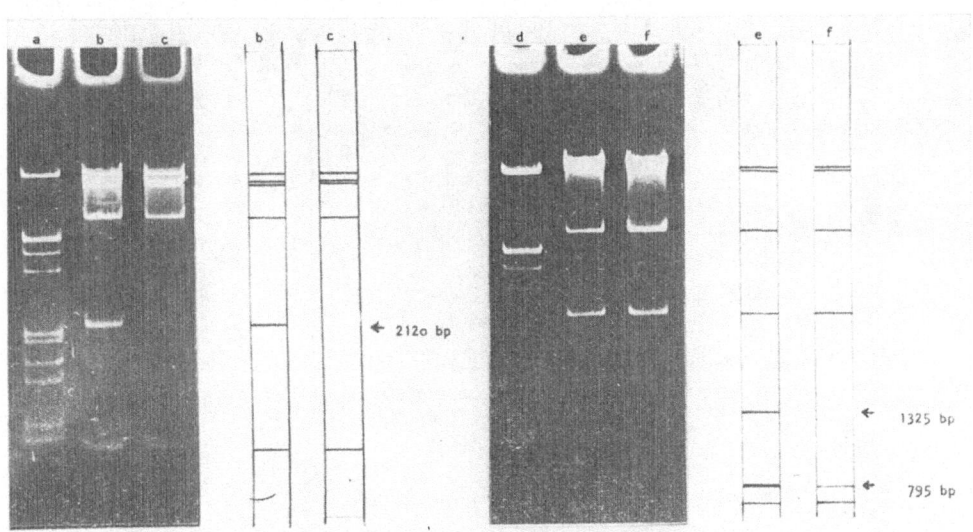

Fig. 2. SmaI and PstI digestion of R68.45 and R68 DNA.
R68.45 and R68 DNA was digested by the restriction endonucleases SmaI and PstI. The fragments were separated by agarose gel electrophoresis. λ fragments generated by an EcoRI/HindIII double digestion were used as reference markers for molecular weight determination. The lengths of interesting fragments are given in bp. The lanes show:

a) λ *EcoRI/Hind*III   d) λ *EcoRI/Hind*III
b) R68.45   SmaI   e) R68.45   PstI
c) R68   SmaI   f) R68   PstI

in the PstI digest of R68.45 two new fragments of 1.325 bp and 795 bp appear. The smaller new PstI fragment is exactly as long as one of the PstI fragments of plasmid R68. Since R68.45 and R68 differ only in one DNA segment shown by the heteroduplex experiment, we conclude that the additional SmaI and PstI restriction sites are located on this DNA segment.

All R68 derived Cma plasmids carry the same additional DNA segment

Plasmids with Cma derived from R68 can be detected in *P. aeruginosa* as follows: In crosses with auxotrophic parental strains R68 is used for transfer of chromosomal markers. Prototrophic exconjugants appear at a very low frequency. From such exconjugants, plasmids with enhanced chromosome mobilization derived from R68 can be isolated. Other plasmids with Cma obtained from independent experiments are termed pM047, pM061, pM062, pM090, pM091, pM091, pM092, pM093 and pM094. There is one exception: pM060 is a natural isolate from Japan which carries the same antibiotic

31

resistance markers as R68 but in addition shows the enhanced mobilization ability of R68.45. From all these plasmids DNA was isolated and digested by SmaI and PstI. Surprisingly, the SmaI and PstI restriction pattern was absolutely identical with that of R68.45 shown in Figure 2. We conclude that all these different Cma plasmids derived from R68 are identical and possess the same DNA segment characterized by one SmaI and two PstI restriction sites.

It was now interesting to speculate about the origin of the additional DNA segment present in Cma plasmids. As all Cma plasmids including pM060, the natural isolate from Japan, carry the same DNA segment, we assume that this special segment is of

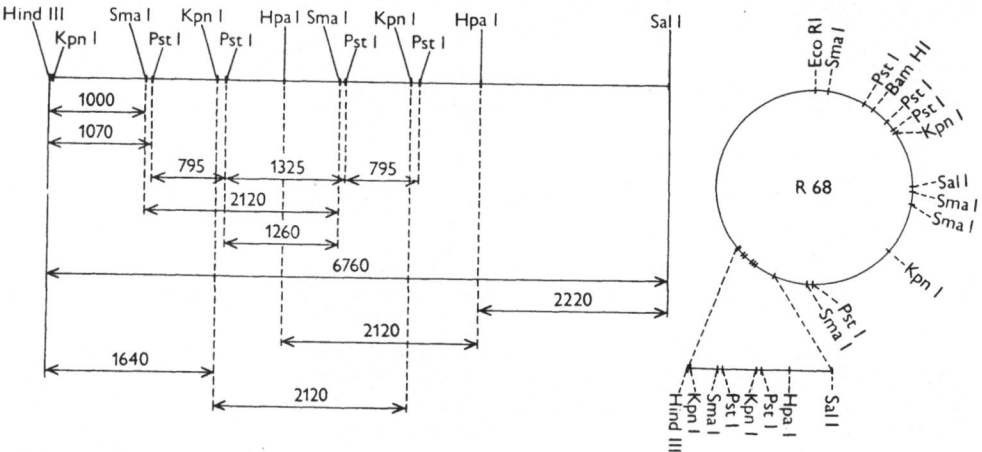

Fig. 3. The restriction map of R68.45 and R68.

The restriction sites for the enzymes HindIII, SmaI, PstI, KnpI and HpaI on the HindIII/SalI fragment of R68.45 were determined. The lengths in bp of the fragments identified in appropriate digests are indicated. A DNA segment with the following sequence of restriction sites seems to be duplicated: *SmaI – PstI – KpnI – PstI – HpaI*. In addition the R68 restriction map is shown. All restriction sites of R68, except for KpnI sites are taken from the RK2 map[7]. The KpnI sites are identical with those of the RP4 map[8]. R68, RP4 and RK2 are identical[6].

R68 origin, for instance a duplication of a preexisting DNA region. Another possibility would be that an insertion-like element present in *P. aeruginosa* has a special integration site on R68. The PstI digestion experiment shown in Figure 2 supports the first hypothesis as the 795 bp fragment of R68 is found twice in R68.45. In order to test the duplication hypothesis we tried to find more restriction enzymes cutting in the presumed duplicated region.

The additional DNA segment of R68.45 is a duplicated region of R68

The heteroduplex experiment showed that the additional DNA segment of R68.45 is located close to the kanamycin resistance gene which carries the single HindIII site of R68. From an EcoRI/HindIII double digestion experiment of R68 and R68.45 DNA we learned that the additional DNA segment is integrated into the larger EcoRI/HindIII fragment (Figure 3). We therefore tried to determine the sequence of different restriction sites downstream from the HindIII site. Using the restriction enzymes HindIII, KpnI, SmaI, PstI, HpaI and SalI, we could establish the restriction map shown in Figure 3.

Evidently a DNA segment preexisting in R68 with the following sequence of restriction sites seems to be duplicated: SmaI, PstI, KpnI, PstI, HpaI. The duplication hypothesis is confirmed by fragment length measurements: In SmaI, KpnI and HpaI digestions a fragment of 2120 bp always appears. The same length can be calculated for the distances of the PstI sites in the tandem region. The length of the duplicated region can be estimated from Figure 3. The minimum value is 2120 bp, whereas a maximum value of 2820 bp can be calculated taking into account that the HpaI/SmaI fragment of the tandem region is 700 bp in size.

CONCLUDING REMARKS

In contrast to R68, R68.45 is able to mobilize chromosomal genes in many bacterial species. We have demonstrated this for *E. coli*. By molecular characterization of plasmid DNA we were able to show that R68.45 differs from its parent R68 only by a duplicated region. We therefore assume that this duplicated region is responsible for the chromosome mobilizing ability of R68.45. But so far nothing is known about the molecular mechanism of chromosome mobilization. We hope to obtain more information about this mechanism by isolation and characterization of R68.45 prime plasmids carrying segments of the *E. coli* chromosome.

This work was supported from a grant of Deutsche Forschungsgemeinschaft (Pu 28/8)

REFERENCES

1. HAAS D. and B. W. HOLLOWAY (1976): Mol. Gen. Genet. 144, 243–251.
2. KONDOROSHI A. et al. (1977): Nature, 268, 525–527.
3. BERINGER J. E. and D. A. HOPWOOD (1976): Nature, 264, 291–293.
4. JOHNSTON A. W. B. and J. E. BERINGER (1977): Nature, 267, 611–613.
5. SISTROM W. R. (1977): J. Bact., 131, 526–532.
6. BURKARDT H. J. et al. (1979): J. Gen. Microbial (in press).
7. MEYER R. et al. (1977): in DNA, Insertion Elements, Plasmids and Episomes CSH Laboratory.
8. A. DE PICKER et al. (1977): in DNA, Insertion Elements, Plasmids and Episomes CSH Laboratory

G. R., Inst. Mikrobiologie, D. 852 Erlangen, Egerlandstr. 7, B.R.D.

MAPPING OF REPLICATION GENES OF PLASMID RP4

R. SIMON and A. P. PÜHLER

Institute of Microbiology and Biochemistry, University of Erlangen, F.R.G.

INTRODUCTION

Considerable emphasis in current plasmid research has been directed to the nature o-plasmid-borne functions required for controlled replication. In this connection the drug-resistant plasmid RP4 has attracted special interest as it can be transferred to any Gram-negative bacterial species that has been tested (1, 2). RP4 is a conjugative P-incompatibility group plasmid that has a molecular weight of about 57 kb and that specifies resistf ance against ampicillin (Ap), kanamycin (Km) and tetracyclin (Tc). The drug-resistance markers of RP4 have been mapped by Tn7 and Tn76 insertion mutagenesis (3). The origin of vegetative replication (oriv) is known for the plasmid RK2 (4). RK2 and RP4 are identical (5). Hence results from studies with RK2 can also be used for RP4 and vice versa. Three non-contiguous regions of RK2 have been identified as being necessary for replication of this plasmid (6). In this paper we present a method for construction of replication-defective mutants of RP4, their characterization in a transcomplementation system and their mapping by heteroduplex analyses.

At the moment we still know too little about the sophisticated mechanisms enabling a RP4 plasmid to be stably inherited. Thus we use the synonym "replication-defectiveness" to describe phenomena which might affect functions of replication-initiation, plasmid segregation or other maintenance factors.

RESULTS

1) *The Plasmid System*

Two derivatives of RP4 were used throughout this work. The plasmid to be mutated in replication genes is RP4-2 (Ap^s). RP4-2 carries an insertion (ISR1) and an adjacent deletion in the ampicillin-transposon Tn1 (7). The plasmid which should complement replication defective RP4-2 mutants is pSR100. pSR100 is a deletion mutant which has been isolated from RP4 by Mu-mutagenesis (8). The deletion has removed most of the transfer functions of RP4 and the kanamycin-resistance gene.

Important additional features of the plasmids are: pSR100 has lost the rlx-locus (the site of relaxation nicking) (6), which makes it unable to be mobilized by RP4 (Mob$^-$). It has also lost the porperty of surface exclusion (Sex$^-$). Thus, it does not prevent the entry of another P-type plasmid. We have also studied the two plasmids with regard to their incompatibility interactions. The incompatibility was tested in different experiments. In one test Ap^s or Tc^s derivatives of PR4 were introduced by conjugation into a strain already carrying pSR100. The incoming plasmid was selected using Km-containing agar. All transconjugants tested (more than 3000) have lost pSR100. In another test, co-selection was used for both the incoming RP4-Tc^s plasmid (Km) and the residing PSR100 (Tc). On plates containing Km and Tc the resulting transconjugant colonies are very small. The cells from 10 single colonies isolated from the co-selection plates were mixed and retested. It was found, that all cells tested still carried the new plasmid (RP4-Tc^s)

35

whereas only 1 out of 10^4 cells could be regrown on Tc-medium indicating the presence of pSR100. Again those colonies grew very slowly and contained pSR100 as well as the RP4-Tc^s plasmid.

Similar results have been obtained in reversed incompatibility tests using pSR100 transducing lysates of the phage P1. We therefore conclude that the incompatibility between RP4 and its mini-RP4 derivative pSR100 is very strong and asymmetric in that pSR100 is always lost. This particular property of pSR100 can be called "Sin$_p$" which stands for "susceptibility for P-type incompatibility" (9).

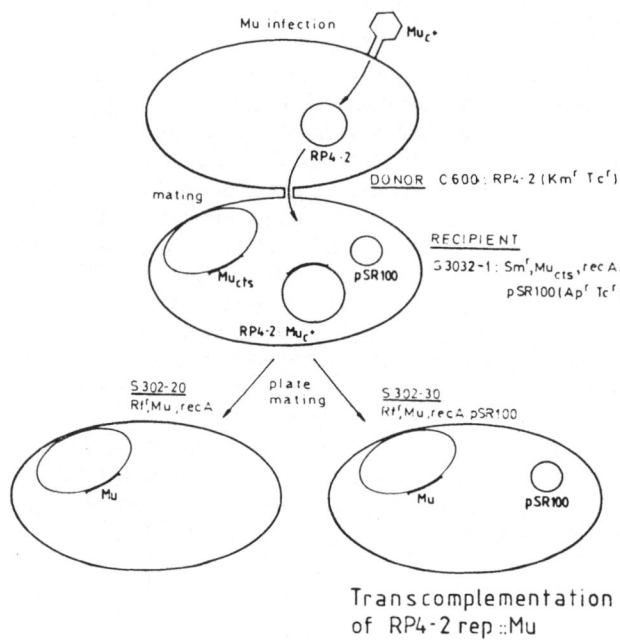

Fig. 1. Scheme of Mu-mutagenesis and screening for RP4-2 *rep*: :Mu plasmids.
Strains: C 600 (*thi, thr, leu, lac, su*$_{III}$); S3032-1, S302-20 and S302-30 are derivatives of CSH52 (*thi, pro, recA*, Sm$_r$) with the additional phenotypes given in the figure. For explanation of the procedure see the text.

The phenotypic feature of the plasmids used in the following chapters are summarized as follows:

RP4-2   (Ap^s, Km^r, Tc^r, Tra$^+$, Inc$_p$$^+$)
pSR100   (Ap^r, Km^s, Tc^r, Tra$^-$, Mob$^-$, Sex$^-$, Sin$_p$$^+$).

2) *Mu-Mutagenesis and Screening for Replication-Defective RP4-2 Mutants*

The temperature *E. coli* bacteriophage Mu was used to construct insertion mutants of RP4-2. Mu shows the extraordinary property of being able to insert its genome at random into the host DNA during lysogenization of a cell (for review see 10). Mu-insertions into an operon cause very stable polar mutations (10). Mu can also be inserted into any transmissible plasmid of an *E. coli* strain.

The mutagenesis procedure for RP4-2 is outlined in Fig. 1. A donor strain carrying RP4-2 is infected with Muc+ (c$^+$ = wild-type repressor gene) leading to random

36

insertion of the Mu-DNA into the plasmid. Since the infected cells will lyse within a short period of time, the plasmids have to be rescued by mating with a Mucts lysogenic recipient. RP4-2::Mu$_c$+ hybrid-plasmids can be detected in the Muc ts background by selection for Km^r at 42°C. Recipients normally will lyse due to heat-induction of the Mucts prophage unless they obtain an RP4-2::Mu$_c$+ plasmid (8). Obviously, the wild-type repressor molecules made by the Mu$_c$+ prophage of the new hybrid-plasmid are transdominant over the thermolabile repressors from the Muc ts. prophage of the recipients. Assuming that the Mu-DNA has been inserted into a trans-acting replication gene of RP4-2, those RP4-2::Mu hybrids should be able to survive in the recipient

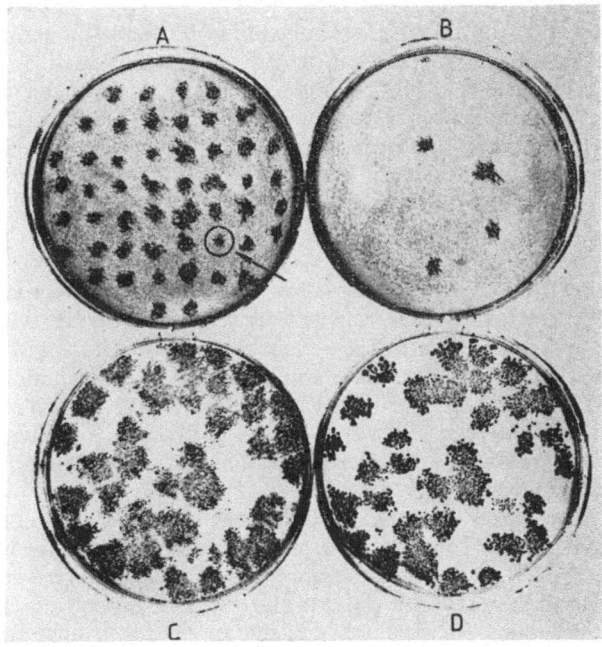

Fig. 2. Screening for RP4-2 rep::Mu by replica plating experiments.

The procedure is described in chapter 2. It can be summarized as follows:

Plate	Medium	Test
A	ApKmSm, 42°C (masterplate)	co-selection for pSR100 and RP4-2::Mu$_c$+; donors for C/D
B	ApSm	stability of pSR100
C	KmRf	transconjugants with S302-20 (*rec*A, Rf, Mu)
D	KmRf	transconjugants with S302-30 (*rec*A, Rf, Mu, pSR100)

Antibiotics were used in the following concentrations: Ampicillin (Ap) 100 µg/ml; Kanamycin (Km) 30 µg/ml; Streptomycin (Sm) 200 µg/ml, Rifampicin (Rf) 50 µg/ml.

S3032-1. The resident pSR100 in S3032-1 should transcomplement the destroyed replication function. This assumption can be tested as follows.

Primarily the transconjugants of the first mating are grown on plates containing Ap, Km and Sm at 42°C (Fig. 2, plate A). These conditions select for both the presence of RP4-2::Mu$_c$+ and pSR100 in the recipients. The clones are retested for stable inheritance of pSR100 by replicating them on Ap-medium. 4 out of 50 colonies tested seemed to contain both plasmids (Fig. 2, plate B). The Km-test plate is not shown, all the 50 colonies grew on it.

Further mating experiments can confirm the existence of transcomplementation of RP4-2 rep::Mu by pSR100. Two new recipients are used, one of them carries pSR100 (S302-30) and one does not (S302-20) (see also Fig. 1). The selection is for transfer of RP4-2::Mu (Km^r). In Fig. 2 the transconjugants of these plate mating experiments are shown. There is one colony (marked on plate A of Fig. 2) which does not transfer its plasmid into S302-20 (plate C of Fig. 2). The plasmid of this clone can only be established in a new recipient with the help of a residing pSR100 (plate D of Fig. 2). Since the transcomplementation phenomenon has been confirmed by this mating experiment, the plasmid of the marked colony is said to be a replication-defective RP4-2 mutant (RP4-2 rep::Mu).

3) *Properties of the PR-4 rep::Mu Hybrid-Plasmids*

Matings performed by replica plating allow only a rough estimation of transfer frequencies. Therefore the mating experiments were repeated with six different rep-mutants by filter membrane crosses. The transfer-frequency of an RP4-2::Mu control-plasmid (containing the Mu-genome in a nonessential region of RP4) is about 10^{-1} with the two test recipients S302-20 and S302-30 (pSR100). The RP4-2 rep::Mu mutants showed about the same transfer-frequency with S302-30 (pSR100), but they were established in S302-20 at a 10^2–10^5 fold lower frequency. This again confirms the existence of the helper function of pSR100.

The stability of the rep-mutants was tested in the following way. From double-selection plates (Ap, Km) single colonies were picked and grown in a medium without antibiotics. At intervals of 5, 10 nad 15 generations single colonies were grown on nutrientagar and tested for their resistance pattern and transfer properties. All tested colonies (50) of each strain retained both plasmids indicated by their resistance against Km and Ap. They also showed the expected transfer properties with the two recipients mentioned above.

Obviously the transcomplementation system described works effectively enough to keep the replication-defective RP4 mutants maintained in their *E. coli* host over at least 15 generations. This observation is puzzling in view of the incompatibility between RP4-2 and pSR 100.

4) *The Problem of Incompatibility*

The foregoing results have primarily focused on identifying replication-defective mutants. In the same mutagenesis procedure as described in 2), we also looked for RP4-2::Mu mutants which became compatible with pSR100 (RP4-2 inc::Mu) but are independent of the presence of pSR100. After mutagenesis and selection for Ap^r and Km^r, RP4-2::Mu transconjugants were tested for their stability of the Ap-resistance by replica plating. As can be seen in Fig. 2 (plate B), only 4 out of 50 colonies were stable Ap^r in the second cycle of growth. One of them has been shown to contain a RP4-2 rep::Mu plasmid (see chapter 2). Surprisingly the plasmids of the other 3 Ap^r clones became transfer-defective despite the fact that the markers of RP4-2::Mu were still present.

It was necessary at this stage to look at the plasmid content of such clones. In a similar experiment we isolated 12 clones which were stable Km^rAp^r and therefore seemed to contain RP4-2 inc::Mu (or RP4-2 rep::Mu) together with pSR100. Fig. 3 shows the results of this experiment. Cleared lysates of the clones were subjected to agarose gel electrophoresis to see how many plasmids they contain and what the approximate sizes of the plasmids are.

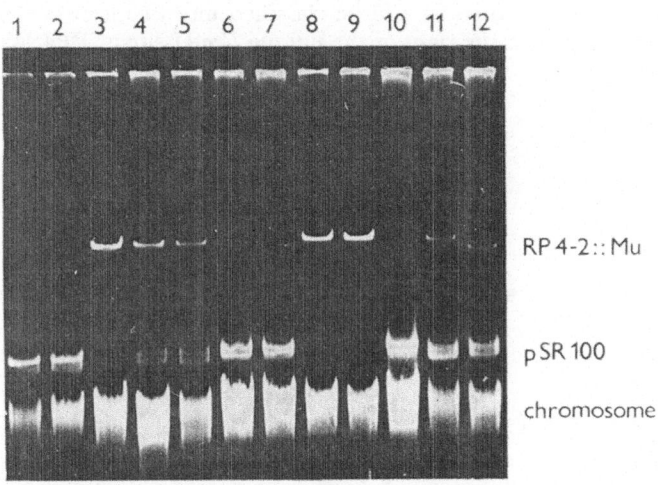

Fig. 3. Agarose-gel-electrophoresis of cleared lysates of presumptive RP4-2 inc::Mu containing clones.
Cleared lysates were prepared from cultures (5 ml) of Ap^rKm^r clones. The lysis was performed by SDS followed by NaCl-precipitation of membrane-chromosome complexes and plasmid-DNA enrichment by PEG-precipitation. The crude DNA preparations were run in a 0.9% agarose gel in borate-buffer for 3 hrs at 120 V. The plasmid content of the strains is explained in chapter 4.

Only strain 4, 5, 11 and 12 contained two independent plasmids, pSR100 and RP4-2::Mu or for strain 12 a deletion mutant of the latter. (It is not yet clear, why pSR100 forms a double band when it is brought together with RP4-2::Mu). These particular strains turned out to contain RP4-2 rep::Mu plasmids rather than pure inc-mutants as could be shown in further transfer tests. All the other so-called inc-mutants clearly inherit only one independent plasmid, either RP4-4::Mu or pSR100. In the first case, represented by strains 3, 8 and 9, the Ap-resistance of pSR100 might have been transposed via Tn1 into the chromosome. The second case, where only pSR100 can be seen in the gel (represented by strains 1, 2, 6, 7 and 10), can most probably be explained by the assumption that RP4-2::Mu has become integrated into the chromosome. Such clones do not transfer RP4::Mu plasmids in plate matings.

From these studies it can be concluded that it is obviously not possible to isolate pure inc-mutants of RP4-2 in a one-step insertion mutagenesis. Stable Ap^r Km^r clones contain either only one plasmid, suggesting that the markers of the other plasmid somehow became integrated into the chromosome, or two plasmids with the condition that the RP4-2::Mu plasmid is a replication-defective one.

The lack of success in isolating pure inc-mutants makes it attractive to consider a model in which the P-type incompatibility is very closely related to, if not identical with, a transacting function needed for replication or maintenance of RP4. Such a hypothesis could be supported by the results reported in chapter 3: RP4-2 rep::Mu hybrid plasmids are stably maintained together with pSR100 even in the absence of any selection pressure. This could mean that they have also lost their inc-properties.

5) *Mapping of Replication Genes*

By restriction analyses with different enzymes it could be shown that the six RP4-2 rep::Mu plasmids examined so far can be classified into two groups. One group maps on the left and the other on the right of the unique *Eco*RI site of RP4. As the *Eco* RIsite

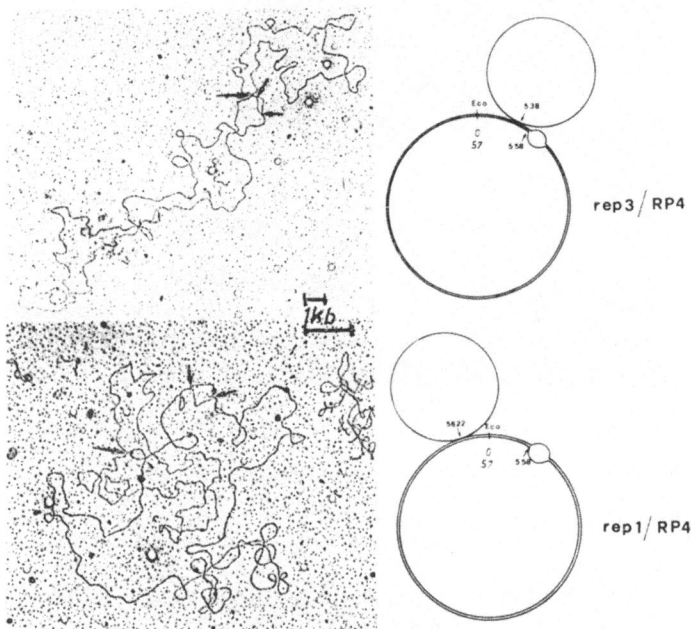

Fig. 4. Heteroduplex analyses of RP4-2 rep::Mu plasmids.
The RP4-2 rep::Mu plasmids were hybridized with RP4-DNA and prepared by formamide spreading for electron-microscopic analyses. The arrows indicate the insertion point of Mu-DNA and the bubble-structure which is used as internal measurement marker. The unique EcoRI restriction site is taken as point 0 and 57 kb (kilobases).

itself does not interrupt an essential operon of RP4 (11), it can be concluded that there are at least two different transacting replication functions on this plasmid. Two members of each class have been subjected to a more defined mapping by means of heteroduplex studies. When hybridized with RP4, RP4-2 gives a characteristic "bubble-structure", which is due to the above mentioned insertion and deletion of RP4-2. This structure is used as an internal marker in the heteroduplex analyses of RP4-2 rep::Mu plasmids. The distances from the Mu-insertions to the bubble-structure in connection with the data from appropriate restriction patterns give the exact map positions of the replication mutations isolated in this work.

Two examples are shown in Fig. 4, one maps very close to the *Eco*RI site (*rep*1 in Fig. 4), the other is distantly located away from the *Eco*RI site close to the ampicillin-transposon Tn1 (*rep*3 in Fig. 4). In summary, the map positions are given in kb taking the *Eco*RI site as point 0 and 57 kb of the RP4 map: in the RP4-2 rep::Mu mutants *rep*1 and *rep*5 the Mu-genome has found to be inserted at 56.7 kb and 56.2 kb respectively (counterclockwise close to *Eco*RI). Mutant *rep*3 carries the Mu-insertion 5.4 kb and mutant *rep*4 4.7 kb clockwise away from the *Eco*RI site (close to Tn1). The distances between *rep*1 and *rep*5 or *rep*3 and *rep*4 respectively are very small. Thus the two pairs of mutations can be regarded to represent two single genes or small operons.

DISCUSSION

By Mu-insertion mutagenesis essential genes of RP4 can be mutated. Immediately after Mu-insertion, the replication defective RP4::Mu hybrids are rescued into a recipient where transcomplementation can occur due to the presence of a mini-RP4 plasmid (pSR100). Further mating experiments confirm the existence of RP4 rep::Mu mutants, as such plasmids can only be established and maintained in new recipients with the help of pSR100. Despite the strong incompatibility between RP4 and pSR100, cells containing a replication defective RP4::Mu plasmid also retain pSR100 even in the absence of any selection pressure. Thus, *rep*⁻ mutants found in the transcomplementation system described also seem to be *inc*⁻. On the other hand, *inc*⁻ *rep*⁺ mutants could not be isolated with Mu-insertions. This suggests that at least the two transcomplementable replication functions of RP4 mapped in this work are very closely related to the incompatibility phenomenon. In other words incompatibility could be an inevitable by-product of essential plasmid functions like replication, maintenance or segregation. In our mutagenesis system probably only such mutants can be found which lost rep- and inc functions simultaneously.

It is interesting to note that C. M. Thomas et al. mapped two transacting replication functions on RK2 : *trf*B close to the *Eco*RI site and *trf*A close to the Tc^r gene and oriv (6). These authors constructed a very small derivative of RK2 capable of autonomous replication by *in vitro* joining of *trf*B with *trf*A and the origin (pCT5). Since RK2 and RP4 are identical (5) those results can be compared with the map positions of the Mu-induced rep-mutants of this paper. The RP4-2 rep::Mu plasmids *rep*1 and *rep*5 are mutated in the region which corresponds to *trf*B.

So far, we could not find Mu-insertions in the region of *trf*A. But, on the other hand, with the mutants *rep*3 and *rep*4 (close to Tn1) we found a DNA-segment which is essential for replication of RP4 but which is obviously dispensable for replication of pCT5. It can be suggested that a normal RP4 needs more transacting information to be stably maintained than its very small derivative pCT5.

V. A. Sakanyan et al. found by deletion mapping of an RP4::ColEI hybrid-plasmid (pAS8) that there are incompatibility functions of RP4 in a DNA-segment with the coordinates 2.1–9.8 kb (12). Our mutants *rep*3 and *rep*4 are located within this section. It should finally be noted that P. T. Barth isolated a RP4 mutant which is not stably inherited in *E. coli* and which also exhibits a reduced host range (3). The mutation is due to insertion of Tn76. Its location coincides with *trf*B (6) and our *rep*1/*rep*5 mutations.

REFERENCES

1. DATTA, N., R. W. HEDGES (1972): J. Gen. Microbiol. *70*, 453.
2. OLSEN R. H., P. SHIPLEY (1973): J. Bacteriol. *113*, 772.
3. BARTH P. T. (1979): in "Plasmids of Medical, Environmental and Commercial Importance", ed. by K. Timmis and A. Pühler, Elsevier, North Holland.
4. MEYER, R. J., D. R. HELINSKI (1977): BBA *478*, 109.
5. BURKARDT H. J., G. RIESS, A. PÜHLER (1979): J. Microbiol. (in press).
6. THOMAS C. M., D. STALKER, D. GUINEY, D. R. HELINSKI (1979): in "Plasmids of Medical, Environmental and Commercial Importance", ed. by K. Timmis and A. Pühler, Elsevier, North Holland.
7. PRIEFER U., A. PÜHLER, this volume.
8. SIMON R., Diplomarbeit, Universität Erlangen (1976).
9. OLSEN R. H., A. L. SHIPLEY (1975): J. Bacteriol. *123*, 28.
10. HOWE M. M., E. G. BADE (1975): Science *190*, 624.
11. MEYER R., D. FIGURSKI, D. R. HELINSKI (1975): Science *190*, 1226.
12. SAKANYAN V. A., L. Z. YAKUBOV, S. I. ALIKANIAN and A. I. STEPANOV (1978): MGG *165*, 331.

R. S., Inst. f. Mikrobiologie,
Univ. Erlangen, Egerlandstrasse 7,
D-852 Erlangen, B.R.D.

TRANSLOCATION OF DRUG RESISTANCE GENES
IN *PSEUDOMONAS AERUGINOSA*

S. IYOBE, H. SAGAI, K. HASUDA and S. MITSUHASHI

Department of Microbiology, School of Medicine, Gunma University, Maebashi, Japan

INTRODUCTION

Conjugative or nonconjugative drug-resistance plasmids have been demonstrated in *P. aeruginosa* strains of clinical origin and they were shown to be important factors in clinical spread of resistant strains (Bryan, Semeka, Van Den Elzen, Kinner, and White-house, 1973, Iyobe, Hasuda, Fuse, and Mitsuhashi, 1974, Korfhagen, Ferrel, Menefee, and Loper, 1976). Conjugative resistance plasmids (R plasmids) in *P. aeruginosa* are mostly transferred only to *P. aeruginosa* strains but not to *E. coli* and other strains of *Enterobacteriaceae*. They have been classified by incompatibility in intraspecies conjugation system (Sagai, Hasuda, Iyobe, Bryan, Holloway, and Mitsuhashi, 1976). On the other hand, nonconjugative resistance plasmids in *P. aeruginosa* are transferred not only to *P. aeruginosa* strains but also *E. coli* srains by mobilization or transformation, and maintained stably in *E. coli* (Iyobe, Hasuda, Kato, Sagai, and Mitsuhashi, 1979).

As the third factor in epidemiological spread of resistance genes, transposable genetic elements, transposons, have been recognized recently (Cohen, 1979), because they encode drug-resistance and are transposable among plasmids, phages, and bacterial chromosome. Transposons carrying the resistance gene(s) to tetracycline (Tc), chloramphenicol (Cm), streptomycin (Sm), sulfanilamide (Su), ampicillin (Ap) or kanamycin (Km) were found in R plasmids isolated from *E. coli* strains. In *P. aeruginosa*, transposons mediating Hg resistance was found in a nonconjugative plasmid (Stanisich, Bennet, and Richmond, 1977) and carbenicillin (Cb) resistance was also mediated by transposons in R plasmids (Cohen, 1976).

In the survey of resistance plasmids in *P. aeruginosa* of clinical origin, we observed that Cm resistance genes(s) (*cml*) in various R plasmids were transposable to lysogenic phages and further translocated to various R plasmids from these phages. We assumed that the *cml* gene in *P. aeruginosa* was carried by transposons as in the case of *E. coli* plasmids, in which the transposon Tn9 carrying *cml* gene was first observed in the formation of active phage P1*cml* (Kondo and Mutsuhashi, 1964). Cm resistance mediated by plasmids was mostly due to the formation of acetyltransferase. But R plasmids mediating Cm resistance incapable of inactivating the drug were often found in *P. aeruginosa* strains (Mitsuhashi, Kawabe, Fuse, and Iyobe, 1975). These two types of the *cml* genes in *P. aeruginosa* were found to be transposable. This paper deals with the genetic characters of these Cm transposons.

Translocation of the cml genes from R plasmids to a phage. In the transduction of R plasmids with a lysogenic phage F116, we found that only Cm resistance was transduced from multiple resistance plasmids but other resistances were not (Table I). We picked up ten colonies and purified them on Cm plate in each case of transduction. It was found that all Cm-resistant transductants had immunity to the F116. The culture filtrates from transductants through membrane filter could transduce Cm resistance at a high frequency and form plaques on the lawn of sensitive cells. Furthermore, most bacterial cells isolated

TABLE I.

Translocation of Cm genes to a phage F116 from R plasmids

R plasmid in F116 lysogenic strain	Transduction of		Existence of F116 in transductants
	Cm	Other genes	
Rms159 (Tc Cm Sm Hg)	+	−	+
Rms161 (Tc Cm Sm Su Km Hg)	+	−	+
Rms171 (Cm Sm Su Hg)	+	−	+
Rms196 (Tc Cm Sm Su Km)	+	−	+
Rms162 (Cm Sm Su Hg)	+	−	+

Donor; ML4600 carrying both F116 and R plasmid.
Recipient; ML4262.

from these plaque centers carried Cm resistance. These facts indicated that the active F116 phages carrying *cml* genes, that is, F116*cml* were formed by translocation of *cml* genes from R plasmids to phage F116.

Further translocation of cml genes to R plasmids from an F116cml phage. We isolated an F116*cml*$_1$ whose *cml* was derived from an Rms159 which belonged to incompatibility group P-2, and examined whether the *cml* gene was further transposable to other replicons. Donor strains for conjugation were prepared by superinfecting R plasmids to the strain carrying F116*cml*, and mated with a recipient ML4731. When we used the resistant mutant of ML4262 (ML4731) to F116 phage as a recipient of R plasmids, it was expected that the R plasmids accepted the *cml*$_1$ gene from F116*cml*$_1$ and could be detected in this recipient by selecting Cm-resistant transconjugants. Various R plasmids belonging to different incompatibility groups were used as acceptors of the *cml*$_1$ gene (Table II). These R plasmids were transferable by conjugation to the recipient strain at frequencies of 10^{-3} to 10^{-1}. When Cm-resistant transconjugants were selected on Cm plate, R plasmids carrying *cml*$_1$ were obtained at lower frequencies than that of R plasmids without *cml*$_1$. The frequency of translocation was expressed as a ratio of the number of R plasmid carrying *cml*$_1$ to that of the original R plasmid. As shown in Table II, the translocation frequencies of the *cml*$_1$ gene were found to be different among acceptor R plasmids ranging from 10^{-4} ro 10^{-2}. A hundred colonies were picked from Cm transconjugants and their genetic properties were examined. Most of the *cml*$_1$-transposed R plasmids carried both the original resistance markers in addition to Cm resistance and conjugal transferability. In rare

TABLE II.

Translocation of cml$_1$ gene to other plasmids from F116cml$_1$

Plasmid in F116 *cml*$_1$$^+$ strains	Group	Resistance	Selection	
			Cm	Plasmid marker
RP4	P-1	Tc Km Cb	10^{-4}	10^{-2}
R1b151	P-3	Sm Su Gm Cb	10^{-4}	10^{-1}
Rms176	P-5	Sm	10^{-6}	10^{-3}
Rms148	P-7	Sm	10^{-4}	10^{-1}
FP2	P-8	Hg	10^{-6}	10^{-2}

Donor; ML4600 carrying both F116 *cml*$_1$ and R plasmid: Recipient; ML4731.

TABLE III.

Rec independent translocation of cml genes between R plasmids

Donor		Freq. of translocation
Plasmids	Rec	
Rms159-1 (Cm Hg)	+	1×10^{-2}
+		
Rms148 (Sm)	−	5×10^{-3}
Rms162 (Cm Sm Su Hg)	+	1×10^{-3}
+		
Rms148 (Sm)	−	9×10^{-4}

Donor; PAO2142 *rif* and its *rec⁻* mutant
Recipient; ML4600

cases, however, the loss of resistance was observed in RP4cml_1 and R1b151cml_1 recombinants, that is, Km resistance was lost from 0.1% of RP4cml_1 and both Sm and Gm resistances were lost from 12% of R1b151cml_1.

Rec independent translocation of cml genes. We selected two transposable *cml* genes originated from Rms159 and Rms162 (Table I) and used for further genetical experiments. We designated *cml* genes of Rms159 and Rms162 as cml_1 and cml_2, respectively. The mechanism of Cm resistance was different in these two *cml* genes, because bacteria carrying Rms162 inactivated Cm by acetylation but that carrying Rms159 did not (Mitsuhashi, Kawabe. Fuse, and Iyobe, 1975). In order to know whether *rec* function of host bacteria is required for translocation of the *cml* genes, the *cml* genes were translocated to RMs148 or to phage P29 in the background of *rec⁺* or *rec⁻* host, e.i., PAO2142 and its *rec⁻* mutant isolated by us (Iyobe, Hasuda, Kato, Sagai, and Mitsuhashi, 1979).

Both donor plasmid and acceptor plasmid of *cml* gene were doubly infected to *rec⁺* or *rec⁻* host to examine the translocation of *cml* gene between R plasmids. The frequency of translocation was expressed as the ratio of the number of transconjugants carrying Rms148*cml* to that carrying Rms148. Table III. shows that the *cml* genes were translocated at about the same frequencies irrespective of *rec* function of their host, to Rms148 from Rms159-1, a derivative of Rms159, or from Rms162. The translocation frequency was 10-fold higher in cml_1 of Rms159-1 than in cml_2 of Rms162.

TABLE IV.

Rec independent translocation of cml genes to a phage P29 from R plasmids

Donor		Freq. of Cm transduction	P29 *cml* in transconjugants
Plasmid and phage	Rec		
Rms159-1 (Cm Hg) + P29	+	2×10^{-6}	+ (10/10)
	−	7×10^{-7}	+ (10/10)
Rms162 (Cm Sm Su Hg) + P29	+	1×10^{-7}	+ (10/10)
	−	5×10^{-8}	+ (10/10)

Donor; PAO2142 *rif* and its *rec⁻* mutant: Recipient; PAO2142.

The *rec* independency of *cml* translocation was also indicated when phage P29 was used as an acceptor of the *cml* genes from Rms159-1 or Rms162 (Table IV.). The *rec+* or *rec−* strain carrying P29 in lysogenic state was superinfected with Rms159-1 or Rms162, and transduction experiment was carried out using bacteria-free filtrates obtained from the cultures. In this case, the frequency of Cm transduction per p.f.u. of P29 phage in the filtrate was expressed as the translocation frequency. The Cm-resistant transductants were obtained at a frequency of about 10^{-6} or 10^{-7} from the filtrates from bacterial cultures carrying Rms159-1 or Rms162, respectively. These frequencies were not changed whether *rec+* or *rec−* strain was used as the host of donor. We picked up 10 colonies of Cm-transductants in each case, and confirmed that all of them had immunity to P29 phage and liberated active P29 phage. We further obtained filtrates from each culture of transductants and isolated plaques on the lawn of P29 sensitive strains after an appropriate dilution of the filtrate. The P29 lysogenized cells in the plaque centers were found to carry *cml* genes. These facts indicate that transductants carried P29cml_1 or P29cml_2, active phages showing Cm resistance. Several percent of lysogenic cells with phages obtained from strains carrying P29cml_1 or P29cml_2 had not accepted Cm resistance, indicating that elimination of the *cml* genes from P29cml occurred when phage particles were formed in bacterial cells or after infection to P29-sensitive cells.

Characterization of the Cm segments translocated on P29 phage. For further characterization of cml_1 or cml_2 gene, we translocated *cml* genes from P29cml to RP4. Donors carried both P29cml and RP4, and 33-72b was used as a recipient which was a resistant strain to the P29 phage. It was found that about half of 10 P29cml phages could translocate the *cml* gene. Electron microscopy of heteroduplex between phage DNAs of P29 and P29cml revealed the formation of insertion roops of single-strand DNA corresponding to translocated segments, and in some cases, the formation of substitution-like roops was observed. The latter type of heteroduplex was frequently shown in P29cml DNA whose *cml* gene could not be further translocated. Agarose gel electrophoresis showed that the transposition of *cml* into P29 phage DNA yielded a new cleavage site for *Eco*R1, *Pst*1 or *Bam*H1 enzymes. These cleavage sites were also detected in RP4cml DNA, indicating that the transposable segments had the sites recognized by these three enzymes. It is known that the RP4 has only one cleavage site for each *Eco*RI and *Bam*H1 DNA (Barth, Grinter and Bradley, 1978). The translocation of the cml segments yield a new cutting site by these enzymes. The location of *cml* segments on RP4 was assumed using the physical map of the RP4 DNA, by estimating the size of the newly appeared segment on agarose gel after digestion of RP4cml DNA by *Eco*R1 or *Bam*H1. Using RP4cml DNAs from independently isolated cells carrying RP4cml_1 or RP4cml_2, we found that both cml_1 and cml_2 transposons were incorporated into different sites of RP4 DNA.

REFERENCES

BARTH P. T., N. J. GRINTER and D. E. BRADLEY (1978): Conjugal transfer system of plasmid RP4: analysis by transposon 7 insertion. J. Bacteriol. 133, 43–52.
BRYAN L. E., S. D. SEMEKA, H. M. VAN DEN ELZEN, J. E. KINNER and R. L. S. WHITEHOUSE (1973): Characteristics of R931 and other *Pseudomonas aeruginosa* R plasmids. Antimicrob. Ag. Chemother. 3, 625—637.
COHEN S. N. (1976): Transposable genetic elements and plasmid evolution. Nature 263, 731–738.
IYOBE S., K. HASUDA, A. FUSE and S. MITSUHASHI (1974): Demonstration of R plasmids from *Pseudomonas aeruginosa*. Antimicrob. Ag. Chemother. 5, 547–552.
IYOBE S., K. HASUDA, T. KATO, H. SAGAI and S. MITSUHASHI (1979): Drug-resistance plasmids in *Pseudomonas aeruginosa*. Microbial Drug Resistance (Mitsuhashi, S., ed.) 2, 187–192.

KONDO E. and S. MITSUHASHI (1964): Active transducing bacteriophage PlCM by the combination of R factors with bacteriophage Pl. J. Bacteriol. 88, 1266–1276.

KORFHAGEN T. R., J. A. FERREL, C. L. MENEFEE and J. C. LOPER (1976): Resistance plasmids of *Pseudomonas aeruginosa*: change from conjugative to nonconjugative in a hospital population. Antimicrob. Ag. Chemother. 9, 810–816.

MITSUHASHI S., H. KAWABE, A. FUSE and S. IYOBE (1975): Biochemical mechanisms of chloramphenicol resistance in *Pseudomonas aeruginosa*. Microbiol Drug Resistance (Mitsuhashi, S. and Hashimoto, H. ed.). 1, 515–523.

SAGAI H., K. HASUDA, S. IYOBE, L. E. BRYAN, B. W. HOLLOWAY and S. MITSUHASHI (1976): Classification of R plasmids by incompatibility in *Pseudomonas aeruginosa*. Antimicrob. Ag. Chemother. 10, 573–578.

STANISICH V. A., P. M. BENNETT and M. H. RICHMOND (1977): Characterization of a translocation unit encoding resistance to mercuric ions that occurs on a nonconjugative plasmid in *Pseudomonas aeruginosa*. J. Bacteriol. 129, 1227–1233.

S. I., Dept. of Microbiology,
School of Medicine, Gunma Univ.
Maebashi, Japan

CARBENICILLIN RESISTANCE AND MULTIRESISTANCE PLASMID IN *PSEUDOMONAS AERUGINOSA*

Y. MICHEL-BRIAND, M. J. DUPONT, I. CHARDON-LORIAUX
and J. M. LAPORTE

Laboratory of Bacteriology-Virology,
Faculty of Medicine, Besançon, France

INTRODUCTION

Pseudomonas aeruginosa used to be susceptible to Carbenicillin, a β-lactam antibiotic. However resistance appeared and the first extrachromosomic resistance mechanism in *P. aeruginosa* was reported in 1969 by Lowbury.

The mechanism involves a β-lactamase able to split the β-lactam ring. In *Pseudomonas aeruginosa* several, but not all β-lactamases may be responsible for the Carbenicillin resistance (Table I.). Other mechanisms (e.g. permeability barrier) do not seem to play a role in plasmidic resistance.

This paper reports: (1) the percentage of resistant strains isolated in eastern France over a period of several years; (2) the rate of plasmidic localisation and (3) the description of a multiple resistance plasmid (see Michel-Briand, Stanisich, Jouvenot, 1977 for a report on single resistance plasmid).

TABLE I.

Main characteristics of β-lactamases from Pseudomonas aeruginosa
(from Sawada 1975, Sykes 1976).

	Pen	Amp	Carb	Clox	Cephr	Cepht	Pi
Penicillinase							
constitutive							
chromosome							
Dalgleish	100	90	150	1	40	5	5.3
plasmid							
HL	100	85	89		37		5.3
Rms 139	100	98	105	2	20		5.9
Rms 149	100	101	253	3	10		7.05
Cephalosporinase							
inducible							
chromosome							
Sabath	100	10	0	0	400		7.7
GN 918	100	10	5	5	770	50	8.7
Peni-cephalosp.							
constitutive							
plasmid							
Rms 165	100	108	10	2	125		5.25
Tem 1	100	150	10	0	120	20	5.4
Tem 2	100	150	10	0	120	20	5.6

RESULTS

(1) *Carbenicillin resistant Pseudomonas aeruginosa strains*

A *Pseudomonas aeruginosa* strain was considered Carbenicillin resistant when the minimal inhibitory concentration (MIC) is equal to or greater than 128 μg/ml (Disc method with Mueller Hinton medium). The proportion of Carbenicillin resistant strains since 1972, has been 25.6% of 3,154 strains. 25% of the strains were resistant in 1972

TABLE II.

Percentage of strains carbenicillin sensitive Cb^s or resistent Cb^r associated with aminoglycosides resistance streptomycin Sm, kanamycin Km, paromomycin Pm, gentamicin gm, tobramycin Tm). Only the most frequent associations are indicated.

% strains resistant to:

	Pm Km Pm	Sm Km Pm	Sm Km Pm Gm Tm
Cb^s	19.8	15.6	5.9
Cb^r	2.1	0	18.2

and 29% in 1978. In 1979, 18.2% of Carbenicillin (Cb) resistant strains were also resistant to streptomycin (Sm), kanamycin (Km), paromomycin (Pm), gentamicin (Gm) and tobramycin (Tm) (Table II.). We cannot say if all these strains are different as pyocin or phage typing was not performed on them.

(2) *Plasmidic localisation of Carbenicillin resistance*

One hundred and eighty six highly carbenicillin resistant strains (MIC $\geq$ 500 μg/ml) were tested for plasmid transfer by at least two methods: (i) mating on solid medium: wild strains were grown for 6 hours on Brain Heart agar and then replicated on selective medium previously inoculated with the recipient strain PAO 2635 F^- *trp-54 rif-3* (Stanisich and Ortiz 1976) (ii) mating in liquid medium: wild and recipient strains (5 . 10^8 bacteria/ml) were mixed in Brain Heart broth and incubated overnight. Then 10 μl of the mixture was dropped on selective medium.

In all cases, colonies obtained on selective media were identified as transconjugants after purification and were checked for their genetic characteristics.

Other methods were also used in part of this work: determination of the β-lactamase involved in the resistance by studying its substrate profile or isoelectric point. It seems that the determination of the characteristics of a β-lactamase cannot prove its chromosomic or plasmidic control (Jouvenot, Michel-Briand and Laporte, 1978). Of 186 strains tested, only 10 transferred carbenicillin resistance.

(3) *Main characteristics of a multiresistance plasmid isolated from Pseudomonas aeruginosa Mo 19.*

Serial transfer of carbenicillin resistance

Wild strain Mo 19 transferred carbenicillin resistance (MIC $\geq$ 1.000 μg/ml) to other

Pseudomonas aeruginosa PAO 2635, PAO 8 *F⁻ met-28 ilv-202 str-1* (Isaac and Holloway, 1972) PAO 1670 *F⁻ ade-136 leu-8 chl-3 rif-1* (Krishnapillai, 1974) with a transfer frequency (TF) 10^{-5} to 10^{-7}. Resistance to aminosides (streptomycin, spectinomycin (Sp), neomycin (Nm)/paromomycin, lividomycin (Lv), ribomycin (Rm), kanamycin, gentamicin, sisomycin (Si), and to other antimicrobial agents [chloramphenicol (Cm), tetracyclin (Tc), Sulfonamide (Su), mercury (Hg)] was also transferred. These resistances are then carried by a plasmid that we call pYMB1 according to the Novick nomenclature (1976).

Pseudomonas aeruginosa transconjugants can transfer all these resistances including carbenicillin, to *Escherichia coli* J5 F⁻ lac⁺ met pro az (Datta and Hedges 1972). However, after storage at 4°C for 3 weeks this plasmid bearing *E. coli* strain lost resistance to Sm, Sp, Gm and Si.

TABLE III.

Substrate profile of the pYMB1 β-lactamase. Enzyme activity was measured by the microacidimetric method. The substrate concentration of β-lactamase saturation was 0.24 mM; it was 6.58 mM for cefaloridin. All figures are relative to the Vmax for benzyl panicillin (arbitrarily = 100).

	Benzyl penicillin	Ampi cillin	Carbeni cillin	Oxa cillin	Cloxa cillin	Cefalo ridin
PA08 (pYMB1)	100	73–105	11–13	0	0	163–181

Plasmid pYMB1 was compared with RP1 by transferring pYMB1 from PAO 2635 $(5 \cdot 10^8$ bacteria/ml) to PAO 8 and PAO 8 (RP1-2) (which is a carbenicillin sensitive derivative of PAO 8 RP1, Curtis, Richmond and Stanisich 1973) $(2.5.10^9$ bacteria/ml). Transfers of pYMB1 or RP1 (control) to PAO 8 (RP1-2) from respectively PAO 2635 (pYMB1) and PAO 2635 (RP1) were unsuccessful (TF 5.10^{-8}). These plasmids, however, could be transferred to PAO 8 (TF $= 5.10^{-4}$ to 5.10^{-6}). Plasmid pYMB1 then seems very similar to RP1.

Resistance to carbenicillin

Carbenicillin resistance was due to a β-lactamase. The activity of this enzyme was measured by the pH-stat method (Jouvenot, Michel-Briand and Laporte, 1978).

Table III. gives a substrate profile and suggests that this enzyme is TEM-like. The isoelectric point (pI) of the β-lactamase was determined in polyacrylamide gel (5%) and ampholines (2.4%) with a pH gradient between 4 and 6.5. β-lactamase of pYMB1, RP1 and RP4 are similar (pI = 5.6) and differs from β-lactamase TEM1 (pI = 5.4). These results suggest that β-lactamase of pYMB1 is TEM2-like.

Isolation of the plasmid

The plasmid was obtained using lysis of bacteria in TRIS 0.05M buffer, with lysozyme (1.15 mg/ml) and 5% sodium dodecyl sulfate. After a short (2 mn) alkaline denaturation (pH = 12), the lysate was chromatographied on Sepharose 4B, then through a nitro-cellulose column, retaining denatured chromosomal DNA strands and allowing the passage of plasmidic DNA in the first fractions. An electron microscopic examination was performed on this material (formamide technique, shadowing with Au/Pd). After 40 determinations the pYMB1 DNA contour length was 25 ± 0.5 μm (40 measurements), giving a molecular weight of 52 Mdal (taking 1 μ = 2.07 Mdal).

DISCUSSION

The occurrence of plasmidic localisation of drug resistance in *P. aeruginosa* is variable according to different authors. Our percentage of successful transfer seems low, although 26% of bacteria are carbenicillin resistant. In the same species, Bryan (quoted by Dean et al. 1977) has found R plasmids in 2–3% of clinical isolates in Canada (collection including susceptible strains.) Dean, Morgan, Asche, Holloway (1977) observed 2.6% carbenicillin resistant strains (17 strains) in Australia. Only 4 strains could transfer resistance to Tc, Sm, and Su. Iyobe, Hasuda, Sagai and Mitsuhashi (1975) found 15% of carbenicillin resistant *Pseudomonas aeruginosa* strains (MIC $\geq$ 400 μg/ml), 69% of these carrying R plasmids.

These discrepancies may be due to several factors, such as therapeutic pressure which may be very different according to the country, or detection of genetic mechanisms from different levels of resistance (we chose highly resistant bacteria, i.e. we favoured the isolation of Dalgleish enzyme, depending on non-transferable gene). The problem of one type of enzyme associated with one type of genetic localisation should be reconsidered.

Plasmid pYMB1, carrying multiple resistance, belongs to the P1 group (exclusion with RP1 Cb sensitive, transfer to *E. coli*) in which other carbenicillin resistance plasmids have been described (Rm 16b, Japan, RP638 South Africa, R527 Spain).

This P1 group is identical with the *E. coli* P group, which seems very rare in human *E. coli* and more frequent in animal *E. coli*. We could speculate that *Pseudomonas* P1 plasmid was acquired from bacterial species of animal source.

β-lactamase coded by pYMB1 is a TEM2–like enzyme which is prevalent in enteric bacteria and for the corresponding gene it is more realistic to think in terms of transposon.

REFERENCES

DATTA N. and R. W. HEDGES (1972): R factors identified in Paris, some conferring gentamicin resistance, constitute a new compatibility group. Ann. Inst. Pasteur *123*, 849–852.
DEAN H. F., A. F. MORGAN L. V., ASCHE and B. W. HOLLOWAY (1977): Isolates of *Pseudomonas aeruginosa* from Australian hospitals having R-plasmid determined antibiotic resistance. The med. J. Australia *2*, 116–119.
JOUVENOT M., Y. MICHEL-BRIAND et J. M. LAPORTE (1978): Argument en faveur de la localisation plasmidique du gène "β-lactamase" chez une souche de *Pseudomonas aeruginosa* hautement resistante â la Carbenicilline. C. R. Acad. Sc. Paris. *287*, série D 1067–1070.
KRISHNAPILLAI V. (1974): The use of bacteriophages for differentiating plasmids of *Pseudomonas aeruginosa*. Genet. Res. *23*, 327–334.
LOWBURY E. J. L., A. KIDSON, H. A. LILLY, G. A. J. AYLIFFE, and R. J. JONES (1969): Sensitivity of *Pseudomonas aeruginosa* to antibiotics: emergence of strains highly resistant to penicillins. Lancet ii 448–452.
MICHEL-BRIAND Y., V. A. STANISICH and M. JOUVENOT (1977): *Pseudomonas aeruginosa* strain isolated in France that carries a plasmid determining Carbenocillin resistance. Antimicrob. Agents chemother. *11*, 589–593.
NOVICK R. P., R. C. CLOWES, S. N. COHEN, R. CURTISS III, N. DATTA and S. FALKOW (1976): Uniform nomenclature for bacterial plasmids: a proposal. Bacteriol. Rev. *40*, 168–189.
STANISICH V. A. and J. M. ORTIZ, (1976): Similarities between plasmids of the P incompatibility group derived from different bacterial genera. J. gen. Microbiol. *94*, 281–289.
SYKES R. B. and M. MATTHEW (1976): the β-lactamases of gram negative bacteria and their role in resistance to β-lactam antibiotics. J. antimicrob. chemoth. *2*, 115–157.

ATTEMPTS TO TRANSDUCE ANTIBIOTIC RESISTANCE IN *PSEUDOMONAS AERUGINOSA*

H. KNOTHE, V. KRČMÉRY, A. SEČKÁROVÁ and S. MITSUHASHI

Hygiene-Institut, Frankfurt University, F.R.G.
Research Institute of Preventive Medicine, Center of Hygiene, Bratislava,
Czechoslovakia
Department of Microbiology, Gunma University, Maebashi, Japan

INTRODUCTION

In the the years 1977–1978 altogether 17 clinical isolates of *Pseudomonas aeruginosa* have been collected in Frankfurt University Hospitals, resistant to high levels of gentamicin (over 64 mcg/ml, but several strains over 512 mcg/ml), tobramycin, netilmicin and/or amikacin. 11 of them produced visible autoplaques. In 6 strains wild-type phages could be isolated. Two strains, as described previously (Knothe and Krčméry, 1979), No. 70 and 78, were found to transfer amikacin and netilmicin resistance to suitable recipients in mixed cultures. Both strains produce an acetylating enzyme AAC-6' and a phosphorylating one, APH-3' (II) Moreau private communication), but additional inactivating enzymes have been found in these strains (M. Kettner et al., this Symposium).

It would be interesting to see whether the wild-type phages could transduce antibiotic resistance genes as well as fertility factors from their hosts. Such transduction was described previously in strains of *P. aeruginosa* having high-level gentamicin-tobramycin//resistance (Krčméry, Výmola and Mitsuhashi, 1977) and producing hemolytic toxins. Strains investigated in the present study are interesting in that they are resistant to newer aminoglycoside antibiotics, amikacin and netilmicin and that levels of resistance to gentamicin, tobramycin and netilmicin are extremely high in some strains (see Table I.).

MATERIALS AND METHODS

Wild-type phages were isolated from strains *P. aeruginosa*, listed in Table I., as described by Krčméry, Výmola and Mitsuhashi (1977). Visible plaques in a dense inoculum of cultures on MacConkey Agar were picked up with the surrounding growth of bacteria and grown in Nutrient broth DIFCO overnight. Cell-free lysates were then propagated on strain ML 4600 trp^-leu^- which, in contrast to our previous study with strains of *P. aeruginosa* from Czechoslovakia, was found to be most suitable for that purpose. Phage titres up to 10^9 to 10^{11} p.f.u./ml could be obtained with this strain.

Transduction experiments were the same as described in our above paper, except that mutliplicity of infection = 0.05, 0.1 and 0.2 was always used, i.e. each transduction was performed in triplicate. L broth and L agar were used as media. For selection of transductans we used MacConkey Agar with 15 mcg/ml of gentamicin, tobramycin, netilmicin and amikacin, and with 75 mcg/ml of carbenicillin, kanamyci (nor lividomycin) and streptomycin. Mixtures with the phage and the recipient culture were always left at least 2 hours at 37°C for phenotypic expression.

TABLE I.

Minimal inhibitory concentrations of aminoglycosides and carbenicillin in strains of P. aeruginosa from which phages have been isolated (in mcg/ml)

Strain No.	Antibiotic					
	GEN[1]	NET	AMI	TOB	KAN	CAR
70	64	128	64	16	128	512
78	64	128	64	128	512	512
93	64	64	32	16	1024	1024
97	64	64	32	16	1024	64
113	1024	512	64	256	1024	512
120	1024	512	64	256	1024	64

[1] GEN = gentamicin, NET = netilmicin, AMI = amikacin,
TOB = tobramycin, KAN = kanamycin A, CAR = carbenicillin

To see which antibiotic resistance determinants were transduced together with the selected marker, 25–50 transductants were stransfered into minitubes with 0.5 ml of Nutrient broth and incubated for 8h. With the aid of a multiloop applicator the cultures were then simultaneously inoculated onto a series of Petri dishes with minimal media without aminoacids and with aminoacids necessary for the growth of the transductants (see Table II.) and with MacConkey Agar without drugs as well as with individual antibiotics.

To see whether the fertility was also transduced along with resistance determinants, 10 to 15 macrocolonies from various monoantibiotic media were cultivated for 6 hours with shaking at 37°C and then mixed with the appropriate recipient (PAO 1670 or ML 4561 if selected with rifampicin or ML 4262 and 4600 if selected with appropriate aminoacids). The mixtures were incubated for 2 hours at 37°C with gentle shaking and then one standard loop was inoculated on a segment of appropriate media with increasing concentrations of single antibiotics. Exconjugants which appeared were retested for their identity and used for further cycles of transfer in a similar manner.

RESULTS AND DISCUSSION

From Table II. it can be seen that phages AP-70/XI, AP-78 and AP-113 can transduce aminoglycoside as well as carbenicillin resistance determinants to suitable strains of *P. aeruginosa*. Transductants of ML 4600 and PAO 1670 could be isolated after transduction with all three above phages. Amikacin and gentamicin resistance determinants were transduced and directly selected with AP-70/XI and AP-78. Further conjugational experiments with transductants showed that fertility function was sometimes co-transduced in these systems (see arrows in Table II.). We are in the process of separating individual R determinants by conjugation into individual clones for closer enzymatic studies of aminoglycoside inactivation.

Phage AP-113 was interesting to investigate in view of the fact that it was carried by a strain of P. aeruginosa with unusually high-level resistance to gentamicin C_1, C_{1a}, netilmicin and tobramycin (see Table I, and Kettner et al., this volume). With this phage, however, in contrast to AP-70/XI and AP-78, we could obtain transductants mostly on plates with carbenicillin (Table II.). Indirect investigation of non-selected determinants showed, however, that a series of R determinants, i.e. GEN, STR and sometimes TOB

TABLE II.

Phage-mediated transfer of antibiotic-resistance determinants in P. aeruginosa

Recipient strain	Transduced resistatnce determinants by		
	Phage AP-70/XI	Phage AP-78	Phage AP-113
PAO 1670 *rif*[R] *ade⁻*, *leu⁻*	GEN, AMI,[1] (STR)[2]	GEN, TOB (AMI) (STR) → ML 4260 (10% *Tra⁺*)	CAR (GEN) (STR) → ML 4262[3] (20% *Tra⁺*)
ML 4600 *trp⁻*, *leu⁻*	CAR, GEN (STR) → PAO 1670 (10% *Tra⁺*)	CAR, GEN, AMI (STR) (TOB)	CAR (GEN) (TOB) (AMI) (GEN) (TOB) (GEN) (AMI) (GEN) → PAO 1670 (20% *Tra⁺*)
ML 4262 *his⁻*, *ilv⁻*, *met⁻* *trp⁻*	NEGATIVE	NEGATIVE	CAR (KAN) → PAO 1670
ML 4561 *trp⁻*, *his⁻*, *ilv⁻* *arg⁻*, *leu⁻* *rif*[R]	GEN, STR → 0	NEGATIVE	NEGATIVE

[1] Abbreviations as in Table I

[2] Determinants in parenthesis were demonstrated as non-selected markers

[3] Arrows indicate results of transfer experiments with transductants by further conjugational experiments

and AMI, was co-transduced. Moreover, up to 20–30% transductants were found to contain the fertility function as well. Practically unlimited cycles of conjugal transfer of transduced resistance factors could be performed, in which high-frequency exchange of resistance determinants between a system of recipient strains took place. This system of a high-frequency transfer system is represented by strains ML-4262, ML 4600 and PAO 1670 (using minimal media with aminoacids).

Quite recently, Hyde and Streitfeld (1978) demonstrated transfer of erythromycin resistance from clinical isolates of *Streptococcus pyogenes* via their endogenous phages. Our previous (Hušťavová et al., 1977, Krčméry, Výmola and Mitsuhashi, 1977) as well as present experiments suggest that a similar mechanism might be active among clinical strains of *P. aeruginosa*. Transposition of CAR, GEN or AMI resistance determinants to endogenous phages might play a role in this process, too.

REFERENCES

Hušťavová H., Krčméry V., Kánik K., Pažický V., Trojanová M., Výmola F., Mitsuhashi S. (1977): Conjugation in lysogenic, multiple-drug resistant nosocomial strains of *P. aeruginosa* producing a hemolytic toxin. Zbl. Bakt. I. Orig. A 239, 59–61.

Krčméry V., Výmola F., Mitsuhashi S. (1977): Transduction, by phages F116 and G101, of gentamicin-tobramycin resistance and of "autoplaque formation" property in *P. aeruginosa*. Zbl. Bakt. I. Orig. A 239, 361–364.

Hyder S. L., Streitfeld M. M. (1978): Transfer of erythromycin resistance from clinically isolated lysogenic strains of *S. pyogenes* via their endogenous phage. J. infect. Dis. 138, 281–285.

H. K.,
Hygiene-Institut, Paul Ehrlich Str. 40,
Frankfurt/Main, B.R.D.

I. THEORETICAL PART
C) TRANSPONS AND REPLICONS
Chairman: D. KOPECKO

MOLECULAR AND GENETIC ANALYSES OF PLASMIDS RESPONSIBLE FOR LACTOSE CATABOLISM IN *SALMONELLAE* ISOLATED FROM DISEASED HUMANS

D. J. KOPECKO, J. VICKROY, E. M. JOHNSON,
J. A. WOHLHIETER and L. S. BARON

Department of Bacterial Immunology
Walter Reed Army Institute of Research
Washington, D.C., U.S.A.

INTRODUCTION

The utilization of various carbohydrates (e.g. lactose or sucrose) is a key diagnostic aid for the rapid identification and classification of clinically isolated enterobacteria. The usefulness of these biochemical tests in diminished by the occasional isolation of organisms that atypically utilize one or more carbohydrates. Falkow and Baron (1962) and Wohlhieter and coworkers (1964) initially demonstrated the plasmidic nature of lactose utilization in species of the normally lactose-nonutilizing *Salmonella* and *Proteus*, respectively. Over the past 15 years, isolates of a variety of organisms that are normally lactose-negative have been found to metabolize lactose. The lactose trait in these atypical strains has been shown in many cases to exist on a bacterial plasmid (Easterling et al., 1969, Falcao et al., 1975, LeMinor et al., 1976, Reeve, 1976). Because of the clinical significance of lactose-degrading (*lac+*) plasmids, genetic and molecular analyses of several atypical *lac+* salmonellae were conducted, the results of which are summarized below.

RESULTS

The bacterial strains as well as the genetic and molecular techniques employed in these studies have been reported elsewhere (Synenki et al., 1973, Macrina et al., 1978, Kopecko et al., 1979). Initial testing revealed that six of seven atypical *lac+* salmonellae, isolated from the stools of diseased humans and obtained from the Center for Disease Control, could transfer the *lac+* trait to the *S. typhi* WR4204 recipient strain at frequencies of from 10^{-3} to 10^{-5} per donor cell in two hour broth conjugal matings (Easterling et al., 1969). The locale and dates of isolation of the six strains are shown in Table 1. These isolates, obtained over a 10 year period from widely separated geographical regions within the U.S., represent three different species which do not appear to share any common surface antigens. The putative plasmid that is responsible for lactose utilization in each of these strains (see Table 1), after transfer to *S. typhi* WR4204 or *E. coli* K-12 recipients was subsequently transferred to similar bacteria at frequencies of approximately 10^{-3} per donor cell in 2 hours. Additional studies revealed that these lactose plasmids are capable of mediating bacterial chromosomal transfer at a low frequency (5×10^{-6}) from either *Salmonella* or *E. coli* K-12 *lac−* hosts (Easterling et al., 1969). In addition, previous results indicated that *E. coli* K-12 carrying any of these six *lac+* plasmids did not plaque or show a titer increase for the F-specific, male phage R17, indicating that these plasmids are not F-like (Synenki et al., 1973). The finding that the lactose trait could be transferred

TABLE I.

Summary of lactose-fermenting plasmids studied

Plasmid designation	Original host strain[a]		Locale of isolation	Year isolated
pWR 20	Salmonella tennessee	1507–65	California	1965
pWR 22	Salmonella tennessee	545–57	New Jersey	1957
pWR 26	Salmonella tennessee	999–57	New York	1957
pWR 28	Salmonella tennessee	886–57	New Jersey	1957
pWR 32	Salmonella seftenberg	1688–55	Illinois	1955
pWR 50	Salmonella typhimurium	4349–64	Massachusetts	1964

[a] All Bacterial Strains Were Isolated From Human Feces And Were Kindly Supplied by W. H. Ewing of the Center for Disease Control in Atlanta, Ga.

apparently via conjugation at a relatively high frequency plus the observation that occasionally the lac+ phenotype is spontaneously lost from these atypical strains, strongly suggested that a plasmid was reponsible for lactose degradation.

For the preliminary physical characterization, each lac+ plasmid was transferred to a lac− E. coli WR3026 strain. After radiolabelling of the DNA and sarkosyl lysis of each strain, the lysate was analyzed by dye-buoyant density centrifugation in CsCl-ethidium bromide gradients. Only small quantities of radiolabelled DNA in each lysate banded in the position expected for covalently-sealed supertwisted plasmid DNA. No plasmid DNA was detected in strains that had spontaneously lost the lac+ character. Initial estimates of molecular size for these plasmids, determined by sucrose gradient sedimentation, ranged from 30–55 megadaltons (Synenki et al., 1973). As shown below, these estimates are inaccurate both because of the limitations of the technique and the small quantities of plasmid DNA available. However, all six plasmids were observed by centrifugation in CsCl gradients to contain an average nucleotide base composition of 50% guanine plus cytosine, similar to that of the original Salmonella hosts. Although these results demonstrate that the lac+ trait is plasmid-mediated, sufficient amounts of plasmid DNA could not be obtained by the above procedure to study the molecular interrelatedness among these plasmids.

Visible quantities of nonradiolabelled plasmid DNA have not been obtained from the original Salmonella hosts or from E. coli K-12 lac− transconjugant strains carrying these lac+ plasmids, even when using a variety of gentle cell lysis procedures. It appears reasonable to assume that these extrachromosomal elements are moleculary unstable, perhaps due to the rapid integration into and excision from the host chromosome in recombinationally-proficient bacteria. Therefore, these lac+ plasmids were conjugally transferred to E. coli K-12 derivatives that are chromosomally deficient both in lactose utilization and general recombination (i.e. lac−, recA−) or that contain a deletion which includes the entire host lac region (i.e. △lac). Plasmid DNA was obtained in sufficient quantities from these latter E. coli strains by triton X-100 lysis of lysozyme-induced spheroplasts followed by dye-buoyant density centrifugation. Purified plasmid DNA was then examined for molecular size by contour length measurements obtained with the electron microscope. Despite having been initially isolated from several different Salmonella species that had been obtained from widely separated geographical regions of the United States, all six plasmids were found to be identical in molecular size (41 megadaltons, see Table II.). Identical genetic characteristics as well as nucleotide composition and molecular size of these six plasmids indicated a close physical interrelatedness. To determine inter-

TABLE II

Certain available information on most known lactose-degrading (lac+) plasmids is summarized here. pWR1 and pWR2, originally termed Folac and P-lac, respectively, have been described by Falkow and Baron (1964) and Wohlhieter et al. (1964). The remaining plasmids are described elsewhere (Synenki et al., 1973; Le Minor et al., 1976; Reeve, 1976; Efstathiou and McKay, 1977; Cornelis et al., 1978). The molecular weight of pWR1 has been determined by contour length measurements with the electron microscope (Casey and Wohlhieter, unpublished data). Plasmid pWR18 confers upon the bacterial host the ability to utilize both lactose and sucrose (Scr+).

Plasmid Designation	Original Host	Molecular Size Mdal	Molecular Size kb pr.	Incompatibility Group	Nucleotide Composition (%G+C)	Self Trans	Revelant Plasmid Phenotype
pWR 20	*Salmonella tennessee*	41	62	* (not F)	50	+	Lac+
pWR 22	*Salmonella tennessee*	41	62	*	50	+	Lac+
pWR 26	*Salmonella tennessee*	41	62	*	50	+	Lac+
pWR 28	*Salmonella tennessee*	41	62	*	50	+	Lac+
pWR 32	*Salmonella senftenberg*	41	62	*	50	+	Lac+
pWR 50	*Salmonella typhimurium*	41	62	*	50	+	Lac+
pGC 1	*Yersinia enterocolitica*	33	50		44	+	Lac+
MIP 234	*Salmonella typhimurium*	ND	ND	F_1	ND	ND	Lac+, Cam^R
MIP 235	*Salmonella oranienburg*	ND	ND	H	ND	ND	Lac+
MIP 242	*Salmonella newport*	ND	ND	*	50	+	Lac+
pWR 1	*Salmonella typhi*	67	101	F_v	ND	+	Lac+, Scr+
pWR 18	*Salmonella tennessee*	164	248	ND	ND	+	Lac+
MIP 236	*Serratia liquefaciens*	ND	ND	*	ND	ND	Lac+
MIP 237	*Enterobacter hafnia*	ND	ND	*	ND	+	Lac+
MIP 238	*Proteus morganii*	ND	ND	*	ND	ND	Lac+
MIP 239	*Proteus morganii*	ND	ND	*	ND		Lac+
pWR 2	*Proteus mirabilis*	ND	ND	P	50	+	Lac+
RK 5	*Klebsiella pneumoniae*	ND	ND	F	ND	+	Lac+, 7 resist.
F_K lac	*Klebsiella pneumoniae*	ND	ND	F_1	ND	+	Lcc+
pLM 1001	*Streptococcus lactis*	30	45	ND	ND	ND	Lac+, Asa^R, Asi^R, Chr^R
pLM 2101	*Streptococcus lactis*	20	30	ND	ND	ND	Lac+, Asa^R, Asi^R

Molecular weights are given in megadaltons and kilobase pairs.
ND = not determined, * = determined not to be F-like, + = indicates transferability.
Resistance to chloramphenicol (Cam), arsenate (Asa), arsenite (Asi), and chromium (Chr) are noted as additional plasmid traits.

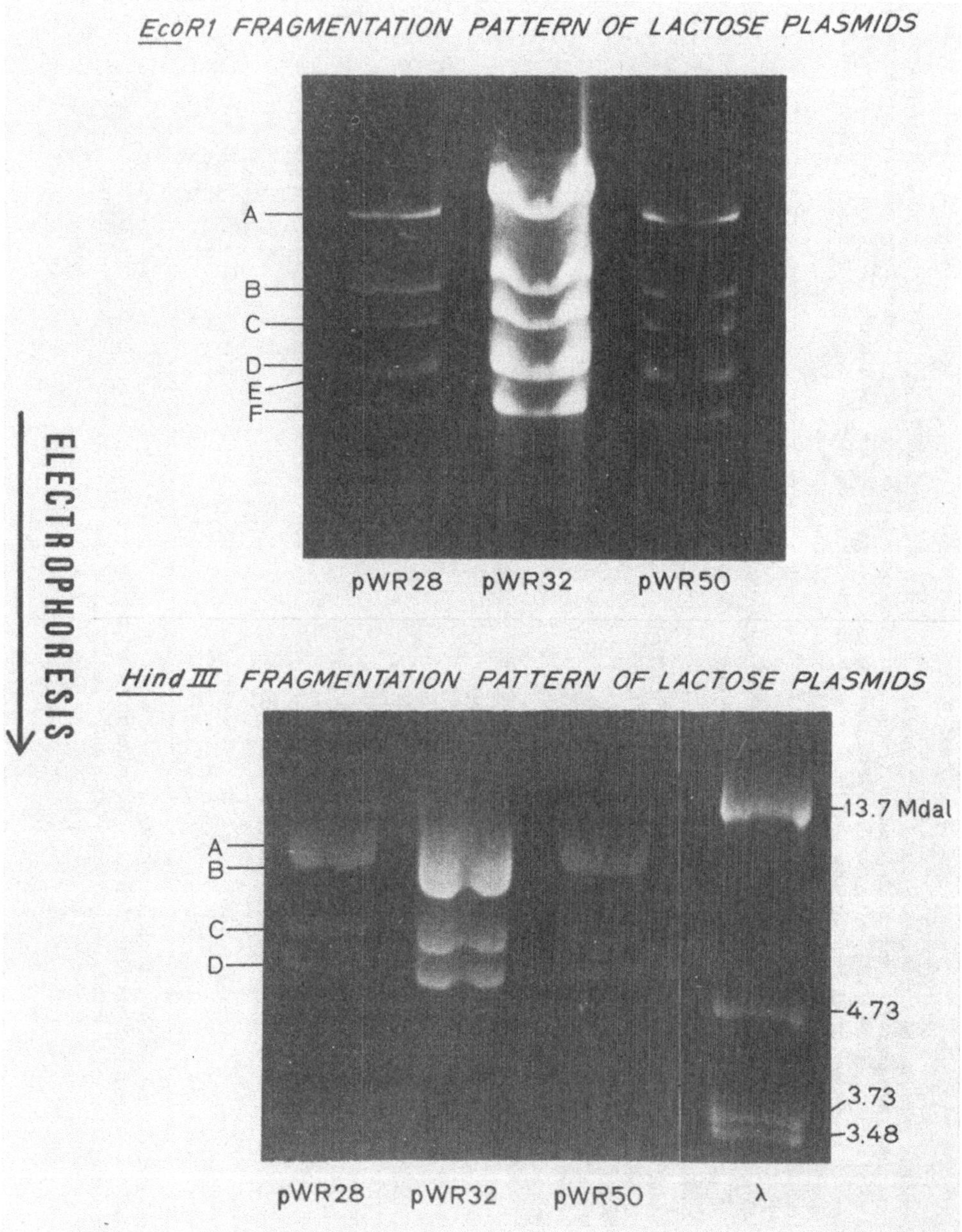

EcoR1 FRAGMENTATION PATTERN OF LACTOSE PLASMIDS

Hind III FRAGMENTATION PATTERN OF LACTOSE PLASMIDS

Fig. 1. EcoRI and HindIII fragmentation patterns of three representative lactose plasmids. Purified plasmid DNAs were digested with either EcoRI or HindIII in standard buffers for 1 hour at 37 C. Fifty to 100 microliters of each digested DNA sample were electrophoresed in Tris-borate buffer through 0.7% agarose vertical slab gels at 125 V for 1–1.5 hours, as described earlier (Kopecko et al., 1979). Gels were stained in ethidium bromide, illuminated with ultraviolet light, and photographed through a Wratten #9 filter using Polaroid type 55P/N film. Linear fragments of known molecular weight of lambda phage were included as size reference molecules in each gel.

62

molecular homology, these plasmids were cleaved with either the restriction endonuclease *Eco*RI or *Hind*III and the specific fragments generated by digestion were separated by electrophoresis on 0.7% agarose vertical slab gels, as shown in Figure 1. Although only three representative examples are shown in this figure, all plasmids were cleaved into six apparently identical *Eco*RI fragments. Two *Eco*RI fragments (labeled D and E in Figure 1) had overlapping mobilities under these electrophoretic conditions. The lower half of Figure 1 shows the fragmentation patterns of each *lac*+ plasmid observed after digestion with *Hind*III. Each plasmid contains four seemingly identical fragments, two of which (A and B) are not well separated. However, these results indirectly demonstrate that these *lac*+ plasmids share most or all of their nucleotide sequences.

DISCUSSION

Genetic and molecular characterization of the *lac*+ trait of six atypical salmonellae indicates that this character is mediated in all strains by the same conjugative plasmid of 41 megadaltons in size. The *lac*+ plasmid contains an average base composition of 50% guanine plus cytosine, transfers itself at a frequency of $\sim 10^{-3}$ per donor cell in 2 hours in broth matings, and mobilizes the host chromosome at a low frequency. Relatively large yields of plasmid DNA can be obtained in *E. coli* hosts deficient in general recombination or deleted for the host chromosomal lactose region, either of which presumably prevents the *lac*+ plasmid from integrating into the chromosome. Restriction enzyme fragmentation patterns obtained with two restriction endonucleases show that all six plasmids are highly homologous, suggesting that between 1955 and 1965 in the United States *lac*+ ability in *Salmonella* could be attributed to a widely disseminated plasmid. Studies are underway to see if the same plasmid presently exists in *Salmonella* in the U.S. Although thought to be infrequent, the occurrence of *lac*+ plasmids in many organisms is probably not detected by most clinical laboratories. Despite the difficulties in detecting *lac*+ plasmids, these extrachromosomal elements have now been observed in *Salmonella*, *Proteus*, *Erwinia*, *Klebsiella*, *Serratia*, *Yersinia*, *Enterobacter*, and in the gram-positive streptococci (Falkow and Baron, 1962, Wohlhieter et al., 1964, LeMinor et al., 1976, Efstathiou and McKay, 1977, Cornelis et al., 1978). However, comparison of the genetic and molecular properties of the six *lac*+ plasmids reported here with the properties of other known plasmids that are responsible for lactose catabolism in a variety of bacteria does not indicate any obvious physical similarity between these six plasmids and other *lac*+ elements. These data are summarized in Table II.

Most of the *lac*+ plasmids shown in Table II. were isolated from bacteria that were obtained from diseased humans or animals. In fact, *lac* Salmonellae have been implicated infrequently as the causative agents of human and animal diarrheal disease outbreaks in the U.S. and Australia, and are presently endemic in parts of Brazil (Falcao et al., 1975, see Hall et al., 1978). Though some bacterial plasmids confer virulence-enhancing capability (e.g. enterotoxin production, intestinal attachment antigen synthesis, or serum resistance factors) upon their host, no such trait has yet been correlated with the *lac*+ plasmids. However, because of their association with diseased patients, one might suspect that the *lac*+ plasmids are responsible for some virulence enhancing capability. Perhaps these plasmids have an, as yet, uncharacterized virulence factor(s), that is intimately involved in, but not essential to the diseased state. Alternatively, or in addition, since most adult humans are lactase deficient, but consume large quantities of dairy products, the *lac*+ plasmids might offer their hosts a selective growth advantage in the intestine because of high lactose concentrations. Studies to assess these possibilities are being conducted.

ACKNOWLEDGEMENTS

We appreciate the technical help of Thomas Casey and thank Patricia Guerry-Kopecko for critically reviewing the manuscript.

REFERENCES

CORNELIS G., D. GHOSAL and H. SAEDLER (1978): Tn 951: a new transposon carrying a lactose operon. Molec. gen. Genet. *160*, 215–224.

EASTERLING S. B., E. M. JOHNSON, J. A. WOHLHIETER and L. S. BARON (1969): Nature of lactose-fermenting *Salmonella* strains obtained from clinical sources. J. Bacteriol. *100*, 35–44.

EFSTATHIOU J. D. and L. L. MCKAY (1977): Inorganic salts resistance associated with a lactose-fermenting plasmid in *Streptococcus lactis*. J. Bacteriol. *130*, 257–265.

FALCAO D. P., L. R. TRABULSI F. W., HICKMAN and J. J. FARMER, III. (1975): Unusual *Enterobacteriaceae*: lactose-positive *Salmonella typhimurium* which is endemic in Sao Paulo, Brazil. J. Clin. Microbiol. *2*, 349–353.

FALKOW S. and L. S. BARON (1962): Episomic element in a strain of *Salmonella typhosa*. J. Bacteriol. *84*, 581–589.

HALL M. L. M., E. J. THRELFALL, B. ROWE, J. A. PINEGAR and G. L. GIBSON (1978): Lactose-fermenting *Salmonella indiana* from turkeys in Britain. Lancet , 1197–1198.

KOPECKO D. J., J. HOLCOMBE and S. B. FORMAL (1979): Molecular characterization of plasmids from virulent and spontaneously occurring avirulent colonial variants of *Shigella flexneri*. Infect. immun. *24*, 580–582.

LEMINOR I., C. COYNAULT, Y. CHABBERT, G. GERBAUD and S. LEMINOR (1976): Compatibility groups of metabolic plasmids. Ann. Microbiol. Inst. Pasteur *127B*, 31–40.

MACRINA F. L., D. J. KOPECKO, K. R. JONES, D. J. AYERS and S. M. MCCOWEN (1978): A multiple plasmid-containing *Escherichia coli* strain: convenient source of size reference plasmid molecules. Plasmid *1*, 417–420.

REEVE E. C. R. (1976): The lactose system of *Klebsiella aerogenes* V9A. Genet. Res. Camb. *28*, 61–74.

SYNENKI R. M., J. A. WOHLHIETER, E. M. JOHNSON, J. R. LAZERE and L. S. BARON (1973): Isolation and characterization of circular deoxyribonucleic acid obtained from lactose-fermenting *Salmonella* strains. J. Bacteriol. *116*, 1185–1190.

WOHLHIETER J. A., S. FALKOW, R. V. CITERELLA and L. S. BARON (1964): Characterization of DNA from a *Proteus* strain harboring an episome. J. Mol. Biol. *9*, 576–588.

D. J. K., Dept. of Bacterial Immunology,
Walter Reed Army Medical Centre,
Washington, D. C. 20012, U.S.A.

THE INFLUENCE OF THE *recA* MUTATION ON THE INTRAMOLECULAR GENE AMPLIFICATION OF Tn1771 IN *ESCHERICHIA COLI*

F. SCHÖFFL

Institute of Microbiology and Biochemistry,
University of Erlangen, F.R.G.

INTRODUCTION

Strains of *E. coli* harboring the conjugative 26-Mdals-plasmid pFS202 that carries the tetracycline transposon Tn1771 were found to be adaptable to very high tetracycline concentrations (Schöffl and Pühler, 1979). The molecular analysis of plasmids isolated from strains with enhanced drug resistance levels revealed an intramolecular amplification

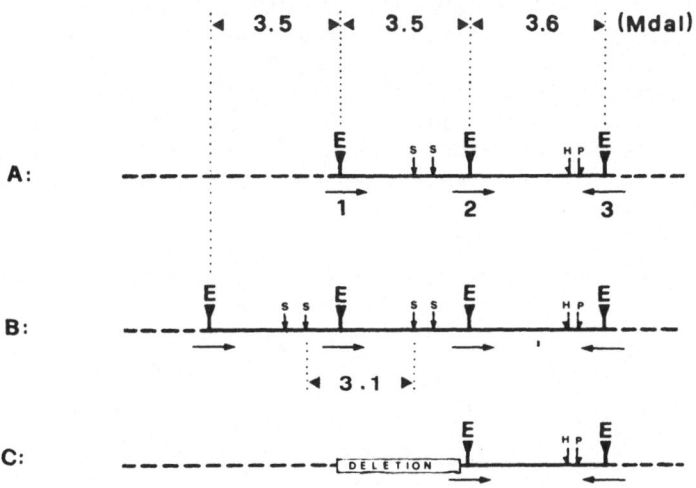

Fig. 1. Models of the physical structure of transposon Tn1771 (Schöffl et al., 1979)
A: normal structure, present on non-amplified pFS202
B: amplified structure, present on pFS202(+)
C: reduced structure found on Tc^S plasmid variants of pFS202
Restriction endonuclease cleavage sites are: E: *Eco*RI, S: *Sma*I, H: *Hind*III, P: *Pst*I. The arrows 1, 2, 3 indicate the positions and orientations of small repetitive DNA segments.

of the resistance determinant comprising a 3.5 Mdals DNA-segment of the transposon. The amplified structure of Tn1771 is characterized by a tandem repetition of the 3.5 Mdals *Eco*RI-fragment (Fig. 1B). The second half of the transposon, a 3.6 Mdals *Eco*RI-fragment, is not affected by the gene amplification process. Spontaneous deletion of exactly the 3.5 Mdals r-determinant-fragment occurs also very frequently in *E.coli*, leading to Tc^S variants of pFS202 which carry only a reduced structure of Tn1771 (Fig. 1C).

In analogy to the models developed for the amplification of certain r-determinants of NR1 in *Proteus mirabilis* (Rownd et al., 1975) and pAMα1 in *Streptococcus faecalis* (Yagi and Clewell, 1977), it is suggested that the intramolecular multiplication of Tn1771 genes on pFS202 is due to recombinational events, occurring at small directly repeated DNA sequences flanking the amplifiable region. These repeats probably map at the positions of the *Eco*RI sites on Tn1771 (Fig. 1A). The existence of a third DNA-repeat, with inverse orientation to the direct repeats, was deduced from electron microscopic observations (Schöffl and Burkardt, 1979).

According to this model, Tn1771 seems to be a genetic element where probably the same target sites are involved in both the transposition and gene amplification process. As transposition obviously occurs by a $recA^+$-independent mechanism, it was examined whether the gene amplification is influenced by the cellular, $recA^+$-dependent recombination system of *E. coli*. This paper describes the results of comparative gene amplification studies using $recA^+$ and $recA^-$ strains of *E. coli* harboring the plasmid pFS202.

RESULTS

E. coli strains used in these studies are derivatives of CSH51 ($recA^+$) and CSH52 ($recA^-$) (Miller, 1972) both of them harboring pFS202. The tetracycline resistance levels exhibited by pFS202 are very similar for both strains, showing a significant reduction of the growth rates in liquid medium containing 50 μg/ml tetracycline or more. However, extension of incubation time leads to an outgrowth of the bacterial cultures even in the presence of very high drug concentrations. Phenotypically $recA^+$ and $recA^-$ strains differ significantly in the extension of the growth lag phase. When cultured in media containing 185 μg/ml tetracycline, the outgrowth of the $recA^-$ strain is delayed for 8 h more compared to the $recA^+$ strain.

TABLE I.

Examination of intramolecular gene amplification on plasmid pFS202 isolated from tetracycline treated E. coli recA$^+$/recA$^-$ strains.

tetracycline (μg/ml)[a]		3	50	185
pFS202(+) – formation (MW: >26 Mdals)[b]	$recA^+$	—	+	+
	$recA^-$	—	—	+/—
generation of 3.1 Mdals *Sma*I-fragment[c]	$recA^+$	—	+	+
	$recA^-$	—	—	+
buoyant density of plasmid DNA[d] (g/ml)	$recA^+$	1.705	1.707	1.714
	$recA^-$	1.705	1.705	1.705/1.714

[a] $recA^+$/$recA^-$ strains were cultured up to stationary phase in liquid media containing the various concentrations of tetracycline shown in the table.

[b] ccc-molecules were isolated by dye buoyant density gradient centrifugation (Bazaral and Helinski, 1968) of sarcosyllysates of cells (Radloff et al., 1968). The purified plasmid fractions were run in an agarose gel electrophoresis, using marker plasmids for the MW-determination (Schöffl et al., 1979).

[c] plasmid DNA samples were digested by *Sma*I-enzyme and run in an agarose gel electrophoresis (see Fig. 2B).

[d] purified ccc-DNA was centrifuged in CsCl-gradients in an analytical ultracentrifuge. Marker DNA (ρ = 1.731) was used for the determination of DNA density.

In previous gene amplification studies of pFS202 it was found that the first step of the amplification process in recA⁺ strains, a tandem duplication of the r-determinant, is selected at a concentration of 50 µg/ml tetracycline. Investigations were focused on this phenomenon in order to clarify the role of the recA⁺-dependent recombination system in the initial step of the gene amplification. On the other hand it was also interesting to learn more about the molecular basis of enhanced drug resistance of cells, derived from

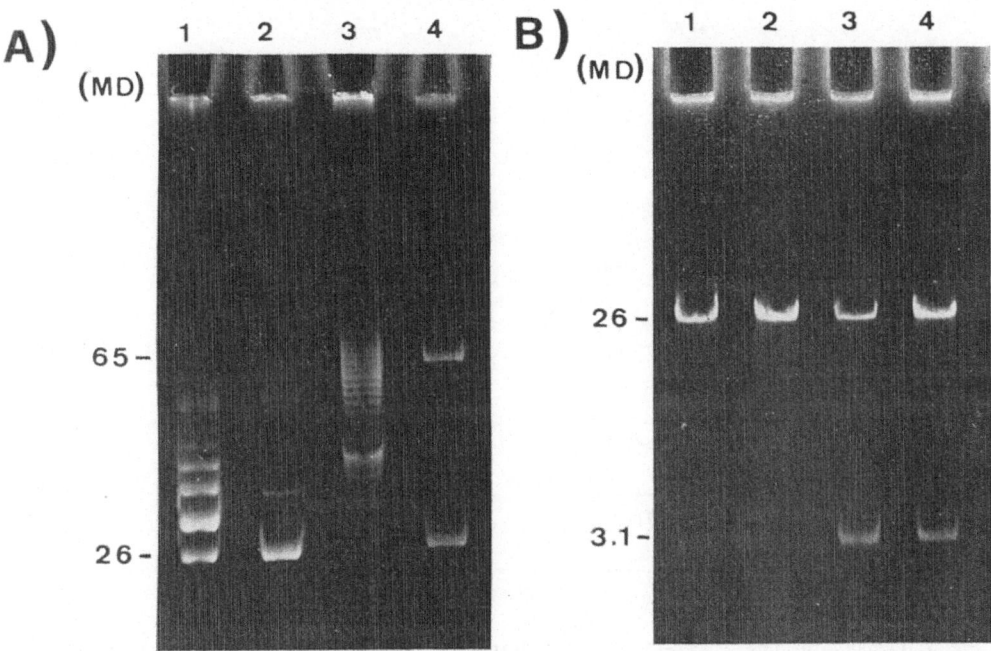

Fig. 2. Agarose gel electrophoresis of plasmid DNA isolated from recA⁺ and recA⁻ strains amplified by different tetracycline concentrations.
Plasmid fractions are from: recA⁺, 50 µg/ml Tet (1), recA⁻, 50 µg/ml Tet (2), recA⁺, 185 µg/ml Tet (3), recA⁻, 185 µg/ml Tet (4)
A: Migration patterns of ccc-molecules in an 0.5% agarose gel according to the method described by Schöffl and Pühler (1979).
B: Fragmentation patterns of SmaI digested plasmid DNA in an 1% agarose gel (Schöffl and Burkardt, 1979).

the recA⁻ strain at high tetracycline concentrations (185 µg/ml). Therefore, the intramolecular gene amplification on PFS202 was studied in response to various tetracycline selection conditions, also applied to recA⁺ and the recA⁻ strains as well. The production of multiple r-determinant copies was examined using the following criteria:
1) Enhanced molecular weights of plasmids (>26 Mdals), forming distinct size classes with differences of 3.5 Mdals between them (Schöffl and Pühler, 1979). Amplified plasmid fractions are designated pFS202(+).
2) Production of a characteristic 3.1 Mdals SmaI-fragment, indicating at least tandem duplication of the r-determinant region (see Fig. 1).
3) Increasing DNA-density of plasmids in CsCl (ρ >1.705 g/ml), resulting from a significant density difference between host plasmid DNA and transposon DNA (1.700 g/ml/1.720 g/ml).

The results of the comparative molecular analysis of plasmids are summarized in Table I. It is indicated that the intramolecular gene amplification of pFS202 starts at 50 μg/ml tetracycline in recA⁺ but not in recA⁻ cells. This is illustrated in Fig. 2, showing the migration of ccc-plasmids (Fig. 2A) and their corresponding SmaI-generated fragment patterns (Fig. 2B) after agarose gel electrophoresis. The physical properties of plasmids isolated from the recA⁺-strain fit the criteria for intramolecular amplification. The DNA-

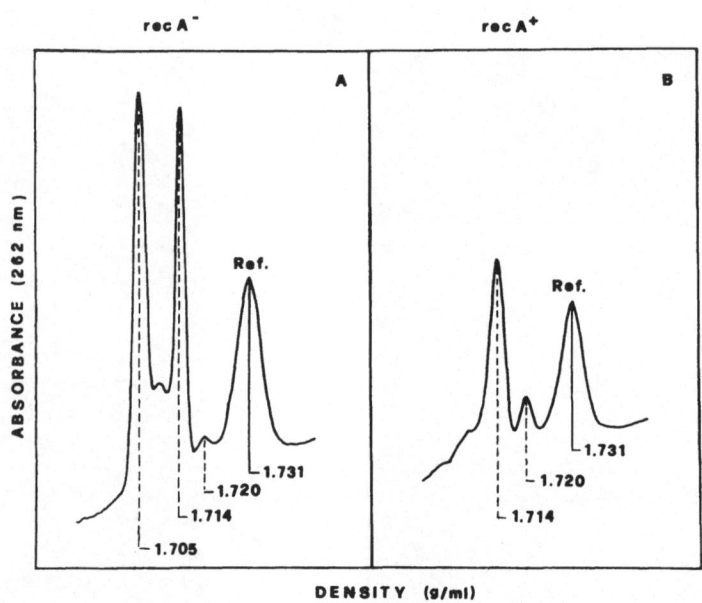

Fig. 3. Density profiles of ccc-DNA isolated from *E. coli* strains with enhanced tetracycline resistance levels (185 μg/ml).
A: Plasmids from *recA*⁻ strain CSH52 (pFS202)
B: Plasmids from *recA*⁺ strain CSH51 (pFS202)
$\varrho = 1.705$ g/ml: non-amplified plasmids (pFS204: multicopy number variant of pFS202)
$\varrho = 1.714$ g/ml: amplified plasmids (pFS202+)
$\varrho = 1.720$ g/ml: r-determinant-DNA (Tn1771)
$\varrho = 1.731$ g/ml: reference DNA (M. lysodeiktikus)
ccc-DNA was isolated by dye buoyant density gradient centrifugation and run afterwards in CsCl gradients in an analytical ultracentrifuge (Beckman, Model E, equipped with a photoelectric scanning system).

density of this plasmid fraction (1.707 g/ml) reflects a duplication of the 3.5 Mdals r-determinant region on an average. In contrast, the plasmids of the *recA*⁻ strain show the physical characteristics of nonamplified pFS202 (MW: 26 Mdals, ρ: 1.705 g/ml). From these results it can be concluded that the initiation of gene amplification is obviously a *recA*⁺-dependent process. The question, whether the enhanced gene dosage in *recA*⁺ is compensated in *recA*⁻ by any other mechanism remains unanswered.

The results obtained for the gene amplification of pF202S in *recA*⁻ after selection with 185 μg/ml tetracycline indicate an intramolecular multiplication of the r-determinant for only one part of the plasmid fraction. The second part of this fraction represents non-amplified pFS202 molecules. The results, obtained by the agarose gel electrophoresis of undigested plasmids (Fig. 2A, lane 4) and by CsCl-gradient centrifugation of the same

DNA-fraction (Fig. 3A), demonstrate the existence of the two main classes of molecules. The physical properties of one class (MW: 26 Mdals, ρ: 1.705 g/ml) fit with the non-amplified pFS202, whereas the second class, pFS202($+$), shows the properties of homogeneously amplified plasmids (MW: 65 Mdals, ρ: 1.714 g/ml) carrying 11–12 r-determinant-copies. The density profile obtained for the plasmids isolated from the recA$^+$-strain (Fig. 3B) shows only one prominent DNA peak at 1.714 g/ml covering the plasmids of all the size classes shown in the agarose gel (Fig. 2A, lane 3). It was calculated that a DNA density of 1.714 g/ml reflects an average copy-number of 11–12 r-determinants/ /molecule. By further analysis of the non-amplified plasmids isolated from the recA$^-$ strain, it was found that this plasmid class represents a plasmid copy-number-mutant of pFS204. This mutant plasmid, called pFS204, stably inherits its high ccc-copy-number of 10–15 molecules/chromosome also after transfer to other E. coli recipient strains. Compared to its single-copy parental plasmid pFS202, the more than 10-fold increase of pFS204 copies/chromosome causes a gene dosage effect for tetracycline resistance similar to an intramolecular amplification. Selection of the mutant plasmid pFS204 by tetracycline treatment of recA$^-$ cells indirectly supports the hypothesis of a recA$^+$-dependent intramolecular gene amplification. In contrast, the production of pFS202($+$), obtained simultaneously during pFS204 selection contradicts this statement. An explanation for the generation of pFS202($+$) might be the existence of residual (or temporary) recombination activities in cells of the recA$^-$ strain. The high selection pressure applied in this experiment may also lead to an enrichment of mutants which have allowed recA$^+$-independent gene amplification. The investigation of this phenomenon is in progress.

Another phenomenon, which is probably closely related to the intramolecular gene amplification, is the generation of Tc^S-deletions on pFS202, comprising the 3.5 Mdals EcoRI-fragment of Tn1771 (see Fig. 1C). Our previous investigations have shown that Tc^S cell are generated at frequencies between 10^{-2} and 10^{-3} in E. coli (pFS202) recA$^+$-cultures grown in drug-free medium. All these segregants were found to contain plasmid variants of pFS202, all lacking the r-determinant region (Schöffl and Pühler, 1979). Analogous investigations using E. coli recA$^-$-cultures revealed a stable inheritance of the Tc^R phenotype without segregation of Tc^S variants. Segregation frequencies up to 10^{-4} were tested in these experiments. On the basis of deletion phenomena, observed for an amplifiable 2.65 Mdals fragment of pAMα1 in Streptococcus faecalis, models for spontaneous gene amplification were developed, proposing simultaneous generation of amplified and deleted plasmids by recombination (Yagi and Clewell, 1977). According to this model, the gene amplification on pFS202 in E. coli recA$^+$-strains results simultaneously in production of deleted plasmid variants. A recA$^+$-dependence of this process can be deduced from the stable inheritence of the Tc^R-phenotype in recombination deficient cells.

CONCLUDING REMARKS

The participation of the recA$^+$-dependent recombination system of E. coli in the intramolecular amplification of the r-determinant of Tn1771, residing on plasmid pFS202, is indicated by the following results:
1) Under non-selective conditions (growth in tetracycline-free medium) deletion formation of the 3.5 Mdals fragment was observed in recA$^+$ but not in recA$^-$ strains.
2) cultural growth in the presence of 50 μg/ml tetracycline selects tandem duplication of the r-determinant region only in recA$^+$-cells, the plasmid of the recA$^-$ strain remains non-amplified.

3) Under high selection pressure (growth in the presence of 185 μg/ml tetracycline) amplified pFS202 molecules (containing multiple r-determinants) are selected in recA$^+$-cells, whereas cells of the recA$^-$ strain predominantly select a mutant plasmid (pFS204) which shows a 10-fold increased plasmid copy number.

The results of the amplification studies focused on the initial step of the gene amplification process (deletion formation — gene duplication) points toward a strictly recA$^+$-dependent mechanism. It is suggested that gene duplication occurs spontaneously during plasmid replication by recombinational events acting on small direct repeats. The molecular structure of Tn1771 implies an involvement of direct repeats (see Fig. 1) in the recA$^+$-dependent multiplication of the r-determinant. The same target sites might be used in combination with the third, inverted repeat in the transposition process. A similar model is described for the tetracycline transposon Tn1721 by Schmitt et al. (1979). The properties of this transposon bear striking resemblance to Tn1771. The identity of these elements is currently being investigated. Similarities to other gene amplification systems are based on the existence of direct repeats, flanking the DNA-segments of repetition. Examples of such direct repeats are the IS1-elements on NR1 (Hu et al., 1975, Ptashne and Cohen, 1975) and the RS-sequences on pAMα1 (Yagi and Clewell, 1977). These direct repeats are probably the structural precondition for very frequent recombination at these sites.

The enhanced tetracycline resistance levels of E. coli cells harboring pFS202 are generally correlated with an enhanced copy number of the r-determinant region of Tn1771. In recA$^+$-cells this is due to intramolecular amplification whereas in recA$^-$ the same gene dosage can be achieved by multiple plasmid-copies of pFS204. The question whether recA$^+$-independent gene amplification can also be selected remains unanswered.

REFERENCES

BAZARAL, M. and HELINSKI, D. R. (1968): Circular DNA forms of colicinogenic factors E1, E2 and E3 from *Escherichia coli*. J. Mol. Biol. 36, 185–194.

HU, S., E. OHTSUBO, N. DAVIDSON and H. SAEDLER (1975): Electron microscope heteroduplex studies of sequence relations among bacterial plasmids: identification and mapping of the insertion sequences IS1 and IS2 in F and R plasmid. J. Bacteriol. 127, 764–775.

MILLER, H. J. (1972): Experiments in Molecular Genetics (ed.: J. H. Miller) Cold Spring Harbor Laboratory.

PTASHNE, K. and S. N. COHEN (1975): Occurrence of insertion sequence (IS) regions on plasmid deoxyribonucleic acid as direct and inverted nucleotide sequence duplications. J. Bacteriol., 122, 776–781.

RADLOFF, R., W. BAUER and J. VINOGRAD (1967): A dye-buoyant-density method for the detection and isolation of closed circular duplex DNA. The closed circular DNA in *Hela* cells. Proc. Nat. Acad. Sci. U.S.A. 57, 1541–1520.

ROWND, R. H., D. PERLMAN and N. GOTO (1975): Structure and replication of R-factor deoxyribonucleic acid in *Proteus mirabilis*. In: Microbiology-1974 (Schlessinger, D., ed.), 76–94. American Society for Microbiology, Washington, D.C.

SCHMITT, R., E. BERNHARD and R. MATTES (1979): Characterization of Tn1721, a new transposon containing tetracycline resistance genes capable of amplification. Molec. Gen. Genet. 172, 53–65.

SCHÖFFL, F. and H.-J. BURKARDT (1979): Intramolecular amplification of the tetracycline resistance determinant of transposon Tn1771 in E. coli. In: Plasmids of medical, environmental and commercial importance (ed. K. Timmis) Amsterdam: Elsevier/North Holland, (in press).

SCHÖFFL, F. and A. PÜHLER (1979): Intramolecular amplification of the tetracycline resistance determinant of transposon Tn1771 on E. coli. Genet. Res. (in press).

F. S., *Inst. of Microbiology
and Biochemistry,
University of Erlangen
8520 Erlangen, F.R.G.*

GENE EXPRESSION OF TRANSPOSABLE ELEMENT DETERMINING AMPICILLIN RESISTANCE

T. YAMAMOTO, R. KATOH and S. YAMAGISHI

Faculty of Pharmaceutical Sciences, Chiba University, Chiba, Japan

A large number of drug resistance determinants on plasmid are recognized to be located in descrete DNA segments, termed transposons, which are capable of translocating to other DNA replicons (2). The sequential transposition of such segments may be the principal mechanism by which multiple drug resistance determinants are accumulated on a plasmid. In spite of the fact that transposon is important in understanding the development of R plasmids, little is known about the transposition mechanism of these segments or the control mechanism of gene expression in transposons.

By using two transposons which mediate ampicillin resistance (TnA), we investigated the correlation among the extent of transposon-specific transcription, the penicillinase (PCase) activity in host bacteria, and their transposition frequency. The two transposons named as Tn101 and Tn102 were originally carried by R plasmids RGN14 (now termed Rms212) and RGN823 (now termed Rms325), respectively. Both the transposons specify the so-called type I(TEM type) PCase which is the most common PCase of R plasmid (6, 8, 9). Tn102 always mediates 10 to 20 fold higher PCase activity per host cell than Tn101. It was concluded from our previous study (10) that the notably higher PCase activity in the cells harboring Tn102 could be ascribed to the superiority of the number of PCase molecules per host cell as compared with Tn101. Such a difference in the ability for PCase production between Tn101 and Tn102 offered an interesting clue to elucidate the control mechanism of transposon-specific transcription.

TRANSPOSITION OF TnA TO BACTERIOPHAGE LAMBDA AND pMK1 PLASMID

The lambda phage derivatives containing TnA were isolated from lysates of the R plasmid-carrying *Escherichia coli* strains infected with lambda phage cI857, s7, xis6, b515, b519 (abbreviated λbb). These phages which obtained TnA will be referred to as λbb:Tn101#1, λbb:Tn101#6, and λbb:Tn101#7 obtained from lysate of strain C600 λbb/RGN14 and as λbb:Tn102#1, λbb:Tn102#3, and λbb:Tn102#8 obtained from lysate of strain C600 λbb/RGN823.

Fig. 1 shows the agarose gel electrophoresis profiles of EcoRI-digested fragments of those phage DNAs and the heteroduplex analysis of the DNAs. The results indicated that the insertions of those TnA occurred at several sites of the phage DNA and were not accompanied by significant deletion of the phage genome. The length of insertion DNA of Tn101 and Tn102 was estimated to be approximately 4800 to 5000 base pairs. From the *E. coli* λbb: Tn101 and the *E. coli* λbb: Tn102 harboring the plasmid pMK1 which was a ColE1 derivative carrying kanamycin resistance gene (personal communication by H. Ogawa), the pMK1 derivatives which contained Tn101 or Tn102 were isolated. The existence of TnA in the resulting plasmid was confirmed by agarose gel electrophoresis of the endonucleases-digested fragments. Using these systems, the extent of the expression of genes encoded by Tn101 and Tn102 was compared.

TABLE I.

TABLE I.

Specific activities of PCase determined by Tn101 and Tn102

Strain	units/mg of total cellular protein[a]
C600λbb/RGN14	1.40
C600λbb/RGN823	19.9
C600λbb:Tn101	0.67
C600λbb:Tn102	8.26
C600/pMK1:Tn101	9.47
C600/pMK1:Tn102	85.3

[a] one unit being equivalent to the micromoles of aminobenzyl penicillin hydrolyzed in 1 min. at 30°C.

The log phase cultures of strains containing TnA cultivated at 37°C or at 30°C in case of lysogenic strains were harvested with centrifugation and washed twice with phosphate buffered saline containing 0.01 M of $MgSO_4$. The aliquots of cells were suspended in 0.1 M of phosphate buffer (pH 7.0) and treated for 4 min. at 20 Kc with an ultrasonic disintegrator. The sonically disrupted cell suspensions were centrifuged at 30 000 rev./min. for 20 min. The supernatants were used as sources for determination of PCase activity.

TABLE II.

Frequency of transposition of Tn101 and Tn102

experiment	mating[a]	frequency[b]
	Tn101 # 1/R388	2.5×10^{-4}
	Tn102 # 8/R388	1.2×10^{-2}
A	Tn101 # 1/R6-5	1.7×10^{-4}
	Tn102 # 8/R6-5	2.6×10^{-3}
	Tn101 # 1/F'lac	3.4×10^{-5}
	Tn102 # 8/F'lac	2.5×10^{-4}
	Tn101 # 1/R388 & RGN823Ap^s	3.3×10^{-2}
B	Tn101 # 1/R388	3.0×10^{-4}
	Tn102 # 8/R388 & RGN823Ap^s	5.3×10^{-2}
	Tn102 # 8/R388	1.0×10^{-2}

[a] Log phase culture of AB2463λbb:Tn101 (or Tn102) containing plasmid indicated in Table as a donor strain and C600^{nal} as a recipient strain were mixed at 1 : 1. The cell mixture was incubated for 2h. at 30°C and dilutions of this mixture were then plated on appropriate selective agar. The following antibiotics were used for selection as appropriate: ampicillin, 50 μg/ml and 200 μg/ml for Tn101 and Tn102, respectively; nalidixic acid, 50 μg/ml; chloramphenicol, 25 μg/ml; trimethoprim, 50 μg/ml. MacConkey agar (Eiken, Tokyo) and Müller Hinton agar (Difco) were used.

[b] Transposition frequency was expressed as the number of ampicillin-resistant transconjugants per number of R plasmid$^+$ (or F plasmid$^+$) transconjugants.

PENICILLINASE PRODUCTIVITY SPECIFIED BY Tn101 AND Tn102

The levels of PCase activity in the host cells harboring Tn101 and Tn102 were compared.

As shown in Table I, Tn102 always conferred about 10 to 20-fold greater degree of PCase activity than did Tn101 regardless of the kinds of replicons carrying TnA. These results suggested the presence of some control mechanism which causes the difference in PCase gene-expression between both the transposons.

The possibility that TnA itself encodes a protein(s) required for its transposition has been reported (5, 7). We measured the transposition frequency by using the *E. coli* strain AB2463(*recA*) lysogenized with λbb:Tn101 or λbb:Tn102.

In a typical example, the strain AB2463 λbb: Tn101 carrying F'lac plasmid was mated with *E. coli* C600[nal] and the progeny was selected according to the procedure described in the legend of Table II. The transposition frequency of Tn102 from the strain λbb:T102

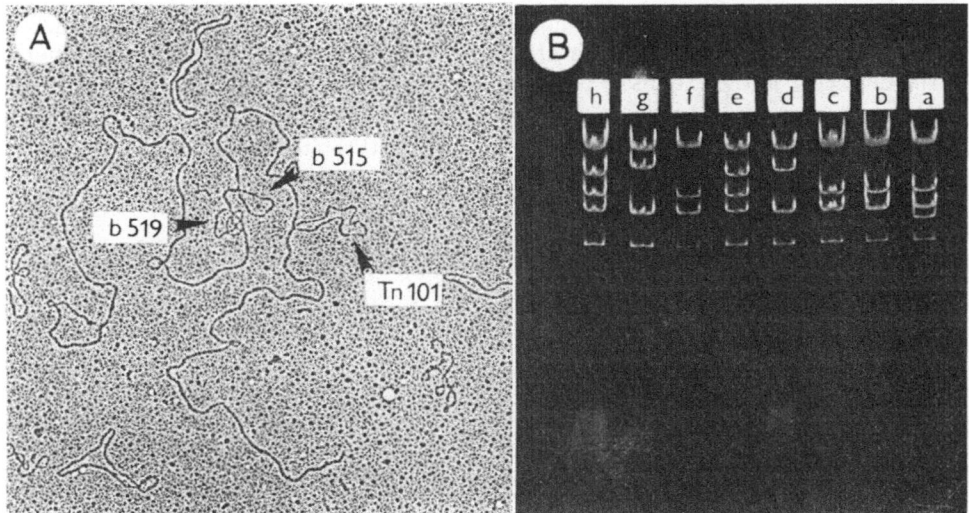

Fig. 1. (A) Heteroduplex of λbb:Tn101#7 DNA and wild phage DNA and (B) agarose gel electrophoresis of EcoR1 digested fragments of various derivatives containing TnA. The samples are (a) λ, (b) λbb, (c) λbb:Tn102#3, (d) λbb:Tn102#1, (e) λbb:Tn102#8, (f) λbb:Tn101#6, (g) λbb:Tn101#7, (h) λbb:Tn101#1.

was 10–50 times greater than that of Tn101 from the strain λbb:Tn101 regardless of the variety of recipient replicons. Furthermore, as shown in the experiment B of Table 2, the transposition frequency of Tn101 was stimulated with the coexistence of Tn102. That is to say, the transposition frequency of Tn101 was increased when the strain AB2463 λbb: Tn101 carrying two plasmids, R388 and RGN823Ap[s] was mated with C600[nal]. RGN823Ap[s] is a mutant of RGN823, which lost the ability for the production of PCase activity by a point mutation in the structural gene of PCase. On the basis of the observation by other workers (7, 5) that the transposition frequency of a transposon closely depends on transposon-specific proteins(s), it is assumed that the protein for transposition encoded by Tn101 and Tn102 is functionally identical and that the number of the protein molecules in Tn102[+] cells is higher as compared with that in Tn101[+] cells.

We have identified presumptive TnA specific proteins which were synthesized in mini cells harboring pMK1: Tn101 or pMK1: Tn102 plasmid by using SDS-polyacrylamide gel electrophoresis. Both the TnAs encoded at least four proteins having molecular weights of 30 500, 29 000, 28 000 and 19 000, respectively, in mini cells. One of those proteins was identified to be PCase. Furthermore, the results of fluorography indicated that three proteins (molecular weight 30 500, 29 000, and 28 000) of those were predominantly synthesized in the Tn102$^+$ mini cells as compared with the Tn101$^+$ mini cells.

These results suggested that the gene products encoded by the two Tn shared almost identical proteins, and that the gene(s) required for transposition as well as the PCase gene were preferentially expressed in the Tn102$^+$ cells as compared with the Tn101$^+$ cells.

TRANSCRIPTION OF Tn101 AND Tn102 DNAS *IN VIVO*

During the course of the investigation mentioned above, it should be examined whether the difference in the productibility of those products between the two TnAs is attributed to the difference in the extents of transcription of genes.

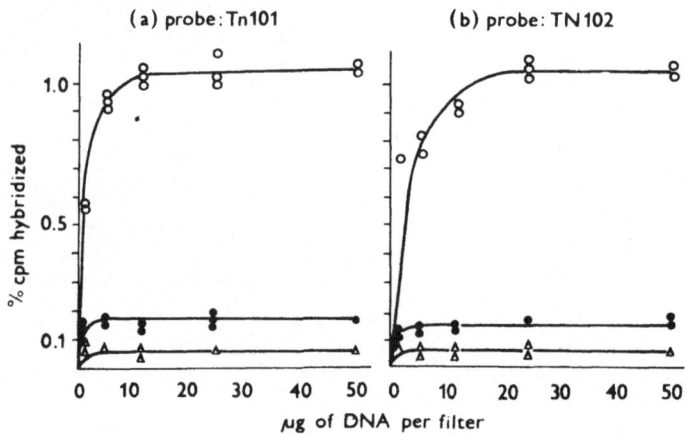

Fig. 2. Hybridization of pulse-labelled RNAs with increasing amounts of unlabelled λbb:TnA DNA. Ten ml of cultures grown exponentially in K medium were labelled 30 μCi of ^{3}H-uridine (specific activity 25 Ci/mmol) per ml. The RNA was purified by a modified procedure of Yamamoto and Imamoto (11). DNA-RNA hybridization was performed by the procedure of Gillespie and Spiegelman (4). (a) The DNA of λbb:Tn101#1 of (b) λbb:Tn102#8 was immobilized on nitrocellulose filters. Input amounts of all RNAs were constant at 5 μg and at 10 000 cpm. The percentage of the cpm hybridized represents the percentage of the input counts that were recovered as RNase-resistant hybrids on nitrocellulose filters.

symboles: ○, C600/pMK1:Tn102; ●, C600/pMK1:Tn101;
△, C600/pMK1

Prior to the determination of extents of transposon-specific mRNA, DNA-DNA sequence homology of total DNA of Tn101 and Tn102 was estimated according to the method of DNA-DNA filter hybridization described by Denhardt (3). By using ^{3}H-labelled pMK1: Tn101 (or Tn102) and pMK1: Tn102 (or Tn101), it was concluded that Tn101 and Tn102 share high DNA sequence homology, more than 90%.

We then determined the extent of hybridization of the λbb: Tn101 DNA or the λbb: Tn102 DNA with ^{3}H-labeled RNA. The ^{3}H-labelled RNA was prepated from the *E. coli* cells carrying pMK1: Tn101 or pMK1: Tn 102 which were previously pulse-labelled with ^{3}H-uridine. The results are illustrated graphically in Fig. 2.

The saturation curves show that there is a very small degree of homology between the transcripts from the pMK1 DNA containing no TnA and the TnA DNA. When the λbb: Tn102 DNA was used as the probe, the hybridized counts of RNA from the Tn102+ cells were seven fold greater than those from Tn101+ cells. Whenever λbb: Tn101 DNA was used as the probe, the hybridized counts of labelled RNA from Tn102+ cells were significantly greater as compared with those from Tn101+ cells. These results suggest that the difference in rates of transcription between two transposons plays a key role in determining the difference in PCase activity and transposition frequency between Tn101 and Tn102. And it can be speculated that the expression of the PCase gene and the gene for transposition may be transcribed co-ordinatively under control of a common regulatory region.

CONCLUSION

The work described here demonstrated that (i) the Tn101 and Tn102 specifying type Ia and type Ib PCase, respectively, share a very high degree of DNA sequence homology (more than 90%), (ii) the transcription rate of Tn102 DNA is several fold greater as compared with that of Tn101, and (iii) the difference in the extent of transcription between Tn101 and Tn102 is responsible for significant differences in PCase productivity and transposability between the two TnAs.

Furthermore, it could be postulated that the genes for transposition and the PCase gene are transcribed co-ordinatively under control of a common regulatory region.

Recently, Calame et al. have demonstrated that two promoter regions were located on the TnA Tn1701, and that the strong promoter located at the right of the BamH1 site was proximal to PCase gene and the transcription from the promoter proceeded in the right direction in vitro (1). In addition the deletion analysis of a plasmid containing Tn3 has suggested that Tn3 gene(s) required the left part of BamH1 site for its transposition (7). On the basis of these facts, it is assumed that the promoters of the gene for transposition and PCase gene are independent.

In order to define the possibility that the co-ordinative transcription of the gens is affected by the common regulatory region as presented in this report, the mode of transcription of Tn101 and Tn102, especially in vivo, is now under investigation.

REFERENCES

1. CALAME, K. L., Y. AAMDA, S. H. SHANBBLATT, and D. NAKADA (1979): Location of promoter sites on plasmid NTP1 which contains the ampicillin transposon Tn1701. J. Mol. Biol. 127:397–409.
2. COHEN, S. N., A. C. Y. CHANG, and L. HSU (1972): Nonchromosomal antibiotic resistance in bacteria: Genetic transformation of Escherichia coli by R-factor DNA. Proc. Nat. Acad. Sci. U.S.A. 69:2110–2114.
3. DENHARAT, D. L. (1966): A membrane-filter technique for the detection of complementary DNA. Biochem. Biophys. Res. Commun. 23:641–646.
4. GILLESPIE, D. L., and S. SPIEGELMAN (1965): A quantitative assay for DNA-RNA hybrids with DNA immobilized on a membrane. J. Mol. Biol. 12:829–842.
5. GOEBEL, W., W. LINDENMAIER, F. PFEIFER, H. SHREMPF, and B. SCHELLE (1977): Transposition and insertion of intact, deleted and enlarged ampicillin transposon Tn3 from mini-R1(Rsc) plasmids into transfer factors. Molec. gen. Genet. 157:119–129.
6. HEDGES, R. W., N. DATTA, P. KONTOMICHALOU, and J. T. SMITH (1974): Molecular specificities of R factor-determined Beta-lactamases: Correlation with plasmid compatibility. J. Bacteriol. 117:56–62.

7. HEFFRON, F., P. BEDINGER, J. J. CHAMPUX, and S. FALKOW (1977): Deletions affecting the transposition of an antibiotic resistance gene. Proc. Nat. Acad. Sci. U.S.A. 74:702–706.
8. SAWAI, T., K. TAKAHASHI, S. YAMAGISHI, and S. MITSUHASHI (1970): Variant of penicillinase mediated by R factor in *Escherichia coli*. J. Bacteriol. 104:620–629.
9. YAMAGISHI, S., K. O'HARA, T. SAWAI, and S. MITSUHASHI (1969): The purification and properties of penicillin B-lactamases mediated by transmissible R factors in *Escherichia coli*. J. Biochem. 66:11–20.
10. YAMAGISHI, S., T. SAWAI, and T. KOBAYASHI (1974): The mode of highly active penicillinase synthesis by an R factor p101–113. In S. Mitsuhashi, and H. Hashimoto (ed.), Microbial drug resistance, Tokyo, Univ. Tokyo.
11. YAMAMOTO, T., and F. IMAMOTO (1975): Differential stability of *trp* messenger RNA synthesized originating at the *trp* promoter and P– promoter of lambda *trp* phage. J. Mol. Biol. 92:289–309.

T. Y., Faculty of Pharmaceut. Sci.
Chiba University,
1–33 Chiba, Japan

NEW RECOMBINANT PROPHAGES BETWEEN BACTERIOPHAGE P1 AND THE R PLASMID NR1

K. MISE

Department of Microbiology, Institute of Public Health,
Tokyo, Japan

It has been shown that P1 phage transduces the R plasmids by two different mechanisms, that is, by specialized transduction and by generalized transduction (Kondo and Mitsuhashi, 1964, Mise and Arber, 1976). In generalized transduction, the transducing particles contain only a fragment of R plasmid DNA without measurable amount of phage-specific DNA (Ikeda and Tomizawa, 1965). On the other hand, specialized transducing phages contain both R plasmid- and phage-specific DNA covalently joined. These hybrid phages transduce the relevant genetic markers at a high frequency by lysogenic conversion. It has been suggested by Iida and his associates (Iida and Arber, 1977, Iida et al., 1978) that the presence of an IS1 element on the P1 genome and the common existence of Tn elements on the R plasmid genome are responsible for the formation of the P1-R hybrid phages.

In the previous paper (Mise and Nakaya, 1977), we reported the occurrence of three kinds of P1-R hybrid prophages in P1 mediated transductants for the R plasmid NR1. Further characterization of these hybrid prophages has been carried out in our laboratory, and it is shown here that: 1. The proportion of specialized transductants among all transductants was 3–7% in the P1-NR1 transduction system. 2. These P1-R hybrids were divided into 4 groups on the basis of the presence or absence of 8 genes of NR1. 3. The *rec*A$^+$ gene product is helpful for the formation of P1T$c$tra$^+$ phage.

Isolation of P1-R hybrids in P1-mediated transductants for NR1.

A one step growth lysate of P1 . WA921 (NR1) transduced the *Cm* or *Tc* marker of NR1 at a frequency of 1–2 × 10^{-5}/PFU at a MOI of 0.3–0.6. More than 90% of the transductants carried all the drug-resistance markers of NR1. However, two types of segregants were obtained at a reduced rate as reported earlier (Nakaya et al., 1960). They were the Cm^{r}Sm^{r}Su^{r}Hg^r and Tc^r types. In Cm^{r}Sm^{r}Su^{r}Tc^{r}Hg^r transductants, only 0.3–2% of the transductants were lysogenic for P1Cm Sm Su Tc Hg prophages. The remaining transductants were derived from generalized transduction of the R plasmid. All the prophages produced tiny plaques and low PFU titers upon UV irradiation. Unlike the Cm^{r}Sm^{r}Su^{r}Tc^{r}Hg^r transductants, most of the Cm^{r}Sm^{r}Su^{r}Hg^r transductants were lysogenic for P1Cm Sm Su Hg prophages. 10–40% of the Tc^r transductants were lysogenic for P1Tc phages which could be divided into two groups on the basis of their plaque-forming ability, size of plaques and PFU titer in the lysates. The 1st group of P1Tc, which consisted of the great majority, made tiny plaques on WA921. Titres of PFU of the lysates obtained from the prophages were very low, at most 10^7 PFU/ml. No plaque-forming, Tc^r-transducing phages were isolated from these lysates as far as tested. On the other hand, the 2nd group of P1Tc produced much bigger plaques than the 1st group. The titer of PFU in the 2nd group lysate was higher than that in the 1st group. Plaque formers of the 2nd group carried the *Tc* marker of NR1 when determined by plaque-center test (Kondo and Mitsuhashi, 1964). The fraction of specialized transductants and Mitsuhashi,

1964). The fraction of specialized transductants among all transductants in the P1-NR1 transduction system was calculated as 3–7%. The fraction is highly dependent upon the MOI in the transduction mixture. All P1-R prophages so far tested were stably maintained during growth of the lysogens.

Conjugal transferability of P1CmSmSuTcHg and P1Tc prophages.

In order to identify the occurrence of genetic markers other than the drug resistance markers of RN1 in P1-R prophages, we first tested conjugal transferability of the P1-R prophages. In each experiment, a P1-resistant strain of *E. coli* was used as a recipient to

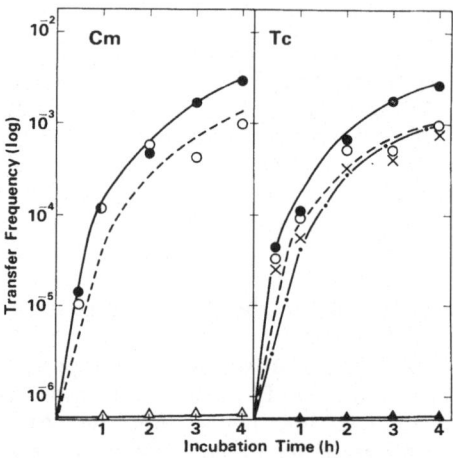

Fig. 1. Kinetics of conjugal transfer of P1-R hybrid prophages. Donor strains used were WA921 (P1CmSmSuTcHg*tra*+12) [○ −−−−−− ○], WA921 (P1Tc*tra*+235) [X−−−·−−−X], WA921 (P1Tc234) (▲ −−−−−−− ▲], WA921 (P1CmSmSuHg50) [△ −−−−−−− △] and WA921 (NR1) [● −−−−−− ●]. The recipient strain used was W3623/P1.

exclude the possibility that the resistance markers were transferred by lysogenic conversion of specialized transducing particles spontaneously produced from the P1-R lysogens. Surprizingly, the resistance markers, as well as the P1 genome, of all P1CmSmSuTcHg prophages were transferred by conjugation at almost the same frequency as NR1 to the P1-resistant cells (about 10^{-4}/hr/donor). Similarly, all the 1st group P1Tc prophages transferred the *Tc* marker and P1 genome at the same frequency as NR1. In contrast to the 1st group, the 2nd group P1Tc did not transfer the *Tc* marker at all by conjugation. The conjugal transferability of the resistance marker was not observed either in any P1CmSmSuHg prophages. The observation that the presence of deoxyribonuclease in the conjugation mixture did not decrease the frequency of transfer strongly indicates that transformation by free DNA did not play an important role in the transfer of these prophages to the recipient cells.

The kinetics of conjugal transfer of the P1-R*tra*+ prophages in comparison with NR1 is shown in Fig. 1. The *tra*+ genes of the P1-R*tra*+ prophages were cotransduced with resistance genes at high frequencies at a high MOI by P1-R*tra*+ lysates. These results strongly indicate that the *tra*+ genes are covalently joined to the P1-R genome.

Occurrence of the fin+ and inc+ genes in P1-Rtra+ prophages.

The above mentioned observations that P1CmSmSuTcHg and P1Tc prophages carried the *tra*+ genes suggest that their prophages might contain other genetic markers in the Tc-Tra region (= RTF) of NR1. As shown in Fig. 2, both Pl Cm Sm Su Tc Hg*tra*+ and Pl Tc*tra*+ carried the *inc*+ marker of NR1. What is more, these two prophages carried the *fin*+ genes. The *fin*+ and *inc*+ markers of Pl Cm Sm Su Tc Hg*tra*+ and Pl Tc*tra*+ were generally cotransferred with drug resistance markers. On the other hand, both the 2nd group Pl Tc and PlCmSmSuHg prophages lacked the *inc*+ and *fin*+ genes.

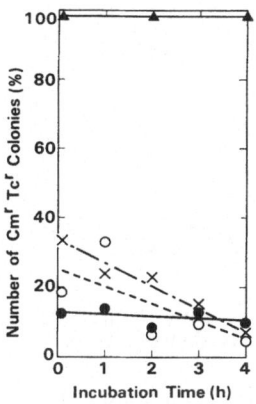

Fig. 2. The occurrence of the *inc*+ marker in P1CmSmSuTcHg*tra*+12 and P1Tc*tra*+235 prophages. One colony of WA921 (P1-R + NR1-101) grown on PAB agar containing Cm and Tc was suspended in 5ml of L broth, and the suspension incubated at 37°C with aeration. At the appointed times, 0.1 ml of aliquot was taken and after appropriate dilution plated on PAB agar with no antibiotics. Segregation of the *Cm* and *Tc* markers was tested by the replica method. WA921 (P1Tc*tra*+235 + NR1-101) [X———·———X], WA921 (P1Tc234 + NR1-101) [△———△], WA921 (NR1-101 + NR1-102)[●———●] and WA921 (P1Tc*tra*+1201 + NR1-101) [○----○]. Prophage P1Tc*tra*+1201 is a derivative of P1CmSmSuTcHg*tra*+12. NR1-101 is resistant to CmSmSuHg, while NR1-102 is resistant to Tc only.

RecA+ *protein is necessary for formation of* Pl Tc*tra*+ *phages from P1 and NR1.*

It is well known that most of hybrids between different incompatibility groups of plasmids or of plasmids and phages are formed in the absence of *recA*+ gene products (Hedges and Jacob, 1974, Kopecko and Cohen, 1975). In the R plasmid-P1 phage system, two cases are known where P1-R hybrids were formed in the *recA* condition (Iida and Arber, 1977, Mise and Nakaya, 1977). Here, we tested the possibility that the formation of Pl Tc*tra*+ is dependent upon the function of *recA*+ protein.

Contrary to our expectation, Pl Tc*tra*+ phages were not obtained from Pl . *recA* (NR1) in either of two independent experiments. All 46 Pl Tc phages isolated lacked the *tra*+, *fin*+ and *inc*+ markers and belonged to the 2nd group of Pl Tc. This result is remarkable in view of the fact that the 1st group Pl Tc was more easily isolated from Pl . *recA*+ (NR1) than the 2nd group was, both a one step growth lysate and multistep lysate of Pl . *recA*+ (NR1) contained much more Pl Tc*tra*+ than Pl Tc without the *tra*+ genes. These observations lead to the conclusion that, unlike other P1-R hybrids, the formation of Pl Tc*tra*+ is highly *recA*+ dependent. The mechanism underlying this phenomenon remains unknown.

DISCUSSION

The occurrence of 8 genes of the R plasmid NR1 in representative P1-R hybrid prophages is summarized in Table I. Apparently, P1 Cm Sm Su Tc Hg *tra*⁺ prophage carries all the genetic markers of NR1. This hybrid seems to be a cointegrate between P1 and NR1. P1Tc*tra*⁺ phage carries all the genetic markers of the RTF region of NR1 flanked by two IS*l* elements (Fig. 3 and Table I), suggesting that this hybrid is P1-RFT.

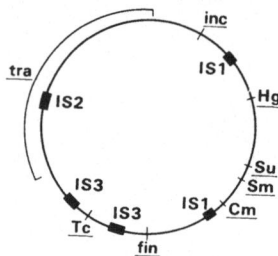

Fig. 3. Genetic map of NR1 (=R100). Only IS elements and genetic markers tested in this paper are shown. See also Table I.

Restriction cleavage analysis carried out in W. Arber's laboratory indicates that these hybrids are indeed P1-NR1 and P1-RTF cointegrates (S. Iida and W. Arber, unpublished results). P1 Cm Sm Su Hg prophage seems to carry the whole gene of the r-determinant of NR1 as is the case with P1 Cm Sm Su81 isolated and characterized by Iida, Arber and their coworkers (Iida et al., 1978, Arber et al., Cold Spring Harbor Symp. Quant. Biol.

TABLE I.

Occurrence of various genetic markers of the R plasmid NR1 in representative P1-R prophages

P1-R prophages	Cm	Sm	Su	Hg	inc	tra	Tc	fin
P1 Cm Sm Su Tc Hg *tra*⁺ 12	+	+	+	+	+	+	+	+
P1 Cm Sm Su Hg 50	+	+	+	+	−	−	−	−
P1 Tc *tra*⁺ 235	−	−	−	−	+	+	+	+
P1 Tc 234	−	−	−	−	−	−	+	−

in press). The most likely explanation for the formation of P1Tc234 is that this phage was produced by transposition of the Tn*10* element to the P1 genome as was the case with P1Tcl. The clear clustering of P1-R hybrids obtained from P1 transductants into 4 groups suggests that only a very limited number of types of the hybrids are obtainable from P1 and NR1 in a single recombinational step. It is assumed that IS elements of the R plasmid and P1 phage may have an important role in the formation of these phages.

In addition to P1-R hybrids, several phage-R plasmid hybrids have been isolated in many laboratories (Dubnau and Stocker, 1964, Coetzee, 1974, Inoue and Mitsuhashi, 1976). By spontaneous induction, these hybrids produced particles that transduced the relevant resistance marker at a high frequency. These observations, together with ours, provide indirect support for the idea suggested earlier (Kondo and Mitsuhashi, 1964) that the specialized transduction, as well as conjugation, might have an important role for the spread of drug resistance genes between different genera of bacteria in nature.

ACKNOWLEDGEMENTS

We acknowledge Prof. Werner Arber and Dr. Shigeru Iida (Biocenter, Basel University) for their kind communications throughout this work. We are grateful to Dr. Laurence Cousins (University of California) for reading the manuscript.

REFERENCES

COETZEE, J. N. (1974): High frequency transduction of kanamycin resistance in *Proteus mirabilis*. J. Gen. Micobiol. 84, 285–296.

DUBNAU, E., and B. A. D. STOCKER (1964): Genetics of plasmids in *Salmonella typhimurium*. Nature 204, 1112–1113.

HEDGES, R. W., and A. E. JACOB (1974): Transposition of ampicillin resistance from RP4 to other replicons. Mol. Gen. Genet. 132, 31–40.

IIDA, S., and W. ARBER (1977): Plaque-forming transducing phage P1: Isolation of P1CmSmSu, a potential precursor of P1Cm. Mol. Gen. Genet. 153, 259–269.

IIDA, S., J. MEYER, and W. ARBER (1978): The insertion element IS*l* is a natural constituent of coliphage P1 DNA. Plasmid 1, 357–365.

IKEDA, H., and J. TOMIZAWA (1965): Transducing fragments in generalized transduction by phage P1. I. Molecular origin of the fragments. J. Mol. Biol. 14, 85–109.

INOUE, M., and S. MITSUHASHI (1976): Recombination between S1 and the Tc resistance gene on *Staphylococcus aureus* plasmid. Virology 72, 322–329.

KONDO, E., and S. MITSUHASHI (1964): Drug resistance of enteric bacteria. IV. Active transducing bacteriophage P1CM produced by the recombination of R factor with bacteriophage P1. J. Bacteriol. 88, 1266–1276.

KOPECKO, D., and S. COHEN (1975): Site-specific *recA*-independent recombination between bacterial plasmids: Involvement of palindromes at the recombinational loci. Proc. Natl. Acad. Sci. U.S.A. 72, 1373–1377.

MISE, K., and W. ARBER (1976): Plaque-forming transducing bacteriophage P1 derivatives and their behaviour in lysogenic conditions. Virology 69, 191–205.

MISE, K., and R. NAKAYA (1977): Transduction of R plasmids by bacteriophages P1 and P22: Distinction between generalized and specialized transduction. Mol. Gen. Genet. 157, 131–138.

NAKAYA, R., A. NAKAMURA, and Y. MURATA (1960): Resistance transfer agents in *Shigella*. Biochem. Biophys. Res. Commun. 3, 654–659.

K. M., Inst. of Public Health,
Shirokanedai 4–6–1,
Minatoku, Tokyo 108, Japan

PROMOTERS OF REPLICATION DETERMINANTS OF PLASMID R6-5

S. ELY, P. M. SLOCOMBE, I. ANDRÉS and K. N. TIMMIS

Max-Planck-Institut für Molekulare Genetik, Berlin-West

INTRODUCTION

The replication of R6-5 and other large, low copy number plasmids is strictly controlled. Replication control is mediated by the *cop* gene which, for plasmids from the *inc*FII group (Uhlin and Nordström, 1975, Danbara and Timmis, 1979), and others (Cabello et al., 1976) probably specifies a repressor of replication. This repressor is assumed to inhibit the initiation of replication. Plasmids require an RNA primer at the replication origin for initiation. Control of initiation of replication could therefore be achieved by repression of (a) the synthesis of an RNA primer, (b) the synthesis of a positive-acting plasmid-encoded protein that participates in the initiation of replication, (c) the activity of the DNA replication complex at the RNA-DNA switch point or (d) transcription events that "activate" the origin structure.

An understanding of the control of initiation of plasmid DNA replication requires characterization of plasmid origin sequences and the proteins and controlling elements

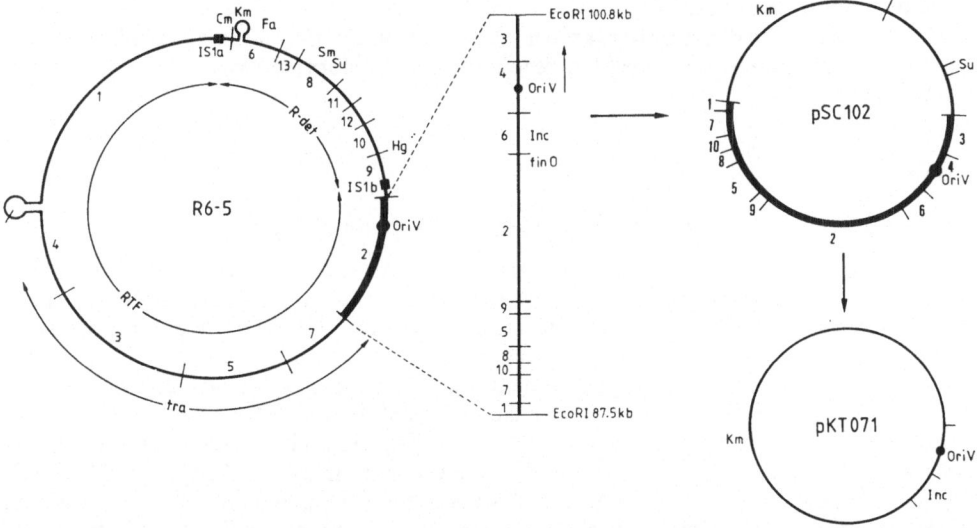

Fig. 1. Plasmid R6-5 and its "mini" derivatives pSC102 and pKT071. *Eco*RI sites are shown on the R6-5 map; *Eco*RI fragment 2, designating the *Rep*A replication region or *Eco*RI-*Rep*A (Timmis et al., 1975; Timmis et al., 1978a), is shown in detail as a fine structure *Pst*I map. The *Pst*I fragments 6 and 4 together constitute a 2.7kb region capable of autonomous replication. Replication proceeds undirectionally from *ori*V towards the R-determinant region. In the pSC102 map, *Eco*RI sites are depicted as lines extending to both sides of the map whereas *Pst*I sites are shown as lines extending only to the outside. The derivation of the mini-plasmids and mapping of the *Eco*RI-*Rep*A fragment have been reported previously (Andrés et al., 1979).

that interact with one another and with these sequences and analysis of the patterns of synthesis of these replication components. At the outset this type of study requires identification of promoters of the replication determinants and analysis of their activities. In this communication we describe the identification of RNA polymerase binding sites in the *Rep*A replication region of the *inc*FII antibiotic resistance plasmid R6–5 (Fig. 1) and show, by *in vitro* and *in vivo* transcription studies, that these sites represent active promoters.

MATERIALS AND METHODS

Bacterial strains, plasmids and conditions for electrophoresis through agarose gels have been described (Andres et al., 1979). RNA polymerase binding assays and *in vitro* transcription reactions are reported in detail elsewhere (Slocombe et al., 1979).

In vivo transcription: Plasmid-containing minicells were purified by a procedure similar to that described by Levy (1971). After penicillin treatment to reduce cellular contamination, minicells were washed in Low Phosphate Medium (Studier, 1973), adjusted to a concentration of 1.0 optical density unit/ml in the same medium, and incubated 60–90 minutes at 37°C. Minicells were labelled with ^{32}Phosphorous (Amersham-Buchler) at a final concentration of 100–250 µCi/ml during overnight incubation at 37°C with aeration. They were subsequently washed once with Low Phosphate Medium, resuspended in cracking buffer (Studier, 1973), boiled 3 minutes, phenol extracted 2–3 times, and then precipitated. The cracking step prevents losses of high molecular weight transcription products which result from direct phenol extraction of intact minicells (S. Ely, unpublished results).

DNA Plotting and Hybridization: Transfer of digested DNA to cellulose nitrate sheets and subsequent hybridization reactions were performed essentially as described by Southern (1975).

RESULTS

1. *RNA polymerase binding ites in the R6-5 Rep*A *replication region.*

RNA polymerase binding reactions were intially performed with *Pst*I-digested DNA from either the miniplasmid pSC102 or the purified *Eco*RI-*Rep*A fragment. Seven of the *Pst*I fragments generated from pSC102 were capable of binding RNA polymerase (Fig. 2A), fragments *Pst*I-1 (Km resistance fragment), *Pst*I-2, 3, 4 (*ori* fragment) and *Pst*I-6 (*inc* fragment) bound RNA polymerase strongly, whereas *Pst*I-7 and *Pst*I-8/9 bound polymerase only weakly. These results enabled assignment of RNA polymerase binding sites to specific *Pst*I subfragments of *Eco*RI-*Rep*A (Fig. 2C). The location of these binding sites were then more precisely identified in experiments in which *Hae*II, *Pvu*II, *Sma*I, *Bgl*I or *Ava*I subfragments of *Eco*RI-*Rep*A were used in the polymerase binding reaction. These experiments together with reactions using double and triple digests of individually cloned *Pst*I subfragments, showed that both the *Pst*I-2 fragment and the *Pst*-I4 (*ori*) fragment contained more than one RNA polymerase binding site.

2. *In vitro transcription of Rep*A *region DNA sequences.*

The identified RNA polymerase binding sites were subsequently shown by *in vitro* transcription studies to be active promoters. When purified the *Eco*RI-*Rep*A fragment was used as a template for *in vitro* transcription, the resultant labelled RNA hybridized back to almost all *Pst*I subfragments of *Eco*RI-*Rep*A. This was probably due to transcription initiating at a given promoter on one fragment and continuing into an adjacent fragment, and to the fact that correct termination of transcripts may be lacking in this

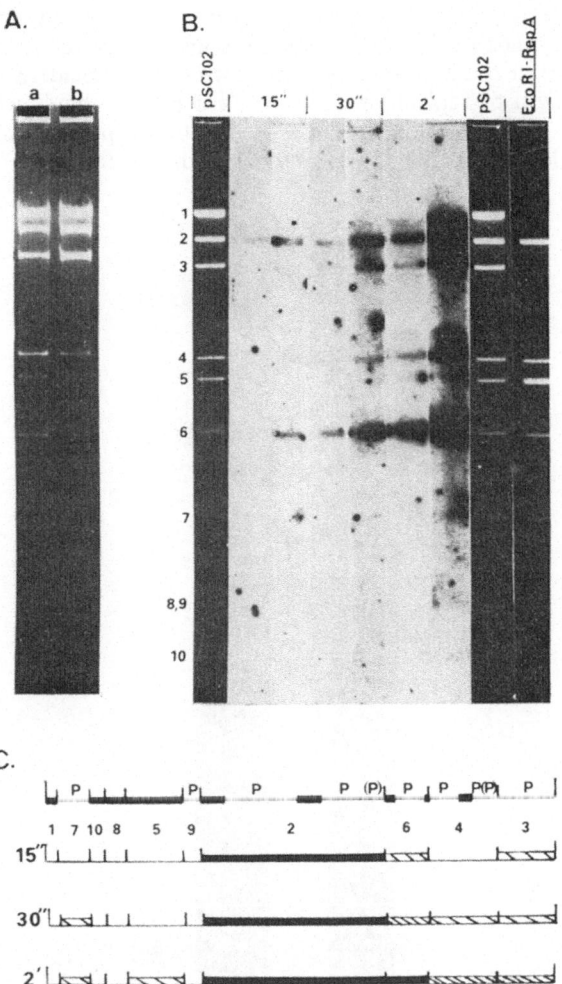

Fig. 2. RNA polymerase binding sites and *in vitro* transcription in the *Eco*RI-*Rep*A fragment of R6-5. 2A: RNA polymerase binding to pSC102 digested with *Pst*I, a) control digest, b) binding of RNA polymerase to the digested DNA. 2B: Limited *in vitro* transcription from the *Eco*RI-*Rep*A fragment template. RNA transcribed during 15″, 30″ or 2′ reactions was hybridized to *Pst*I digests of pSC102. Two autoradiographic exposures aer shown for each time point. The numbers of the *Pst*I fragments of pSC102 are shown on the left of the figure and are used in the maps. A *Pst*I digestion of purified *Eco*RI-*Rep*A fragment is shown on the extreme right of the figure. Plasmid pSC102 was used in this experiment in order to distinguish between the *Eco*RI-*Rep*A *Pst*I fragments 3 and 4 which run as a double . 2C: Summary of RNA polymerase binding data and *in vitro* transcription data. Positions of active promoters are indicated by the symbol P, whereas RNA polymerase binding sites which have not been confirmed as promoters are indicated by (P). The black bars indicate regions where no RNA polymerase binding has been detected.

in vitro system. However, in limited transcription reactions (i.e. with low UTP concentration and short-term labelling) in which short RNA transcripts are produced, it was possible to identify specific *Eco*RI-*Rep*A subfragments carrying active promoters (Fig. 2B). The promoter locations determined in these *in vitro* transcription experiments are consistent with the mapped RNA polymerase binding sites.

Detailed analysis of RNA transcribed *in vitro* from purified *Pst*I-4 (*ori*) fragment template indicated that there are at least two promoters located within the *ori* fragment. The location of one of these promoters was shown to be coincident with the established location of the origin of replication (Fig. 3). A deletion mutant which removes about 400 bp of DNA around the origin results in the loss of this promoter. The other promoter was shown to be located in the left-hand third of the *ori* fragment (Fig. 3). Preliminary

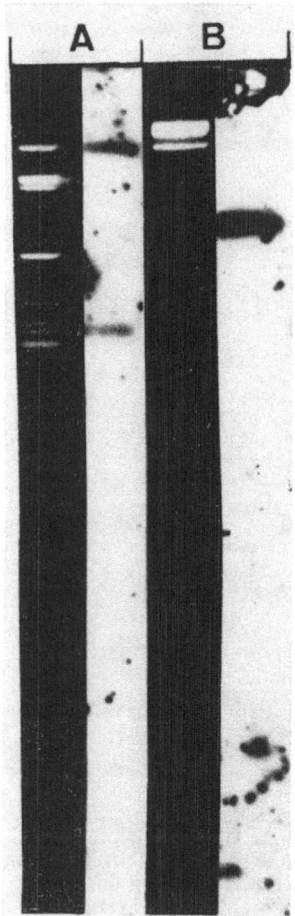

Fig. 3. RNA transcribed from the hybrid plasmid pBR322-*Pst*I-4 (pKTO39) in minicells. The right-hand tracks show Southern hybridizations of *in vivo* ^{32}P-labeled RNA and the left hand tracks show the respective DNA digestion patterns before blotting. A: pSC102 DNA digested with *Hae*II; B: pSC102 DNA digested with *Pst*I. The only bands which bind label, i.e. pSC102/ /*Hae*II bands 1 and 8 (bands 2 & 3 run as a double) and pSC102/*Pst*I band 4 (bands 1 & 2 are not well separated here) constitute *ori* fragment DNA.

results from both RNA polymerase binding and *in vitro* transcription experiments have suggested the possible existence of a third, rather weak promoter at the extreme right-hand end of the *ori* fragment (Fig. 3, Slocombe et al., 1979).

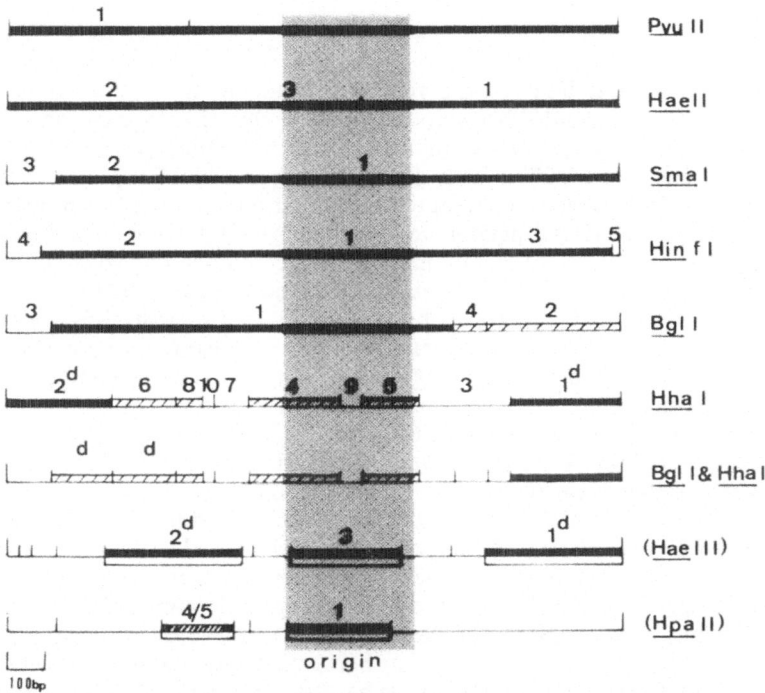

Fig. 4. Summary of RNA transcribed *in vitro* from purified *Pst*I-4 (*ori*) fragment template. Digestions of *ori* fragment DNA were run on agarose gels, transferred to cellulose nitrate strips by the Southern technique (Southern, 1975) and hybridized with *in vitro* labelled RNA. The extent of hybridization to restriction fragments is indicated on the map by the density of shading. The open bars below the *Hae*III and *Hpa*II maps indicate fragments which bind RNA polymerase. The *Bgl*I-2 fragment also binds RNA polymerase, although weakly (Slocombe et al., 1979). The letter "d" indicates fragments which run as doubles on gels. Although the *Hae*III and *Hpa*II maps are incomplete, the fragments shown have been accurately positioned on the basis of mapping by appropriate double and triple restriction endonuclease digestions. The origin of replication is depicted as a stippled region: the left border of this region is the position of the origin as determined by electron microscopy of replicative intermediates (Synenki et al., 1979), whereas the right border is the estimate obtained by combining gel electrophoresis and electron microscope measurements.

3. *In vivo transcription of RepA region DNA sequences.*

To confirm that the *in vitro* transcription data accurately reflect *in vivo* transcription, we analyzed RNA synthesized in minicells which contained pBR322 hybrid plasmids carrying individual *Pst*I subfragments of *Eco*RI-*Rep*A, or the R6-5 miniplasmids pKT071 or pSC102. The pBR322-*Pst*I-4 (*ori*) hybrid plasmid, pKT039, directed the synthesis of RNA which had sequence homology with the *ori* fragment DNA (Fig. 4). As expected, the mini R6-5 plasmid pKT071 directed the synthesis of RNA which had sequence homology to the *inc*, *ori*, and Km fragments. Miniplasmid pSC102-containing minicells

87

synthesized RNA homologous to subfragments PstI-2, 4, 5, 6 & 9. These preliminary experiments indicate that results from *in vitro* transcription experiments do reflect patterns of *in vivo* transcription events.

DISCUSSION

All determinants required for the regulated autonomous replication of the 100kb antibiotic resistance plasmid R6-5 were shown to be clustered together on a 13kb *Eco*RI fragment which formed the basis of mini R6-5 plasmids constructed by *in vitro* cloning methods (Timmis et al., 1975). This *Eco*RI fragment, *Eco*RI-*Rep*A, is located within the RTF part of R6-5, directly adjacent to the R-determinant. Subsequently, it was shown that the essential replication functions were located on a small segment of the *Eco*RI-*Rep*A fragment that is only 2.7kb in length and that it is composed of two adjacent *Pst*I fragments, *Pst*I-4 and *Pst*I-6 (Timmis et al., 1978b, Andrés et al., 1979).

Replication origin mapping studies have demonstrated that the R6-5 origin of replication is located in the middle of the *Pst*I-4 fragment and that replication is unidirectional in the direction of the R-determinant (Synenki et al., 1979). The gene responsible for the control of initiation of replication/incompatibility was shown by cloning experiments to be located entirely within the *Pst*I-6 fragment (Timmis et al., 1978b), but another gene that is essential for plasmid replication (K. N. Timmis and F. Cabello, unpublished results), probably identical to the *Rep*A gene identified by Yoshikawa (1974), appears to be located at the *Pst*I-4/*Pst*I-6 junction, as sequence continuity of this junction region is essential for autonomous replication of R6-5 (Andrés et al., 1979). The expression of these replication determinants and their interaction with one another and with host replication factors is responsible for the ordered replication and inheritance of plasmid R6-5 in dividing bacteria. Study of the expression of these determinants first requires identification of the promoters involved in their transcription and of the activity of such promoters. In this series of experiments we have located eight RNA polymerase binding sites on the *Eco*RI-*Rep*A fragment of R6-5 and have shown that most or all of these sites contain active promoters of transcription. Three promoters are located within the 2.7kb essential region, one on the *Pst*I-6 fragment, and two on the *Pst*I-4 fragment. It is tempting to speculate that these three promoters are involved in the functional expression of the three R6-5 replicative determinants that have thus far been identified, namely the *cop* gene, the *Rep*A gene, and *ori*V.

The finding of a promoter on the *Pst*I-6 fragment is consistent with the expression of incompatibility by this fragment when inserted in either orientation in the pBR322 vector plasmid (Andrés et al., 1979). An analogous, but smaller *Pst*I fragment of the R1 plasmid expresses incompatibility only in one orientation (Danbara et al., this volume) and presumably the promoter for this gene is located on an adjacent DNA sequence that is included in the R6-5 *Pst*I-6 fragment.

It is highly likely that the *Rep*A gene is situated at the *Pst*I-4/*Pst*I-6 junction and the finding of a promoter on the *Pst*I-4 fragment, close to this junction, suggests that this might be the *Rep*A gene promoter. In this case, transcription must be right to left, or in the direction opposite to that of replication.

The finding of a promoter within a 350 base pair sequence that contains the R6-5 origin of replication indicates that this promoter may play a role in the initiation event, either by providing an RNA primer for initiation of replication or by "activation" of the origin in a manner analogous to transcriptional activation of the bacteriophage λ origin. This possibility would appear to be inconsistent with the recent finding of Womble and

Rownd (1979) that replication of plasmid NR1 is inhibited to the same extent by chloramphenicol and rifampicin, but firm conclusions must await more rigorous experiments.

ACKNOWLEDGEMENTS

We wish to thank H. Mayer and K. Yoshinaga for gifts of *Pst*I and of RNA polymerase, respectively, and D. Vogt for valued technical assistance. P.M.S. and I.A. were supported by fellowships from the European Molecular Biology Organization and the Spanish Ministry of Education, respectively.

REFERENCES

ANDRÉS, I., SLOCOMBE, P. M., CABELLO, F., TIMMIS, J., LURZ, R., BURKARDT, H. J. and K. N. TIMMIS (1979): Plasmid Replication functions. II. Cloning analysis of the *RepA* replication region of antibiotic resistance plasmid R6-5. Mol. gen. Genet. **168**, 1–25.
CABELLO, F., TIMMIS, K. N. and S. N. COHEN (1976): Replication control in a composite plasmid constructed by *in vitro* linkage of two distinct replicons. Nature **259**, 285–290.
DANBARA, H. and K. N. TIMMIS (1979): Submitted for publication.
LEVY, S. B. (1971): Physical and functional characteristics of R-factor deoxyribonucleic acid segregated into *Escherichia coli* minicells. J. Bacteriol. **108**, 300–308.
SLOCOMBE, P. M., ELY, S. and K. N. TIMMIS (1979): Promoters in the replication region of plasmid R6-5. Submitted for publication.
SOUTHERN, E. M. (1975): Detection of specific sequences among DNA fragments separated by gel electrophoresis. J. Mol. Biol. **98**, 503–517.
STUDIER, F. W. (1973): Analysis of bacteriophage T7 early RNAs and proteins on slab gels. J. Mol. Biol. **79**, 237–248.
SYNENKI, R. M., NORDHEIM, A. and K. N. TIMMIS (1979): Plasmid replication functions. III. Origin and direction of replication of a "mini" plasmid derived from R6-5. Molec. gen. Genet. **163**, 27–36.
TIMMIS, K. N., CABELLO, F. and S. N. COHEN (1975): Cloning, isolation and characterization of replication regions of complex plasmid genomes. Proc. Natl. Acad. Sci. USA **72**, 2242–2246.
TIMMIS, K. N., CABELLO, F. and S. N. COHEN (1978a): Cloning and characterization of *Eco*RI and *Hind*III restriction endonuclease-generated fragments of antibiotic resistance plasmids R6-5 and R6. Molec. gen. Genet. **162**, 121–137.
TIMMIS, K. N., ANDRÉS, I. and P. M. SLOCOMBE (1978b): Plasmid incompatibility · cloning analysis of an *inc*FII determinant of R6-5. Nature **273**, 27–32.
UHLIN, B. E. and K. NORDSTRÖM (1975): Plasmid incompatibility and control of replication: Copy mutants of the R-factor R1 in *Escherichia coli* K-12. J. Bacteriol. **124**, 641–649.
WOMBLE, D. D. and R. H. ROWND (1979): Effects of chloramphenicol and rifampicin on the replication of R plasmid NR1 deoxyribonucleic acid in *Escherichia coli*. Plasmid **2**, 79–94.
YOSHIKAWA, M. (1974): Identification and mapping of the replication genes of an R factor, R100-1, integrated into the chromosome of *Escherichia coli* K-12. J. Bacteriol. **118**, 1123–1131.

S. E., Max-Planck-Institut
für Molekulare Genetik,
1000 Berlin-West 33.

RP4 MUTANTS GENERATED BY INSERTION ELEMENT ISR1

U. B. PRIEFER, P. SPITZBARTH, H. J. BURKARDT and A. PÜHLER

Institute of Microbiology, University of Erlangen, Erlangen, F.R.G.

INTRODUCTION

The conjugative Inc P plasmid RP4 carries genes for ampicillin (Ap^r), kanamycin (Km^r) and tetracycline (Tc^r) resistance (1). Because of its broad host range RP4 can be transferred to nearly all gram$^-$ bacteria. In strains of Rhizobium lupini, described by W. Heumann (2, 3, 4), RP4 shows a behaviour different from that found in *E. coli*:

- Fertility inhibition
  An R. lupini donor strain carrying RP4 is no longer capable of transferring chromosomal genes. This inhibition is reversible: after loss of RP4 the donor cells regain their fertility (5).
- Instability of RP4
  Plasmid RP4 is lost at high frequencies (50–60%) in strains which have been cultivated in antibiotic free medium (5).
- Increased rate of spontaneous mutations
  In addition to the loss of the whole RP4 very often transfer defective RP4 plasmids (Tra$^-$) can be detected (up to 10%). At a frequency of 1–2%, one can isolate spontaneous mutants which are sensitive to ampicillin (Ap^s), to kanamycin (Km^s) or to both drugs simultaneously (5).

In this paper we report that the high rate of spotaneous mutation is due to an insertion element, designated ISR1, which very frequently integrates into the RP4 molecule and causes deletions.

RESULTS

Properties of RP4 mutants isolated in Rhizobium lupini

In Rhizobium lupini the following RP4 mutants can be isolated: RP4:Ap^s, RP4:Km^s and RP4:Ap^{s}Km^s. In addition, each of these drug sensitive mutants can be found in combination with Tra$^-$. These RP4 mutants are generated at a high frequency. They can be detected without special selection procedures by replica plating single colonies on antibiotic-containing agar. No tetracycline sensitive RP4 mutants were found.

As RP4 mutants are as unstable in *R. lupini* as the parental RP4 plasmid, mutant plasmids were transferred to *E. coli*, where they proved to be absolutely stable. In *E. coli* we found neither complete loss of RP4 mutants nor additional loss of the remaining resistance genes.

Stability of various RP4 mutants was also tested in *Rhizobium leguminosarum* and *Rhizobium meliloti* strains. No plasmid loss or occurrence of further mutations were found in these strains.

For molecular analysis of RP4 mutants we therefore transferred these plasmid mutants to *E. coli* and isolated plasmid DNA from this host.

91

RP4 mutants carry DNA insertions

Plasmid DNA of an Ap^s and a Km^s RP4 mutant was investigated in the electron microscope. Contour length measuremets indicated that the mutated plasmids were slightly larger than RP4, but there was no striking difference in length.

More information about the molecular background of the mutations was obtained by heteroduplex experiments with plasmid RP8, which consists of the total RP4 DNA and an additional insertion of about 33 kb.

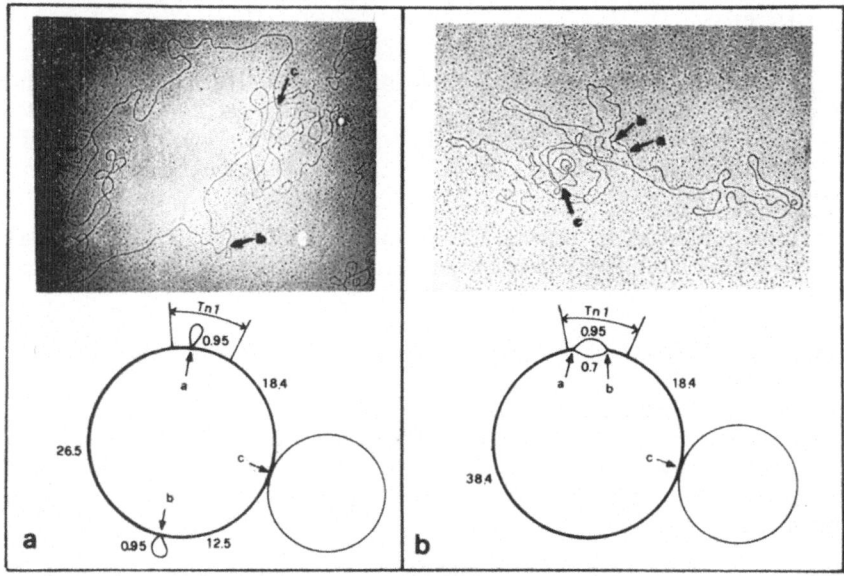

Fig. 1. Electron micrograph of heteroduplex molecules between RP8 and two different RP4 mutants.

A heteroduplex molecule between a Km^s RP4 mutant and RP8 is shown in Fig. 1a. Arrows a and b mark two single stranded loops, arrow c indicates the RP8 insertion within the RP4 molecule. In Fig. 1b a heteroduplex molecule between an Ap^s RP4 mutant and RP8 is shown. In this case arrows a and b mark a heterologous region, while arrow c again indicates the RP8 insertion. All distances are given in kb.

Fig. 1a shows a heteroduplex molecule between a Km^s RP4 mutant and RP8. The two single-stranded regions looping out (arrows a and b) are similar in length (0.95 kb) and represent the insertions of additional DNA into the mutant molecule. Length measurements excluded the possibility that the loops are caused by deletions. The location of the insertions could be mapped by means of their distances to the RP8 insertion (arrow c). Insertion (a) was found to be integrated within the ampicillin transposon Tn1, insertion (b) is located in the region that codes for kanamycin resistance. In this case integration of additional DNA leads to kanamycin sensitivity, whereas the insertion within Tn1 does not affect ampicillin resistance. Figure 1b shows the heteroduplex between RP8 and the Ap^s RP4 mutant. The heterologous region (between arrows a and b) again is located within Tn1 and can be interpreted as an insertion into transposon Tn1 correlated with a subsequent deletion.

The results of this study indicate that the mutant RP4 plasmids carry DNA insertions.

92

The integration of such DNA fragments, however, does not necessarily generate drug sensitivity. In addition, there is evidence that these insertions can cause deletions.

The insertions in RP4 mutants are identical

The interpretation of the electron microscopical data was confirmed by the restriction analysis of different mutant RP4 plasmids, using the enzymes *Hind*III, *Eco*RI, *Pst*I, *Sma*I and *Bam*HI. In all RP4 mutants tested so far, a DNA element was found to be inserted into *Pst*I fragment 4, just to the left of the unique *Bam*HI site. The map of plasmid RP4, given in Fig. 2c, demonstrates that this *Pst*I fragment is part of the ampicillin transposon Tn1. As the DNA fragment is present in both Ap^s and Ap^r plasmids, the insertion per se does not affect the expression of drug resistance.

Except for one, all RP4 mutants showed a second insertion integrated near the region that codes for kanamycin resistance. In this region the integration site did not prove to be as specific as in Tn1. DNA was found to be inserted either within the *Pst*I fragment 5 or between the single *Hind*III and the *Pst*I site to the right.

We suppose that the DNA fragments present in RP4 mutants are dentical. There is good evidence for this assumption:

– the inserted DNA elements have the same length (1.1–1.2 kb)
– the insertions introduce additional cleavage sites for *Bam*HI, *Hind*III and *Pst*I into the molecule
– the order of these restriction sites and their distances are always identical, namely

$$Hind\text{III} - - - - - - Pst\text{I} - - - - - Bam\text{HI}$$
$$\text{0.5 kb} \qquad \text{0.2 kb}$$

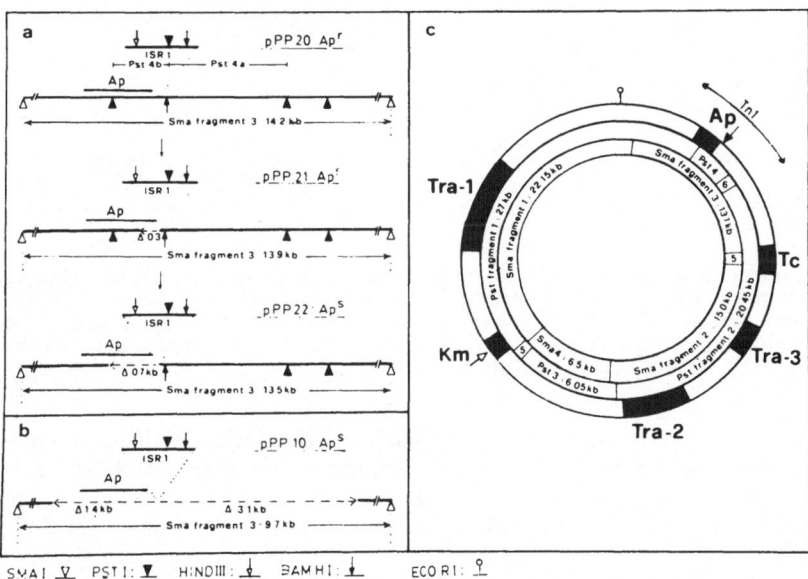

Fig. 2. Demonstration of the deletion forming process within Tn1, generated by ISR1. Part a shows the formation and progress of deletion in two Ap^s RP4 mutants originating from pPP20. In part b a deletion is depicted that extends in both directions, starting at the ends of ISR1. Fig. 2c shows the genetic and physical map of plasmid PR4. Location of structural genes is taken from P. Barth (6, 7).

93

These results suggest the existence of an insertion sequence with a high affinity for the RP4 plasmid. As this element was found in *Rhizobium lupini* strains, we called it ISR1 (insertion sequence of *Rhizobium lupini* 1).

It should be mentioned that, whenever we found two ISR1 elements integrated into one RP4 molecule, they were inserted in the opposite orientation. This could be determined from the sequence of restriction sites located on ISR1.

The insertion element ISR1 can cause deletions

The integration of ISR1 usually occurs in the neighbourhood of the genes for ampicillin and kanamycin resistance. Thus the insertion per se does not generate mutations of drug resistances. Investigation of sensitive RP4 mutants revealed that the loss of antibiotic resistance is due to deletions caused by ISR1.

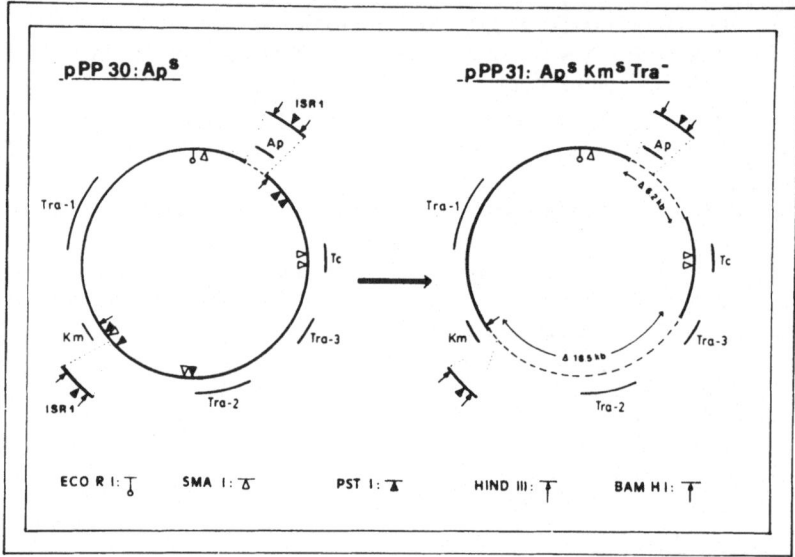

Fig. 3. Generation of an RP4 deletion mutant induced by ISR1 insertions. pPP30 carries two ISR1 elements and has lost Ap^r by deletion of the ampicillin resistance gene. pPP31 is generated by two simultaneous deletion events: the parental deletion within Tn1 progresses in both directions; in addition the ISR1 element close to the Km^r determinant causes a considerable deletion which extends in both directions inactivating Km^r and Tra$^+$.

The first evidence for the existence of such deletions was obtained by studying the heteroduplex molecule between an Ap^s RP4 mutant and plasmid RP8 (Fig. 1b).

These deletions are no unique event. They occur continually. Thus the described drug-sensitive RP4 mutants represent only intermediates in this continuous deletion forming process.

This is demonstrated in Fig. 2a. pPP20 is an Ap^r RP4 mutant carrying ISR1 within *Pst*I fragment 4. The *Pst*I site introduced by the insertion element cuts *Pst*I fragment 4 (2.6 kb) into two pieces of 2.3 kb (fragment 4a) and 1.4 kb (fragment 4b) length. Two Ap^s derivatives, pPP21 and pPP22, were isolated independently from an R. lupini strain carrying pPP20. Gels pPP21 exhibit a somewhat shortened *Pst*I fragment 4b. This suggests

94

that the plasmid has suffered a small deletion in this region, starting as the left end of ISR1 and extending into the ampicillin resistance gene.

In pPP22 PstI fragment 4b is completely absent: the deletion has progressed and removed a great deal of the Ap^r gene including the *Pst*I site.

Fig. 2b shows that the deletions generated by ISR1 can extend in both directions. pPP10, an Ap^s RP4 mutant, shows a rather small SmaI fragment 3 (only 9.7 kb) and neither PstI fragment 4 nor PstI fragment 6 can be detected on gels. ISR1 is proven

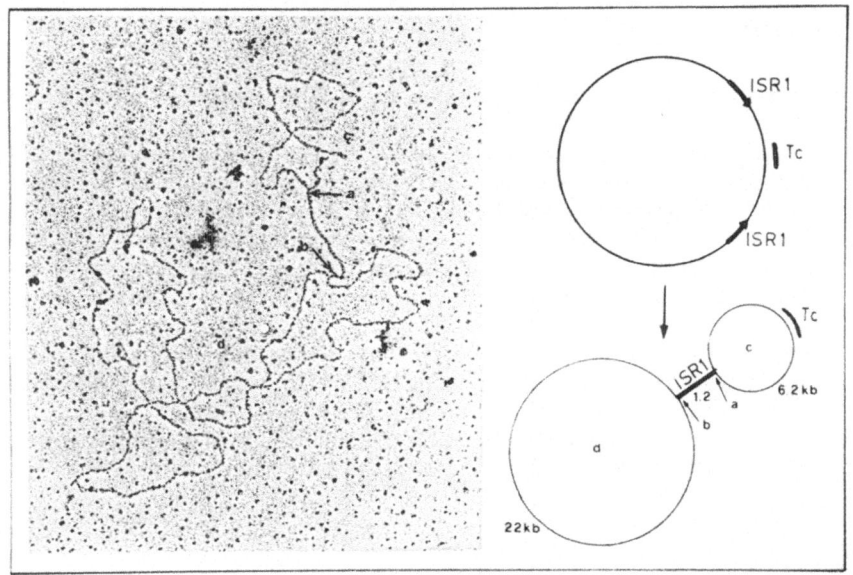

Fig. 4. Electron micrograph of the homoduplex molecule of pPP31. Arrows a and b mark the double stranded region formed by the two inverted ISR1 elements. c and d represent the single stranded parts of the mutant RP4 molecule.

to be present and supposed to be inserted into the usual locus, namely to the left of the parental *Bam*HI site. Thus we require a deletion extending in both directions, starting at the ends of ISR1 and deleting the three *Pst*I sites characteristic for Tn1.

All results obtained so far in connection with ISR1 are summarized in a further example given in Fig. 3. pPP30 shows a structure typical for an Ap^s mutant: ISR1 is integrated near the ampicillin resistance gene of Tn1 which has been inactivated by a subsequent deletion. Another ISR1 element is inserted in the opposite orientation close to the kanamycin resistance determinant, not affecting the expression of the resistance genes.

pPP31 was isolated as a derivative of pPP30 from an R. lupini strain carrying this plasmid. It has additionally lost Km^r and Tra^+, thus only retaining the Tc^r marker. Molecular analysis of pPP31 revealed that both ISR1 insertions have generated additional deletions. The parental deletion within Tn1 has progressed in both directions. In addition the plasmid has suffered a considerable deletion, starting at both ends of the second ISR1 element. There is a short deletion in the Km^r determinant causing kanamycin sensitivity, and a very large one in the other direction, extending through Tra-2 into the Tra-3 region. The plasmid is shortened to nearly one half of the original RP4 length.

The structure of pPP31 derived from restriction analysis was confirmed by electron microscopic investigations. Fig. 4 shows the homoduplex molecule of pPP31: since the two ISR1 elements are identical and integrated in opposite orientations (Fig. 3) they can hybridize and form a short double stranded region (between arrows a and b). This double stranded region divides the molecule into two single stranded loops of different size. The small loop (c) represents the short RP4 region carrying the tetracycline resistance gene, the large loop (d) is due to residual RP4 DNA.

CONCLUSION

In Rhizobium lupini, plasmid RP4 proves to be highly unstable and undergoes enhanced mutations.

Our investigations revealed that this behaviour is due to the insertion of a DNA fragment, designated ISR1. So far, we found integration of ISR1 into two preferred regions, one within Tn1 and the other close to the kanamycin resistance determinant. The two elements were found to be always integrated in opposite orientations.

The insertion of ISR1 per se does not affect drug resistance. Loss of antibiotic resistance is due to deletions of various extent, generated by ISR1. These deletions can progress bidirectionally, starting at both ends of ISR1.

We suggest that the deletion mutants isolated in R. lupini represent intermediates of a total plasmid degradating process. Insertion of ISR1 correlated with subsequent deletions might be an explanation for the instability of RP4 in *R. lupini*.

ISR1 can be used to generate RP4 deletion mutants which are of considerable interest for the analysis of the molecular structure of broad host range plasmids.

REFERENCES

1. DATTA, N., HEDGES, R. W., SHAW, E. J., SYKES, R. B. and RICHMOND, M. H. (1971): J. Bact. 108, 1244–1249.
2. HEUMANN, W., (1968): Molec. gen. Genet., 102, 132–144.
3. HEUMANN, W., PÜHLER, A. and WAGNER, E. (1971): Molec. gen. Genet., 113, 308–315.
4. HEUMANN, W., PÜHLER, A. and WAGNER, E. (1973): Molec. gen. Genet., 126, 267–274.
5. PÜHLER, A. and BURKARDT, H. J. (1978): Molec. gen. Genet., 162, 163–171.
6. BARTH, P. T., GRINTER, N. J. and BRADLEY, D. E. (1978): J. Bacteriol., 133, 43–52.
7. BARTH, P. T.: Proceedings of the Spitzingsee Symposium on "Plasmids of Medical, Environmental and Commercial Importance", Elsevier, in press (1979).

U.B.P., Institute of Microbiology,
University of Erlangen,
Erlangen, F.R.G.

A MUTANT AFFECTING THE DELETION
OF RESISTANCE-DETERMINANT(S) ON R PLASMIDS

H. WATANABE, H. HASHIMOTO and S. MITSUHASHI
*Department of Microbiology, School of Medicine, Gunma University,
Maebashi, Japan*

It has been shown that IS elements play significant roles in chromosomal rearrangement such as deletion, transposition, and duplication, since IS elements were first found as insertions in the genes of several bacterial operons (Starlinger et al., 1972). In particular, the IS1 element was found to enhance the deletion frequency of adjacent DNA sequences as much as 1,000-fold in *E. coli* strains harboring this element in the *gal*T gene of the *gal* operon (Rief et al., 1975).

These IS elements are present as repeated sequences in bacterial plasmids, such as the fertility factor F and R plasmid (Hu et al., 1975). Recent electron microscope heteroduplex studies on R plasmid DNA molecules showed that the ends of IS1 elements which flanked the r-determinants of FII R plasmids also acted as hot spots in the formation of deletions. It is known that there are two types of this deletion. One is caused by legitimate recombination between the two IS1 sequences, which was shown in the dissociation of the R plasmids; during the amplification of r-determinants in *P. mirabilis* (Rownd et al., 1978), during the generation of an r-determinant molecule in integratively suppressed strains of *E. coli* (Chandler et al., 1977), or the growth of R100 in *S. typhimurium* (Rownd et al., 1978). This dissociation is dependent on *rec*A gene function.

The other is caused by illegitimate recombination between one IS1 a nda nonhomologous second site, which was shown in the generation of miniplasmids from pRR12, a copy mutant of R100 (Mickel et al., 1977), or from pKN102, a copy mutant of R1*drd*-19 (Ohtsubo et al., 1978).

The exact mechanism of these deletion processes is unknown, but a first step to solve the mechanism was made by the isolation of an *E. coli* mutant causing the reduction of IS1-mediated deletion formation from the *gal*T gene. Nevers et al. (1978) designated this mutation *del*. In this communication, we describe the isolation and the properties of a novel *S. typhimurium* mutant affecting the deletion of r-determinant(s) on various R plasmids. This mutation was designated *dor* (deletion of r-determinants) and distinct from both *del* and *rec*A.

RESULTS

Isolation and properties of the dor mutant. Rms312 confers resistance to Tc, Cm, Sm, Su, and Mer. In *S. typhimurium*, Rms312 loses either (Tc)- or (CM.SM.Su.Mer)-determinants at a frequency of 80% or 20% of the total cells, respectively, after 10 generations of the growth of Rms312+ cells. Rms312 was transferred from *E. coli* K12 W3630 to *S. typhimurium* LT2 ML4910, and ML4910 (Rms312)+ was cultured to the stationary phase in L-broth containing Tc(25 µg/ml). Then, the culture was diluted 10-fold by fresh L-broth containing Cm(25 µg/ml) and the diluted culture was grown to the stationary phase. This process was repeated 20 times without any mutagenesis, and we selected

TABLE I.

Stability of r-determinant of various R plasmids in ML4912

R plasmid	Incompatibility group	Tested markers	Dor[a] of host	Deletion frequency of r-determinants
Rms312	FII	Tc	+	0.8
			−	0.008
		Cm.Sm.Su.Mer	+	0.2
			−	<0.001
R100	FII	Cm.Sm.Su.Mer	+	0.18
			−	0.001
		Tc	+	<0.001
			−	<0.001
R1	FII	Am.Cm.Sm.Su	+	0.1
			−	<0.002
R7	0	Ap.Tc	+	0.005
			−	0.002

[a] Dor+, ML4910; Dor−, ML4912.

the strains stably maintaining resistance determinants of Rms312. Eight clones were obtained independently.

Three out of eight clones were found to be the host cell mutants according to the following experiments. When R plasmids in three clones were conjugally transferred to a wild type host *S. typhimurium* ML4910, the r-determinants of plasmids became unstable and were segregated at the same frequency as that of Rms312, indicating that the mutation did not affect the R plasmid. The R plasmids were cured from three clones and Rms312 was conjugally transferred to the R− cells, resulting in the stable existence of Rms312 in these cells, different from Rms312 in ML4910. One of three mutants was selected for further study and denoted as ML4912. The mutation was designated *dor* (deletion of r-determinants).

S. typhimurium ML4912 and ML4912 carrying R plasmid did not show any growth deficiency, compared with a wild strain ML4910. The levels of drug resistance in ML4912 R+ strain and the copy number of R plasmid in ML4912 were the same as those in its parent ML4910.

Stability of r-determinants of various R plasmids in ML4912. We examined the stability of r-determinants of various R plasmids in ML4912. As shown in Table I., Tc-determinant and (Cm.Sm.Su.Mer)-determinants in Rms312, (Cm.Sm.Su.Mer)-determinants in R100 and (Ap.Cm.Sm.Su)-determinants in R1 became stable and their deletion frequencies in ML4912 were found to be about 100 times less than in ML4910. But the mutation did not affect the deletion frequency of r-determinants such as Tc-determinant in R100, and (Ap.Tc)-determinants in R7, which were lost at a low frequency even in a wild strain ML4910.

Effect of the dor mutation on generalized genetic recombination. A *dor* mutant strain ML4912 was more sensitive to UV-light than ML4910, but not so sensitive as a *recA* strain ML4920. We determined the frequency of generalized genetic recombination by the following system. SA540(Hfr) was crossed with ML4910 *met*, ML4912 *dor met*, or ML4920 *recA met*, respectively, and *met+* recombinants were selected. Frequency of *met+*

TABLE II.

Effect of the dor mutation on generalized genetic recombination

Donor	Recipient	Recombination frequency of $metA^+ purE^+$
SA540[a]	ML4910 *metA*	1.7×10^{-5}
(Hfr *purE*)	ML4912 *dor metA*	1.7×10^{-5}
	ML4920 *recA metA*	10^{-8}

[a] Donor and recipient were mixed at 1:4 volume ratio. After 150 min of incubation at 37°C, the culture was spread on *pur+ met+* selective media. Frequency of recombination was determined by the number of recombinants per input donor cell.

recombinants per input donor cell was found to be the same in both ML4910 and ML4912. Recombination frequency in ML4920 distinctly decreased, compared with that in ML4912 (Table II.). This result shows that the *dor* mutation has no effect on generalized genetic recombination and *dor* is distinct from *recA*.

Effect of the dor mutation of IS1-mediated deletion. R-determinants (Cm.Sm.Su.Mer) of R100 are flanked by two IS1 sequences (Hu et al., 1975). These IS1 elements acted as hot spots in the formation of deletion (Mickel et al., 1977). We estimated the deletion frequency by the use of streptomycin-dependent strain (LD1 Sm^d), in which only streptomycin-sensitive R plasmid could survive. R100 was transferred from *S. typhimurium* to *E. coli* LD1 Sm^d strain and the R100+ transconjugants were selected in the medium containing both Sm(200 μg/ml) and Cm(25 μg/ml), or containing both Sm(200 μg/ml) and Tc(25 μg/ml). Most transconjugants obtained by (Sm.Cm)-selection lost (Sm.Su.Mer)-determinants (designated as Sm^{s}Cm^r in this report). Most transconjugants by (Sm.Tc)-selection lost (Cm.Sm.Su.Mer)-determinants (designated as △r in this report.) When ML4910 was used as the donor, the frequency of △r was 3.7×10^{-1}, which was coincident with the frequency of spontaneous loss of (Cm.Sm.Su.Mer)-determinants in ML4910. As shown in Table III, the loss frequency of △r in ML4912 (*dor*) was 70 times less than that in its parent ML4910. But, the frequency of Sm^{r}Cm^s in ML4912 (*dor*) was the same as that in ML4910. The *dor* mutation affected

TABLE III.

Effect of the dor mutation on IS1-mediated deletion

Donor	Frequency of transconjugants showing[a]	
	Sm^{s}Cm^r [b]	△r [c]
S. typhimurium		
ML4910 (R100)+	3.4×10^{-3}	3.7×10^{-1}
ML4912 *dor* (R100)+	3.3×10^{-3}	5.3×10^{-3}
E. coli		
W3630	2.1×10^{-4}	4.3×10^{-3}

[a] *E. coli* K12 LD1 Sm^r or LD1 Sm^d (streptomycin dependent) strain was used as the recipient.
[b] Frequency of the deletion of (Sm.Su.Mer)-determinants (Sm^{s}Cm^r) was calculated by the number of Cm^r transconjugants in LD1 Sm^d per that of Cm^r transconjugants in LD1 Sm^r.
[c] Frequency of the deletion of (Cm.Sm.Su.Mer)-determinants (△r) was calculated by the number of Tc^r transconjugants in LD1 Sm^d per that of Tc^r transconjugants in LD1 Sm^r.

TABLE IV.
Genetic constitution of the recombinants

| Cross | | Selection | Frequency of Hfr allelic genes (%) | | | | | | | |
Donor	Recipient		metE (84)[a]	serA (63)	thyA (62)	glyA (57)	dor	(U)[b]	trpB (34)	purE (12)
A SA540 (Hfr)	ML4912 (dor)	thyA+purE+	2		100		54	(54)	20	0
ML4918 (Hfr dor)	ML4910	thyA+purE+	0		100		50	(50)	29	0
B ML4918 (Hfr dor)	JE564	glyA+purE+		42		100	82	(82)		0

[a] ( ) showed the map positions of *S. typhimurium* described by K. E. Sanderson (1978).

[b] The *dor* marker of the recombinants was scored by the following way. Rms312 was transferred into 100 recombinants, and the transconjugants were purified by two successive single colony isolations. Ten colonies out of each clone were picked and streaked on the medium containing Cm (25 μg/ml) or Tc (25 μg/ml). Clone carrying *dor* maintained Tc^r in 9–10 out of 10 colonies but clone carrying *dor*+ lost Tc^r in all of 10 colonies.

only the generation of △r. The frequency of △r in ML4912 (*dor*) decreased to the same level as that in *E. coli* (Table III.). △r is caused by legitimate recombination between two IS1 sequences, and Sm^{s}Cm^r is caused by illegitimate recombination between one IS1 and a non-homologous second site (T. Miki, personal communication).

Mapping of the dor mutation. It has been reported that *S. typhimurium* LT2 strain has a cryptic plasmid (Oliver et al., 1974). We must testify that the *dor* mutation does not exist on a cryptic plasmid but on the host chromosome. Reciprocal transfer of the Dor marker was performed, one transfer was from SA540 Hfr *dor*+ *pur*E to ML4912 *dor thy*. Another transfer was from ML4918 Hfr *dor pur*E to ML4910 *dor*+ *thy*, ML4918 was made by the transfer of *dor* gene from ML4912 to SA540 in order to obtain Hfr *dor*. Both donors and recipients had the same markers except the Dor marker. Both crosses were selected by *thy*+*pur*E+, and *thy*+ recombinants were examined for unselected markers. The results are shown in Table IVA. The frequency of Hfr allelic genes emerging in the recipient was similar in both crosses. This result shows that the *dor* mutation exists on the host chromosome.

ML4918 was crossed with JB564 *serA glyA* recipient and *glyA*+ recombinants were scored for both *dor* and UV sensitivity characters. The linkage of *dor* and UV sensitivity with *glyA* was about 80% and UV sensitivity did not segregate from *dor* (Table IVB). *Dor* was cotransduced with P22 phage at about 10% linkage with *glyA*. The *glyA* gene was mapped to a position at 57 min of the *S. typhimurium* chromosome. Then, *dor* is considered to be within one minute from *glyA*.

DISCUSSION

In this paper we described the isolation and characterization of *dor* mutation in *S. typhimurium*, which affects the deletion formation of the r-determinants on R plasmids. The *dor* mutation has been mapped to a position close to *glyA* at about 57 min of *S. typhimurium* map, near which any genes responsible for recombination and repair have not been observed. Recently, Nervers et al., (1978) reported the *E. coli* mutant defective in IS1-mediated deletion formation (designated *del*). The *del* mutation reduced 100-fold as much as the deletion frequency caused by *gal*T: IS1 system (Nervers et al., 1978). The *dor* mutation is distinct from the *del* mutation in the following three points. Deletion frequency of Tc-determinant of Rms312 was not reduced in the *del* mutant, compared with that of wild strain (unpublished data), in spite of the reduction of the deletion frequency of Tc-determinant in the *dor* mutant. As shown in TableIII., the *dor* mutation did not affect IS1-mediated deletion formation but affected the deletion formation caused by legitimate recombination between two IS1 sequences. The *del* mutation has been mapped to a position close to *lys*A and *gal*R at about 61 min of *E. coli*, but this position is distinct from *gly*A.

The *dor* mutation reduced the deletion frequency of (Cm.Sm.Su.Mer)-determinants on R100 and (Cm.Sm.Su.Ap)-determinants on R1, which are flanked by directly repeated two IS1 sequences (Hu et al., 1975). In Rms312, (Cm.Sm.Su.Mer)-determinants are flanked by IS1, but Tc-determinant is seemed to be flanked by other IS sequences except IS1 (unpublished data). The *dor* mutation also affects the deletion of Tc-determinant of Rms312. This suggests that the deletion formation in *dor*+ strain is not due to the recognition of IS1 specific site but due to the affect on one step of common processes of the deletion formation caused by legitimate recombination between regions with some homology. This is considered to be different pathway from the generalized genetic recombination, because the *dor* mutation has no affect on this recombination.

REFERENCES

CHANDLER, M., ALLET, B., GALLAY, E., BOY de la TOUR, E., and CARO, L. (1977): Involvement of IS1 in the dissociation of the r-determinants and RFT components of the plasmid R100-1. Molec. gen. Genet. 153, 289–295.
HU, S., OHTSUBO, E., DAVIDSON, H., and SAEDLER, H. (1975): Electron microscope heteroduplex studies of sequence relations among bacterial plasmids: identification and mapping of the insertion sequences IS1 and IS2 in F and R plasmids. J. Bacteriol. 122, 764–775.
MICKEL, S., OHTSUBO, E., and BAUER, W. (1977): Heteroduplex mapping of small plasmids derived from R factor R12 in vivo recombination occurs at IS1 insertion sequences. Gene. 2, 193–210.
NERVERS, P. and SAEDLER, H. (1978): Mapping and characterization of an *E. coli* mutant defective in IS1-mediated deletion formation. Molec. gen. Genet. 160, 209–214.
OHTSUBO, E., ROSENBLOOM, M., SCHREMPF, W., GOEBEL, W., and ROSEN, J. (1978): Site specific recombination involved in the generation of small plasmids. Molec. gen. Genet. 159, 131–141.
OLIVER, D., MANIS, T. J., and WHITFIELD (1974): Evidence for a composite state of an F'his, gnd element and a cryptic plasmid in a derivative of *Salmonella typhimurium* LT2. J. Bacteriol. 119, 192–201.
REIF, H. J. and SAEDLER, H. (1975): IS1 is involved in deletion formation in the *gal* region of *E. coli* K-12. Molec. gen. Genet. 137, 17–28.
ROWND, R. H., MIKI, T., GRENGERG, J., LUCKOW, V., EASTON, A., HUFFMAN, G., MILLER, J., and BARTON, C. (1978): Structure, amplification, and replication of composite R plasmid DNA. EMBO workshop. Plasmids and other extrachromosomal genetic elements. p. 86.
STARLINGER, P. and SAEDLER, H. (1972): Insertion mutation in microorganism. Biochimie. 54, 177.

H. W., Dept. of Microbiology,
School of Medicine, Gunma Univ.
Maebashi, Japan

101

I. THEORETICAL PART
D) GENETICS AND MOLECULAR BIOLOGY
Chairman: P. KONTOMICHALOU

PLASMID R1 CONTAINS TWO DISTINCT
FUNCTIONAL REPLICONS

H. DANBARA, J. K. TIMMIS and K. N. TIMMIS
Max-Planck-Institute of Molecular Genetics, Berlin-West.

INTRODUCTION

A fundamental property of cellular genetic elements is their stable inheritance in dividing cells. Replicons that are present in only a few copies per cell, i.e. chromosomes and plasmids of low copy number, must utilize specific partition systems for their ordered inheritance (see Timmis, 1979, for a discussion of this subject). Although membrane binding sites (Jacob et al, 1963) and folded chromosomes (Kline and Miller, 1975) have been proposed as partition structures involved in the segregation of plasmid molecules in dividing bacteria, no direct evidence in support of these hypotheses has yet been obtained. It was recently suggested that plasmid partition mechanisms could be responsible for the phenomenon of plasmid incompatibility (Novick and Schwesinger, 1976), the inability of closely related plasmids to be stably co-inherited by dividing bacteria. We and others have shown that incompatibility results primarily from plasmid copy control systems (Uhlin and Nordström, 1975, Timmis et al, 1978a; Danbara and Timmis, 1979). However, if plasmid incompatibility, even to a small extent, does result from partition systems, it could prove to be a highly useful phenotypic trait in the study of such systems.

We have shown that the copy control (Cop) incompatibility (Inc) gene of plasmid R6-5 is located on one of the two *Pst*I fragments that comprise the essential replication region of this plasmid (Timmis et al., 1978a). The identification of an incompatibility function outside of, but adjacent to, the essential replication region of the closely related plasmid R1 (Kollek and W. Goebel, to be published), indicated that this second incompatibility function could result from an R1 plasmid partition determinant.

In order to investigate this possibility we have undertaken cloning analysis of the replication and incompatibility determinants of plasmid R1. In our experiments we have detected two incompatibility determinants on this plasmid, one located in the essential *Rep*A region, which corresponds to the copy control gene of R6-5 identified earlier, and another situated between this essential region and the R-determinant. The generation of miniplasmids from R1 demonstrated the existance of a second replication region on this plasmid, designated RepD, which is capable of autonomous replication. This region contains the newly-identified incompatibility determinant, *Inc*D.

MATERIALS AND METHODS

Most materials and methods, including the purification of plasmid DNA, its analysis by restriction endonuclease cleavage and agarose gel electrophoresis, and its transformation and manipulation by recombinant DNA techniques, have been described previously (Timmis et al., 1978b, Andres et al., 1979). Only bacterial derivatives of *E. coli* K-12 C600 were used in this study. CR34 is a Thy⁻ derivate of C600 and YC259 is a Thy⁺

*rec*A derivate of CR34 that was obtained by conjugation with KL16–99 Hfr Thy⁺ *rec*A and that was provided by M. Yoshikawa. Plasmid R1 was obtained from N. Datta and R100drd△Tc from M. Achtman.

Incompatibility tests

Plasmids were introduced by transformation into strain YC259 and YC259 (R100drd △Tc). Transformation mixtures were plated on media selective for a marker on the DNA used for transformation (Method 1) and on media selective for both the transformation and resident (R100drd△Tc) DNA. In both methods, individual transformant clones were picked to nutrient broth and incubated without selection for about 13 generations. Cultures were then diluted and plated on a non-selective medium for single colonies, which were subsequently transferred by replica plating onto appropriate antibiotic-containing plates, which indicated the continued carriage of each of the original plasmids (i.e. Cm for R100, Km for R1 miniplasmids and Tc for pBR322 hybrids). Loss of one or the other plasmid during the incubation period is indicated in Table II. as + incompatibility. The transformation incompatibility index is the ratio of YC259 (R100drd△Tc) transformants obtained on double selection plates to YC259 transformants on single selection plates.

RESULTS

1. *Generation of a mini R1 plasmid*

Plasmid R1 DNA was digested with *Bgl*II endonuclease, treated with T4 DNA ligase, and the ligation mixture used to transform competent C600 bacteria. A large number of transformant clones were obtained on plates containing kanamycin (Km) and ampicillin (Ap) but few on plates containing streptomycin (Sm), suplhonamide (Su) or chloramphenicol (Cm). Plasmid DNA prepared from Km^r Ap^r transformant clones and analyzed by cleavage with *Bgl*II, followed by agarose gel electrophoresis, demonstrated that all clones contained a plasmid composed of a single *Bgl*II–generated fragment of R1. Comparison with the published cleavage map of R1 (Blohm and Goebel, 1978) of the digestion patterns produced by various restriction endonucleases from one such mini R1 plasmid, designated pKT300, allowed us to precisely locate the R1 DNA segment carried by the mini plasmid (Fig. 1). This segment starts at the *Bgl*II site previously mapped in R6-5 just to the left of the RepA replication region, but very close to Cop/Inc, and ends in the R-determinant region between Ap and Su. If a second Inc determinant located between the RepA and R-determinant regions exists, then it should be expressed by pKT300. We therefore cloned subfragments of pKT300 and analysed these for replication and incompatibility functions. It is known for both R1 (Kollek et al., 1978) and R6-5 (Timmis et al., 1978a) that *Pst*I endonuclease cleaves the RepA and surrounding DNA region into conveniently-sized fragments. We have therefore mapped the *Pst*I cleavage sites in plasmid pKT300 and have cloned *Pst*I-generated fragments.

2. *Generation of mini pKT300 plasmids*

Plasmid pKT300 DNA was cleaved with *Pst*I endonuclease, treated with T4 ligase, and the ligation mixture used to transform competent C600 bacteria. Plasmid DNA was prepared from a number of purified Km^r transformant clones and analysed by cleavage with *Pst*I endonuclease and agarose gel electrophoresis. As can be seen in

Fig. 2A and Table I., the majority of miniplasmids, e.g. pKT311, contain the P2 (Km; Danbara et al., 1979), P6 (OriV; Ohtsubo et al, 1978; Andres et al, 1979) and P8 (Cop/Inc; see below) fragments and are therefore analogous to the mini R6-5 plasmid pKT071 isolated by Timmis et al (1978a). Surprisingly, one mini pKT300 plasmid that is designated pKT312 did not contain *Rep*A region fragments P6 and P8 but instead contained

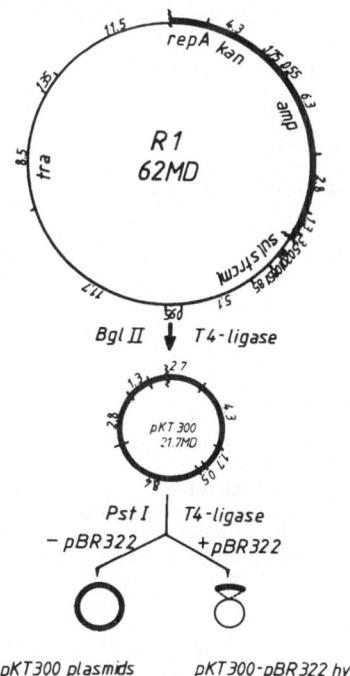

Fig. 1. Cloning scheme for the generation of mini R1 plasmids and pBR322-R1 hybrid plasmids. Straight and wavy cross bars indicate respectively the locations of *Eco*RI and *Bgl*II cleavage sites. The locations of the *Eco*RI sites are taken from Blohm and Goebel (1978).

fragments P2, P3 and P4 (Fig. 2B). Fragments P3 and P4 are not contiguous (Fig. 2D) and it should not be assumed at this stage that both contain essential replication determinants. Because fragment P3 contains an incompatibility determinant (see below) we assume that it carries all or most of the replication determinants of pKT312 (the adjacent Km fragment may also contribute functions); fragment P4 on the other hand is contained within the Tn3 element. We have designated this second series of functions for autonomous replication, *Rep*D (*Rep*A, *Rep*B and *Rep*C regions have all been previously described: Yoshikawa, 1974; Timmis et al., 1975; Timmis et al., 1978 b), and have tentatively placed *Rep*D on the P3 fragment. The pKT312 plasmid is somewhat unstable which explains why only one such isolate was obtained in this experiment (Table I.).

3. *Identification of Incompatibility determinants on plasmid pKT300*

The *in vitro* construction described above in section 2 was repeated except that *Pst*I-cleaved pBR322 DNA was included in the ligation reaction, and Tc[r] Ap[s] transform-

TABLE I.
Composition and properties of mini pKT300 plasmids

Group	Plasmids	No. of Isolates	P8	P6	P3	P2	P5	P4	P7	P1	Km	Ap	Stability*
Parent	pKT300	1	+	+	+	+	+	+	+	+	R	R	Stable
I	pKT301	1	+	+	+	+	+	−	+	−	R	S	N.T.
	pKT302	1	+	+	+	+	+	−	−	−	R	S	N.T.
	pKT303	4	+	+	+	+	−	+	−	−	R	S	Stable
	pKT304	2	+	+	+	+	−	−	−	+	R	S	N.T.
	pKT305	6	+	+	+	−	−	−	−	−	R	S	Stable
II	pKT306	1	+	+	−	+	+	+	−	+	R	R	N.T.
	pKT307	3	+	+	−	+	+	+	−	−	R	R	Stable
	pKT308	1	+	+	−	+	−	−	+	+	R	S	N.T.
	pKT309	2	+	+	−	+	+	−	−	−	R	S	Stable
	pKT310	1	+	+	−	+	−	−	−	+	R	S	Stable
	pKT311	13	+	+	−	+	−	−	−	−	R	S	Stable
III	pKT312	1	−	−	+	+	−	+	−	−	R	S	Unstable

* Plasmid stability was tested by growth of plasmid-carrying clones for 13 generations in the absence of Km selection. After dilution and plating of the cultures for single colonies on nutrient agar, the Km-resistance of about 100 colonies was tested by transfer to Km plates. "Stable" plasmids were found in 99% or more of the colonies tested whereas "unstable" plasmids were found in only 70% of colonies tested. N.T. = not tested.

TABLE II.
Identification of PstI fragments of pKT300 that carry incompatibility determinants

Group	Plasmid	P8	P6	P3	P2	P5	P4	P7	P1	pBR 322	Tc	Km	Ap	Transformation Incompat. Index	Incompat. Method 1	Incompat. Method 2	Ability to Replicate in PolA Bacteria
Parents	pKT300	+	+	+	+	+	+	+		−	S	R	R	0.69	+	+	+
	pBR322	−	−	−	−	−	−	−	−	+	R	S	R	1.58	−	−	−
I	pKT350	+	+	+	+	−	−	+	+	+	R	R	S	0.01	+	+	+
	pKT351	+	+	+	+	−	−	+	+	+	R	R	S	<0.01	+	N.T.	+
	pKT352	+	−	+	−	−	+	+	−	+	R	S	S	0.002	+	+	−
	pKT353	+	+	+	+	−	−	−	−	+	R	R	S	0.02	+	+	+
	pKT354	+	−	+	−	−	−	−	+	+	R	S	S	<0.002	+	N.T.	−
II	pKT355	+	+	−	+	−	−	+	−	+	R	R	S	0.45	+	+	−
	pKT356	+	−	−	−	−	−	−	−	+	R	S	S	0.01	+	+	N.T.
	pKT357	+	−	−	−	−	−	−	−	+	R	S	S	0.12	−	−	N.T.
III	pKT358	−	−	+	+	−	−	−	−	+	R	R	S	0.91	−	−	N.T.
	pKT359	−	−	+	−	−	−	−	−	+	R	S	S	0.039	+	−	N.T.

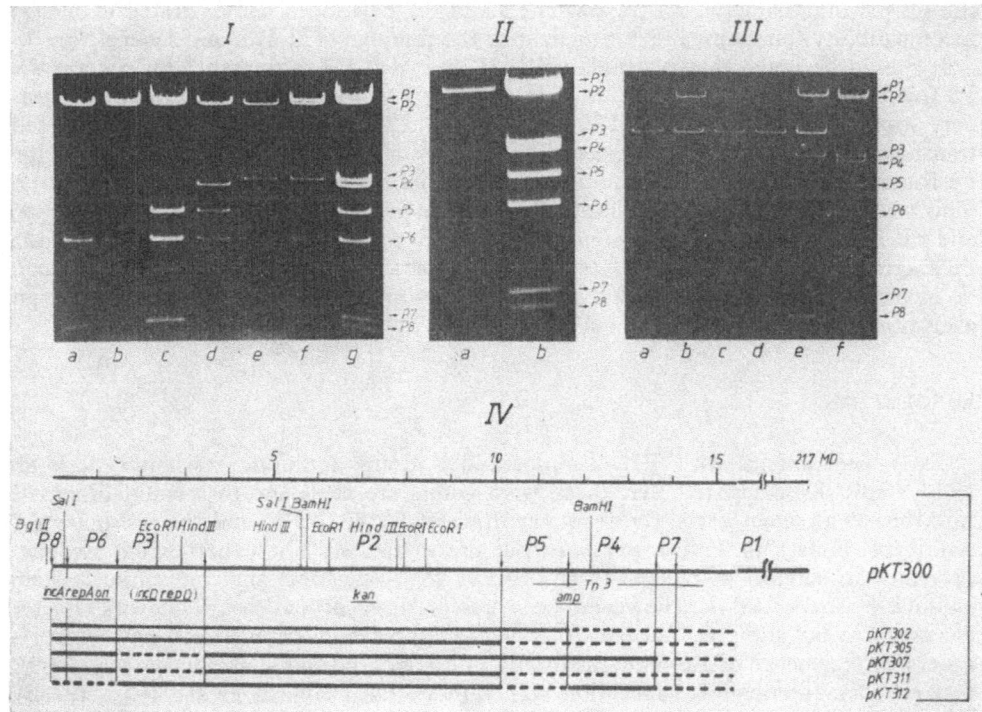

Fig. 2. R1 DNA fragment composition of mini plasmids and pBR322-R1 hybrid plasmids. Plasmid DNA preparations were treated with *Pst*I endonuclease and analysed by electrophoresis through agarose gels (0.8% in tris-brorate buffer), *I*: a, pKT311; b, pKT310; c, pKT309; d, pKT307, e, pKT305; f, pKT304; g, pKT300; II: a, pKT312; b, pKT300; III: a, pBR322; b, pKT360 (this pBR322-P2 hybrid plasmid expresses resistance to Km); c, pKT359; d, pKT356, e, pKT354; f, pKT350; IV: Schematic representation of the structure of the new miniplasmids. Solid lines indicate DNA segments carried by the plasmids, dotted lines indicate DNA segments not carried. Vertical arrows indicate the locations of *Pst*I cleavage sites and the P numbers indicate the *Pst*I-generated fragments in order of size that are generated from the pKT300 mini R1 plasmid.

ant clones were isolated. Plasmid DNAs were prepared from a number of representative clones that had been tentatively identified by a rapid screening procedure (Danbara et al., 1979) as carrying plasmids that express incompatibility towards plasmid R100, and were transformed into competent cells of strains YC259 and YC259 R100drd△Tc to analyse their incompatibility properties. Table II. shows that, as expected, pBR322 is entirely compatible with R100 whereas the mini R1 plasmid, pKT300, is incompatible. pBR322-pKT300 hybrid plasmids that express incompatibility towards R100 fall into three classes. Class I consists of plasmids that carry fragments P8 and P3 and exhibit severe incompatibility, class II consists of plasmids that carry fragment P8 and also exhibit strong incompatibility, and class III consists of plasmids that carry fragment P3 and exhibit weak incompatibility. The degree of incompatibility as measured by transformation frequencies obtained on double selection plates (transformation incompatibility index) and by segregation rates obtained under non-selective conditions (methods I and II), in general were mutually consistent. Thus two distinct regions of the R1 plasmid, namely *Pst*I fragments P3 and P8, have been demonstrated to encode incompatibility functions, in agreement with the findings of Kollek and Goebel.

It should be noted that plasmids pKT356 and pKT357, although both contain the P8 fragment, exhibited quite different incompatibility properties: that of pKT356 was very strong, whereas that of pKT357 was very weak and only detectable by its reduced transformation incompatibility index. This result suggested that the orientation of the P8 fragment in the vector plasmid might be crucial for expression of its Inc property. This was tested in two ways. Firstly, the orientations of the P8 fragments in pKT356 and pKT357 were analysed and found to be different. Secondly, the orientation of the P8 fragment in pKT357 was reversed and the new derivatives tested for expression of incompatibility. All derivatives having P8 in reversed orientation expressed high evels of incompatibility (Danbara et al., 1979).

DISCUSSION

Early experiments with FII incompatibility group antibiotic resistance plasmids (R1, R6, R100) suggested that these were composite replicons that could dissociate into their component parts, the resistance transfer factor (RTF) and the resistance determinant (R-det), in *Proteus mirabilis*, but not in *E. coli*. The experimental evidence presented in support of the existence of more than one functional replication system on such plasmids has not, however, been universally accepted (see Falkow, 1975, for discussion of this subject). The construction of mini plasmids from *Eco*RI- and *Hind*III-generated fragments of R6-5 revealed only one region of this plasmid, namely RepA, that encodes autonomous replication ability in *E. coli* (Timmis et al., 1975, 1978b). This region is contained within the RTF segment of the plasmid, as are two other regions that encode replication functions, RepB and RepC, from which it has so far not been possible to contruct autonomous replicons (Timmis et al., 1978b). Additional evidence for multiple replication systems on IncFII plasmids has been provided by Warren et al. (1978) who identified 4 distinct origins of replication that were used by a copy number mutant of plasmid NR1 in *Proteus mirabilis*.

The experiments described in this paper now provide conclusive evidence for the existence of two distinct DNA segments on the R1 plasmid that contain functions for autonomous replication. One of these segments contains the RepA replication functions that have already been subjected to analysis in several laboratories (Kollek and Goebel, 1978; Timmis et al., 1978a; Andres et al., 1979; Mølin and Nordström, submitted for publication). The other region, which we have designated RepD, is adjacent to the *Rep*A region and appears to be situated either on DNA fragment P3, which means that it is contained within the RTF part of R1, or on both the P3 and P2 DNA fragments and thereby is contained in both the RTF and R-det parts of R1 (the *Pst*I cleavage site that is common to P3 and P2 is located within the IS1b sequence). The generation of new *Rep*D mini plasmids containing a selection fragment other than Km fragment P2 will clarify this situation.

The finding of *Eco*RI and *Hind*III cleavage sites within the P3 fragment, i.e. within the DNA segment provisionally identified as carrying the *Rep*D functions, provides a plausible explanation for previous failures to detect more than one replication region on the R6-5 and R1 plasmids, by *in vitro* construction of miniplasmids from *Eco*RI and *Hind*III fragments, and for the success of Kollek and Goebel and us by *in vivo* generation of miniplasmids and the *in vitro* construction of miniplasmids from *Bgl*II fragments, respectively. Recently, unstable miniplasmids were produced from the R1 plasmid by endonuclease cleavage and transformation of the unligated fragments into *E.coli* (D. Blohm, personal communication) but it is not known which region of R1 these miniplasmids carry.

Although the *Rep*A and *Rep*D regions of plasmid R1 are physically and functionally distinct replication regions, each of which is able to form miniplasmids, their functions may well interact to accomplish replication of the R1 replicon. We have found that the miniplasmid driven by the *Rep*D region is much less stable than are miniplasmids containing both replication regions and Mølin and colleagues (personal communication) have found that miniplasmids driven by the *Rep*A region are less stable than those also carrying the P3 fragment. This suggests that the *Rep*A and *Rep*D systems complement one another in the processes of stable replication and inheritance of the R1 plasmid. The nature of possible interactions between *Rep*A and *Rep*D is now being investigated in this laboratory.

The cloning of *Pst*I-generated fragments of the pKT300 plasmid onto pBR322 allowed the detection of two distinct incompatibility determinants on the R1 plasmid. These are located on DNA fragments that are contained within the two distinct replication regions of R1 and we have consequently designated them *Inc*A (in *Rep*A) and *Inc*D (in *Rep*D). *Inc*A is known to be the copy control gene of R6-5/R1, *Inc*D could either be a second copy control gene or could cause incompatibility by some other mechanism, possibly a partition mechanism. This is now under investigation.

We have shown that the *Pst*I fragment of R6-5 which encodes copy control/incompatibility contains the Cop promoter (see Ely et al., this volume) and hence expresses incompatibility in both possible orientations when cloned into pBR322 (Andres et al., 1979; our previous suggestion that expression of Inc is orientation dependant, Timmis et al., 1978a, was incorrect). This *Pst*I fragment of R6-5 is 0.7 mD in length and corresponds to *two* adjacent *Pst*I fragments of plasmid R1, due to the existence of an additional *Pst*I site in this region of R1. Our present studies have shown that the expression of incompatibility by the P8 fragment is orientation dependant when cloned into pBR322 and probably requires transcription from the β-lactamase promoter. This indicates that the Inc/Cop promoter is located to the left of the R1 P8 fragment and that transcription is rightward (Fig. 2), although other explanations are possible. Our finding of a P3 fragment-carrying hybrid plasmid that does not express incompatibility would also suggest that expression of the *Inc*D incompatibility of this fragment, when cloned into pBR322, is orientation dependant.

ACKNOWLEDGEMENTS

We thank H. Mayer for generous gifts of *Pst*I endonuclease, W. Goebel, R. Kollek and D. Blohm for stimulating discussions, and D. Vogt for valuable technical assistance. H. D. acknowledges with gratitude receipt of a postdoctoral fellowship from the Alexander von Humbolt Stiftung.

REFERENCES

ANDRES, I., P. M. SLOCOMBE, F. CABELLO, J. K. TIMMIS, R. LURZ, H. J. BURKARDT and K. N. TIMMIS (1979): Plasmid replication functions. II. Cloning analysis of the *Rep*A replication region of antibiotic resistance plasmid R6-5. Molec. Gen. Genet. 168, 1–25.
BLOHM, D. and W. GOEBEL (1978): Restriction map of the antibiotic resistance plasmid R1drd19

and its derivatives pKN102 (R1drd19B2) and R1drd16 for the enzymes *Bam*HI, *Hind*III, *Eco*RI and *Sal*I. Molec. Gen. Genet. 167, 119–127.

DANBARA, H., J. K. TIMMIS and K. N. TIMMIS (1979): Plasmid replication functions. V. Plasmid R1 consists of two replicons each capable of autonomous replication in *Escherichia coli*. Submitted for publication.

DANBARA, H. and K. N. TIMMIS (1979): Plasmid replication functions. VI. Genetic analysis of a plasmid copy control mechanism. In preparation.

FALKOW, S. (1975): Infectious multiple drug resistance, Pion Limited, London.

JACOB, F., S. BRENNER and R. CUZIN (1963): On the regulation of DNA replication in bacteria. Cold Spring Harbor Symp. Quant. Biol. 28, 329–348.

KLINE, B. C. and J. R. MILLER (1975): Detection of nonintegrated plasmid deoxyribonucleic acid in the folded chromosome of *Escherichia coli*: physicochemical approach to studying the unit of segregation. J. Bacteriol. 121, 165–172.

KOLLEK, R., W. OERTEL and W. GOEBEL (1978): Isolation and characterization of the minimal fragment required for autonomous replication ("Basic Replicon") of a copy mutant (pKN102) of the antibiotic resistance factor R1. Molec. Gen. Genet. 162, 51–57.

NOVICK, R. P. and M. SCHWESINGER (1976): Independance of plasmid incompatibility and replication control functions in *Staphylococcus aureus*. Nature 262, 623–626.

OHTSUBO, E., M. ROSENBLOOM, H. SCHREMPF, W. GOEBEL and J. ROSEN (1978): Site specific recombination involved in the generation of small plasmids. Molec. Gen. Genet. 159, 131–141.

TIMMIS, K. N. (1979): Mechanisms of plasmid incompatibility. *In* Plasmids in bacteria of medical, environmental. and commercial importance, Eds. K. N. Timmis and A. Pühler, Elsevier, Amsterdam, in press.

TIMMIS, K. N., I. ANDRES and P. M. SLOCOMBE (1978a): Plasmid incompatibility: cloning analysis of an *inc*FII determinant of R6-5. Nature 273, 27–32.

TIMMIS, K. N., I. ANDRES, P. M. SLOCOMBE and R. M. SYNENKI (1979): Plasmid-encoded functions involved in the replication and inheritance of antibiotic resistance plasmid R6-5. Cold Spring Harbor Symp. Quant. Biol. 43, in press.

TIMMIS, K. N., F. CABELLO and S. N. COHEN (1975): Cloning, isolation, and characterization of replication regions of complex plasmid genomes. Proc. Nat. Acad. Sci. U.S.A. 72, 2242–2246.

TIMMIS, K. N., F. CABELLO and S. N. COHEN (1978b): Cloning and characterization of *Eco*RI and *Hind*III restriction endonuclease-generated fragments of antibiotic resistance plasmids R6-5 and R6. Molec. Gen. Genet. 162, 121–137.

UHLIN, B. E and K NORSTRÖM (1975): Plasmid incompatibility and control of replication: copy mutants of the R-factor R1 in *Escherichia coli* K-12. J. Bacteriol. 124, 641–649.

WARREN, R. L., D. D. WOMBLE, C. R. BARTON, A. M. EASTON and R. H. ROWND (1978): Multiple origins for DNA replication of FII composite R plasmids in *Proteus mirabilis*. In Microbiology 1978. Ed. D. Schlessinger, American Society for Microbiology, Washington D. C. pp 96–98.

YOSHIKAWA, M. (1974): Identification and mapping of the replication genes of an R factor, R100.1, integrated into the chromosome of *Escherichia coli* K-12. J. Bacteriol. 118, 1123–1131.

H. D., Max-Planck-Inst. für Molekulare
Genetik, Ihnestrasse 63/73,
D-1000 Berlin-West

CHROMOSOMAL MUTATION AFFECTING
THE EXPRESSION OF PLASMID R6K △1 *IN ESCHERICHIA COLI* K-12

J. NEŠVERA and J. HOCHMANNOVÁ
*Institute of Microbiology, Czechoslovak Academy of Sciences, Prague,
Czechoslovakia*

INTRODUCTION

R-plasmids conferring resistances to different antibiotics in their host bacteria have both theoretical and epidemiological significance. Their maintenance in cells as well as their expression are controlled by cooperation of genes located on plasmids and on bacterial chromosomes (Falkow, 1975). The plasmid-determined resistance to individual antibiotics is accomplished by different mechanisms, namely by inactivation of an antibiotic via chemical modification or by affecting its penetration into the bacterial cell. Plasmid-determined resistance to aminoglycoside antibiotics (e.g., to streptomycin) seems to represent a combination of both of these mechanisms (Davies and Kagan, 1977). The study of phenotypic expression of the same antibiotic resistance determinant in different bacterial strains provides important information on the mechanism of drug resistance carried by a plasmid.

The present communication describes the effect of a thermosensitive chromosomal mutation on the expression of resistance to streptomycin determined by plasmid R6K△1, a deletion mutant (Nešvera, Hochmannová, and Štokrová, 1978) of plasmid R6K (Kontomichalou, Mitani, and Clowes, 1970).

MATERIALS AND METHODS

Bacterial strains. *Escherichia coli* K-12 J5-3 (pro met) harboring plasmid R6K△1, which determines resistances to ampicillin (Ap) and streptomycin (Sm) and which has a molecular weight of 21×10^6 (Nešvera, Hochmannová, and Štokrová, 1978) was used as the original strain. The thermosensitive (ts) mutant M15 (pro met trp) was isolated after nitrosoguanidine mutagenesis of *E.coli* K-12 J5-3 (R6K△1). The recipient strains in conjugation experiments were *E.coli* IV-28 (leu thi thy str-r) and *E.coli* JC5455 (his trp, kindly supplied by Dr. N. S. Willetts, University of Edinburgh). The donor strain in mapping experiments was *E.coli* Hfr H (thi).

Mutagenesis and selection of mutants. Exponential cells of *E.coli* J5-3 (R6K△1) grown in Brain Heart Infusion (BHI, Difco) were treated with N-methyl-N'-nitro-N--nitrosoguanidine (Koch-Light, 300 μh/ml) at 30°C for 2 h in the presence of chloramphenicol (Spofa, 30 μg/ml). After mutagenesis, the cells were washed twice in BHI, diluted and plated on dishes with Tryptose Blood Agar Base (TBAB, Difco). Colonies grown after 2 days at 30°C were tested for resistance to ampicillin (Pentrexyl, Bristol Italiana) and to streptomycin (streptomycin sulfate, Medexport, U.S.S.R.) at both 30°C and 42°C.

Characterization of mutant M15. Growth of the M15 mutant at 30°C and 42°C under shaking was measured in Penassay Broth (PAB, Difco). Segregation of plasmid R6K△1 was tested after growth of M15 in PAB overnight at the respective temperatures by plating the culture on TBAB dishes containing Ap or Sm. The survival of cells grown in PAB with Ap, Sm and erythromycin (Serva) for 24 h at 30°C and at 42°C was determined by colony-counting on TBAB dishes after 2 days at 30°C. The lethal effect of rifampicin (Rifadin, NMB Drug Factory, Bucharest) was tested by colony-counting on TBAB dishes containing this drug using the colony count on drug-free plates as a reference.

Assay for streptomycin phosphotransferase. Crude enzyme extracts from cells of both original and mutant strains grown at 30°C and at 42°C were prepared and assayed for their streptomycin-inactivating activity at 37°C according to Lundbäck and Nord-ström (1974b). The residual streptomycin was estimated by paper disk bioassay (Lund-bäck and Nordström, 1974a) using *E.coli* JC5455 as an indicator strain and plotting the standard curve for streptomycin in each experiment. ATP (adenosine-5′-triphospho-ric acid disodium salt) was purchased from Reanal, Budapest. One enzyme unit is defined as the amount of enzyme that inactivates 1 μmol substrate per min. The protein amount was estimated by the method of Lowry, Rosenbrough, Farr, and Randall (1951).

Conjugation and mapping. The conjugation was carried out as published previously (Šrogl, Hochmannová, Nešvera, Štokrová, and Klégr, 1977), using 30°C instead of 37°C for incubation. In the mapping experiments, the time of conjugation mixture incubation was prolonged to 140 min. The presence of a ts mutation in selected *pro⁻* and *trp⁺* recombinants was demonstrated by their sensitivity to erythromycin in PAB during 24 h at 42°C. The position of the detected ts mutation on *Escherichia coli* K-12 chromosome map was calculated according to Verhoef and de Haan (1966).

RESULTS AND DISCUSSION

The M15 mutant isolated after nitrosoguanidine mutagenesis of *E.coli* K-12 J5-3 (R6K△1) exhibited normal growth on TBAB plates with Ap (4 000 μg/ml) or Sm (30 μg/ml) at 30°C, partially reduced growth on TBAB without drugs or with Ap at 42°C, and significantly reduced growth on TBAB with Sm at 42°C. It was assumed that the detected ts mutation could affect Ap and Sm resistances due to defects in maintenance or in expression of plasmid R6K△1. The segregation of these characters was therefore tested to decide between these two possibilities. Since the normal resistances to both antibiotics at 30°C were preserved after overnight growth of mutant M15 in PAB without antibiotics at 42°C, the possibility of plasmid elimination at 42°C due to

TABLE I.

Activity of Sm-phosphotransferase in cell-free extracts of E. coli J5-3 (R6K△1) and of mutant M15 grown at 30°C and at 42°C

Strain	Temperature (°C)	Activity of Sm-phosphotransferase	
		enzyme unit/mg protein	enzyme unit/viable cell
J5-3 (R6K△1)	30	4.7×10^{-3}	1.4×10^{-13}
	42	2.4×10^{-3}	6.3×10^{-14}
M15	30	1.4×10^{-3}	1.8×10^{-13}
	42	1.9×10^{-4}	1.7×10^{-13}

defects in its replication could be excluded. On the other hand, the survival on TBAB with Sm (but not with Ap) was significantly reduced at 42°C, regardless of preincubation temperature. It follows from these results that the ts mutation affects the expression of the Sm determinant located on the plasmid.

Lundbäck and Nordström (1974b) showed that plasmid R6K codes for streptomycin phosphotransferase. The activities of this streptomycin-inactivating enzyme were therefore estimated in cell-free extracts of *E.coli* J5-3 (R6K△1) and of the M15

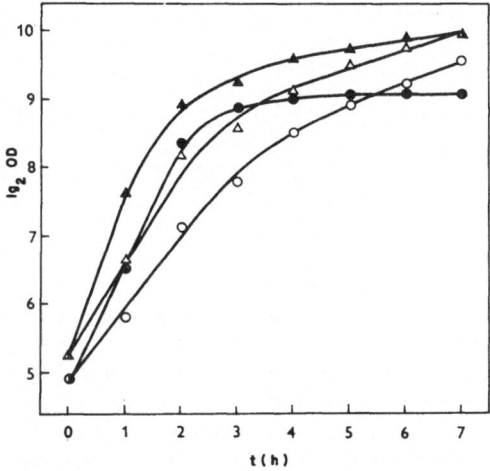

Fig. 1. Growth of *E. coli* J5-3 (R6K△1) and of mutant M15 in PAB at 30° C and at 42° C. — △——△—, J5-3 (R6K△1) at 30° C; —▲——▲—, J5-3 (R6K△1) at 42° C; —○——○—, M15 at 30°C; —●——●—, M15 at 42°C.

mutant, both grown at 30°C and at 42°C (Table I). The activity of Sm-phosphotransferase from M15 grown at 42°C represents only 13.6% of this activity assayed for the same strain grown at 30°C, if both values are referred to one milligram of protein. As the number of viable cells of M15 grown at 42°C was ten times lower than that of M15 grown at 30°C, and because Sm-phosphotransferase activity could be diminished (if not zero) in dead cells, the enzyme activities per one viable cell were also calculated. Respecting both points of view and also taking into consideration the fact that the obtained data (Table I.) do not include the possible difference in protein content between viable and dead cells, it is not possible to correlate the values of Sm-phosphotransferase activity assayed for M15 grown at 30°C and at 42°C with the presence of ts mutation in an unambiguous way. Therefore, additional characters of M15 had to be checked.

At 42°C the M15 mutant exhibits slower growth and altered morphology of colonies on solid media without antibiotics. Partial thermosensitivity was observed during its growth in a liquid medium without drugs (Fig. 1). The growth of M15 stops at 42°C much earlier than at 30°C and the corresponding yield of biomass is lower at this temperature. The total number of M15 cells grown during 24 h at 42°C represents about 15% of the total cell number of M15 grown at 30°C, in addition, a majority of these cells form filaments, suggesting defects of cell division in this mutant at 42°C. In the same sample, the number of viable M15 cells at 42°C corresponds to only 2.6% of viable

TABLE II.

Lethal effect of antibiotics on E. coli J5-3 (R6K△1)
and on mutant M15 at 30°C and at 42°C

Antibiotic (μg/ml)	Surviving cells (%) [a] of strain			
	J5-3 (R6K△1)		M15	
	at 30°C	at 42°C	at 30°C	at 42°C
Ampicillin (1000)	77.0	84.1	90.0	94.4
Streptomycin (30)	85.7	76.9	94.0	13.8
Erythromycin (20)	32.0	15.2	34.0	0.05
Rifampicin (3)	59.0	54.5	54.6	12.8

[a] Number of viable cells grown in/on drug-free media at 30°C and at 42°C was set as 100%.
STR ±222

cells of this strain grown at 30°C. These results point to a pleiotropic effect of the ts mutation.

While the M15 mutant was found to be resistant to Ap at both 30°C and 42°C, its resistance to Sm was proved only at 30°C (Table II.). Streptomycin penetrates across biological membranes very poorly under normal conditions (Nielsen, 1978), the observed increased sensitivity of the M15 mutant to Sm at 42°C could therefore result from an increased penetration of this drug into cells at the given temperature due to a decreased efficiency of the permeability barrier. To test this possibility, the lethal effect of rifampicin (Rf) and erythromycin (Em), which both penetrate rather poorly into *E.coli* cells (Riva, Fietta, Silvestri, and Romero, 1972) was determined. The significantly increased sensitivities of mutant M15 to Rf and especially to Em at 42°C in comparison with the effect of these drugs at 30°C (Table II.), indicate that the ts mutation can cause a defect in the outer penetration barrier. This conclusion seems to be much more probable than the concept of decreased activity of Sm-phosphotransferase due to ts mutation at 42°C, as in *E.coli* K-12 (R6K△1) cells Rf and Em cannot be affected by this modifying enzyme. The possibility that the activity of Sm-phosphotransferase could be lowered e.g. by leakage of this enzyme from the periplasmic space due to defects in the cell envelope, also does not seem to be probable, as the Sm-phosphotransferase coded by plasmid R6K was shown to be localized mainly in the cytoplasm (Lundbäck and Nordström, 1974b).

As proved by a conjugation experiment, the mutation responsible for the thermosensitive phenotype of mutant M15 is located on the chromosome. Transconjugants harboring plasmid R6K△1 transferred from mutant M15 exhibit the same antibiotic resistance as the original *E. coli* J5-3 (R6K△1) strain. As all the other tested characters in a selected transconjugant *E.coli* JC5455 (R6K△1 from M15) were found to be the same as in the original strain, they must also be caused by mutation(s) located on the chromosome. Most likely, a single-point chromosomal mutation is responsible for the pleiotropic phenotype of mutant M15, because a spontaneous revertant reveals all the tested characters identical with those of the original strain (data not shown).

A cross between *E.coli* Hfr H and mutant M15 was carried out to determine the position of the ts mutation on the chromosome map of *E.coli* K-12. After Hfr H × M15 mating, the pro⁺ and trp⁺ recombinants were selected and tested for the presence or absence of ts mutation. The order of genes *pro – trp – ts* was determined by a three-point cross. The thermosensitive phenotype was missing in 4.5% of pro+ recombinants and in 33.4% of trp+ recombinants. The distance of the ts mutation from the trp operon was calculated as 11.5 min clockwise. The studied ts mutation is therefore probably located near 39 min on the *E.coli* K-12 chromosome map (Fig. 2). For more exact location of this ts mutation, its linkage to chromosomal markers located in the neighbourhood has to be tested by transduction mapping.

It could be concluded that the defective expression of plasmid R6K△1-determined Sm resistance in mutant M15 is due to chromosomal ts mutation being very probably responsible for the increased penetration of Sm into cells at 42°C. The activity of Sm-

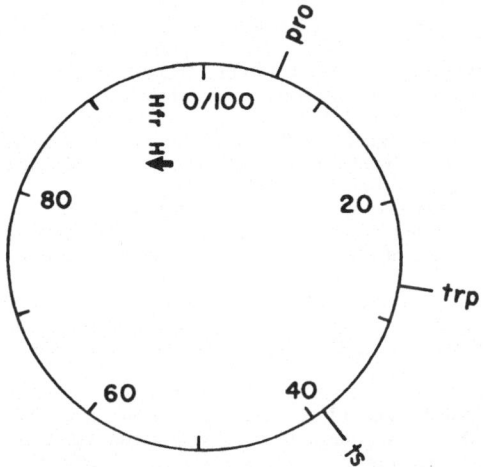

Fig. 2. Position of the described ts mutation on the *E. coli* K-12 simplified chromosome map. Hfr H origin and direction of chromosome transfer is indicated by an arrowhead.

phosphotransferase does not seem to be high enough to inactivate the higher amount of Sm which had penetrated into M15 cells at this temperature. Such a conclusion is consistent with the recent models proposed for Sm uptake by bacterial cells and for the role of aminoglycoside-modifying enzymes in resistance to streptomycin (Davies and Kagan, 1977, Dickie, Bryan, and Pickard, 1978, Nielsen, 1978). Although these models are based on the results referring to Sm-adenylyltransferase localized in the periplasmic space, rules for Sm-phosphotransferase coded by plasmid R6K and present in the cytoplasm (Lundbäck and Nordström, 1974b) need not be different. Following the uptake of radioactively labelled Sm by mutant M15, further information on the mechanism of its changed Sm resistance could be obtained.

Taking in consideration the sensitivity of M15 to Rf and to Em at 42°C (Table II.) as well as the structure of these antibiotics which is different from Sm, it appears less probable that the increased penetration of Sm into M15 at this temperature could be due to increased activity of the transporting system specific for Sm. On the other hand, alterations in the cell wall which represents the barrier(s) against penetration of Sm (Nielsen, 1978) and also of Rf and Em (Riva, Fietta, Silvestri, and Romero, 1972) may influence the Sm resistance in M15. Biochemical comparison of the cell envelope components of mutant M15 grown at 30°C and at 42°C could therefore be valuable for locating the defect in the expression of plasmid R6K△1 caused by the described chormosomal ts mutation.

117

REFERENCES

DAVIES, J. and S. A. KAGAN (1977): What is the mechanism of plasmid-determined resistance to aminoglycoside antibiotics? In: R-factors: Their properties and possible control (J. Drews, and G. Högenauer, eds.), pp. 207–215. Springer-Verlag, Wien.

DICKIE, P., L. E. BRYAN and M. A. PICKARD (1978): Effect of enzymatic adenylation on dihydro-streptomycin accumulation in Escherichia coli carrying an R-factor: Model explaining amino-glycoside resistance by inactivating mechanisms. Antimicrob. Ag. Chemother. 14, 569–580.

FALKOW, S. (1975): Infectious multiple drug resistance. Pion Ltd., London.

KONTOMICHALOU, P., M. MITANI and R. C. CLOWES (1970): Circular R-factor molecules controlling penicillinase synthesis, replicating in Escherichia coli under either relaxed or stringent control. J. Bacteriol. 104, 34–44.

LOWRY, O. H., N. J. ROSEBROUGH, A. L. FARR and R. J. RANDALL (1951): Protein measurement with Folin phenol reagent. J. Biol. Chem. 193, 265–275.

LUNDBÄCK, A. and K. NORDSTRÖM (1974a): Effect of R-factor-mediated drug-metabolizing enzymes on survival of Escherichia coli K-12 in presence of ampicillin, chloramphenicol, or streptomycin. Antimicrob. Ag. Chemother. 5, 492–499.

LUNDBÄCK, A. K. and K. NORDSTRÖM (1974b): Mutations in Escherichia coli K-12 decreasing the rate of streptomycin uptake: Synergism with R-factor-mediated capacity to inactivate strepto-mycin. Antimicrob. Ag. Chemother. 5, 500–507.

NEŠVERA, J., J. HOCHMANNOVÁ and J. ŠTOKROVÁ (1978): Application of R-plasmid DNA's from Escherichia coli minicells in genetic transformation. Folia Microbiol. 23, 278–285.

NIELSEN, P. L. (1978): Uptake of streptomycin by the bacterial cell. Arch. Pharm. Chem., Sci. Ed. 6, 41–55.

RIVA, S., A. M. FIETTA, L. G. SILVESTRI and E. ROMERO (1972): R-factor determined changes in permeability of E. coli towards rifampicin and other antibiotics. In: Bacterial plasmids and anti-biotic resistance (V. Krčméry, L. Rosival, and T. Watanabe, eds.), pp. 343–348. Springer-Verlag, Berlin.

ŠROGL, M., J. HOCHMANNOVÁ, J. NEŠVERA, J. ŠTOKROVÁ and M. KLÉGR (1977): Transforming activity of plasmid and chromosomal DNA in Escherichia coli. Folia Microbiol. 22, 353–359.

VERHOEF, C. and P. G. DE HAAN (1966): Genetic recombination in Escherichia coli. I. Relation between linkage of unselected markers and map distance. Mutat. Res. 3, 101–110.

J. N., Institute of Microbiology,
Czechoslovak Academy of Sciences,
142 20 Prague 4, Czechoslovakia

ISOLATION AND CHARACTERIZATION OF TEMPERATURE-SENSITIVE MUTANTS FOR REPLICATION OF COMPOSITE PLASMID Rms201

Y. IKE, H. HASHIMOTO, K. MOTOHASHI, N. FUJISAWA
and S. MITSUHASHI

Department of Microbiology, School of Medicine, Gunma University, Maebashi, Japan.

INTRODUCTION

Cuzin and Jacob(Cuzin and Jacob, 1967) first isolated temperature-sensitive(*ts*) F'lac' mutants and postulated that the replication of plasmids was positively controlled by a protein coded by its own genome. To clarify the control mechanism of plasmid replication several *ts* mutants for replication have been isolated (Hashimoto and Sekiguchi, 1977; Kingsbury and Helinski, 1973; Kretschmer, Chang and Cohen, 1975).

Extensive analyses of composite R plasmid were done with FII group R plasmids, i.e., NR1, R1, and R6, which are strigently controlled and have similar DNA buoyant density. Recently, Gustafsson and Nordström (1978) isolated temperature-dependent copy number mutants from Rldrd-19 by a protein. However, temperature-sensitive mutants for replication have not been isolated from these composite R plasmids.

We have studied the control system of the replication of a conjugative plasmid Rms201. Rms201 plasmid belongs to the FII incompatibility group and encodes resistance to five drugs, i.e., ampicillin(Ap), tetracycline(Tc), chloramphenicol(Cm), streptomycin (Sm), and sulfanilamides(Su). Rms201 is strigently controlled and its molecular size is 60×10^6 daltons (Ike, Hashimoto and Mitsuhashi, 1976; Odakura, Hashimoto and Mitsuhashi, 1974). In this paper, we describe the isolation of temperature sensitive Rms201 mutants for replication.

RESULT

Selection of R plasmid mutants. Isolation of ts mutants followed the method of Baumberg (1976). P1*kc* lysate of *E. coli* W3630 *mal* carrying an Rms201 plasmid was prepared and used for transduction to *E.coli* ML1410 *met* Nal^r (resistance to nalidixic acid) after treatment of transducing lysate with hydroxylamine. The cells were plated on BTB lactose agar plates containing Cm(25 μg/ml) and incubated at 30°C for 40 hr. For the selection of *ts* mutants, transductants were picked with a toothpick and inoculated onto two BTB lactose agar plates without drug. One plate was incubated at 30°C and the other at 42°C. Cells grown at 30°C or 42°C were replicated, respectively, to drug-free plate and agar plates containing each of the four drugs individually, and they were incubated at 30°C and 42°C, respectively. Drug plates contained each of four drugs, i.e., Cm(25 μg/ml), Tc(12.5 μg/ml), Sm(6 μg/ml) and Ap(25 μg/ml). Clones that produced colonies at 30°C, but did not at 42°C in the presence of each of the drugs were selected as *ts* mutants. We examined about 1.000 transductants and obtained two *ts* mutants for replication. The mutant R plasmids were conjugally transferred to W3630, and then they were again transferred from W3630 transconjugants to ML1410. After the transfer, the mutant R plasmids were denoted Rms201*ts*14 and Rms201*ts*69.

Transformation analysis. The plasmid CCC-DNAs were isolated from ML1410

119

(Rms201ts14)⁺, ML1410(Rms201ts69)⁺ and ML1410(Rms201)⁺, and were used for transformation to calcium chloride-treated ML1410. After the phenotypic expression of resistance at 25°C for 10 hr in L broth, the cells were plated on drug-containing BTB lactose agar plates and incubated at 25, 30, 37 and 42°C, respectively. Drug plates contained each of 4 drugs, i.e., Cm(25 μg/ml), Tc(12.5 μg/ml), Sm(6 μg/ml) and Ap(25 μg/ml). Colonies formed were counted as transformants. *E. Coli* with the DNA

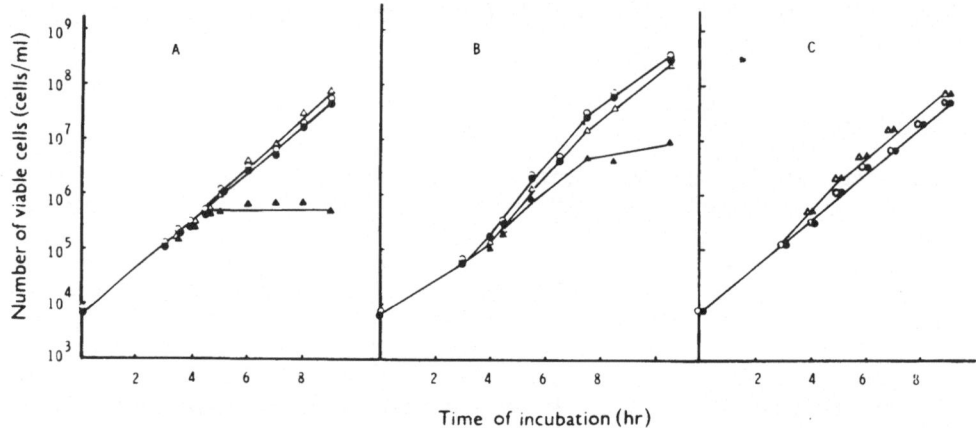

Fig. 1. Growth curves of *E. coli* ML1410 met Nal^r (Rms201ts14)⁺ or M11410 met Nal^r (Rms201ts69)⁺. *E. coli* ML1410 carrying Rms201ts14, Rms201ts69 or Rms201 was incubated for 16 hr at 30°C in L broth. The cultures were diluted 10⁵-fold with L broth and incubated at 30°C. After 3 hr of incubation, the cultures were divided into two portions and were incubated at 30°C and 42°C. At time indicated in the figure, the total number of cells was counted on BTB lactose agar and the number of antibiotic-resistant cells was counted on BTB lactose agar containing Cm (25 μg/ml). (A) ML1410 *met* Nal^r(Rms201ts14)⁺, (B) ML1410 *met* Nal^r(Rms201ts69)⁺, and (C) ML1410 *met* Nal^r(Rms201). An arrow indicates the shift of temperature from 30 to 42°C. Symbols: △, total number of cells at 42°C; ▲, antibiotic resistant cells at 42°C; ○, total number of cells at 30°C, and ●, antibiotic-resistant cells at 30°C.

from Rms201ts14 or Rms 201ts69 formed colonies on the selective plates incubated at 25, 30, and 37°C, at a frequency of 10³/μgDNA, but could not form colonies on the selective plates incubated at 42°C. *E.coli* with the DNA from Rms201 formed colonies on selective plates at indicated temperatures, even at 42°C, at a frequency od 10³/μgDNA.

Segregation of R⁻ cells at high temperature. Segregation kinetics of ML1410(Rms201ts 14)⁺ or ML1410(Rms201ts69)⁺ was tested at various temperatures. When the incubation temperature of ML1410(Rms201ts14)⁺ was shifted to 42°C from 30°C in L broth, the increase in the number of antibiotic-resistant cells ceased at 90 min after the temperature shift (Fig. 1). However the total number of cells continuously increased at almost equal growth rate at both 30°C and 42°C, and the Rms201ts14 plasmid was lost at a frequency of 97% after 5 hr of incubation at 42°C. When ML1410(Rms201ts69)⁺ was grown at 42°C in L broth, antibiotic-sensitive cells appeared at a high frequency 2 hr after temperature shift, and the Rms201ts69 was lost at a frequency of about 60% 5 hr after incubation at 42°C. These results indicated that the replication of Rms201ts14 or Rms201ts69 was affected at 42°C. But the plasmids did not interfere with the host cell growth at both 30 and 42°C. By contrast, the cells harboring a wild type Rms201 did not segregate drug-sensitive cells even at 42°C (Fig. 1)

Number of covalently closed circular plasmid DNA(CCC-DNA) per chromosome equi-

120

valent. Bacteria carrying plasmids were labelled with [³H] thymidine at 30°C for 90 min. Cleared lysate was prepared and analyzed by cesium chloride-ethidium bromide equilibrium centrifugation. The amounts of CCC-DNA of Rms201*ts*14, Rms201*ts*69 and Rms201 plasmids per chromosome equivalent were 3.8, 5.3 and 6.3%, respectively.

Synthesis of plasmid DNA at high temperature. The effect of temperature on plasmid DNA synthesis was examined in double-labelling experiments. Bacteria carrying plas-

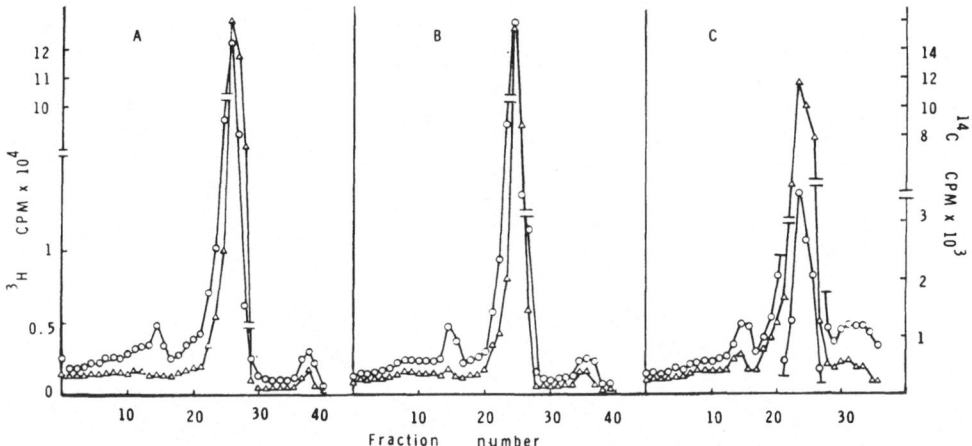

Fig. 2. Temperature effect on the replication of *ts* mutant plasmid DNA. *E. coli* ML1410 carrying plasmids were labelled with [³H] thymidine (5 μC/ml) for 90 min at 30°C, washed with fresh medium, and suspended in 5 ml of L broth. After incubation for 10 min at 30°C, the temperature of the culture was shifted to 42°C. [¹⁴C] thymidine (final concentration, 2 μC/ml) was added 5 min after the temperature shift, and the labelling was continued for 30 min at 42°C. The cleared lysate was analyzed by cesium chloride-ethidium bromide equilibrium centrifugation at 38,000 rpm for 50 hr at 25°C in Hitachi Rp65 rotor. Symbols: ○, DNA prelabelled with [³H] at 30°C and △, DNA labelled with [¹⁴C] at 42°C. (A), Rms201*ts*14, (B), Rms201*ts*69 and (C), Rms201.

mids were labelled with [³H] thymidine at 30°C for 90 min, and then labelled at 42°C with [¹⁴C] for 30 min. Cleared lysates were prepared and analyzed by cesium chloride-ethidium bromide equilibrium centrifugation. Centrifugation profile of DNAs from cells carrying plasmids are shown in Fig. 2. No incorporation of radioactive thymidine into CCC-DNA in Rms 201*ts*14 was observed at 42°C, although a significant amount of [³H]-labelled DNA in the plasmid precultured at 30°C, was found, indicating no replication of Rms201*ts*14 at 42°C. In the case of Rms201*ts*69, a little incorporation of radioactive precursor into CCC-DNA(0.66% per chromosome) was observed at 42°C. On the other hand, active incorporation of radioactive precursor into CCC-DNA(3.8% per chromosome equivalent) of Rms201 plasmid was observed even at 42°C.

The conjugal transferability of Rms201ts14 and Rms201ts69. The transferability of Rms201*ts*14 and Rms201*ts*69 was then examined. Transfer frequency of Rms201*ts*14, Rms201*ts*69 and Rms201 was 3.4×10^{-7}, 3×10^{-6} and 2.5×10^{-5} at 30°C, respectively, and was less than 10^{-8}, 2.8×10^{-7} and 2.5×10^{-5} at 42°C, respectively.

Isolation of mini-plasmid from Rms201ts14. We reported that deletion mutants were isolated from Rms201 at a high frequency after incubation of the R⁺ cells for a long period at 37°C. From the combination of the deleted resistance markers of Rms201 and its copy number mutants, a circular genetic map of Rms201 was constructed (Ike,

121

Hashimoto and Mitsuhashi, 1979). The gene order of Rms201 was, *-tet-tra-amp-rep-sul-str-cml-* (Ike et al. 1979). The replication gene *rep* and the gene *amp* on Rms201 were included in a miniplasmid pMs201, and its molecular weight was about 5×10^6 daltons (Ike et al. 1979) (Fig. 4).

ML1410(Rms201*ts*14)⁺ cells were incubated in L broth for 10 days at 30°C and drug-resistant patterns of the survived cells were examined. Thus we obtained a clone called

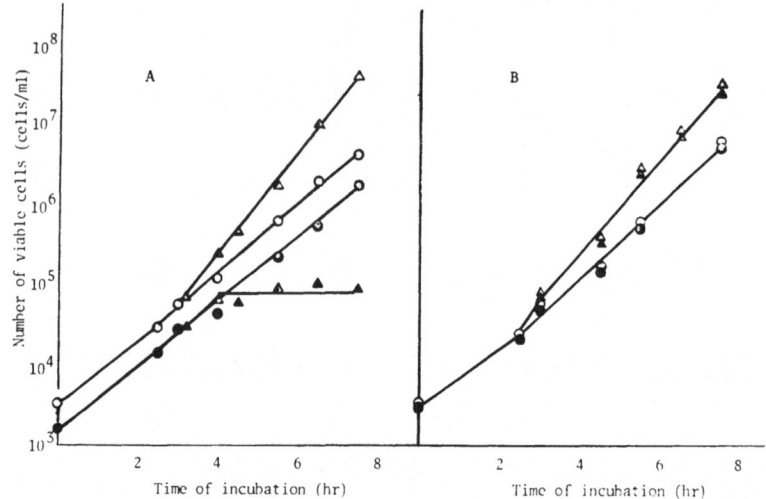

Fig. 3. Growth curves of the cells harboring pMSts214 and pMS201. *E. coli* ML1410 carrying pMSts214 or PMS201 was incubated for 16 hr at 30°C in L broth. The cultures were diluted 10⁵-fold by L broth and incubated at 30°C. After 2.5 hr of incubation, the cultures were divided into two portions. One portion was incubated at 30°C and the other at 42°C. At the time indicated in the figure, the total number of cells was counted on BTB lactose agar and the number of antibiotic-resistant cells was counted on BTB lactose agar containing Ap(25 μg/ml). (A), cells harboring pMSts214 and (B), cells harboring pMS201. An arrow indicates the shift of temperature from 30°C to 42°C. Symbols: ○, total number of cells at 42°C; ●, antibiotic-resistant cells at 42°C; △, total number of cells at 30°C; antibiotic resistant cells at 30°C.

Rms201*ts*14Ap which encodes resistance only to Ap. The plasmid CCC-DNA were isolated from ML1410(Rms201*ts*14Ap)⁺ and ML1410 (pMs201)⁺, and were used for transformation to CaCl₂-treated ML1410. The transformants were selected on BTB lactose agar plate containing Ap(25 μg/ml) and incubation temperatures were 25, 30, 37 and 42°C. *E.coli* with the DNA from pMS201 formed colonies on the selective plate incubated at 25, 30, 37 and 42°C, at a frequency of 10³/μg DNA. On the other hand, *E.coli* ML1410 with the DNA from Rms201*ts*14Ap formed colonies on the selective plates incubated at 25, 30, and 37°C, at a frequency of 10³/μg DNA, but could not form colonies on the selective plate incubated at 42°C. One transformant obtained from the selective plate incubated at 25°C was denoted as ML1410 (pMSts214). The molecular weight of pMSts214 was about 5×10^6 daltons and the plasmid contained a single *EcoR1*-cutting site(Fig. 4).

Segregation of Ap-sensitive cells from ML1410(pMSts214)⁺. When ML1410(pMSts214)⁺ was incubated at 42°C, Ap-sensitive cells were segregated. Fig. 3 illustrates the segreg-

122

ation kinetics of pMS*ts*214. When the incubation temperature of ML1410(pMS*ts*214⁺) was shifted to 42 from 30°C in L broth, increase in the number of antibiotic-resistant cells ceased at 90 min after temperature shift. The number of total cells increased continuously, and the pMS*ts*214 plasmid was lost at a frequency of 99% after 5 hr of incubation at 42°C. By contrast, the cells harboring pMS201 did not segregate drug--sensitive cells even at 42°C.

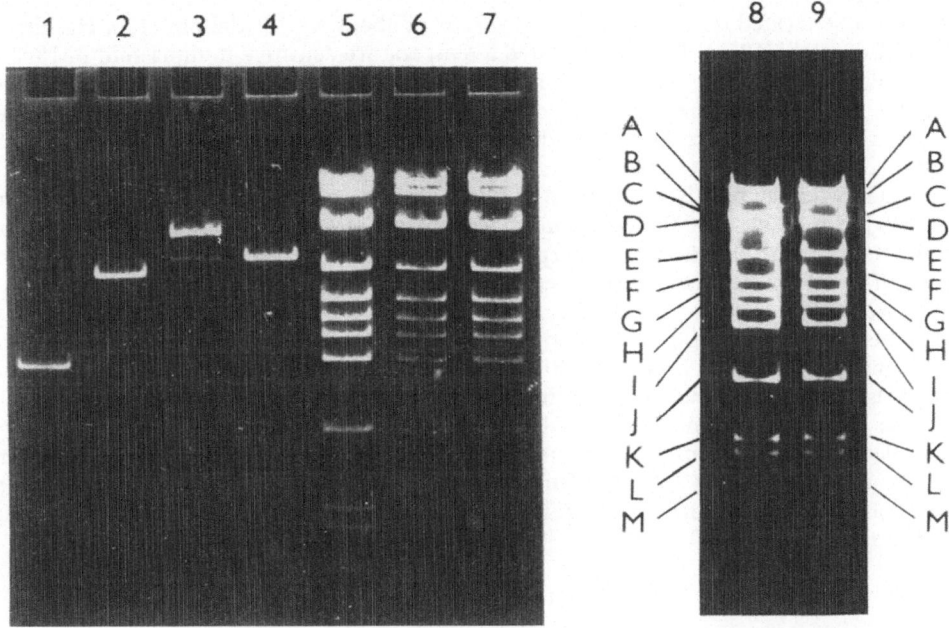

Fig. 4. Agarose-ethidium bromide gel electrophoresis of EcoR1 restriction endonuclease fragments of Rms201*ts*14 DNA, Rms201 DNA and miniplasmids DNA. The DNAs were purified from crude lysate by preparative ethidium bromide-cesium chlolide density gradient centrifugation. The patterns are of: (1), undigested pMS201 (shown in order of decreasing mobilities are the CCC and open circular (OC) (2); EcoR1 digested pMS201; (3), undigested pMS*ts*214 (shown in order of decreasing mobilities are the linear form and open circular (OC); (4), EcoR1 digested pMS*ts*214; (5), EcoR1 digested Rms201; (6), EcoR1 digested Rms201*ts*14; (7), EcoR1 digested Rms201*ts*69; (8), EcoR1 digested NR1 and (9), EcoR1 digested Rms201.

EcoR1 digestion of the plasmid DNA. Fragments of *EcoR1* digestion of Rms201*ts*14 or Rms201*ts*69 were the same as those of Rms201 (Fig. 4). Rms201 DNA yielded 13 major fragments after *EcoR1* digestion. The 13 fragments of Rms201 DNA were the same as those of NR1 except that the fragment B of NR1 migrated slightly faster than that of Rms201 in agarose gel. Calculated difference was 3.2×10^6 daltons, indicating that *EcoR1* fragment B of Rms201 contained gene *amp* (unpublished data). The miniplasmid pMS*ts*214 derived from Rms201*ts*14 and pMS201 derived from parent Rms201 contained a single *EcoR1* site and the molecular weight of both plasmids was about 5×10^6 daltons.

123

DISCUSSION

The results described in this article indicate that Rms201ts14 and Rms201ts69 are the temperature-sensitive(ts) mutants for replication. The existance of temperature-sensitive mutation for replication in Rms201 plasmid indicates that the control system for the replication of the plasmid Rms201 contains an element with a positive function and this element is a protein.

A gene concerning the autonomous replication of Rms201 and an *amp* gene are also found to be located on a miniplasmid pMS201 of about 5×10^6 daltons (Ike, Hashimoto and Mitsuhashi, 1979). We have isolated a temperature-sensitive miniplasmid pMSts214 for replication from Rms201ts14. The miniplasmid pMSts214 was found to be about 5×10^6 daltons and encoded single resistance to Ap. These results indicate that the mutation which results in temperature sensitivity for replication of plasmid Rms201 resides on the miniplasmid.

A model for the positive regulation of DNA replication of a replicon has been presented by Jacob et al. (1963). Cuzin and Jacob first isolated temperature-sensitive F'lac' mutants and postulated that the replication of plasmids is positively controlled by a protein coded by its own. The temperature-sensitive mutants for replication of R plasmid were isolated by several workers (Kretschmer et al.1975). However temperature-sensitive mutants for the replication of composite R plasmid have not been isolated in laboratory experiment. Rms201ts14 and Rms201ts69 are temperature-sensitive mutants for replication isolated in laboratory experiment from a composite plasmid Rms201.

The cessation of Rms201ts14 DNA synthesis after temperature shift was almost complete. However, antibiotic-sensitive cells were segregated after a certain lag time of about 90 min when ML1410(Rms201ts14)$^+$ was grown at 42°C. This lag of period was considered to be the time required for diluting out of the plasmids existing prior to the temperature shift.

REFERENCES

BAUMBERG, B. (1976): Isolation of amber mutant of a P group R factor after hydroxylamide mutagenesis of a transducing lysate. J. General Microbiol. 94, 425–429.

CUZIN, F. and F. JACOB (1967): Mutations de l'épisome F d'*Escherichia coli* K12. II. Mutants a replication thermosensible. Ann. Inst. Pasteur Paris. 112, 398–418.

GUSTAFSSON, P. and K. NORDSTRÖM (1978): Temperature-dependent and amber copy mutants of plasmid R1drd-19 in *Escherichia coli*. Plasmid. 4, 134–144.

IKE, Y., H. HASHIMOTO and S. MITSUHASHI (1976): In S. Mitsuhashi, L. Rosival and V. Krčméry (ed.), Plasmids medical and theoretical aspects. Avicenum and Springer Verlag. P. 277–283.

IKE, Y., H. HASHIMOTO and S. MITSUHASHI (1979): Deletion mutants from R plasmids of increased copy number. Microbiology and Immunology. Vol. 23, No. 8, in press.

JACOB, F., S. BRENNER and F. CUZIN (1963): On the regulation of DNA replication in Bacteria. Cold Spring Harbor Symp. Quant. Biol. 28, 329–348.

KINGSBURY, D. T. and D. R. HELINSKI (1973): Temperature-sensitive mutants for the replication of plasmid in *Escherichia coli*. I. Isolation and specificity of host and plasmid mutations. Genetics. 74, 17–31.

KRETSCHMER, P. J., A.C.Y. CHANG and S. N. COHEN (1975): Indirect selection of bacterial plasmids lacking identifiable phenotypic properties. J. Bacteriol. 124, 225–231.

ODAKURA, Y., H. HASHIMOTO and S. MITSUHASHI (1974): R-factor mutant capable of specifying hypersynthesis of penicillinase. J. Bacteriol. 120, 1260–1267.

I. Y., Dept. of Microbiology,
Gunma University, Maebashi, Japan

GENETIC AND PHYSICO-CHEMICAL ANALYSIS OF ANTIBIOTIC RESISTANCE PLASMIDS OF CLINICAL STRAINS

S. B. VAKULENKO, L. E. BODUNKOVA, M. M. GARAEV, I. P. FOMINA
and S. M. NAVASHIN

National Research Institute of Antibiotics, Moscow, U.S.S.R.

Antibiotic resistance plasmids are suitable objects for investigation of the mechanisms of DNA replication and segregation in bacteria. The fact that many of the R plasmids consist of a resistance transfer factor (RTF) and resistance determinants capable of autonomous replication under certain conditions indicates the presence of at least two origins at which replication can be initiated (Rownd R. H. et al., 1975). There are some data on dissociation of transferable R plasmids with formation of miniplasmids capable of autonomous replication. The dissociation may be due to the factors depending on the properties both of the plasmids themselves (Goebel W. R. et al., 1975) and of the host strain (Humphreys G. O., 1977).

Data on dissociation of R plasmids isolated from a clinical strain of *E. coli* and *E. coli* CSH-2 (R1drd-19) on transformation of E. coli C600 rif^r with plasmid DNA from these strains are presented.

MATERIALS AND METHODS

Bacterial strains: 32 strains of *E. coli* isolated from surgical patients, *E. coli* K-12: CSH-2 (R1drd-19) and *E. coli* K-12 C600 ($r^-, m^-, thr, leu, thi, lac^-$) resistant to rifampicin and nalidixic acid (SV-10) were used.

Media. L broth was used for the growth of the bacteria. Agar plates consisted of L broth plus 2% Difco agar. If necessary ampicillin (Ap)-50 µg/ml, streptomycin (Sm)-20 µg/ml, chloramphenicol (Cm)-25 µg/ml. kanamycin (Km)-25 µg/ml, tetracycline (Tc)-10 µg/ml, rifampicin (Rif)-50 µg/ml, nalidixic acid (Nal)-50 µg/ml were added to the medium.

Bacterial Matings. Strains were grown overnight at 37°C in L broth. Equal volumes of the donor and recipient (SV-10) were mixed and diluted 10 fold. After incubation for 18 h at 37°C the colonies were selected on nutrient agar plates, containing appropriate antibiotics at the concentrations indicated above.

Preparation of Plasmid DNA. Cells lysis and DNA preparation for the agarose gel electrophoresis were performed as described by Guerry P. et al., (1973). The cleared lysis procedure used for isolation of plasmid DNA for transformation, electron microscopy and endonucleasis digestion is described by Timmis K. N. et al., (1978).

Agarose gel electrophoresis of Plasmid DNA. The plasmid DNA was identified and characterized by the method of Meyers S. A. et al., (1976).

Transformation. Transformation of *E. coli* K-12 C600 rif^r with bacterial plasmids was performed according to a procedure described by Cohen N. S. et al., (1972). Determination of the number of the plasmid copies was carried out by the procedure described by Nisioka T. et al., (1970).

Electron microscopy. Electron microscopy was carried out according to the procedure of Davis R. W. et al., (1971).

RESULTS

For elucidation of the plasmid presence in the strains of *E. coli* an electrophoretic analysis of plasmid DNA isolated from these strains was carried out. The plasmids were detected in 30 out of 32 strains and 80 per cent of the strains carried more than one plasmid. Some strains contained 5–6 plasmids. Two strains resistant to Tc contained no plasmid DNA which could be connected with the chromosomal nature of their resistance. On crossing of the isolated strains with *E. coli* SV-10 conjugation transfer of R plasmids was shown in 18 strains (56 per cent) (Table I). In all clases plasmids carrying the determinants of drug resistance with molecular weights of 30 to 70 Mdal were transmitted.

For detection of nontransferable plasmids governing the antibiotic resistance, 12 strains of *E. coli* were analyzed by means of transformation. The transformation analysis showed that transfer of all or a part of the resistance markers took place in 8 strains of *E. coli* (Table II).

TABLE I

Conjugative properties of E. coli strains

	Phenotype of the donors	Phenotype of the transconjugants
1.	Km	Km
2.	Ap	Ap
3.	Tc	Tc
4.	Sm Tc	Sm
5.	Ap Tc	Ap
6.	Ap Sm	Ap
7.	Cm Km Tc	Cm Km Tc
8.	Cm Km Tc	Cm Km
9.	Ap Cm Tc	Ap Cm Tc
10.	Ap Cm Km	Ap Cm Km
11.	Ap Km Tc	Ap Km Tc
12.	Ap Cm Tc	Ap
13.	Ap Sm Tc	Ap Sm Tc
14.	Ap Cm Km Tc	Ap Cm Km Tc
15.	Ap Cm Km Tc	Ap
16.	Ap Cm Km Tc	Ap Cm Km Tc
17.	Ap Cm Sm Km Tc	Ap Cm Sm Km Tc
18.	Ap Cm Sm Km Tc	Ap Cm Sm Km Tc

TABLE II.

Results of Transformation of E. coli C600 rif^r with plasmid DNA from E. coli strains

	Phenotype of the donors	Phenotype of the transformants
1.	Tc	Tc
2.	Tc	Tc
3.	Tc	—
4.	Tc	—
5.	Sm	Sm
6.	Sm	Sm
7.	Ap Tc	Ap Tc
8.	Tc Sm	Tc; Sm; Tc Sm
9.	Tc Sm	Sm
10.	Tc Cm	—
11.	Ap Cm Sm Tc	—
12.	Ap Sm Tc	Ap; Sm; Ap Sm

126

The transformation and conjugation analyses of R plasmids revealed that 82 per cent of the *E. coli* strains carried the antibiotic resistance plasmids.

On transformation of the cells of *E. coli* K-12 C600 rif^r with DNA of the 60 Mdal R factor isolated from a clinical strain of *E. coli* 278 resistant to ampicillin, streptomycin

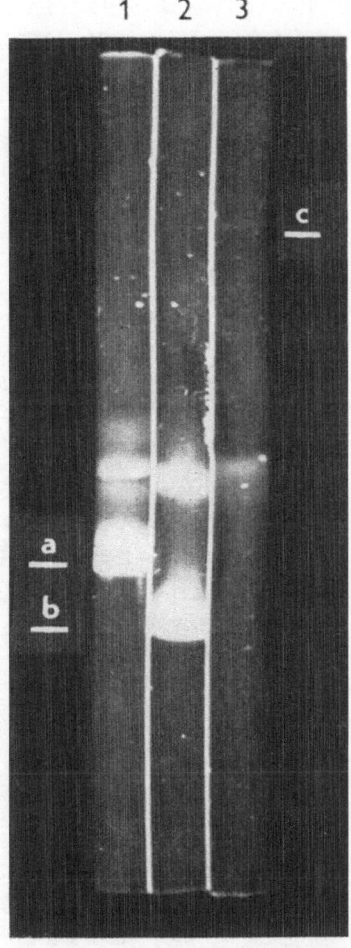

Fig. 1. Agarose gel electrophoresis of DNA from *E. coli* 278 (3) and transformants: resistant to Ap (1), resistant to Sm (2)

        a – plasmid conferring resistance to Ap
        b – plasmid conferring resistance to Sm
        c – plasmid, conferring resistance to Ap, Sm, Tc.

and tetracycline, it was demonstrated that transformants of three classes are formed: (1) those carrying the same antibiotic resistance markers as the donor strain, (2) those resistant only to Ap and (3) those resistant only to Sm. The electrophoretic analysis of DNA isolated from these transformants revealed the presence of the plasmid DNA with molecular weights of 60, 4–6 and 3–4 Mdal respectively (Fig. 1).

An analogous phenomenon of the plasmid dissociation was observed in the trans-formants obtained on transformation of the cells of *E. coli* C600 rif^r with DNA of plasmid R1drd-19 (Table III). As is evident from the data presented in Table III at least 5 classes of transformants differing in the antibiotic resistance markers are formed. Most of the

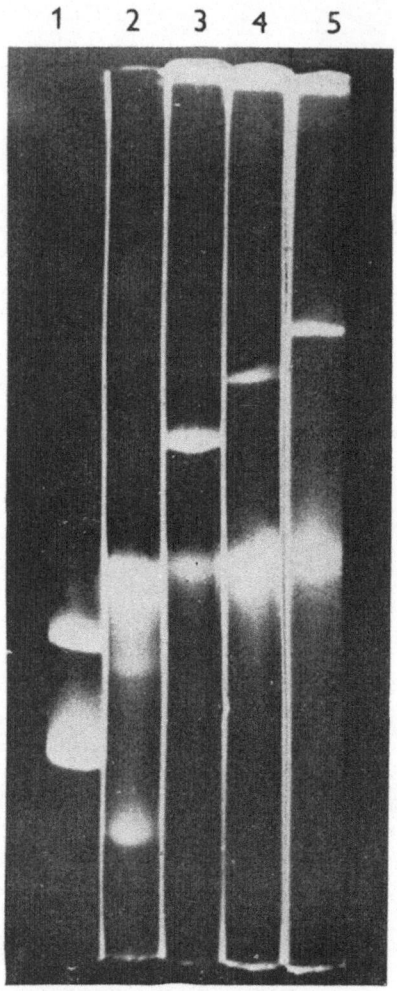

Fig. 2. Agarose gel electrophoresis of plasmids: (1) p AP; (2) p SM; (3) p KM; (4) p AK; (5) R1drd-19.

transformants were resistant to Ap, Cm, Sm, Km, and had a plasmid of similar size as that carried by the host strain *E. coli* K-12: CSH-2 (R1-19).

Three independent clones were taken from every transformant class for analysis. The frequency of Ap resistant clones was the highest. Those transformants had a higher level of resistance to Ap as compared to *E. coli* K-12: CSH-2 (R1-19).

Three Ap resistant transformants were examined by agarose gel electrophoresis for the

TABLE III.

Properties of the R1-19 derivative plasmids

Classes of of transformants	MIC (mcg/ml)				Conjugation transfer frequency	Plasmids	Molecular weight (Mdal)
	Ap	Km	Cm	Sm			
Ap	1000	<2	<2	<2	<10^{-8}	pAP	5.3 ± 0.2
Sm	<2	<2	<2	50	<10^{-8}	pSM	3.6 ± 0.5
Km	<2	100	<2	<2	$5 . 10^{-3}$	pKM	28 ± 1.4
Ap Km	100	100	<2	<2	$5 . 10^{-2}$	pAK	38 ± 1.5
Ap Km Sm Cm (Control)	100	100	100	50	$5 . 10^{-2}$	RI-19	62 ± 2.0

presence of extrachromosomal. DNA. The lysates of these isolates gave a single plasmid band in the gel. All plasmids showed similar mobility. The plasmid present in the transformant strains was named p Ap. The copy number of plasmid p Ap determined in the sucrose alkaline gradient was equal to ~50 per chromosome. Ap resistance was not transferable to *E. coli* C600 rif[r] strain by co-cultivation (Table III).

However, DNA isolated from the representatives of this class of the transformants had high ability to transform *E. coli* C600 rif[r] to Ap resistance. The molecular weights of p Ap plasmid determined with the methods of electron microscopy and electrophoresis in agarose gel were equal to 5.3 ± 0.2 Mdal (Fig. 2). Plasmid p Ap was not restricted by endonucleases *Eco*R1, *Bam*H1, *Sal*1. Three fragments of p Ap DNA were produced by restriction endonucleas Bgl 1.

Analysis of three clones of the Sm resistant transformants showed that, in the resistance level to this antibiotic, they did not differ from *E. coli* K-12: CSH2 (R1drd-19). Analysis of the DNA preparations revealed the presence of the plasmid in all the clones tested. All plasmids (named p SM) had similar molecular weights of 3.6 Mdal (Fig. 2) and a single site for restriction endonuclease *Eco*R1.

All Km resistant transformants had the same resistance level to Km as strain *E. coli* K 12: CSH2 (R1drd-19) Still they had a decreased capacity for conjugation transfer of the Km resistance marker as compared to the parent plasmid R1drd-19 (Tabl. III). Electrophoresis of the DNA samples from Km resistant transformants showed the presence of extrachromosomal DNA. This plasmid was designated p Km. The estimation of the plasmid mass from the extent of DNA migration revealed a molecular weight of about 28 Mdal.

The clones of the transformants resistant to both Ap and Km had a plasmid with a molecular weight of about 38 Mdal (Fig. 2), which provided the same resistance levels to Ap and Km as plasmid R1drd-19. The transfer frequency of this plasmid (designated p AK) was the same as that of plasmid R1drd-19.

Therefore, transformation of *E. coli* C600 strain with DNA of R1drd-19 and DNA isolated from a clinical strain of *E. coli* resulted in appearance of several discrete classes of plasmids, capable of autonomous replication. These plasmids carried only some of the antibiotic resistance markers as compared to those of the parental strains.

Their formation is probably associated with segregation of the parental plasmid fragments conditioned by the presence of "hot spots" on R plasmids possibly providing recombination events with high frequency. Comparison of the data obtained by us with the genetic map R1drd-19 plasmid suggests that the "hot spots" of recombination are connected with the presence of insertion sequences in the structure of plasmid R1drd-19.

REFERENCES

COHEN S. N et al (1972): Nonchromosomal antibiotic resistance in bacteria genetic transformation of *Escherichia coli* by R-factor DNA. Proceedings of the National Academy of Sciences of the United States of America 69, 2110–2114.

DAVIS R. W. et al. (1971): Electron microscope heteroduplex methods for mapping regions of base sequence homology in nucleic acids. Methods in Enzymology 21, 413–429.

GOEBEL W. et al. (1975): Transition of the R-factor NR1 in *Proteus mirabilis*: level of drug resistance of nontransitioned cells. J. Bacteriol. 123, 56–68.

GUERRY P. et al. (1973): General method for the isolation of plasmid DNA. J. Bacteriol. 116, N 2, 1064–1066.

HUMPHREYS G. O. et al. (1977): The molecular nature of R-factors in different bacterial hosts. R-factor Prop and Possible Contr. Symp. Baden near Vienna, 1977. Wien–New York, 277–293.

MEYERS S. A. et al. (1976): J. Bacteriol., 127, 1529–1537.

NISIOKA T. et al. (1970): Molecular recombination between R-factor deoxyribonucleic acid molecules in *Escherichia coli* host cells. J. Bacteriol, 103, 1, 166–177.

ROWND R. H. et al. (1975): Dissociation and transition of R plasmid in *Proteus mirabilis*. Microbial drug resistance. Edited by S. Mitsuhashi and H. Hashimoto p. 3–25.

TIMMIS K. N. et al. (1978): Cloning and characterization of *Eco* R 1 and *Hind* III restriction endonuclease-generated fragments of antibiotic resistance plasmid R 6-5 and R 6. Mol. Gen. Genet., 162, N 2, 121–137.

S. M. N., National Research Inst.
of Antibiotics,
Nagatinskaya 3a,
Moscow M-105, U.S.S.R.

THE PROPERTIES OF *ESCHERICHIA COLI* (R+) STRAIN WITH A MUTATION IN THE R1drd-19 PLASMID

H. BRANÁ, J. HUBÁČEK, O. NAVRÁTIL and O. BENDA

Institute of Microbiology Czechoslovak Academy of Sciences and Faculty of Natural Sciences, Charles University, Prague, Czechoslovakia

INTRODUCTION

The population of R+ cells includes mutants with a phenotypically high resistance to streptomycin (Sm) (Figure 1); these are selected at a frequency of 10^{-5} in a medium with antibiotic concentration higher than that inhibiting the growth of R+ cells (Pearce and Meynell, 1968). The resistance is the result of a synergistic action of two determinants, the R plasmid Sm-determinant and a mutation which determines the low level of Sm resistance (*lrs*, Hubáček, Braná, and Čejka, 1975). The mapping of the *lrs* gene has not supported its chromosomal localization, at least in the tested regions of the *E. coli* chromosome (Braná and Hubáček, 1979).

It was thus tempting to investigate the role of the plasmid DNA in the *lrs* phenomenon. Nordström and coworkers anticipated the direct relationship between the plasmid copy number and the permeability of the cell for Sm (Nordström et al., 1977). Therefore we tried to isolate *E. coli* mutants with an increased chloramphenicol (Cm) resistance, to choose bacteria with a higher plasmid copy number or other alterations of the plasmid molecule among them and to find the relationships between such an alteration and the Sm resistance level.

RESULTS AND DISCUSSION

E. coli JC5455 (R1drd-19) strain was used throughout the experiments. The plasmid is a mutant of R1 derepressed with respect to transfer (Meynell & Datta, 1967) which has a molecular size of ca. 25 μm and confers resistance to Cm, Ap, Sm and Su. The

TABLE I.

Activity of CAT and ABL and the frequency of Sm-r mutants in E. coli

Strain	Plasmid	Enzyme		Frequency of Sm-r mutants
		CAT akat	ABL akat	
JC5455	R1 drd-19	0.7	12.7	4.8×10^{-5}
JC5455-1	pON 1800	1.4	54.2	—
JC5455-2	pON 5300	3.0	60.3	7.8×10^{-7}
JC5455-3	pON 5301	—	—	5.3×10^{-5}

CAT was determined spectrophotometrically according to Shaw and Brodsky (1967), ABL by the microiodometric assay of Novick (1962), the frequency of Sm-r mutants was estimated as described previously (Braná and Hubáček, 1979).

mutants isolated were induced by treating the cells with N-methyl-N′-nitro-N-nitroso-guanidine (MNNG). Short-term mutagenesis was carried out according to Kingsbury & Helinski (1973). After the period of mutation expression, the bacteria were spread on M9 agar plates with Cm (600 μg/ml). A number of 56 clones was selected and tested for the stability of the Cm-resistance phenotype by growing the cells in a drug-free M9 medium as described previously (Hubáček, Braná, and Čejka, 1975). In 12 stable clones

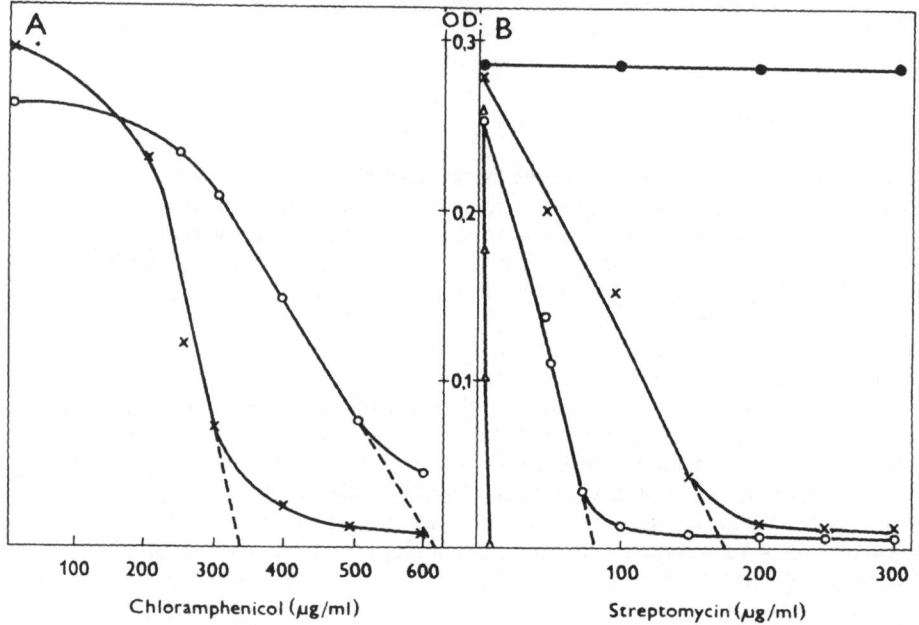

Fig. 1. Determination of the minimal inhibitory concentration. △——△ JC5455; ×——× JC5455 (R1drd-19); ○——○ JC5455-2 (pON5300); ●——● JC5455 (R1drd-19) strain selected on Sm medium (300 μg/ml).

the activities of drug-inactivating enzymes, Cm-acetyltransferase (E.C.2.3.1.28, CAT) and ampicillin-β-lactamase (E. C. 3.5.2.6, ABL), were assayed. In two strains, designated JC5455-1 and JC5455-2, the increase in the activity of both enzymes was found (Table I.). It is interesting to note that the JC5455-2 variant lost the ability to mutate to a higher Sm resistance with the 10^{-5} frequency on LA with Sm (300 μg/ml). As shown in Table I., the frequency is by about two logs lower in order to have the parental JC5455(R1) strain. Some properties of the mutant were further analyzed to elucidate the nature of this phenomenon.

To compare the resistance level in the parental JC5455(R1) and mutant JC5455-2 (pON5300) strains the minimal inhibitory concentration (MIC) of Cm and Sm was determined. As is evident from Figure 1, the mutant strain exhibited twice enlarged MIC value of Cm, while the resistance to Sm was unambiguously decreased. The sensitivity to rifampicin was determined according to Nordström et al. (1977) and a 10 percent survival of exponentially growing bacteria on LA plates with 2 μg of the antibiotic per ml was observed in comparison with the parental strain. The pleiotropic effect of the mutation, i.e. the increased activities of drug-inactivating enzymes and the corresponding levels of resistance to these antibiotics and, on the other hand, the higher sensitivity of the mutant to Sm and rifampicin could be easily explained by a change in the plasmid copy number or some structural alterations of the plasmid molecule.

The plasmid DNA pON5300 was isolated and visualized in an electron microscope (Fig. 2). The average length and molecular weight of the original R1drd-19 plasmid corresponds to the experimentally determined molecular weight published for this

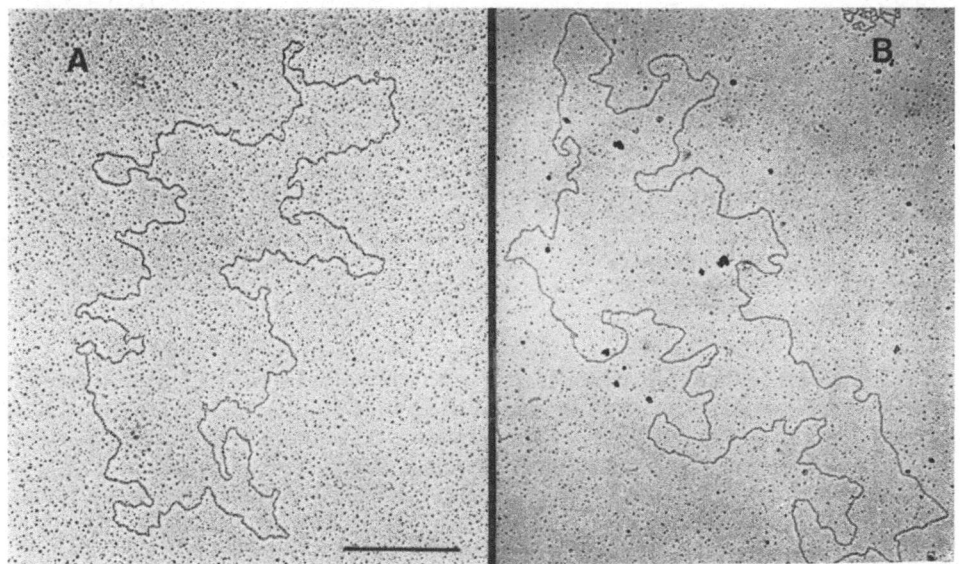

Fig. 2. Electron-micrographs of the R1drd-19 (1) and the pON5300 (B) DNA molecules. Magnification 30.600, the bar represents 1 μm.

plasmid by Kopecko and Cohen (1975) and Nešvera, Hochmannová and Štokrová (1978). The plasmid pON5300 exhibits a significant increment in the contour length of about 2.5 μm which might be due to amplificattion of some resistance determinants (Table II.).

When the plasmid pON5300 from the mutant strain was transferred to the JC5455 *trp*+ recipient to yield the R factor for a strain that had not been MNNG treated, the frequency of highly Sm-r colonies (as illustrated in Table I, strain JC5455-3) as well as MIC of Cm were found to be the same both in the Cm-r colonies with the acquired R plasmid and in the parental JC5455 (R1drd-19) strain. After the transfer of the pON5300 to the JC5455 *trp*+ recipient the amplified pON5300 molecules were not observed (Table II., pON5301). This finding evokes a question of transferability of the pON5300 plasmid. A part of the modified pON5300 molecule might be excised during the conjugation process or already in the donor cell and only the residual pON5301 plasmid could be transferred. The other assumption necessitates contemporary occurrence of two plasmids, the original R1 and the non-transferable pON5300. However, although the dispersion of the molecular lengths is more conspicuous in the pON5300 than in the original plasmid, the collection of the analyzed pON5300 molecules did not fall into two different groups.

So far, no direct evidence concerning the relationship between the enlargement of the pON5300 molecule and the altered frequency of highly Sm-r mutants in the JC5455-2

TABLE II.

Electron-microscopic measurements of the plasmid DNA molecules

Plasmid DNA	Number of molecules	Mean contour length ± S. D. μm	Molecular weight ± S. D. Mg mol⁻¹
R1drd-19	23	24.67 ± 0.61	51.06 ± 1.27
pON 5300	30	26.23 ± 0.98	54.29 ± 2.02
pON 5301	24	24.90 ± 0.63	51.54 ± 1.31

Plasmid DNA was isolated from cleared lyzates obtained by the procedure of Meyers et al. (1976) and Guerry et al. (1973) which were treated with polyethylenglycol 6000 (Humphreys and Anderson, 1975) and further purified in CsCl gradient. After dialysis the amount of plasmid DNA was calculated from the absorbance at 260 nm. For electron-microscopic length measurements the plasmid DNA was spread according to the modified aqueous technique (Davis et al., 1971). Samples were rotatory shadowed with Pt + Ir (80:20) from an angle of 8°. The grids were viewed through a Tesla BS613 electron microscope and photographed at an original magnification of 10,200 × on ORWO EU02 electron-microscope plates. The resulting negatives were enlarged 3 times in a DURST photoprinter. The length of plasmid DNA molecules was measured in a Hewlett-Packard Digitizer 9864A.

strain has been obtained. Nevertheless, the sensitivity of the strain to Sm and rifampicin hints at a possible impairment of the permeability barrier for these antibiotics (Nordström et al., 1977). The changes in the permeation of Sm might be responsible for the decrease in the frequency of highly Sm-r mutants in the JC5455-2 strain.

We wish to thank Dr. M. Jílek from the Institute of Microbiology for his valuable advice on statistical calculations.

REFERENCES

BRANÁ, H. and J. HUBÁČEK (1979): High-level resistance to streptomycin in *Escherichia coli* with R1 plasmid and the analysis of genetic determinants. Folia Microbiol. 24, 136–143.

GUERRY, P., D. J. LEBLANC and S. FALKOW (1973): General method for the isolation of plasmid deoxyribonucleic acid. J. Bacteriol. 116, 1064–1066.

DAVIS, R. W., M. SIMON and N. DAVIDSON (1971): Electron Microscope Heteroduplex Methods for Mapping Regions of Base Sequence Homology in Nucleic Acids, p. 413–428 in L. Grossman and K. Moldave (Eds): Methods in Enzymology XXI. Nucleic Acids, Part D. Academic Press, New York – London.

HUBÁČEK, J., H. BRANÁ and K. ČEJKA (1975): Mechanism of increased rate of appearance of highly streptomycin-resistant and chloramphenicol-resistant variants in *E. coli* R+, p. 171–177 in S. Mitsuhashi, L. Rosival, V. Krčméry (Eds): Drug Inactivation Enzymes and Antibiotic Resistance. Avicenum, Prague, and Springer-Verlag, Berlin–Heidelberg–New York.

HUMPHREYS, G. O., G. A. WILLSHAW and E. S. ANDERSON (1975): A simple method for the preparation of large quantities of pure plasmid DNA. Biochem. Biophys. Acta 383, 457–463.

KINGSBURY, D. T. and D. R. HELINSKI (1973): Temperature-sensitive mutants for the replication of plasmids in *Escherichia coli*. I. Isolation and specificity of host and plasmid mutations. Genetics 74, 17–31.

KOPECKO, D. J. and S. N. COHEN (1975): Site-specific recA independent recombination between bacterial plasmids: Involvement of palindromes at the recombinational loci. Proc. Nat. Acad. Sci. USA 72, 1373–1377.

MEYERS, J. A., D. SANCHEZ, L. P. ELWELL and S. FALKOW (1976): Simplex agarose gel electrophoretic method for the identification and characterization of plasmid deoxyribonucleic acid. J. Bacteriol. 127, 1529–1537.

MEYNEL, E. and N. DATTA (1967): Mutant drug resistance factors of high transmissibility. Nature, Lond. 214, 885–887.

134

NEŠVERA, J., J. HOCHMANNOVÁ and J. ŠTOKROVÁ (1978): Application of R-plasmid DNA's from *Escherichia coli* minicells in genetic transformation. Folia Microbiol. 23, 278–285.

NORDSTRÖM, K., B. ENGBERG, P. GUSTAFFSON, S. MOLIN and B. E. UHLIN (1977): Effect of Plasmids on Cell Division and on the Cell Envelope of *Escherichia coli*, p. 221–237 in J. Drews and G. Högenauer (Eds.): R-Factors: Their Properties and Possible Control. Springer-Verlag, Wien–New York.

NOVICK, R. P. (1962): Microiodometric assay for penicillinase. Biochem J. 83, 236–240.

PEARCE, L. E. and E. MEYNELL (1968): Mutation to high-level streptomycin-resistance in R$^+$ bacteria. J. gen. Microbiol. 50, 173–176.

SHAW, W. V. and R. F. BRODSKY (1968): Characterization of chloramphenicol acetyltransferase from chloramphenicol-resistant Staphylococcus aureus. J. Bacteriol. 95, 28–36.

H. B., Institute of Microbiology,
Czechoslovak Academy of Sciences,
142 20 Pargue 4, Czechoslovakia

INVESTIGATION OF AN EVENTUAL CORRELATION BETWEEN GENERATION TIME OF *E. COLI* INCREASED BY R PLASMIDS AND THEIR MOLECULAR SIZE

G. LEBEK and P. ZÜND

Institute of Hygiene and Medical Microbiology,
University of Berne, Switzerland

INTRODUCTION

It was often observed that *E. coli* strains grow slower when harbouring certain resistance plasmids (1, 4).

This was a reason for us to measure the generation time of an *E. coli* K12 strain bearing different R-factors. For that purpose we transferred about one hundred freshly isolated R-plasmids into *E. coli* K12 921 and determined its speed of growth with the method of

TABLE I.

Generation time of E. coli K12 921 harbouring different plasmids

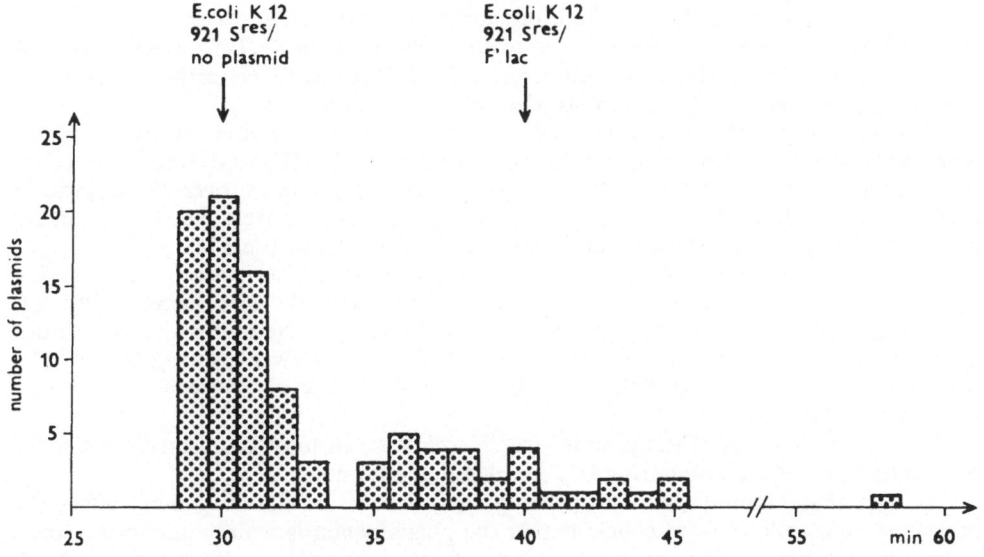

The auxotroph hoststrains' generation time without plasmid is 30 min. About one forth of the 101 tested R-factors prolong generation time up to over 35 min.

turbidity-measuring at 580 nanometers in a flow-through chamber. A definite number of cells per ml from an overnight culture was incubated in columbia broth at 37°C and its extinction was measured every minute. From the growth curve obtained by this results it is possible to calculate the generation time (G.T.) of the strain. This standardized

TABLE II.

Disc susceptibility test of E. coli K12 bearing resistance-determinant-free plasmid F'lac

Antibiotic	No plasmid E. coli 921 GT: 30'	F'lac GT: 40'	Rel. s. dev. — max. in %
Tetracyclin	19	30	3.5
Chloramph.	23	28	2.9
Kanamycin	17	20	7.8
Gentamycin	20	25	6.7
Tobramycin	18	24	5.9
Amicacin	15	19	2.4
Ampicillin	17	20	3.5
Carbenicillin	24	29	2.6
Cefalothin	18	23	3.3
Co-Trimoxazol	28	35	4.1

* WHO Expert Committee on Biological Standardization. 28th Report. Tech. Rep. Series 610, 1977
Inhibitory zones of the test strain harbouring the generation time prolonging plasmid F'*lac* (G.T. =
= 40 min) are distinctly enlarged with all tested antibiotics.

method is very precise; multiple measurings of the same samples gave differences of only ± 30 sec. With this method we obtained the results of Table I.

Our host strain for all the tests was *E. coli* K12 921 *lac⁻*, *leu⁻*, *threo⁻*, *met⁻*, S^{r-s}. Its G.T. without plasmid is 30 min.

From our 101 tested R-factors about one forth makes the G.T. go up to over 35 min and approximately half of them have no distinct effect on the G.T. This increasing of the generation time can be drastically long, pGL 207 for instance, makes its host cell grow only half as fast (G.T. 58 min) as when it is not present.

When checking if those R-factors which strongly prolong, possess many resistance determinants, we saw that this is not the case (see also Table III). Furthermore we could not discover any accumulation of the strongly prolonging plasmids (over 35 min) within certain incompatibility groups. Those which could be classified were distributed among the same groups that we usually find in our region (Inc FI and Inc FII). (5)

Subsequently the effect of prolonged G.T. on resistance testing was examined. We analysed our host strains, bearing different plasmids, with the disc-susceptibility test according to the WHO recommendations. As with this test the antibiotics diffuse out of the discs into the medium, one would expect that with slower growing bacteria the inhibitory zones increase since the antibiotics have more time to diffuse. We found that this is obviously true.

The inhibitory zones of the plasmid-free *E. coli* were in most cases smaller than the ones of the same strain harbouring G.T.-prolonging plasmids.

The fact that plasmids with no prolonging functions do not in general affect the inhibitory zones allows us to conclude that the phenomenon is mainly due to the speed of growth and not to other qualities of the R-factors. In order to eliminate very small possible resistance that the R-factors could confer to their host cells, we did the same experiments with the F'*lac* plasmid that contains no resistance determinants and found a corresponding increase of the zone diameters. (Table II.). These findings mean, that a G.T.-prolonging R-factor can cause wrong results in resistance testings with susceptibility-discs because the inhibitory zones of a resistant strain might become so large that it is wrongly classified as non-resistant.

After that, minimal inhibitory concentrations for R-infected cells were measured

TABLE III.

Generation time — molecular size — resistance determinants

Plasmid	Resistance	G.T. (min)	Size (Kb)	Copies
pGL 207	Tc, Sm, Su	58	115	+
pGL 1468	Tc, Cm, Sm, Km, Su	45	>150	+
pGL 113	Tc, Ap, Su, Tm	44	43	+++
pGL 229	Cm, Sm, Km, Ap, Su, Tm	42	85	+
F'lac		40	150	+
pGL 250	Tc, Cm, Sm, Ap, Su	40	110	+
pGL 104	Tc, Sm, Ap, Su	40	50; 3, 1	+
pGL 231	Tc, Sm, Cm, Km, Su, Tm	39	100	+
pGL 267	Tc	39	85	+
pGL 242	Sm, Su, Tm	38	>150	+
pGL 220	Cm, Sm, Su	37	120	×
pGL 611	Tc, Cm, Km, Ap	37	110	+
Al	Tc, Su	36	136	+
pGL 247	Tc, Cm, Sm, Su	36	115; 90; 4,2; 4,0; 0,74	+
pGL 118	Tc, Ap, Su, Tm	36	60	+
RP4	Tc, Cm, Km, Nm	33	57	+
pGL 228	Tc, Sm, Su	32	65	+
pGL 369	Tc, Sm, Ap, Su, Tm	31	110	+
pGL 251	Tc, Cm, Km, Su, Tm	31	100	+
pGL 204	Tc, Sm, Su	31	75	+
pGL 372	Tc, Cm, Sm, Ap, Su, Tm	30	120	+
pGL 217	Sm, Cm, Su	30	100	+
pGL 108	Tc, Sm, Ap, Su, Tm	30	50	++
Sa	Sm, Cm, Km, Gm, Su, Tm	30	40	++
pGL 126	Tc, Sm, Ap, Su	29	110	+
pGL 127	Cm, Km, Ap	29	65	+
pGL 111	Tc, Sm, Km, Ap, Su, Tm	29	40	++

E. coli K12 921 bearing one plasmid larger than 80 Kb, one multicopy-plasmid or several different plasmids have in most cases a generation time over 35 min. The number of resistance determinants has no influence on generation time.

quantitatively. To our surprise, the sensitivity that we found to be increased with the disc-test *in general* was only increased with *certain* antibiotics in the M.I.C.-test. We found that all the R-factors which prolong G.T. up to over 35 min render their host cells two to four times more sensitive to Aminoglycosid antibitotics, but only to them. No such effect could be noticed with Ampicillin and Chloramphenicol. In these cases the plasmid harbouring strains were even a bit less sensitive, which we believe is the result of a very small possible resistance against these antibiotics. Since the therapeutical range of the Aminoglycosid antibiotics is very small, we find these results rather meaningful.

To find a possible reason for R-factors to increase G.T., we discussed the possibility that those which prolong it might have more genes than those which do not.

Therefore we determined the size of a selection of prolonging and not prolonging plasmids by agarose gel electrophoresis (2, 3). To estimate the size of our R-factors we used a number of plasmids whose size has been measured by electron microscopy as F'lac (150 Kb), R1 (98 Kb), RP4 (57 Kb), Sa (40 Kb), M47 (33 Kb, 14,7 Kb), 4,2 Kb) and *Hind* II digested Lambda phage.

It is apparent that our suggestion has a certain validity. Among the plasmids that increase G.T. for more than 5 min one can mainly find the larger ones (±80 Kb) except

139

in cases where more than one plasmid are in a strain or where one plasmid exists in multi-copies.

The number of copies/cell was estimated roughly by diluting the amount of plasmid DNA obtained from 1 ml bacterial suspension and by comparing its intensity on agarose gel with the intensity of a plasmid whose copy number is known (plasmid R6, two copies/cell) (6, 7).

TABLE IV.

Generation time versus molecular size

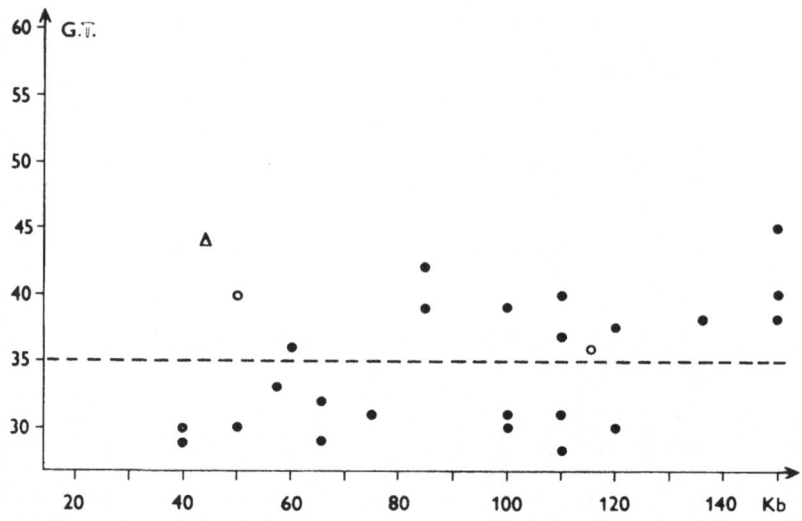

Among 17 R-factors larger than 80 Kb 12 increase generation time up to over 35 min. Among 8 plasmids smaller than 80 Kb only one does (multicopy- and multiplasmid-strains excepted). (●) hoststrain harbouring one plasmid, (○) hoststrain with several plasmids, (△) hoststrain with more than 20 copies/cell.

A possible proportionality between molecular size and generation time that R-factors confer to their host cells would become visible when generation time is plotted versus molecular size (Table IV.). This is obviously not the case. However, it becomes apparent that the majority of R-factors with a size over 80 Kb increases its host strain's generation time up to over 35 min and the majority of R-plasmids smaller than 80 Kb do not.

Investigations to understand these phenomena are at work.

ACKNOWLEDGEMENTS

We thank R. Binggeli and D. Moser for technical assistance.

Plasmids A1, R1, RP4, Sa were generously provided by N. Datta, F'*lac* by J. Miller and Strain M 46 by B. Wiedemann.

This work was in part supported by the Swiss National Foundation.

REFERENCES

1. LEBEK, G. (1969): Die infektiöse (plasmidische) bakterielle Antibiotikaresistenz. Verlag Hans Huber, Bern.
2. MEYERS, J., D. SANCHEZ, L. ELWELL and S. FALKOW (1976): Simple Agarose Gel Electrophoretic Method for the Identification and Characterisation of Plasmid DNA. J. Bacteriol. 127.
3. SAEDLER, M.: unpublished data.
4. TONOMURA, M. and S. YOSHIKAWA (1975): Host Chromosome Replication Control by Plasmid. Part 2. JPN. J. Bacteriol. 30.
5. ZÜND, P. and G. LEBEK (1978): Classification of fi$^+$ R-factors from Enterobacteriaceae Isolated from Medical Material. Int. Soc. Chemoth. (ed.). Current Chemotherapy 1, 451–453.
6. LEBEK, G. (1963): Ueber die Enstehung mehrfachresistenter Salmonellen. Ein experimenteller Beitrag. Zbt. f. Bakt. *176*.
7. DAVIES, J. E. and R. ROWND (1972): Transmissible drug resistance in Enterobacteriaceae. Science *176*.

G. L., Hygiene-Inst.
Friedbrühlstrasse 51,
CH-3008 Bern, Schweiz

LOSS OF THE MULTI-COPY RESISTANCE PLASMID pBR 325 FROM *ESCHERICHIA COLI* GY 2354 pBR 325 DURING CONTINUOUS CULTIVATION

M. ROTH, G. MÜLLER and D. NOACK

Central Institute of Microbiology and Experimental Therapy, Academy of Sciences of GDR, Jena G.D.R.

INTRODUCTION

In a recent paper we reported the results of investigations on the stability of the antibiotic forming capacity of the turimycin producer Streptomyces hygroscopicus during continuous cultivation (Roth and Noack, 1978). During cultivation of this strain at glucose limitation and low dilution rates in a chemostat an increasing number of clones which did not produce the antibiotic appeared. In addition, these clones did not form aerial mycelium and were sensitive to turimycin. At high growth rates, segregation of the strain into two types of clones was not observed. Because of the stability of the turimycin non-producing clones which appear with high frequency during continuous cultivation at low dilution rates and glucose limitation it is assumed that this segregation is caused by the loss of extrachromosomal genetic information. In both the original producing strain and the isolated non-producing clones a fraction of extrachromosomal DNA was detected (Zippel and Noack, 1978). Characterization of these DNA's has not yet been possible. There are difficulties in obtaining sufficient amounts of plasmid DNA from this Streptomyces strain.

In order to investigate the kinetics of the loss of extrachromosomal DNA from bacterial cells during continuous cultivation under the conditions mentioned above, a well-characterized model system was used. The results expected from a favourable model system can be useful for the further interpretation and evaluation of the experimental results obtained on *Streptomyces hygroscopicus*. A suitable plasmid for this purpose should not be self-transferable, to rule out the reinfection of cured cells. Furthermore, it should be a multi-copy plasmid to simulate the corresponding situation in streptomycetes. These bacteria growing in the chemostat form small mycelium pieces which contain about 20 to 100 chromosomes and eventually a comparable number of plasmid copies.

Therefore we chose the plasmid pBR325 which was constructed by F. Bolivar (1978) as a cloning vehicle. This plasmid codes for the resistance to the antibiotics ampicillin (Ap^r), tetracycline (Tc^r) and chloramphenicol (Cm^r). It is a relaxed replicating element and carries the replication origin of the ColE1 plasmid. As host we used *E. coli* GY2354 (arg his thi mal lac spc-339).

In this paper we describe first results concerning the stability of the pBR325 plasmid in *E. coli* GY2354 during continuous cultivation in a chemostat at different growth rates and limiting substrates.

MATERIAL AND METHODS

Strain. The *E. coli* K12 derivative GY2354 F'*lac* (arg his thi mal lac spc-339) was kindly supplied by R. Devoret. The F'lac plasmid was eliminated by acridin orange curing. GY2354 pBR325 used in the experiments was constructed by transformation of GY2354 with pBR325 DNA. The pBR325 (Ap^r Tc^r Cm^r) plasmid in *E. coli* GM31 (his thr leu thi dcm) was generously provided by F. Bolivar.

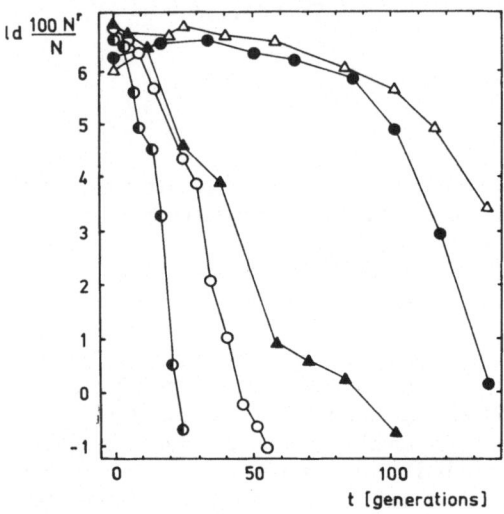

Fig. 1. Kinetics of loss of the pBR325 plasmid from *E. coli* GY2354 during continuous cultivation at different dilution rates (D). The limiting substrate of the culture medium was glucose. The percentage of antibiotic resistant cells in the continuous culture expressed as ld 100N^r/N (N^r, number of cells resistant to ampicillin, tetracycline and chloramphenicol per ml; N, total number of cells per ml) is plotted as function of the cultivation time (t).

Symbols: ●, D = 0.11 h^{-1}, τ = 6.3 h; ○, D = 0.15 h^{-1}, τ = 4.6 h; ▲, D = 0.41 h^{-1}, τ = 1.7 h; ●, D = 0.49 h^{-1}, τ = 1.4 h; △, D = 0.55 h^{-1}, τ = 1.25 h.

Media. The continuous cultivation was carried out in glucose mineral salt minimal medium containing 2.72 g KH$_2$PO$_4$, 7.16 g Na$_2$HPO$_4$. 12 H$_2$O, 5 g NaCl, 1 g Na$_2$SO$_4$, 41 mg MgCl$_2$. 6 H$_2$O, 5.4 mg FeCl$_3$, 3.9 mg MnCl$_2$, 50 mg of histidine, arginine and vitamin B$_1$ per 1. For cultivation at glucose limitation 0.5 g glucose and 0.5 g NH$_4$Cl and at NH$_4$Cl limitation 2 g glucose and 0.08 g NH$_4$Cl per 1 were added. Plating was done on NB agar (0.8% Bacto nutrient broth, 0.4% NaCl, 1.5% Bacto agar) and on NB agar supplemented with oxytetracycline, chloramphenicol (20 µg/ml) and ampicillin (70 µg/ml).

Cultivation conditions. *E. coli* GY2354 pBR325 was cultivated under aeration in a chemostat. The volume v of the culture vessel was 80 ml. The relation between the influx rate F of fresh medium and the dilution rate D is given by D = F/v. The specific growth rate is equal to D (Herbert et al., 1956) The generation times is τ = 1n 2/D. The limiting substrate of the medium was glucose or NH$_4$Cl. Samples withdrawn from the chemostat were diluted and plated on plasmid-selective and non-selective media. From the numbers of colonies grown on the respective media the percentage of antibiotic resistant cells carrying the pBR325 plasmid was calculated.

All experiments were performed at 37°C.

RESULTS

The stability of the plasmid pBR325 in *E. coli* GY2354 was examined during continuous cultivation at different dilution rates and limiting substrates in the culture medium. The chemostat cultivation was started in all experiments with a population containing 100% of antibiotic resistant cells carrying the plasmid.

Growth of the strain in medium with glucose limitation resulted in the loss of the plasmid at all dilution rates tested (Fig. 1). The kinetics of the plasmid loss shows a dependence on the growth rate. The lower the growth rate the faster the plasmid is

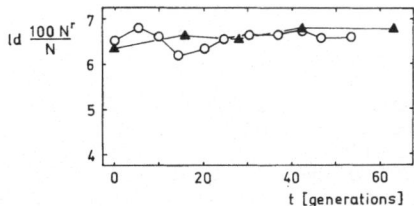

Fig. 2. Kinetics of stability of the plasmid pBR325 in *E. coli* GY2354 during continuous cultivation in medium with NH_4Cl as limiting substrate. The percentage of antibiotic resistant cells in the continuous culture as ld $100N^r/N$ is plotted as function of the cultivation time (t).

Symbols: ○, $D = 0.15 \, h^{-1}$, $\tau = 4.6 \, h$; ▲, $D = 0.41 \, h^{-1}$, $\tau = 1.7 \, h$.

diluted out from the population. At $D = 0.55 \, h^{-1}$, which corresponds to the maximal possible dilution rate, and $D = 0.49 \, h^{-1}$ only a very slow decrease of the percentage of plasmid-containing cells was observed. However within the range of dilution rates between $0.1 \, h^{-1}$ and $0.4 \, h^{-1}$ the portion of antibiotic resistant cells in the culture decreases rapidly.

In contrast to the findings at glucose limitation during continuous cultivation in medium with NH_4Cl as limiting substrate the plasmid was not diluted out from the culture at all dilution rates (Fig. 2).

DISCUSSION

The experimental results suggest that the replication frequency of the pBR325 plasmid in relation to the chromosomal replication resulting in the loss or stable inheritance of this replicon depends on both the growth rate and the limiting substrate of the medium.

For interpretation of these observations it must be pointed out that the replication of the ColE1 plasmid DNA from which the replication origin of pBR325 was derived and the chromosomal DNA is initiated by a common initiator (Goebel, 1970). In contrast to this the DNA polymerase III is not used for the ColE1 replication (Goebel, 1972), whereas the large plasmids (F, R1, etc.) and the chromosome are replicated by this enzyme (Nordström et al., 1974, Goebel, 1972).

Therefore only in the process of replication initiation is a competition between replication of chromosomal and pBR325 DNA to be expected. Noack and Klaus (1973) supposed that the replication of both the chromosomal and episomal DNA is initiated when the concentration of one or several precursors of DNA synthesis in the cell reaches a critical value which may be different for the chromosome and an extrachromosomal replicon. On the basis of this model, it can be supposed that, under energy starvation due to glucose limitation, the concentration of the precursors needed for the initiation of plasmid DNA replication is reached only with a low probability insufficient for the maintenance of a constant plasmid copy number per chromosome. This interpretation is able to explain both the stable inheritance of the plasmid at NH_4-Cl limitation providing a glucose excess and the delayed loss of the plasmid at glucose limitation at high dilution rates because the glucose concentration in the culture medium is then increased significantly compared with low dilution rates.

REFERENCES

Bolivar, F. (1978): Construction and characterization of new cloning vehicles. III. Derivatives of plasmid pBR322 carrying unique *Eco*RI sites for selection of *Eco*RI generated recombinant DNA molecules. Gene 4, 121–136.

Goebel, W. (1970): Replication of the colicinogenic factor ColE1 in two temperature sensitive mutants of Escherichia coli defective in DNA replication. Eur. J. Biochem. 15, 311–320.

Goebel, W. (1972): Replication of the colicinogenic factor E1 (ColE1) at the restrictive temperature in a DNA replication mutant thermosensitive for DNA polymerase III. Nature New Biol. 237, 67–70.

Herbert, D., R. Elsworth and R. C. Telling (1956): The continuous culture of bacteria: a theoretical and experimental study. J. Gen. Microbiol. 14, 601–622.

Noack, D. and S. Klaus (1973): A model for the regulation of replication of bacterial and episomal DNA. Acta Phys. Acad. Sci. Hung. 33, 369–376.

Nordström, U. M., B. Engberg and K. Nordström (1974): Competition for DNA polymerase III between the chromosome and the R-factor R1. Mol. Gen. Genet. 135, 185–190.

Roth, M. and D. Noack (1978): Stability of Streptomyces hygroscopicus concerning the antibiotic forming capacity during continuous cultivation. Proceedings of the 7th Int. Symp. on Continuous Cultivation (Prague, 1978). In press.

Zippel, M. and D. Noack (1978): Personal communication.

M. R., Central Institute of
Microbiology and Experimental Therapy,
Beuthenbergstr. 11
69 Jena, G.D.R.

ISOELECTRIC FOCUSING ANALYSIS OF MONOPLASMIDS CARRIED BY ESCHERICHIA COLI STRAINS

O. SOVA, A. SOKOL, V. KMEŤ

Department of Microbiology and Zoohygiene, University of Veterinary Medicine; Laboratory of Molecular Genetics of Institute of Animal Physiology of Slovak Academy of Sciences; Laboratory of Genetics of Microorganisms, Institute of Experimental Veterinary Medicine, Košice, Czechoslovakia

INTRODUCTION

When R plasmids were discovered in Japan in 1959, the sex factor F of *Escherichia coli* strain was already known. Watanabe's (1964) idea suggesting the extrachromosomal localisation of genetic information of resistance on antibiotics represented by the R plasmids in the F factor aroused hitherto unseen interest in plasmids as such and gave an impulse to the birth of "plasmidology".

1. METHODS OF STUDYING PLASMIDS

The study of plasmids has always been conditioned by the existence of suitable methods used in microbiology and in related fields, such as biochemistry, molecular biology, biophysics, etc., First plasmids were studied by investigating their phenotypical properties. The clasical study of nucleic acids of the plasmids by ultracentrifugation (Freifelder, 1970) was added to these methods. A further step in this research was the introduction of analytical and preparative electrophoresis in agarose gel (Aaij and Borst, 1972).

In spite of these methical improvements that, together with electronmicroscopy (Falkow and al., 1975), provide a fairly clear picture of the plasmid molecule, so that today we have several models of their internal structure and the dynamics of their function and metabolism are known, there are still several facts that require explanation, one of the most important is the discrepancy between the amount of visualised plasmids and their phenotypical phenomena.

The above problems lead our research team to consider the *limiting factors* of the study of plasmids in more detail. Our attention was directed to a modern method often used in biochemistry and molecular biology, to the method of *isoelectric focusing*. This method has been used relatively recently in microbiology, where it was initially used mainly for immunoglobulin subfractionation (Bordenave and al., 1974) and later for bacterial enterotoxin analysis (Yotis et al., 1974). Mouillot and Netter (1977) used it for succesful differentiation of Orthopox type viruses. This indicated that this method might be used, not only for proteins, but also for complexes of proteins and nucleic acids, such as viruses and perhaps directly for nucleic acids. Brive et al. (1977) found that beta-lactamase activity determined by electrophoresis in acrylamide gel showed only one homogeneous fraction, but in isoelectric focusing it appears to be markedly heterogeneous. It should be mentioned that beta-lactamase synthesis is coded by a single plasmid, which has a different molecular weight for the different strains, but it can always be isolated and identified by gel electrophoresis as a homogeneous monoplasmid with exactly definable molecular weight. If we try to explain the heterogeneity of this enzyme by the coding of several *structural genes*, it can be assumed to be due to the simultaneous presence of different DNA segments in the same genome, phenotypically manifested differently, or it may be considered as conformative heterogeneity of the genetic information as a whole. However, these considerations led us to investigate

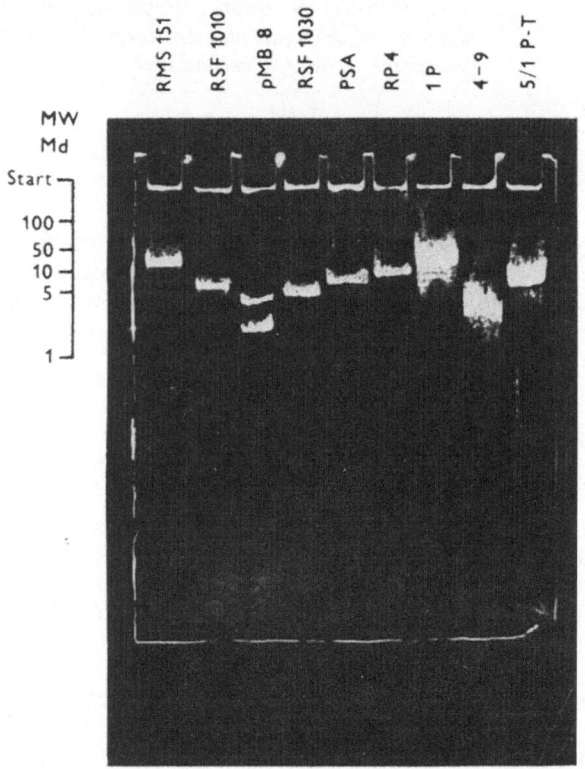

Fig. 1. Agarose-gel electrophoresis of plasmid DNA.

the possible heterogeneity of extrachromosomal genetic elements in plasmids – carried by reference and wild strains of *E. coli* which appeared by the current methods as unambiguously homogeneous (and well characterized) or heterogeneous.

2. THE USE OF ISOELECTRIC FOCUSING TO STUDY THE HETEROGENEITY OF PLASMID DNA

In the experiment we used *exactly defined reference R monoplasmid* carrying *E.coli* strains (C 600 with plasmid RSF 1010 and pMB 8, strain J 5 with plasmid RP 4, strain 1485–1 containing plasmid RSF 1030, strain J 53 with plasmid pSa and strain F 027 with plasmid RMS 151). In addition to these exactly defined reference monoplasmids we also used *undefined multiplasmids* carried by our wild *E. coli* strains 4–9, 1P and 5/1

148

isolated from calves and piglets on breeding farms when a lively circulation of several plasmids was observed.

The plasmids from the reference and wild *E. oli* strains were isolated by classical enzymatic lysis combined with phenol-alcohol extraction according to Mayers et al. (1976). After isolation of every reference plasmid and also the plasmids from wild strains, all of them were subjected to comparative visualization using agarose gel electro-

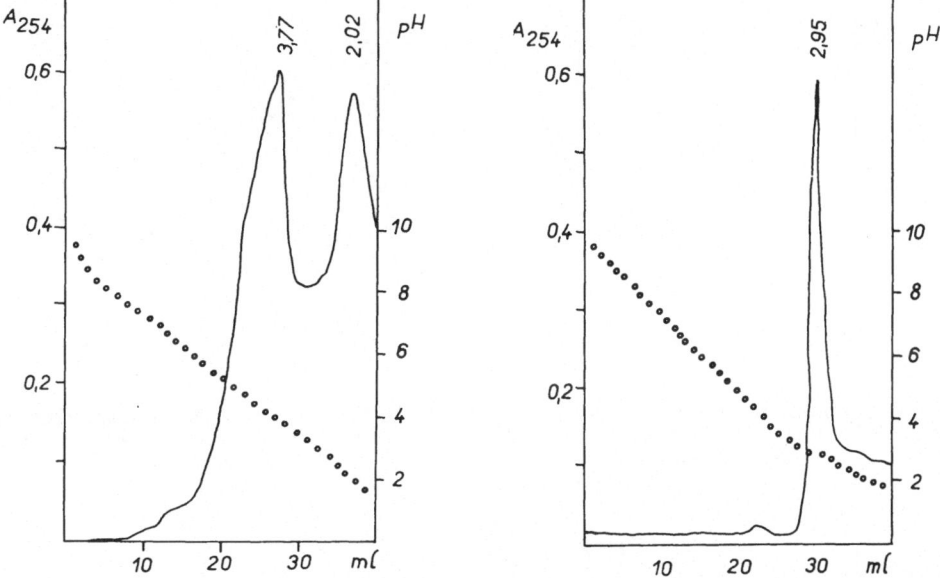

Fig. 2. Record of continual-flow measuring of plasmid DNA Rms 151 at 254 nm after isoelectric focusing.

Fig. 3. Record of continual-flow measuring of the absorbance of the RSF 1010 plasmid DNA at 254 nm after isoelectric focusing.

phoresis. On this way we checked the homogeneity of the reference plasmids and the heterogeneity of the plasmids from the wild strains and determined their molecular weights.

Electrophoresis was performed by the modified method according to Mayers et al. (1976). After the electrophoresis it was found that all the monoplasmids of the reference strains, except plasmid pMB 8, were perfectly homogeneous, while the plasmids from the wild strains were slightly heterogeneous.

Then all the plasmids, except plasmid pMB 8, were subjected to analysis by isoelectric focusing carried out in an apparatus described by Osterman (1970), with a volume of 38 ml, where the content was stabilized by the classical method using 5—45% gradient saccharose. To the basal solution was added 1.6 ml of ampholytic buffer with a pH of 2-20 and 100 μl sample of the plasmidic DNA, i.e. approximately 2-8 μg of pure DNA. The isoelectric focusing lasted for 24 hours at 100 V and an initial value at 0,6 mA.

After this time, the current was stopped and the column was evacuated at a flow rate of 1 ml/min. Simultaneous registration of the absorbance in UV light at 254 um on flow spectrophotometer Uvicord (LKB) was realized and single fractions were united after 1 ml by an automatic fraction collector. In every fraction the pH was precisely maesured by a pH meter (Radelkis) with a combined glass electrode.

149

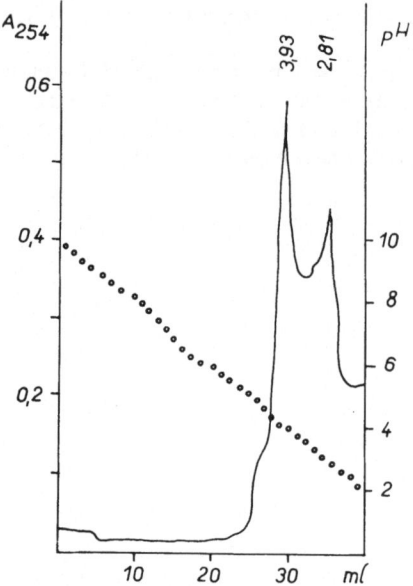

Fig. 4. Continual-flow record of the absorbance of the RSF 1030 plasmid DNA at 254 nm after isoelectrofocusing.

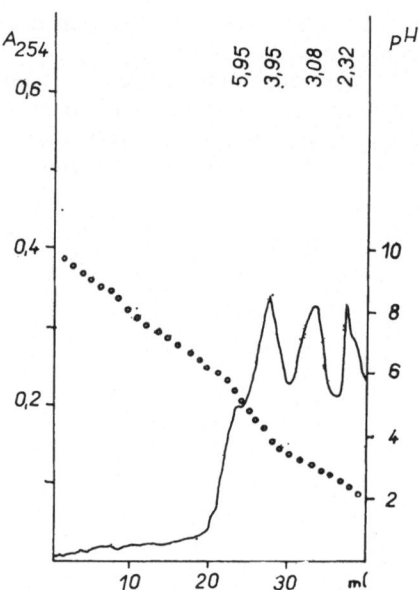

Fig. 5. Continual-flow record of absorbance of pSa plasmid DNA at 254 nm after isoelectric focusing.

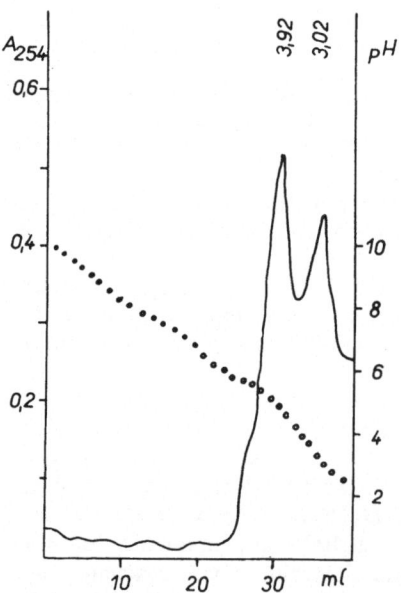

Fig. 6. Continual-flow record of absorbence of RP 4 plasmid DNA at 254 nm after isoelectric focusing.

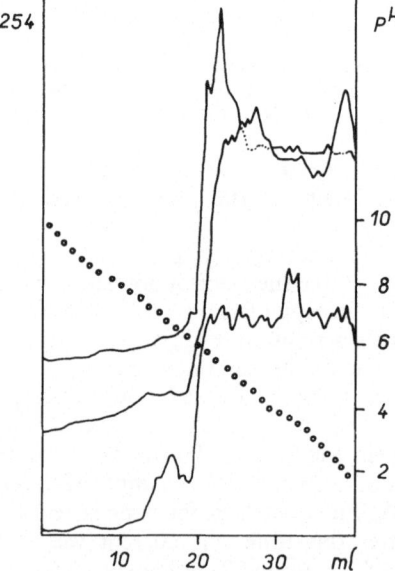

Fig. 7. Continual-flow record of the absorbance of the plasmids DNAs of wild strains 4–9, 1P and 5/1 separated by isoelectric focusing.

Figure 1 presents a photograph of an agarose-gel electrophoretic visualization of 6 reference and 3 wild strains of *E. coli*. It can be seen in the figure that, except plasmid pMB 8 and the plasmids of the wild strains that displayed heterogeneity in the electrophoresis on agarose gel, the plasmids are homogeneous. It should be noted that these plasmids were perfectly characterized already by their original discoverer (Mayers et al., 1976). They determined their molecular weight, which was confirmed in our experiments.

Figure 2 shows the registration of the flow measurement of the absorbance in UV light at 254 nm and the course of the pH gradient after isoelectric focusing of plasmid Rms 151. This plasmid, that was characterized as completely homogeneous by the electrophoresis and electronmicroscopy methods, showed evident heterogeneity in isoelectric focusing as it was divided into two well-defined peaks with different isoelectric points.

Figures 3 to 6 show the characteristics of further reference plasmids. It is interesting that, except plasmid RSF 1010, two to four peaks of evidently heterogeneous fractions were observed.

Figure 7 gives a survey of the characteristics of the plasmids in three wild strains of *E. coli* circulating on breeding farms isolated here. These plasmids, as was seen to the electrophoresis separation, did not appear to be completely homogeneous. The visibly blurred fractions did indicate a tendency to heterogeneity. After isoelectric focusing the plasmids isolated from the wild strains yielded a distinctly wide spectrum of peaks for the individual fractions with different isoelectric points.

All the peaks of isolated plasmid DNA are represented by isoelectric points in the pH range from 2 to 6. None of them has a higher isoelectric point: on the contrary, some – e.g. plasmid pSa, show a tendency to further separation at a lower pH value. It should be mentioned that all the plasmids divided by isoelectric focusing were repeatedly subjected to electrophoresis on agarose-gel, in which they again displayed the same mobility as before focussing. The combined peaks of the referent plasmids were again homogeneous – monoplasmidic under reelectrophoresis.

3. CONCLUSIONS

The results obtained from our experiments can be discussed from the viewpoint of using different separation techniques and also from that of the detectable microheterogeneity of the reference plasmids, that were characterized by classical methods as monoplasmids. This is indicated by the fact that, after repeatedly mixing microheterogeneous focusing zones, these exhibit the same mobility of the monoplasmids as before the focusing.

It unambiguously follows that isoelectric focusing as an individual analytical method may be used for finer characterization of plasmid DNA and provides a further criterion for their differentiation – according to their isoelectric point.

REFERENCES

AAIJ, C. and P. BORST (1972): The gel eletrophoresis of DNA. Biochim. Biophys. Acta 269, 197–200.
BORDENAVE, G., B. A. ASKONAS (1974): Isoelectric Focusing Spectra of Rabbit Antibodies to *Salmonella abortus-equi* Detected by Anti-idiotypic Sera. Immunology 27, 1–9.

Brive, C., M. Barthelemy, D. H. Bouanchaud, R. Rabia (1977): Microheterogenite en electrofocalisation analytique de Beta-lactamases d'origine plasmidique. Ann. Microbiol. 128, 309–317.

Falkow, S., P. Guerry, R. W. Hedges and N. Datta (1975): Polynucleotide sequence relationships among plasmids of the I compatibility complex. J. Gen. Microbiol. 85, 65–76.

Freifelder, D. (1970): Isolation of extrachromosomal DNA from bacteria, p. 153–162. In L. Grossman and K Moldave (ed.), Methods in enzymology vol. 21. Nucleic acids. Academic Press Inc., New York.

Mayers, J. A., D. Sanchez, L. E. Elwell, S. Falkow (1976): Simple Agarose Gel Electrophoretic Method for the Identification and Characterization of Plasmid Deoxyribonucleic Acid. J. Bacteriol. 127, 1529–1537.

Mouillot, L., R. Netter (1977): Identification of Orthopox Virus by Isoelectrofocusing in a Granulated Gel. Ann. Microbiol. 128, 417–419.

Osterman, L. (1970): A Column for Microanalytical Electrofocusing. Science Tools 17, 31–33.

Watanabe, T., H. Nishida, C. Ogata, T. Arai, S. Sato (1964): Episome mediated transfer of drug resistance in *Enterobacteriaceae* VII. Two types of naturally occurring R factors. J. Bacteriol. 88, 716–726.

Yotis, W. W., N. Catsimopolas, M. S. Bergoldi, J. Edward (1974): Scanning Density gradient Isoelectric Focusing of *Staphylococcus aureus* Enterotoxins B and C_1. Inf. a. Immun. 9, 974–976.

O. S. Institute of Animal Physiology
SAV, Palackého 36, Košice,
Czechoslovakia

ENHANCED CHLORAMPHENICOL RESISTANCE OF THE LYSOGENIC STRAIN *ESCHERICHIA COLI* GY 2354 P1CM

CH. HOFFMEIER, D. NOACK, R. GEUTHER and H. BRANÁ

Institute of Microbiology and Experimental Therapy, Jena, GDR and Institute of Microbiology, Czechoslovak Academy of Sciences, Prague, Czechoslovakia.

The genome of the temperate phage P1CM exists as an extrachromosomal replicon within the cell (Ikeda and Tomizawa 1968). It bears the chloramphenicol-resistance gene coding for the synthesis of the chloramphenicol-acetyltransferase. The experiments described here were carried out to obtain strains with enhanced chloramphenicol (CM) resistance and to shed some light on the mechanism responsible for this enhanced resistance level.

In our work *Escherichia coli* K12 GY 2354 strain infected with the P1CM phage from Dr. J. L. Rosner was used. In cultures of this strain with a basal resistance of 200 μg/ml CM a few colonies spontaneously resistant to higher CM concentration (500 μg/ml) could be found. The minimal inhibition concentration (MIC) was measured

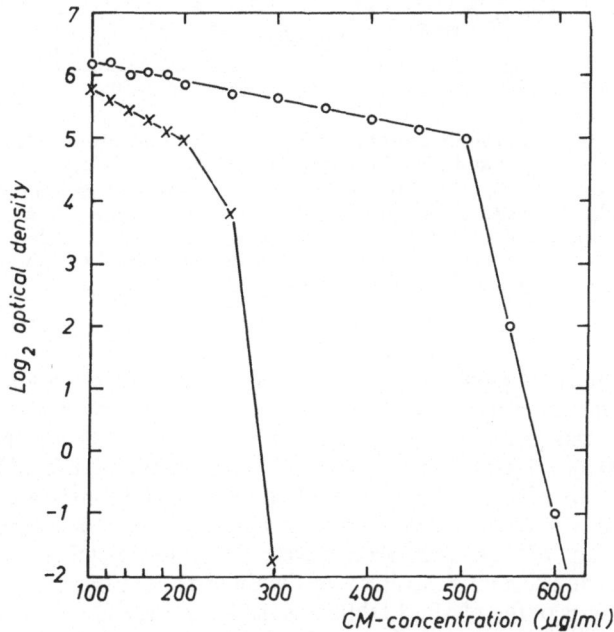

Fig. 1. MIC test of the *Escherichia coli* Gy 2354 P1CM wild type (WT) and the chloramphenicol higher resistant mutant R 56 strains. The MIC was determined in liquid media with different CM concentration, the optical density was read in cultures grown for 20 h at 470 nm on a Pulfrich photometer.

×——×——× = wild type strain
○- —○——○ = mutant strain R 56

153

in these strains and it was approximately 3 times higher compared with the parental strain. The activity of the enzyme chloramphenicol-acetyletransferase was estimated to be $1.05 \times 10^{-1} \mu g$ DTNB (Dithio-bis-nitrobenzoid acid). min^{-1}/mg protein in the parental and $1.23 \mu g$ DTNB . min^{-1}/mg protein in the higher resistant strains. The enzyme assay was performed by the method of Shaw (Shaw 1967) in cell-free extracts prepared as described previously (Braná and Hubáček 1974). Protein was determined according to Lowry (Lowry et al. 1951).

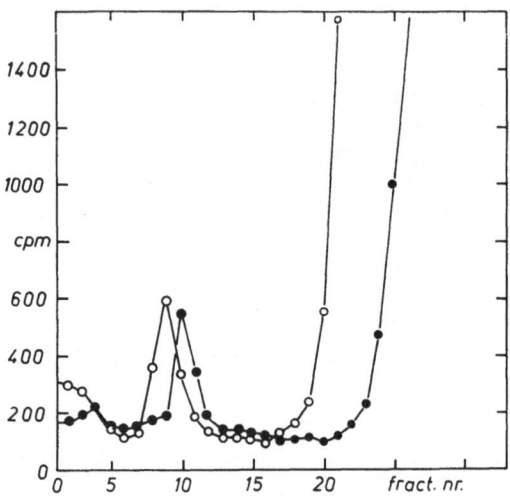

Fig. 2. Alkaline sucrose sedimentation profile of the DNA of *Escherichia coli* GY 2354 P1CM wild type and the CM higher resistant mutant R 56.

The cultures were shaken at 37 °C in M9 medium supplemented with casamino acids (100 μg/ml), thiamine (20 μg/ml), deoxyadenosine (250 μg/ml) and ³H-thymidine 5,5 M Bq for 3 h. Then cleared lysates were prepared after deproteinization with 1M sodium perchlorate and chloroform--isoamylalcohol, the gradient analysis was carried out by the method of Lowett (Lowett 1973).

 = wild type
= R 56

Because the ability of growth at the higher antibiotic concentration was lost during further subcultivations without CM, the mutagenization with N-methy-l-N-nitro-N-nitrosoguanidine was performed (200 μg/ml for 45 min at 37°C) and mutants with resistance to 500 μg/ml CM were selected. Typical result of the MIC test of these mutants is shown in Fig. 1. The activity of the CM-acetyltransferase was determined in 5 mutant strains and the enzyme activity was on average 4–5 times higher than in the wild type strain. All 5 mutant strains were found to be stable with respect to the CM resistance during several subcultivations without the antibiotic.

A gradient analysis of the plasmid DNA was made by centrifugation in alkaline sucrose gradient to estimate the amount of plasmid copies. As shown in Fig. 2 mutant strain R56 contains the same amount of plasmid DNA as compared with the wild type strain. The same picture was obtained with all 5 higher resistant mutants tested.

The increased enzyme activity, which was proved in the mutant strains, was in agreement with the MIC values in these strains. This effect can be explained either by an increase in the enzyme activity or by an enhancement of the enzyme amount within the cell. However so far a CM – acetyltranferase with a higher enzymatic activity could not be found. A higher amount of enzyme within the cell could by produced either due to a higher copy number of the gene for CM-resistance or as a result of changed control of transcription of this gene. The results obtained by alkaline sucrose gradient centrifugation point rather to altered transcription control of the CM gene than to a mutation resulting in an increased copy number of the plasmid.

In seperate experiments it could be shown that the mutated strains are not able to release free P1 phages into the culture medium either spontaneously or after UV irradiation. This observation indicates that, within the P1 genome, a mutation must have occurred affecting the vegetative development of phage growth. Similar results have recently been described by Austin (Austin et al. 1978). These authors investigated mutants of P1CM concerning incompatibility and vegetative phage multiplication. All the mutations could be mapped near the CM gene.

From our experimental results and with regard to related data from other authors it can be concluded that the enhanced CM resistance in our mutated P1CM bearing strains of *E. coli* GY 2354 is the result of an altered transcription control of the CM gene resulting in a higher amount of CM-acetyltransferase within each cell.

REFERENCES

AUSTIN, S., STERNBERG, N. and YARMOLINSKY, Y., (1978): Miniplasmids of bacteriophage P1: I. Stringent plasmid replication does not require elements that regulate the lytic cycle. J. mol. Biol., *120*, 297–309.
BRANÁ, H. and HUBÁČEK, J. (1974): Regulation of synthesis of chloramphenicolacetyltransferase in *Escherichia coli* (Infected with R1 plasmid). Antibiotiki, *5*, 390–394.
IKEDA, H. and TOMIZAWA, J. (1968): Prophage P1, an extrachromosomal unit. Cold Spring Harb. Symp. Quant. Biol. *XXXIII*, 791.
LOWETT, P. S. (1973): Plasmid in Bacillus pumilus and the enhanced sporulation of plasmid negative variants. J. Bacteriol. *115*, 291.
LOWRY, H. O., ROSEBROUGH, N. J., FARR, A. L. and RANDALL, R. (1951): Protein measurements with the folin phenol reagent. J. Biol. Chem. *193*, 265.
SHAW, W. V. and BRODSKY, R. F. (1967): Chloramphenicol resistance by enzymatic acetylation: Comparative aspects. Antimicrobial Agents and Chemotherapy, 1967, 257–263.

*CH. H., Inst. of Microbiology
and Experimental Therapy,
Beuthenbergstrasse 11,
69-Jena, G.D.R.*

REASSORTMENT AND GENE AMPLIFICATION OF R-FACTOR R 1767 FROM *SALMONELLA TYPHIMIRIUM*

F. SCHMIDT, U. van TREECK and B. WIEDEMANN

Institute of Medical Microbiology and Immunology, University of Bonn, Bonn, F.R.G.

A number of composite structures of drug resistance plasmids have been shown by genetic and physical studies. But many R-plasmids occurring in clinical isolates seem to be even more complicated. The study of mechanisms involved in rearrangement of plasmid structures can help to understand the spread of drug resistance among clinically important bacteria. In this context R 1767, a plasmid derived from a strain of *Salmonnella typhimurium*, seems to be a good model. It carries the determinants Ap, Su, Sm, Tc, Cm, ColI and Tra and fragments during conjugal transfer considerably (Richmond and Wiedemann, 1974).

Conjugation and transduction experiments reveal a process of reversal dissociation of RTF, ColI and the different r-determinants.

In *E. coli* transconjugants and transformants so far more than 20 different molecules with molecular weights between 5.2 and 92 Mdal could be demonstrated by electron microscopy. Such a broad variety of different aggregation products complicates a correlation between structure and genotype of each replicon. Thus we isolated some of the smallest molecules and mapped their position in larger aggregates. In our communication we focus on a transconjugant plasmid of R 1767, designated rBP 11, that determines resistance to ampicillin and sulfonamides, the most frequent marker combination in transconjugant plasmids of R 1767, selected on ampicillin. We analysed the plasmid structure and genome organisation by using genetical studies, restriction mapping, gene cloning and homo- and hetero-duplex mapping, demonstrating that rBP 11 might serve as good example for reassortment of plasmid DNA.

PHYSICAL CHARACTERIZATION OF rBP 11

In contrast to the majority of R-factors (Bukhari et al., 1977), rBP 11 is expressed in the *E. coli* host in more than one size class: 78.15, 53.52, and 26.36 Kb. In order to exclude the possibility that all three replicons might be derivated from different ancestors, DNA was suffiently separated by agarose gel electrophoresis. The lowest band, representing the smallest replicon with the size of 26.36 Kb, was extracted and transformed. Recipient cells yield the same pattern of size classes, minimizing the possibility that one of the different plasmids represents a cryptic plasmid, as cotransformation appears to be a rare event.

For further explanation of this phenomenon hairpin-loop structures of rBP 11 were examined in the electron microscope. Fig. 1a shows micrographs of typical snap-back structures of rBP 11, containing both inverted repeat and unique sequences similar to the amphimer molecules described for 2 µg yeast DNA (Hollenberg et al., 1976) and for several types of dv mutants (Chow et al., 1974). All three types of replicons, different in size, show appearantly the same principle of genome organisation: They contain unique sequences of the same length (loop-regions), while the duplicated sequence of the larger entities seems to represent a multiple copy of the smallest unit. Based on this result it appears that formation of larger aggregates apparently arises from variation of the basic replicon without change of genetic information.

Estimation of the number of copies of plasmid DNA per genome equivalent corresponds to this finding. Lysates of *E. coli* cultures harbouring rBP 11 and labelled with ³H-thymidine were analysed by ethidium bromide – CsCl gradient. 17.7% of the amount of the chromosomal DNA was plasmid DNA, being further subjected to sucrose gradient

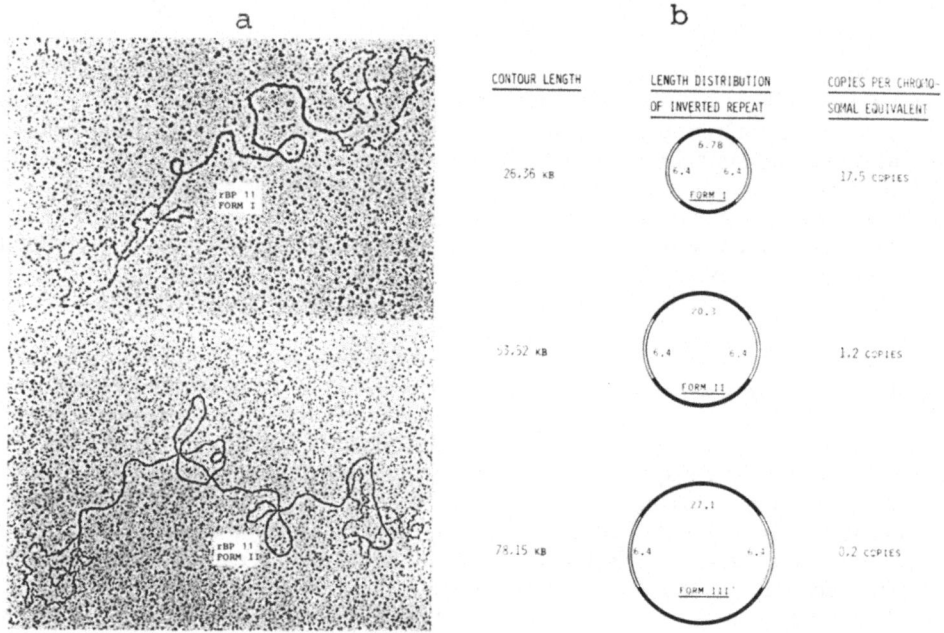

Fig. 1a – Snapbacks observed from rBP 11, form I and II. DNA was denatured and briefly rean-nealed, directly sprited and picked up on parlodion grids. Single and double-stranded forms of Ø X174-phage DNA served as internal standards. 1b – Distribution of inverted repeats, copy number and lenths of different rBP 11-molecules, named form I, II, III, as isolated from a single transformant. The lengths are expressed as units of kilobases (1.000 nucleotide(s)/pairs). ▬▬▬ inverted repeat sequence ═══ unique sequence

centrifugation. Under normal conditions only the smallest unit is under relaxed control with 17 copies per chromosomal equivalent while the larger plasmid structures are present with 1 copy for the 53.52 Kb and 0.2 copies for the 78.15 Kb replicon, thereby being under stringent control (Fig. 1b).

PHYSICAL MAPPING OF rBP 11 BY RESTRICTION ANALYSIS AND LOCALIZATION OF REGION RESPONSIBLE FOR EXPRESSION OF RESISTANCE

Plasmid rBP 11 isolated in the absence of drug pressure was subjected to single and double digestion by six different restriction endonucleases. The fragmentation pattern, as analysed by gel electrophoresis, led to the construction of the restriction map, presented as snap-back structure shown in Fig. 2. Molecules, linearized by digestion with one or the other restriction endonuclease, were denatured, reannealed and examined electron microscopically, ensuring some of the critical restriction data and making an exact localisation and differentiation of the loop structures possible. The small loop-and stem-structures with restriction sites for endonucleases SalI and Pst I were analysed analogously by Sl-hydrolysis of rBP 11 after formation of snap-back structures followed by homoduplex formation of the undigested repeated sequences and located upon evaluation of defective snap-backs (Fig. 2b).

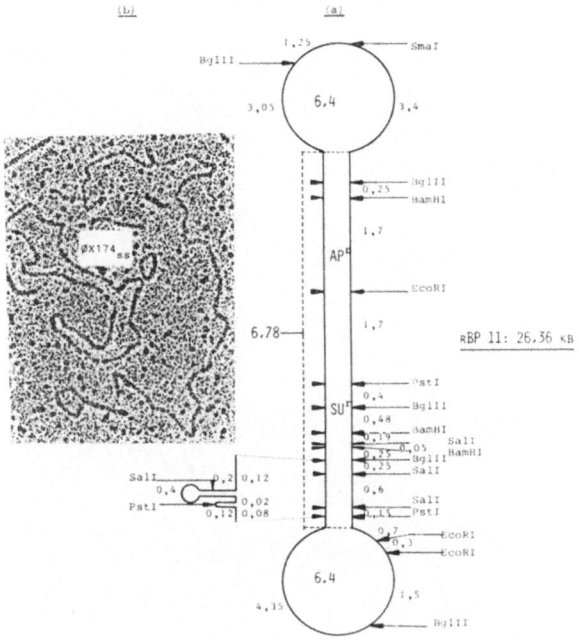

Fig. 2. Restriction map and structure of plasmid rBP 11, presented as snap-back structure, obtained by single and combined digestions with restriction endonucleases (a). Restriction coordinates are given as Kb (Kilobases) relative to each restriction cut. (b) shows defective homoduplex molecule, indicating position of small loop-and stem-structure, left to Su^r-gene.

All distances represent the average numbers of more than 10 measurements each. The left loop with the 2 *Eco*RI-restriction sites does not contain the sites for *Pst*I-fragmentation, since position of PstI-cuts have not yet been obtained.

The 4.2 Mdal plasmid ColE1 contains a single EcoRI site within the gene determining colicin activity. This activity is lost, if an *Eco*RI fragment is ligated into *Col*E1. EcoRI fragments of rBP 11 were used as substrates for ligation (Tanaka and Weisblum, 1975) and transformants were selected on agar plates containing ampicillin and sulfonamide and tested for colicin immunity. Several clones were isolated, further characterized and the orientation of integrated fragments was determined by restriction analysis and homoduplex presentation. Clones with Bam HI – restriction fragments of pSC 105 and rBP 11 were analysed in parallel. By these data position of Ap^r and Su^r genes could be located in the repeat sequence shown in Fig. 2.

159

RESISTANCE LEVELS IN RESPONSE TO AMPICILLIN
AND SULFONAMIDE-STRESS

Cells harbouring rBP 11 were subjected to increasing concentrations of ampicillin and sulfonamide. Drug concentration was raised over a great number of passages for several generations each. Thus MIC (Minimal Inhibitory Concentration) could be increased from 0.52 mg/ml to 32 mg/ml for ampicillin and from 8 mg/ml to 64 mg/ml for sulfonamides.

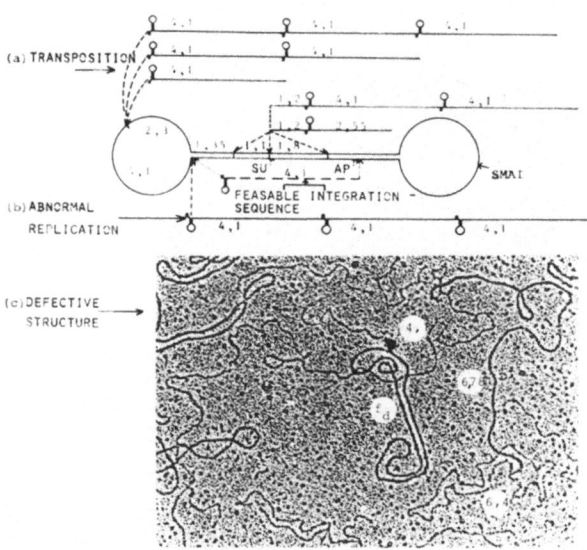

Fig. 3. Electron micrograph of one type of "amplified" rBP 11-replicon from ampicillin-stressed *E. coli.* Drawing of the molecule shows thick-lines corresponding to the double-stranded regions and thin lines representing single-stranded-DNA. Distances are in units of kilobases.

Estimation of values for copy numbers of "amplified" rBP 11 turned out to be difficult as amplified plasmids appear in a broad variation of different size classes. As only 20% of the amount of chromosonal DNA represents the plasmid content of the host cell, copy number functions play a minor role in the mechanism leading to high Ap and Su resistance.

CONTOUR LENGTH, HOMODUPLEX- AND RESTRICTION ANALYSIS OF
AMPICILLIN STRESSED PLASMID DNA

Nonstressed rBP 11 molecules are yielded in three different size classes whereas plasmids from Ap stressed cells form a heterogeneous population with at least 16 different size classes, thus complicating statistically significant determination of contour lengths. Restriction analysis indicates an apparent change in the hairpin loop structure, distal to the resistance genes. Homoduplex analysis confirms this finding and in addition reveals typical structures shown in Fig. 3. The sum of altered plasmid structures resulting from ampicillin drug pressure is listed in Fig. 4. Several integration sites in rBP 11 serve as target for a "jumping" element of 4.1 Kb, bordered at one of its ends by several short inverted repeats. The insertion event involves one single unit or several units of the element. Concerning its dimension and characteristic structure this unit corresponds to a fragment due to the part of the repeat region of rBP 11, consisting of the stem- and loop-sequence, the Ap^r and the Su^r genes (Fig. 4: FEASABLE INTEGRATION SEQUENCE).

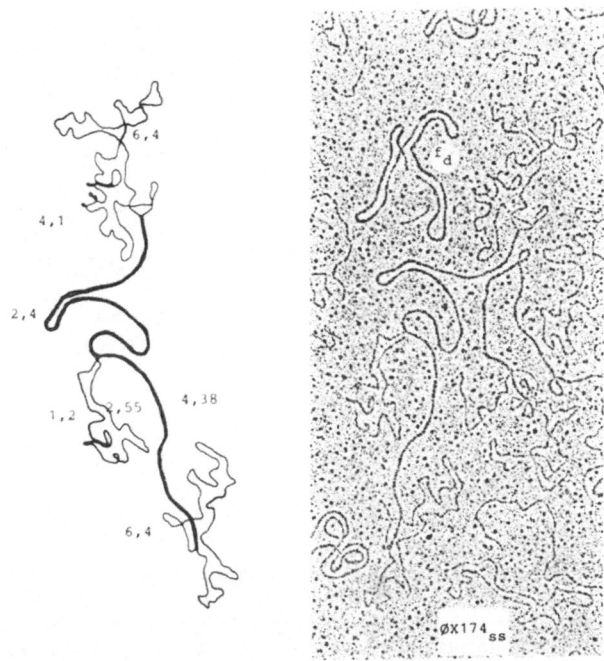

Fig. 4. Summary of integration cointegrates of rBP 11 resulting from gene amplification by ampicillin (25 mg/ml). Sites for integration are indicated by arrows, lengths of integrated fragments are given in Kb. Whereas the size of the stem- and loop-fragment is not listed specially, its overall size has been found to be 1.05 ± 0.02 Kb. As indicated, the multiple segment might correspond to a "feasable integration sequence" of about 5.16 Kb in the inverted repeat, including Su^r- and Ap^{r}genes. rBP 11 is thought to be enlarged by this segment via transposition (a), or abnormal replication (b), as indicated by characterization of defective structure (c).

CONCLUDING REMARKS

I. *Amplification of gene structure in rBP 11*. R factor R 1767 yields a broad varity of plasmid structures, independant from host rec-functions (Richmond and Wiedemann, 1974). Most of these structures contain duplicated sequences, as can be seen by homo- and heteroduplex-studies, thus indicating the presence of sequences, mediating recombinational events. Amplification of rBP 11 by ampicillin might serve as a model for such sequence arrangements. The most likely explanation for drug mediated reorganization of rBP 11 seems to be a transition event as is described for tetracycline resistance-plasmids by Clewell and Yagi, 1974, and by Mattes et al., 1979. It means amplification of resistance genes simply occurs by recombination of directly repeat terminal sequences, resulting in an enlarged replicon, which contains several copies of the resistance determinant region.

Recent experiments prove the existence of at least two independant. transposable elements in R 1767 (manuscript in preparation). One of them, TnA, was derived from rBP 11. The characterization of the gene product yields a β-lactamase with an unknown substrate profile. Together with restriction analysis it can be assumed that it represents

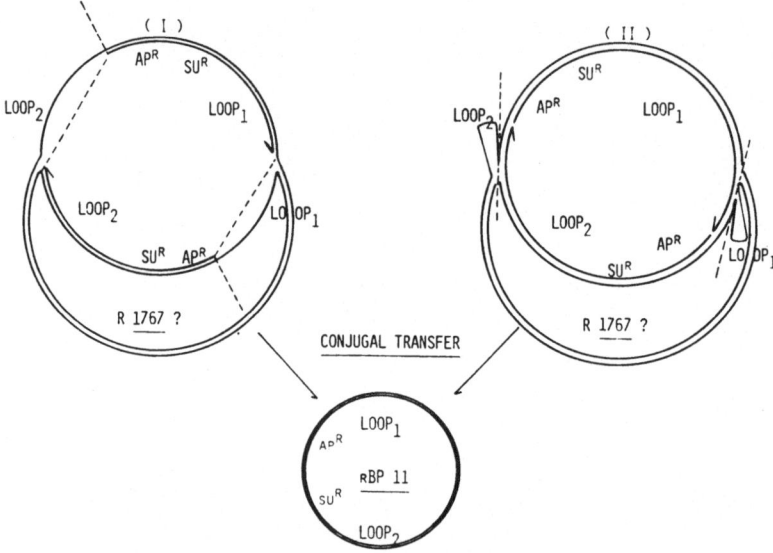

Fig. 5. Illustration of the generation of rBP 11 by illegitime recombination and/or abnormal replication of the growing fork of a parental replicon – R 1767?

In (I) illegitime recombination precedes abnormal replication, in (II) abnormal replication precedes the recombination event, thus in each case forming an "amphimer", a structure with Ap^r and Su^r as inverted repeat (thick lines) and two different loops as unique sequences at both ends. The dashed lines indicate the points connected by recombination.

a so far unknown transposable structure. The other "jumping" sequence was revealed to be a Su-transposon, which might correspond to the Su-sequence in rBP 11. Therefore, it can be assumed that translocation processes of these sequences could be responsible for transition of rBP 11 in drug stressed hosts.

The most frequent composite structure, obtained from conjugal transfer of R 1767 after selection on ampicillin, is plasmid rBP 10 with the determinants Ap, Su, and Tra. This composite structure shows the same transition phenomenon as rBP 11, probably due to its repeat sequences. Formation of underwound loop-structures could be proved in most plasmids, derived from R 1767, mediating functions for recombinational events and so far might be "diagnostic" of inverted duplications and transposon sequences as proposed by Broker et al., 1977.

II. *Formation of rBP 11.* Similar events could be assumed for the formation of rBP11. Probably transposable elements of parental plasmids serve to fuse two replicons thereby producing rBP 11, as predicted for replicon fusion by Shapiro, 1979. It is thought that this process generally underlies the same rules as transposition does. The model implies a cointegrate structure containing a directly repeated copy of a transposable element at each juncture of the two replicons. Up to now there is no proof of directly

162

repeated sequences in rBP 11. On the other hand it cannot be ruled out that rBP 11 might be created during replication of an unknown parental plasmid by the mode of abnormal replication as firstly predicted for formation of τ dv mutants by Chow et al., 1974. As shown in Fig. 5, a hypothetical parental plasmid – R 1767? – might underly illegitime recombination and/or abnormal replication explaining formation of "amphimer" – structures with unique and inverted repeat sequences, thus resulting in a stable replicon – rBP 11.

Furthermore, formation of different size classes of rBP 11 might be assumed as a more rare event in the life cycle of the plasmid analogous to so-called "slippage-replication", as described for formation of IS2-6 by Ghosal and Saedler (1979).

This problem is under work now.

REFERENCES

BROKER, T. R., SOLL, L. and CHOW, L. T. (1977): Underwound loops in self-renatured DNA can be diagnostic of inverted duplications and translocated sequences. J. Mol. Biol., *113*, 579–589.

BUKHARI, A. I., SHAPIRO, J. A. and ADYA, S. L. (eds.). (1977): DNA insertion elements, plasmids and episomes (Cold Spring Harbor Laboratory, Cold Spring Harbor, NY).

CHOW, L. T., DAVIDSON, N. and BERG, D. (1974): Electron microscope study of the structures of λdv DNAs. J. Mol. Biol. *86*, 69–89.

GHOSAL, D. and SAEDLER, H. (1979): DNA Sequence of the mini-insertion IS2-6 and its relation to the sequence of IS2. Nature, *275*, 611–617.

HOLLENBERG, C. P., DEGELMANN, A., KUSTERMANN-KUHN, and ROYER, H. D. (1967): Characterization of 2-μm DNA of Saccharomyces cerevisiae by restriction fragment analysis and integration in an Escherichia coli plasmid. Proc. Nat. Acad. Sci. USA *73*, 2072–2076.

MATTES, R., BURKHARDT, H. J., and SCHMITT, R. (1979): Repetition of tetracycline resistance determinant genes on R plasmid pRSD1 in *Escherichia coli*. Molec. gen. Genet. *168*, 173–184.

RICHMOND, M. H., and WIEDEMANN, B. (1974): Plasmids and bacterial evolution. In: Evolution in the Microbial world. Cambridge University Press, 59–85.

TANAKA, T., WEISBLUM, B. (1975): Construction of a colicin E1 – R factor composite plasmid *in vitro*: means for amplification of deoxyribonucleic acid. J. Bacteriol. *121*, 354–362.

YAGI, Y., and CLEWELL, D. B. (1977): Identification and characterization of a small sequence located at two sites on the amplifiable tetracycline resistance plasmid pAMα1 in *Streptococcus faecalis*. J. Bacteriol. *129*, 400–406.

F. S., Inst. of Medical Microbiology and Imunology, University of Bonn, 5300 Bonn, F. R. G.

CHARACTERIZATION OF A SmSu-PLASMID (rBP1) FREQUENTLY OCCURRING IN CLINICAL ISOLATES OF *E. COLI*

U. VAN TREECK, B. WIEDEMANN and W. KALTHOFEN

Institute of Medical Microbiology and Immunology, University of Bonn, Bonn, F.R.G.

INTRODUCTION

Streptomycin is with 25% one of the most common resistance markers in strains of clinical isolates of Escherichia coli. The high figure is surprising as the use of streptomycin in Middle Europe is restricted to the therapy of tuberculosis. Streptomycin resistance is usually mediated by plasmid coded adenyl- or phosphotransferases. Barth and Grinter (1974) and Grinter and Barth (1976) described several plasmids with linked

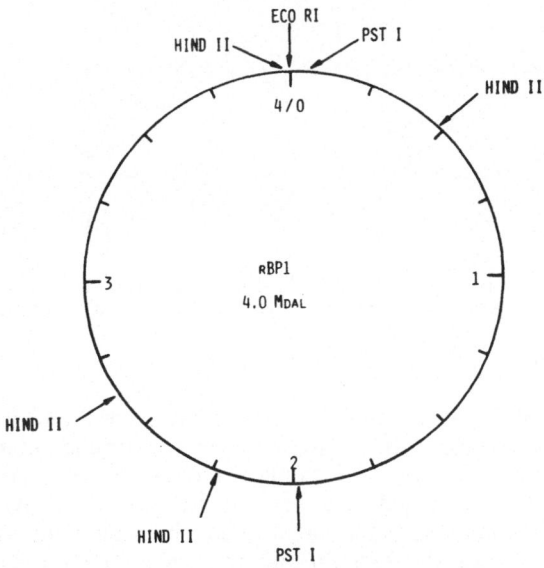

Fig. 1. Circular cleavage map of rBP1.

streptomycin and sulfonamide (SmSu)-resistance belonging to the incompatibility group Inc Q. Ten out of twelve of the plasmids had a molecular mass between 5.5 and 6.3 Mdal and a single cut site for restriction endonuclease Eco RI. DNA-DNA-hybridization showed a very close relation of these plasmids. One of them hybridized with RSF1010, in heteroduplex analysis almost completely as shown by Heffron, Rubens and Falkow (1975).

On the other hand Kawabe, Tanaka and Mitsuhashi (1978) postulated that all non-conjugative plasmids with Sm-resistance inactivate streptomycin by phosphorylation, while the mechanism of inactivation by conjugative plasmids is mostly due to adenylation.

Until now epidemiological data of the surveillance of spezialized SmSu-plasmids are not available. This paper deals with the characterization and the incidence of a plasmid with a molecular weight of 4.0 Mdal, encoding for SmSu-resistance. The Sm-resistance is mediated by a phosphotransferase (APH-(3'')). The plasmid is designated rBP1.

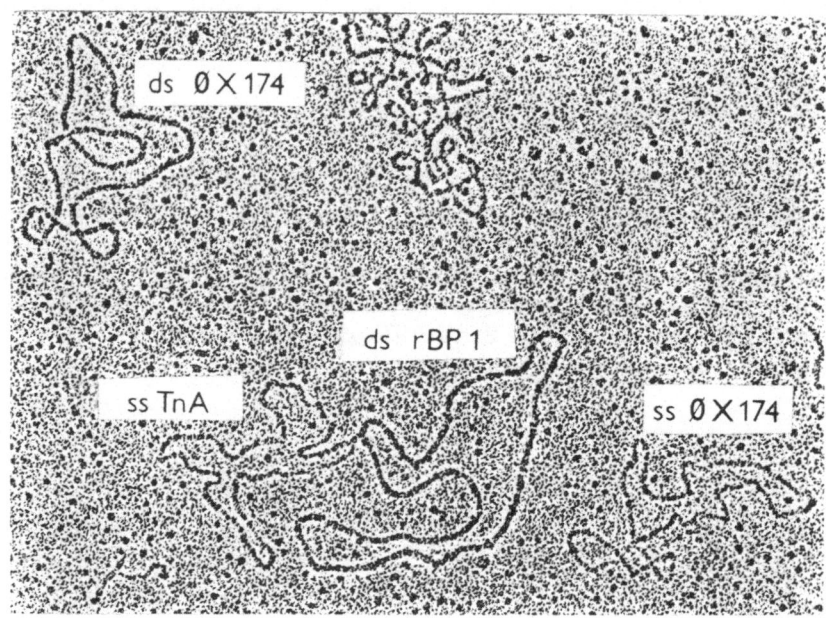

Fig. 2. Heteroduplex of rBP1 with the 7.2 Mdal plasmid (rBP1-1) of strain 019393. Single (ss) and double stranded (ds) forms of phage ⌀X 174 served as internal standards.

Characterization of the plasmid (rBP1):

In faecal strains of *E. coli* isolated from a volunteer we found in different serotypes (017, 057, and 0106) besides other plasmids a small resistance plasmid with resistance determinants to streptomycin and sulfonamides and with a molecular weight of 4.0 Mdal. This ecological study encouraged us to look for the spread of this plasmid in bacteria from a clinical specimen. Therefore a detailed characterization was necessary. The streptomycin resistance was due to a phosphotransferase APH-(3''). This marker enables the bacteria to resist a concentration of 512 μg streptomycin/ml. The MIC of sulfonami-des is 8 mg/ml. The cleavage with restriction endonucleases revealed fragments listed in Table I. Bam H 1 restriction endonuclease was not able to cut the molecule. Digestion with combinations of two enzymes and a partial digestion with the *Hind* II enzyme allowed to construct the restriction map shown in Fig. 1. The *Eco* RI site is the zero point. A partial cleavage was necessary to find the location of the *Hind* II fragments on the plasmid. As we found fragments with 1.85 and 2.15 Mdal we conclude that the fragments 0.5 and 1.35 Mdal as well as 1.75 and 0.4 Mdal we got in the complete digestion with *Hind* II are neighboured.

Incidence of plasmids identical with rBP1 in clinical isolates

130 strains of *E. coli* from clinical specimens (mostly urinary tract infections) were screened for resistance to streptomycin and sulfonamides, and sensitivity to spectinomycin. 8.5% of all strains were resistant to Sm and Su and sensitive to spectinomycin, while 16.9% phenotypically showed the resistance pattern Sm^{r}Su^{r}Sp^r. From 10 strains with the markers Sm^r, Su^r, Sp^s we isolated inactivating enzymes. The activities of these enzymes indicated that all strains harbour a APH-(3″) and no adenyltransferase AAD--(3″).

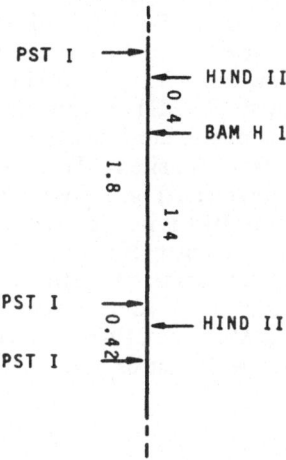

Fig. 3. Linear cleavage map of TnA. The fragment sizes are expressed in Mdal. Total size of TnA is 3.2 Mdal. The distances of the restriction sites to both ends are not yet determined.

These strains were tested for their sensitivity to other chemotherapeutics. The number and molecular size of plasmids was determined with the gel electrophoresis (Tab. II.). In all organisms except one a plasmid with a molecular weight of 4.1 ± 0.1 Mdal could be demonstrated. The strain lacking this sort of molecule harboured another slightly bigger plasmid with a molecular weight of 7.2 Mdal.

The isolated plasmid DNA of these 10 strains was used for a transformation after purifying in CsCl-gradient centrifugation. The transformants (*E. coli* C600) were selected with streptomycin or sulfonamides. The transformants from 9 strains were resistant to streptomycin and sulfonamides and harboured only one plasmid with a molecular mass of 4.0 Mdal. Those derived from strain No. 019393 harboured a 7.2 Mdal plasmid and showed additional resistance to ampicillin.

The digestion of the Sm^r, Su^r plasmids from the transformants with *Eco* RI, *Pst* I and *Hind* II revealed fragments identical with those of rBP1.

Integration of an ampicillin transposon into rBP1

The R-factor with the additional ampicillin resistance marker was cut once by *Eco* RI and *Bam* H1. *Pst* I gave two fragments of 2.05 and one fragment of 1.8, 0.75 and 0.42 Mdal. *Hind* II gave fragments with a molecular weight of 2.6, 1.8, 1.35, 0.52, 0.5 and 0.4 Mdal. As four of these fragments are identical in size we presumed that one part of this plasmid might be related to rBP1.

Heteroduplex between the ApSmSu plasmid and rBP1 showed a homologous region corresponding to the size of rBP1 and an additional single stranded insertion (about 3.2 Mdal) on a short double-stranded stalk (68 kilobases) (Fig. 3). These results indicated the presence of an Ap-transposon (TnA) integrated in the plasmid rBP1.

Genetic evidence for the ability of translocation of the Ap marker from the 7.2 Mdal plasmid to plasmids pUB307 and R751 was shown by mating experiments. About 1% of the transconjugants inserted the Ap marker into either of the plasmids.

Characterization of the Ap transposon

As described above the ApSmSu plasmid has three additional *Pst*I- and two other *Hind* II restriction sites and one site for *Bam* HI, indicating that these sites are present on TnA. The position of the *Hind* II and *Pst*I sites relative to the *Bam* HI site was determined by double digestionswith *Bam* HI/*Hind* II and *Bam* HI/PstI. The 1.8 Mdal *Hind* II fragment was cleaved with *Bam* HI into two fragments with 1.4 and 0.4 Mdal in size. The 1.8 Mdal PstI fragment was cut into a 1.22 Mdal fragment and a 0.58 Mdal fragment. The position of the third PstI site in TnA has to be 0.42 Mdal distinct from the right end of the 1.22 Mdal fragment. Fig. 2 gives the restriction map of TnA.

The integration site of TnA into rBP1 is not yet determined exactly, but it lies between positions 0.5 and 0.75 on the cleavage map of rBP1.

The β-lactamase of TnA was characterized by the isolation of a raw enzyme extract with sonication followed by isoelectric focussing in comparison to the TEM 1 and TEM 2 enzymes described by Sykes and Mathews (1976). We could demonstrate a β-lactamase band in a position corresponding with that of TEM 1 enzyme.

CONCLUSIONS

The nonconjugative resistance plasmid rBP1 confers resistance to sulfonamides and streptomycin in bacteria. The latter marker is expressed by the production of APH-(3″). It has a molecular weight of 4.0 Mdal and one cleavage site for *Eco* RI, two for *Pst*I and four for *Hind* II. These characters enable a clear cut identification of rBP1 like plasmids in bacteria from clinical specimens. It is not identical with one of the SmSu plasmids described by different workers (Grinter and Barth, 1976, Kawabe, Tanaka and Mitsuhashi, 1978, Heffron, Rubens and Falkow, 1975). The epidemiological data presented show that this plasmid is common in clinical isolates of *E. coli*. Although the number of tested strains is still small we conclude from data derived from clinical strains from Egypt (Elkhouly and Wiedemann, 1979, manuscript in preparation) that this plasmid is distributed all over the world. In this study the incidence of rBP1 plasmids was much higher than in our material. This difference in the prevalence is probably due to different drug usage. The linkage of the streptomycin with the sulfonamide marker is presumably the reason for the wide spread of Sm resistance, as sulfonamides are still used quite frequently. The example of rBP1 demonstrates that nonconjugative plasmids can spread easily in nature.

Although we found rBP1 with the integration of a TnA only once and have no idea how common this sort of enlarged rBP1 is, we suggest that other transposons with identical insertion sequences can be integrated in the same manner. We designate the 7.2 Mdal plasmid rBP1 with the integrated TnA as rBP1-1.

The *Hind* II restriction sites of the TnA of rBP1-1 and the presence of TEM 1 enzyme suggests that this transposon is identical with the Tn 3 described by Heffron, Bedinger, Champoux and Falkow (1977).

TABLE I.

Restriction fragments of rBP1 in Mdal after cleavage with restriction endonucleases. PstI abd EcoRI fragments of λ DNA were used as markers in agarose gel electrophoresis.

EcoRI	PstI	HindII	EcoRI/PstI	PstI/HindII	EcoRI/HindII
4.0	2.05	1.75	1.95	1.5	1.75
	1.9	1.35	1.9	1.35	1.35
		0.5	(0.1)	0.42	0.48
		0.4		0.4	0.4
				0.25	(0.02)
				(0.08)	

TABLE II.

Characteristics of the tested E. coli strains from the university hospital of Bonn. Strain 14518 is a faecal E. coli isolated from a volunteer in Frankfurt.

Strain	Resistant to						Number of plasmids	Plasmid sizes (Mdal)
	Ap	Tc	Cm	Sp	Sm	Su		
018438c	×	×	×		×	×	4	32 6.9 4.1 3.1
018488a	×	×			×	×	7	33 4.9 4.0 3.0 2.7 2.0 1.6
018561					×	×	2	31 4.0
018567		×			×	×	3	31 6.2 4.0
018592		×			×	×	2	33 4.1
018617					×	×	4	5.0 4.0 3.2 2.1
007870					×	×	6	32 5.9 4.0 2.9 1.8 1.4
019632	×	×			×	×	4	50 7.0 4.8 4.2
019809					×	×	3	45 4.1 1.4
019393	×	×			×	×	3	39 7.2 3.1
14518					×	×	2	77 4.0 (rBP1)

REFERENCES

BARTH, P. T. and N. J. GRINTER (1974): Comparison of the deoxyribonucleic acid molecular weights and homologies of plasmids conferring linked resistance to streptomycin and sulfonamides. J. Bacteriol. *120*, 618–630.

GRINTER, N. J. and P. T. BARTH (1976): Characterization of SmSu plasmids by restrictionendonuclease cleavage and compatibility testing. J. Bacteriol. *128*, 394–400.

HEFFRON, F., P. BEDINGER, J. J. CHAMPOUX, and S. FALKOW (1977): Deletions affecting the transposition of an antibiotic resistance gene. Proc. Nat. Acad. Sci. U.S.A. *74*, 702–706.

HEFFRON, F., C. RUBENS, and S. FALKOW (1975): Translocation of a plasmid DNA sequence which mediates ampicillin resistance: molecular nature and specifity of insertion. Proc. Nat. Acad. Sci. U.S.A. *72*, 3623–3627.

KAWABE, H., T. TANAKA, and S. MITSUHASHI (1978): Streptomycin and spectinomycin resistance mediated by plasmids. Antimicrob. Agents Chemother. *13*, 1031–1035.

SYKES, R. B., and M. MATTHEW (1976): The β-lactamases of Gram-negative bacteria and their role in resistance to β-lactam antibiotics. J. Antimicrob. Chemother. *2*, 115–157.

U. v. T., Inst. of Medical Microbiology
and Immunology, University of Bonn,
5300 Bonn, F.R.G.

I. THEORETICAL PART

E) PLASMIDS IN THE NATURE

Chairman: W. GOEBEL

REPLICATION FUNCTIONS DETERMINED BY THE BASIC REPLICON OF THE ANTIBIOTIC RESISTANCE FACTOR R1

W. GOEBEL, R. KOLLEK, W. OERTEL, K. J. BURGER and R. RÖLLICH

Institute of Genetics and Microbiology, University of Würzburg, Würzburg, F.R.G.

Recent investigations have indicated that even large plasmids, such as transmissible R-factors, may require rather little of their genetic information for autonomous replication (Eichenlaub et al., 1975, Timmis et al., 1975, Kollek et al., 1978, Stalker et al., 1979). In addition, these studies provide evidence that, with the exception of RP4, the information necessary for the replication function(s) is located on a contiguous stretch of DNA surrounding the origin(s) of replication. However, the complexity of these "basic replicons" appears to be different.

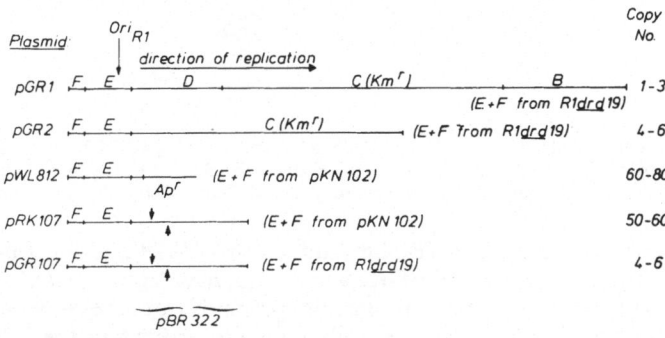

Fig. 1. Physical maps of miniplasmids derived from R1drd-19 and its copy mutant pKN102. Miniplasmids pGR1 and pGR2 were obtained by cleaving R1drd-19 RNA with restriction endonuclease PstI and religating the fragments. Transformation was performed with the obtained mixtures into *E. coli* HB101 and colonies resistant to kanamycin were selected. Miniplasmid pWL812 is a derivative of pKN102 (Kollek et al., 1978). The recombinant plasmids pRK107 and pGR107 were obtained by cloning of the Pst-E and Pst-F fragments from pGR2 and pWL812, respectively, with vector plasmid pBR322. Copy numbers were determined from the ratio of covalently closed circular DNA to total DNA. The copy numbers of pGR107 and pRK107 were determined under polA-negative conditions.

The antibiotic resistance factor R1drd-19 is, like many other transmissible R-factors of *E. coli*, a stringently controlled plasmid which determines resistance to ampicillin, streptomycin/neomycin, chloramphenicol, kanamycin and sulfonamide (Meynell and Datta, 1967). Copy mutants of this plasmid can be obtained, which occur in the cell in an increased copy number (Nordström et al., 1972). Most of these copy mutants are unstable under normal growth conditions and lead to the formation of miniplasmids (Goebel and Bonewald, 1975, Luibrand et al., 1977, Mickel et al., 1977), which have been successfully used for the further investigation of the replication functions of R1 (Kollek et al., 1978, Oertel et al., 1979). Miniplasmids of the wild type R1 factor which have not been observed to occur *in vivo* can be generated in vitro by partial cleavage of R1*drd*-19 DNA with the restriction enzyme *Pst*I, religating the obtained fragments with DNA ligase, transforming the ligated DNA mixture into *E. coli* and selecting for transformants that are resistant to kanamycin. This resistance marker is located

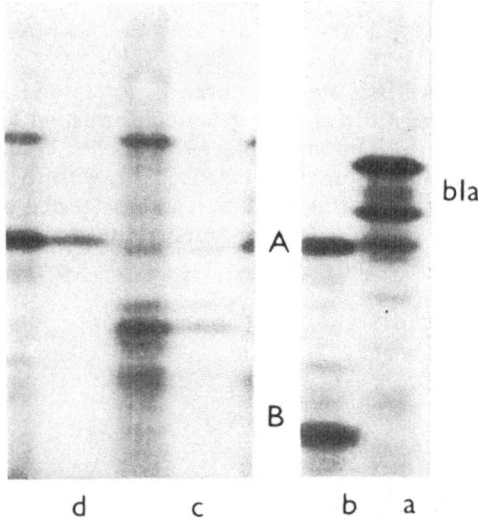

Fig. 2. Proteins expressed in minicells of P678-54 carrying a) pBR322, b) pRK107 (pBR322 . *Pst*-E+F), c) pRK102 (pBR322 *Pst*-F), d) pRK101 (pBR322 . *Pst*-E), *bla* is β-lactamase. For description of proteins A and B see text.

close to the origin of replication (Blohm and Goebel, 1978). Using this method two mini plasmids of R1*drd*-19, termed pGR1 and pGR2, have been isolated. Their physical maps are shown in Fig. 1. In contrast to the mini plasmids from the copy mutant pKN101 (Goebel and Bonewald, 1975) these miniplasmids are present in the cell in low copy numbers, indicating that the genetic information for copy control is still located on these plasmids.

From both types of plasmids a segment of 1.800 base pairs (bp) was obtained by cloning the two *Pst*I fragments E and F (Fig. 1) on pBR322. The two *Pst*I fragments are sufficient for regulated autonomous replication as shown by transforming these recombinant plasmids into a *pol*A^{ts} mutant. As the replication of the vector plasmid pBR322 (ColE1 replicon) but not of plasmid R1 is dependent on high levels of DNA polymerase I, replication of the recombinant plasmids under polymerase I-negative conditions can only start from the R1 origin. As shown in Table I, both recombinant plasmids pGR107 (deriving from wild type R1) and pRK107 (deriving from the copy mutant pKN102) are able to replicate under the restrictive condition in the *pol*A^{ts} mutant MM383 but their copy numbers differ as expected. The low copy number of pGR107 and the high copy number of pRK107 at 43°C indicate that the cloned segment consisting of *Pst*-E and *Pst*-F still contains the essential function(s) for copy control. The larger *Pst*I fragment E carries the origin of replication which has been mapped by electron microscopy of replicative intermediates (Ohtsubo et al., 1978).

174

The proteins A and B which seem to be specific for the replicon part of R1 are expressed in minicells of *E. coli* harbouring the recombinant plasmids pGR 107 and pRK107. Protein A, however, is a fused gene product consisting of the amino acid sequence of the β-lactamase up to the *Pst*I site and the adjacent coding region of *Pst*-E up to the nearest stop codon as shown by proteolytic cleavage of this protein according to Cleveland

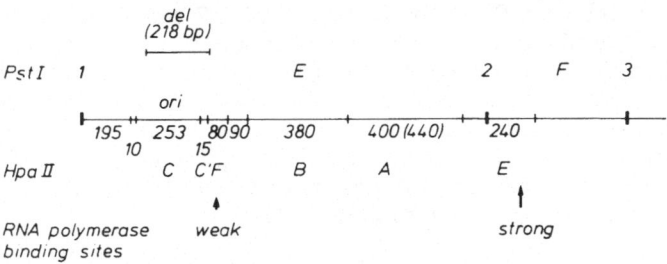

Fig. 3. Fine structure map obtained from *Hpa*II fragments of the replication region of R1 (*Pst*-E+ +*Pst*-F) and location of RNA polymerase binding sites. Transcription from the strong binding site on *Pst*-F runs towards the origin.

et al., 1977 (data not shown). Protein B of pRK107 (mw 11.500 dalton) must be determined by a gene which is cleaved by the *Pst*I site separating *Pst*-E and *Pst*-F of the replication segment, as neither *Pst*-E nor *Pst*-F alone codes for this protein (Fig. 2). Preliminary results indicate that protein B of pGR107 (replicon of wild type R1*drd*-19) differs in size from protein B of pRK107 (replicon of copy mutant pKN102). It is tempting to speculate that the difference in molecular weights of this protein may be related to the difference in copy control of these two replicons. The molecular function of this protein is still unknown.

```
5'-C C G G T C C A G C G C A G A A C C G A A A C C A C A A A G C C C|C T C C C T C A T A A C T G A A A A G C G G|C C C C
   G G C C A G G T C G C G T C T T G G C T T T G G T G T T T C G G G|G A G G G A G T A T T G A C T T T T C G C C|G G G G
                                                                    1              10              20
   G C C C C G G C C C A A A G G G C C G G A A C A G A G T C G C T T|T T A A T T|A T G|A A T G T T G T A A|C T A C A T C
   C G G G G C C G G G T T T C C C G G C C T T G T C T C A G C G A A|A A T T A A|T A C|T T A C A A C A T T|G A T G T A G
              30              40              50              60              70              80
   T T C G T C G C T G T C A G T C T T C T C G C T G G A A G T T C T C A G T A C A C G C T C G T A A G C G G C C C T C G
   A A G C A G C G A C A G T C A G A A G A G C G A C C T T C A A G A G T C A T G T G C G A G C A T T C G C C G G G A G C
              90             100             110             120             130             140
   C G G C C C C G C T A A C G C G G A G A T A C G C C C C G A C T T C G G G T A A A C C C T C G T C G G G A C C A C T C C
   G C C G G G C G A T T G C G C C T C T A T G C G G G G C T G A A G C C C A T T T G G G A G C A G C C C T G G T G A G G
              150             160             170             180             190             200
   G A C C G C G C A C A G A A G|C T C T C T C A T G G C T G A A A G C G G|G T A T G G T C T G G C G A G G G C T G G G G A
   C T G G C G C G T G T C T T C|G A G A G A G T A C C G A C T T T C G C C|C A T A C C A G A C C G T C C C G A C C C C T
              210             220                                          (T T)      (T)         (G)
   T G G G T A A G G T G A A A T C T A T C A A T C A G T A C C G G · · · · · · · C G G G C G G C G G C G G T G T T A C T
   A C C C A T T C C A C T T T A G A T A G T T A G T C A T G G C C · · · · · · · G C C C G C C G C C G C C A C A A T G A
   (T)             (T)                                                          (T)
   C C T G T A T C A T A T G A A G C A A C A G A G T G C C G C C T T C C A T G C C G C T G A T G C A G A T T G T T T G G
   G G A C A T A G T A T A C T T C G T T G T C T C A C G G C G G A A G G T A C G G C G A C T A C G T C T A A C A A A C C

   T A A C G A T A T C T G A A T T G T T A T A C A T G T G T A T A T A C G T G G T A A T G A C A A A A A T A G - 3'
   A T T G C T A T A G A C T T A A C A A T A T G T A C A C A T A T A T G C A C C A T T A C T G T T T T T A T C
```

Fig. 4. Nucleotide sequence of the DNA segment surrounding the origin of replication of R1. The sequence was determined by the Maxam and Gilbert technique.

Pst-F carries one strong binding site for RNA polymerase (Fig. 3), which is likely to be the promotor for gene-determining protein B. In addition, a rather weak binding site has been indetified in the replication region of R1, where the origin of replication has been mapped (Fig. 3). It remains to be seen whether this is the starting point for RNA primer synthesis required for R1 replication (Mayer et al., 1978). The cloned replication segment of R1*drd*-19 and pKN102 also expresses incompatibility function(s).

When the recombinant DNA consisting of pBR322 and *Pst*-E of the copy mutant pKN102 is introduced into *E. coli* K5, a segment of 218 bp is deleted at high frequency when the orientation of the *Pst*-E fragment is reversed within pBR322 relative to its original orientation. The location of the deletion has been mapped by electron micros-copy and restriction enzymes. These analyses show that it is located within the replication region and comprises essential parts (or all) of the replication origin of R1. The deleted segment is covered by three *Hpa*II fragments C, C′ and F (Fig. 3). This area has been sequenced using the Maxam-Gilbert technique (Fig. 4).

5′···CAGAACTGAAACCACAAAGCCC|CTCTCTCATGGCTGAAAGCGGG|TATGG······3′ pRK103

5′···CAGAACTGAAACCACAAAGCCC|CTCTCTCATGGCTGAAAGCGGGT······ ─3′ pRK1031

5′···CAGAACTGAAACCACAAAGCCC|CTCTCTCATGGCTGAAAGCGG······ ─3′ pRK1032

Fig. 5. Nucleotide sequences at the junction sites of *Hpa*II-CF of three independently isolated deletion mutants.

The nucleotide sequence in Fig. 4 indicates that the deletion occurs at a characteristic sequence (21 bp and 22 bp, respectively) which is located as direct repeats on both ends of the deleted segment. The determination of the nucleotide sequence of the arising *Hpa*II-CF fragments of three independently isolated deletion mutants (Fig. 5) suggests that the deletion always occurs precisely at these sites (Kollek et al., to be published). The nucleotide sequence of the deleted 218 bp segment shows some characteristic features (Fig. 4) such as palindromic and hair pin structures (Oertel et al., 1979). The nucleotide sequence of the segment of R1 which has been determined so far is very similar to that of R100 (Rosen et al., 1979), indicating that the replicon of both anti-biotic resistance factors is of the same origin. (Fig. 6).

As indicated above, the *Pst*-E fragment is unable to provide autonomous replication to the recombinant plasmid pKN101 (pBR322-*Pst*-E) under *pol*A-deficient conditions. However, when plasmid pRK101 is propagated in the *pol*A12 mutant MM383, carrying a temperature-sensitive DNA polymerase I (*pol*A), mutants of pRK101 arise at a fre-quency of 10^{-4} to 10^{-5} that are independent of *pol*A in their replication. Similiar obser-vations were also made with pGR101 having the *Pst*-E-fragment from wild type R1*drd*-19 inserted into pBR322 and even with the recombinant plasmids pRK103a and pRK103 b having *Pst*-E-fragment with the 218 bp deletion inserted into pBR322. The mutation(s) which activates a new replicon independent of the original R1 replication origin and of protein B seems to affect the *Pst*-E fragment only, as indicated by the following obser-vation: a) These *Pst*-E- or *Pst*-E$_{del}$-fragments can be ligated *in vitro* to a nonreplicative *Pst*-fragment carrying kanamycin (Km) resistance. Transformation of *E. coli* 5K with this recombinant DNA leads to Km^R transformants which carry the expected recom-binant plasmid. b) *Pst*-E*-or *Pst*-E$_{del}$*-fragments can be recloned on another *Col*E1-type vector plasmid pACYC184. The resulting recombinant plasmids are also *pol*A-independent. c) The restriction enzyme *Sau*3A cleaves *Pst*-E into two almost identical fragments. The 218 bp deletion affects only the right side *Sau*3A fragment. Cloning of the left side *Sau*3A fragment of *Pst*-E$_{del}$ from a mutant plasmid into the *Bam*HI site of pBR322 leads to recombinant plasmids, that are still *pol*A independent. This suggests that a new origin of replication is activated in these mutants which is located on another part of the *Pst*-E-fragment as the original R1 origin. The mutation(s) leading to this event does not change the size of the *Pst*-E-fragment, which suggests that only minor changes

```
5'-C C G G T C C A G C G C A G A A C T G A A A C C A C A A A G C C C CTCCCTCATAACTGAAAAGCGG CCCCG
         5'-C G C A G A A C C G A A A C C A C A A A G C C C CTCCCTCATAACTGAAAAGCGG CCCCG

C C C C G G C C C G A A G G G C C G G A A C A G A G T C G C T T T T A A T T A T G A A T G T T G T A A C T A C T T C A T
C C C C G G C C C A A A G G G C C G G A A C A G A G T C G C T T T T A A T T A T G A A T G T T G T A A C T A C A T C T T

C A T C G C T G T C A G T C T T C T C G C T G G A A G T T C T C A G T A C A C G C T C G T A A G C G G C C C T G A C G G
C G T C G C T G T C A G T C T T C T C G C T G G A A G T T C T C A G T A C A C G C T C G T A A G C G G C C C T C G C G G

C C C G C T A A C G C G G A G A T A C G C C C C G A C T T C G G G T A A A C C C T C G T C G G G A C C A C T C C G A C C
C C C G C T A A C G C G G A G A T A C G C C C C G A C T T C G G G T A A A C C C T C G T C G G G A C C A C T C C G A C C

G C G C A C A G A A G C T CTCTCATGGCTGAAAGCGG G T A T G G T C T G G C A G G G C T G G G G A T G G G T
G C G C A C A G A A G C T CTCTCATGGCTGAAAGCGG G T A T G G T C T G G C A G G G C T G G G G A T G G G T

A A G G T G A A A T C T A T C A A T C A G T A C C G G C T T A C G C C G G G C T T C G G C G G T T T T A C T C C T G T A
A A G G T G A A A T C T A T C A A T C A G T A C C G G · · · · · · · C G G G C G G C G G C G G T G T T A C T C C T G T A
                                                         (T  T)            (T)  ●   (G)        (T)

T C A T A T G A A A C A A C A G A G T G C C G C C T T C C A T G C C G C T G A T G C G G C A T A T C C T G G T A A C G A
T C A T A T G A A G C A A C A G A G T G C C G C C T T C C A T G C C G C T G A T G C G G C A G A T T G T T (T) G G T A A C G A
          (T)    ●                                      (T)          ●(G C A) (C)
                                                                      (C G A)

T A T C T G A A T T G T T A T A C A T G T G T A T A T A C G T G G T A A T G A C A A A A A T A G-3'
T A T C T G A A T T G T T A T A C A T G T G T A T A T A C G T G G T A A T G A C A A A A A T A G-3'
```

Fig. 6. Comparison of the nucleotide sequence surrounding the origin of replication of R1 and R100. Nucleotide sequence of R100 is taken from Rosen et al. (1979).

in the base composition may occur. This new R1 replicon behaves differently from the original R1 replicon in several respects. a) The copy number of the recombinant plasmids carrying this replicon integrated into pBR322 is low under *pol*A-deficient conditions regardless of whether they derive from wildtype R1 or copy mutants. b) It is rather unstable, i. e. plasmids replicating under its control segregate at a high rate. c) In contrast to the "primary" replicon of R1 (*Pst*-E+*Pst*-F), this new R1 replicon appears to be completely compatible with R1.

In conclusion: The replicon of the antibiotic resistance factor R1 consists of a nucleotide sequence of 1.800 bp. This DNA segment carries an origin of replication from which the replication starts unidirectionally. At least one gene product determined by this segment (protein B) that is required for the initiation of replication has been identified. This segment also carries the information for copy control of R1 and the incompatibility function(s). A sequence of 218 bp, which includes essential parts or all of the replication origin, is deleted under certain conditions. The abolishment of the functional R1 origin by this deletion followed by a mutation event in the *Pst*-E fragment activates a new replication origin. This new minimal replicon of R1 is less stable than the original R1 replicon and may be regarded as a "salvage replicon" of R1.

REFERENCES

BLOHM, D. and W. GOEBEL (1978): Restriction map of the antibiotic resistance plasmid R1drd-19 and its derivatives pKN102 (R1drd-19B2) and R1drd-16 for the enzymes *Bam*HI, *Hind*III, *Eco*RI and *Sal*I. Molec. gen. Genet. **167**, 119–127.

EICHENLAUB, R., D. FIGURSKI and D. R. HELINSKI (1975): Bidirectional replication from a unique origin in a mini-F plasmid. Proc. Nat. Acad. Sci **74**, 1138–1141.

KOLLEK, R., W. OERTEL and W. GOEBEL (1978): Isolation and characterization of the minimal fragment required for autonomous replication (("basic replicon") of a copy mutant (pKN102) of the antibiotic resistance factor R1. Molec. gen. Genet. **162**, 51–57.

LUIBRAND, G., D. BLOHM, H. MAYER and W. GOEBEL (1977): Characterization of small ampicillin resistance plasmids (Rsc) originating from the mutant antibiotic resistance factor R1drd-19B2. Molec. gen. Genet. **152**, 43–51.

MAYER, H., G. LUIBRAND and W. GOEBEL (1977): Replication of the mini-R1 plasmid Rsc11 and Rsc11 hybrid plasmids. Molec. gen. Genet. **152**, 145–152.

MEYNELL, E. and N. DATTA (1967): Mutant drug resistance factor of high transmissibility. Nature (London) **214**, 885–887.

MICKEL, S., E. OHTSUBO and W. BAUER (1977): Heteroduplex mapping of small plasmids derived from R-factor R12: *In vivo* recombination occurs at IS1 insertion sequences. Gene **2**, 193–210.

NORDSTRÖM, K., L. C. INGRAM and A. LUNDBÄCK (1972): Mutations in R-factor of *Escherichia coli* causing an increased number of R-factor copies per chromosome. J. Bacteriol. **110**, 562–569.

OERTEL, W., R. KOLLEK, E. BECK and W. GOEBEL (1979): The nucleotide sequence of a DNA fragment from the replication origin of the antibiotic resistance factor R1drd-19. Molec. gen. Genet. **171**, 277–285.

OHTSUBO, E., M. ROSENBLOOM, H. SCHREMPF, W. GOEBEL and J. ROSEN (1978): Site specific recombination involved in the generation of small plasmids. Molec. gen. Genet. **159**, 131–141.

ROSEN, J., H. OHTSUBO and E. OHTSUBO (1979): The nucleotide sequence of the region surrounding the replication origin of an R100 resistance factor derivative. Molec. gen. Genet. **171**, 287–293.

STALKER, D. M., R. KOLTER and D. R. HELINSKI (1979): Nucleotide sequence of the region of an origin of replication of the antibiotic resistance plasmid R6K. Proc. Nat. Acad. Sci **76**, 1150–1154.

TIMMIS, K., F. CABELLO, and S. N. COHEN (1975): Cloning, isolation and characterization of replication regions of complex plasmid genomes. Proc. Nat. Acad. Sci **72**, 2242–2246.

W. G., Inst. für Genetik und Mikrobiologie,
Röntgenring 11,
D-8700 Würzburg, B.R.D.

CLONING OF SYNTHESIS AND TRANSPORT FUNCTIONS OF THE EXTRACELLULAR TOXIC PROTEIN, α-HAEMOLISIN OF *E. COLI*

W. GOEBEL, A. NOEGEL, U. RDEST and W. SPRINGER

Institute of Genetics and Microbiology, University of Würzburg, Würzburg, F.R.G.

INTRODUCTION

Haemolysis, i. e. the ability to cause lysis of erythrocytes, is a rather wide-spread phenomenon among bacteria and other microorganisms. The molecular basis of this property may be different from organism to organism. Extracellular enzymes such as proteases, lipases and other hydrolytic enzymes have been shown to be responsible for haemolysis in some bacteria[1-5], as well as in fungi[6]. In other bacteria "nonenzymatic" proteins, which disrupt the membrane of erythrocytes, have been identified as the causative agents for haemolysis[7-10]. All these agents are termed *haemolysins*. They are often secreted by the cell or are bound to the cellular surface. Hameolysis may be associated with the pathogenic trait exerted by such organisms. In pathogenic *E. coli* strains, causing more or less severe diarrhoea, especially in newborn animals and humans, the capability of haemolysin production is frequently encountered[11-12]. Haemolytic *E. coli* strains are also frequently encountered in infections of the urinary tract. Three types of haemolysins termed α, β and γ, have been identified in *E.coli*[13-15]. The distinction of these three types is mainly based on rather superficial properties, such as more or less free diffusion into the surroundings or different appearance of the lysis zones on blood agar. The α-haemolysin of *E. coli*, which has been shown in some cases to be genetically determined by transmissible plasmids [15-20], is secreted by the producing cells and can be isolated from the supernatant. Despite considerable effort, little is known so far about the nature of this agent[21].

RESULTS

Haemolysin, a plasmid-determined extracellular protein

Recent investigations in our laboratory have shown that the biologically active α-haemolysin of *E. coli* is a protein with a molecular weight of 58.000 dalton (Springer, Rdest and Goebel, in preparation), which can occur in several complex forms in an α-haemolytic *E. coli* strain. Recent data from our laboratory have shown that the genetic information required for the production and secretion of active α-haemolysin is completely plasmid-borne in haemolytic *E. coli* strains isolated from mice.

Most of these strains carry three transmissible plasmids belonging to different incompatibility groups[18] (Noegel, Rdest and Goebel, to be published). Mating of th haemolytic wild type strains with *E. coli* K12 resulted in haemolytic transconjugants, which harboured either all three plasmids (A, B, C), or the two plasmids B and C, either separately or recombined. By transformation of the individual, separated plasmids B and C into *E. coli* HB101 it could be demonstrated that only one of them (B), termed in the following pHly152, carries the determinant for α-haemolysin, whereas the other two plasmids (A and C) are not required for synthesis or excretion of this toxin. The transmissible plasmid pHly152 belongs to the incompatibility group I_2, has a molecular weight of 41×10^6 dalton and exhibits strong restriction to phage λ (plating efficiency of λ in *E. coli* C600 harbouring pHly152 is reduced to 10^{-4}). A physical map of this plasmid was constructed from its *Eco*RI cleavage products (Fig. 1). The eleven *Eco*RI- (and also the eleven *Hind*III-) fragments of pHly152 were cloned using pACYC184 as vector plasmid. The properties of these clones were determined with regard to autonomous replication, incompatibility, transfer, restriction and modification, and haemolysis. A summary of the results is given in Fig. 2. In the following we shall discuss only the haemolytic determinant of plasmid pHly152.

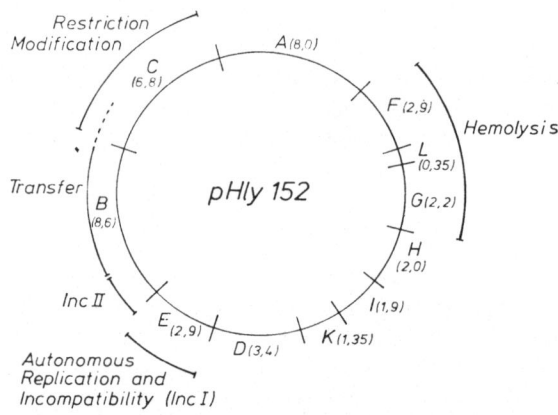

Fig. 1. Physical and functional map of the haemolytic plasmid pHly152. The ordering of the *Eco*RI fragments was performed by standard procedures ([30]). The *Eco*RI fragments were cloned using the vector plasmids RSF2124 ([31]) or pACYC184 ([23]) in *E. coli* HB101. The functions of the recombinant plasmids were analysed with regard to autonomous replication, incompatibility, restriction/modification, transfer and haemolysis. The numbers given represent the molecular weights of the *Eco*RI fragments in 10^6 dalton.

Two pools of active α-haemolysin can be detected in logarithmically growing *E. coli* cells harbouring this plasmid. Haemolysin of pool I exists in a free (*i. e.* not cell-bound) extracellular form, whereas that of pool II is found inside the cell, either in the periplasmic space or bound to the cytoplasmic membrane (Springer, unpublished results). We shall refer in the following to α-haemolysin in these two pools as "external haemolysin" or Hly_{ex} and "internal haemolysin" or Hly_{in}.

Mutagenesis with nitrosoguanidine leads to the isolation of plasmid-specific mutants which are unable to secrete α-haemolysin. The pool of Hly_{in}, however, is not affected in these mutants. The mutations are either constitutive or temperature-sensitive (Table I.). The other type of mutants can no longer synthesize α-haemolysin, *i. e.* neither Hly_{in} nor Hly_{ex} can be detected. Again, constitutive as well as temperature-sensitive mutants of this type were isolated (Table I.).

To further characterize the genetic loci underlying these defects, mutations affecting the synthesis or excretion of α-haemolysin were induced by transposition of the ampicillin transposon Tn3 to the haemolytic plasmid pHly152 using a procedure described previously ([22]). Again both types of mutants (Hly^-_{ex}/Hly^-_{in} and Hly^-_{ex}/Hly^+_{in}) were obtained (Table II.). Cleavage of the mutant plasmids with *Eco*RI indicates that the Tn3 insertions in mutants of type I are located on *Eco*RI fragments F or G, whereas those of type II are always located on *Eco*RI fragment G. The *Eco*RI fragments F and G

Mapping of the Tn3-insertions of the hemolysis-negative mutants

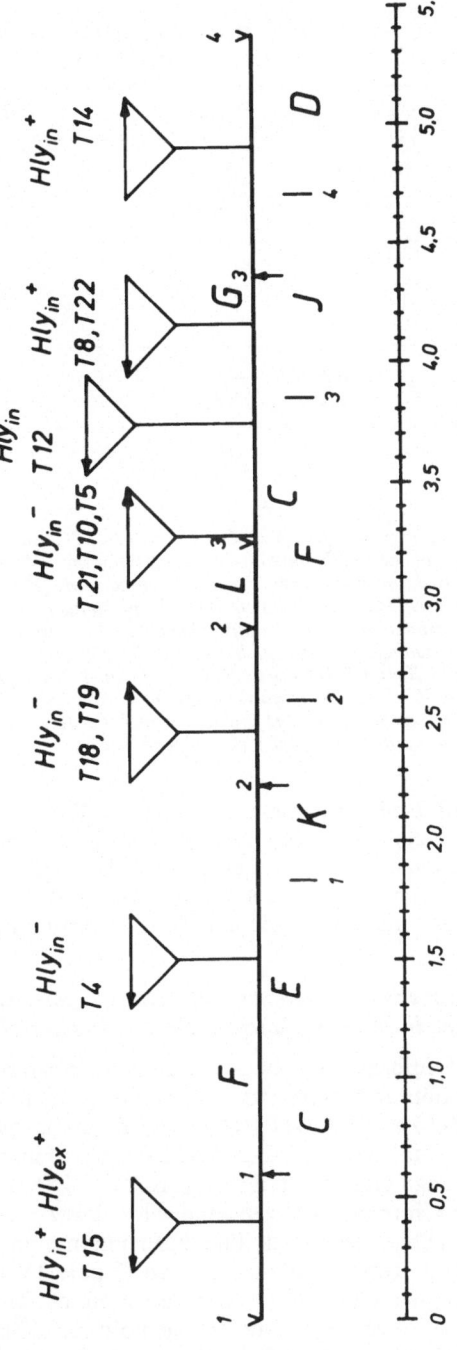

Fig. 2. Mapping of the Tn3-insertions leading to a defect in the synthesis of haemolysin (Hly⁻in) or in the excretion of intracellular haemolysin (Hly⁺in). The determination of the locations of Tn3 within EcoRI fragments E and G and its orientation was performed by isolating the corresponding EcoRI fragment DNA from agarose gels and digesting it with HindIII and PstI. Tn3 is cleaved three times by PstI but not by HindIII. The sites of the Tn3 insertions of mutants T5, T10 and T21 and of mutants T18 and T19 cannot be distinguished by this method. The scale below the map is given in 10⁶ dalton.

181

TABLE I.

Plasmid-specific mutants affected in the synthesis and/or secretion of α-haemolysin

Mutant	Haemolytic activity at 30°C (units*/ml)		Haemolytic activity at 43°C (units*/ml)		Type of mutant
	Hly_{in}	Hly_{ex}	Hly_{in}	Hly_{ex}	
pHly152-365	0.62	—	0.42	—	
-367	0.58	—	0.40	—	$Hly^+{}_{in}/Hly^-{}_{ex}$
-370	0.52	—	0.35	—	
pHly152-363	—	—	—	—	
-366	—	—	—	—	$Hly^-{}_{in}/Hly^-{}_{ex}$
-373	—	—	—	—	
pHly152-30	0.53	0.63	—	—	$Hly^{TS}{}_{in}/Hly^{TS}{}_{ex}$
pHly152-31	0.52	0.25	0.38	—	$Hly^+{}_{in}/Hly^{TS}{}_{ex}$
pHly152 (control)	0.5	1.16	0.41	0.33	

Mutants unable to secrete α-haemolysin (Hly_{ex}) at 30 °C and/or 43 °C were selected after treatment of E. coli K12, pHly152 with N-methyl-N-nitroso-N'-nitroguanidine. Mutants, which were unable to secrete haemolysin, were tested for periplasmic internal haemolysin (Hly_{in}). The plasmid specific character of the mutations was verified by transferring the Hly-plasmids of the mutant strain to a K12 wild-type strain, where they exhibited the same mutant phenotype for haemolysis. Haemolytic activity was measured on the basis of the amount of haemoglobin released from an incubation mixture containing 2% bovine erythrocytes, 10 mM Tris-HC1 buffer pH 7.4, 100 mM NaCl and 10 mM $CaCl_2$. The reaction tubes were incubated at 40 °C for 30 minutes, centrifuged and the amount of haemoglobin was spectrophotometrically determined at 420 nm. One unit of haemolytic activity is defined as the amount that causes the release of 1 μmole haemoglobin under the test conditions.

are separated by the small *Eco*RI fragment L. (Fig. 1). By additional cleavage with *Pst*I of the isolated *Eco*RI fragments containing the Tn3 insertions the precise location of Tn3 within these fragments could be determined. The analyses are summarized in Fig. 2. From these data it is obvious that a rather large stretch of the plasmid DNA comprising at least 3.4×10^6 dalton is necessary for determining the production of extracellular α-haemolysin.

Cloning of α-haemolysin synthesis and excretion functions and complementation or recombination of haemolysis-negative insertion mutants by the recombinant plasmids

To obtain more detailed information about the genetic apparatus and the gene products required for production and excretion of α-haemolysin, the cloned *Eco*RI- and *Hind*III-fragments of pHly152 (Table III.) were tested for haemolytic properties. As expected from the above data, none of the obtained recombinant plasmids carrying single *Eco*RI- or *Hind*III-fragments inserted into pACYC184[23] could express biologically active extracellular or intracellular α-haemolysin. They were, however, able to complement some of the above described Tn3 insertion mutants to full haemolytic activity, *i. e.* production and secretion of haemolysin (Table IV.). Transformation of plasmid pAN250 into mutants defective in the secretion of Hly_{in} always resulted in transformants capable of secreting active α-haemolysin. This indicates that the *Eco*RI fragment G carries gene(s) required for the secretion of α-haemolysin. Plasmid

TABLE II.

Haemolysis-negative mutants obtained by Tn3 insertion

Mutant	Restriction fragments carrying Tn3 insertion in fragments			Property affected by the mutation	
	*Eco*RI	*Pst*I	*Hind*III	Hly$_{in}$	Hly$_{ex}$
pHly 152-T5	G	F	C	—	—
pHly 152-T10	G	F	C	—	—
pHly 152-T12	G	F	C	—	—
pHly 152-T21	G	F	C	—	—
pHly 152-T8	G	J	C	+	—
Hly 152-T22	G	J	C	+	—
pHly 152-T14	G	D	NT	+	—
Hly 152-T4	F	C	E	—	—
Hly 152-T18	F	F	C	—	—
Hly 152-T19	F	F	C	—	—
Hly 152-T15	F	C	NT	+	+

```
 V            F (2.9)                    V  L (0.35)        G (2.15)     EcoRI  V
      ↑       E (1.65)          ↑           C (2.1)       ↑             HindIII
          C (2.7)  |  K (0.57)  |  F (1.4)  |  J (0.84)  |  D (1.55)     PstI
```

Hly$_{in}$ (+) *or* (—) *indicates that mutant produces either active haemolysin, which accumulates in the periplasmic space* (+), *or no active haemolysin* (—). *Hly*$_{ex}$ (—) *indicates that no active haemolysin appears in an extracellular form.*

The numbers given in the physical map of the Hly-region of plasmid pHly152 drawn underneath the table represent molecular weights in $10^6 x$ *dalton.* (NT) = *not tested.*

TABLE III.

Hybrid plasmids containing EcoRI — or HindIII-fragments which cover the Hly-region of pHly152

Hybrid plasmid	Vector plasmid	Fragment inserted	Orientation of the inserted fragment*	Size of the inserted fragment in M dalton
pAN 104	pACYC 184	*Eco*RI-F	A	2.9
pAN 104-1	pACYC 184	*Eco*-RI-F	B	2.9
pAN 250	pACYC 184	*Eco*RI-G	A	2.15
pAN 250-1	pACYC 184	*Eco*RI-G	B	2.15
pAN 202	pACYC 184	*Hind*III-E	A	1.65
pAN 202-1	pACYC.184	*Hind*III-E	B	1.65
pAN 215**	pACYC 184	*Hind*III-C	—	2.1

* A represents the original, B the reverse orientation of the inserted fragment relative to the vector plasmid.
** The reverse orientation could not be obtained.

pAN215 with the *Hind*III fragment C inserted into pACYC184 could not complement either of the "transport" mutants, indicating that the genes required for this process are located on the right half of *Eco*RI fragment G in the map given in Fig 3. As the location of the Tn3 insertions leading to a defect in the secretion of Hly$_{in}$ in the mutants T8, T22 and T14 are 0.7×10^6 dalton apart, it can not be exluded that more than one gene is involved in the secretion of active intracellular α-haemolysin (Hly$_{in}$).

TABLE IV.

Complementation of haemolysis-negative Tn3 insertion mutants by recombinant plasmids

Mutants	Tn3 insertion in	Hly-Phenotype of mutant		Complementation to production of Hly$_{ex}$ by						
		Hly$_{in}$	Hly$_{ex}$	pAN 104	pAN 104-1	pAN 250	pAN 250-1	pAN 202	pAN 202-1	pAN 215
pHly 152-T4	EcoRI F	—	—	+	+	—	—	+	+	—
pHly 152-T18		—	—	—	—	—	—	—	—	—
pHly 152-T19		—	—	—	—	—	—	—	—	—
pHly 152-T5	EcoRI G	—	—	—	—	—	—	—	—	—R
pHly 152-T10		—	—	—	—	—	—	—	—	—R
pHly 152-T21		—	—	—	—	—	—	—	—	—R
pHly 152-T12		—	—	—	—	—	—	—	—	—R
pHly 152-T8	EcoRI G	+	—	—	—	+	—	—	—	—
pHly 152-T22		+	—	—	—	+	—	—	—	—
pHly 152-T14		+	—	—	—	+	—	—	—	—

↓	HindIII-E (pAN202)	↓	HindIII-C (pAN215)	↓
∧	EcoRI-F (pAN104)	∧ ∧ EcoRI-L	EcoRI-G (pAN250)	∧

The description of the recombinant plasmids used in the complementation studies is given in the physical map of the Hly-region drawn underneath the Table and in Table III. Complementation was carried out in C600 recA, r⁻, m⁻. The recombinant plasmids were transformed into C600 recA carrying the indicated Tn3 insertion mutant Hly plasmid. R indicates recombination between the resident recombinant plasmid and the Tn3 mutant Hly plasmid introduced by conjugation, which leads to haemolysis-positive transconjugants in a C600 rec⁺ strain and with decreased frequency in a recA strain.

The hybrid plasmid pAN104 having the *Eco*RI fragment F inserted into pACYC184 could complement the Tn3 insertion mutant T4 which is Hly⁻$_{in}$/Hly⁻$_{ex}$ (Table IV.). The same complementation was obtained when the recombinant plasmid pAN202 with the cloned fragment *Hind*III-E, comprising the middle part of *Eco*RI fragment G is

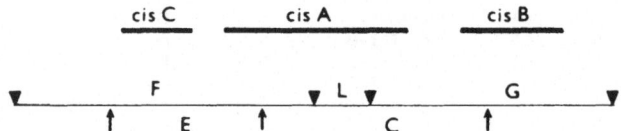

Fig. 3. Preliminary map of the location of the cistrons involved in the synthesis of active α-haemolysin and its excretion. The map is based on the results from the complementation tests performed with the recombinant plasmids and the Tn3 insertion mutants and from the protein studies in minicells carrying these plasmids (see Fig. 4).

transformed into this mutant (Table IV.). The observed haemolytic activity after complementation corresponded to about 25% of the wild type activity.

None of the other mutations caused by Tn3 insertions into the *Eco*RI fragment F, *i. e.* T18 and T19, or the *Eco*RI fragment G. *i. e.* T5, T10, T21 and T12 could be complemented by pAN104, pAN202, pAN215 or pAN250 when the latter recombinant plasmids were transformed into a recA strain harbouring the corresponding insertion mutant plasmid (Table IV.) The Tn3 insertions of these mutants are located on the right half of the EcoRI fragment F and the left half of EcoRI fragment G, respectively (Fig. 2). This suggests that none of the fragments in the recombinant plasmids used for the complementation tests carries the whole information for the gene product(s) which are defective in these Tn3 mutants. However, when the mutant plasmids of T5, T10, T21 or T12 were transferred by conjugation in a rec+ strain already carrying pAN215 as resident plasmid, two types of transconjugants could be selected. Type I (about 30% of the obtained transconjugants) is Hly+in/Hly+ex and carries a recombinant large plasmid whereas type II (about 70% of the transconjugants) is still unable to produce α-haemolysin and contains both plasmids as separate units (Fig. 4). This indicates

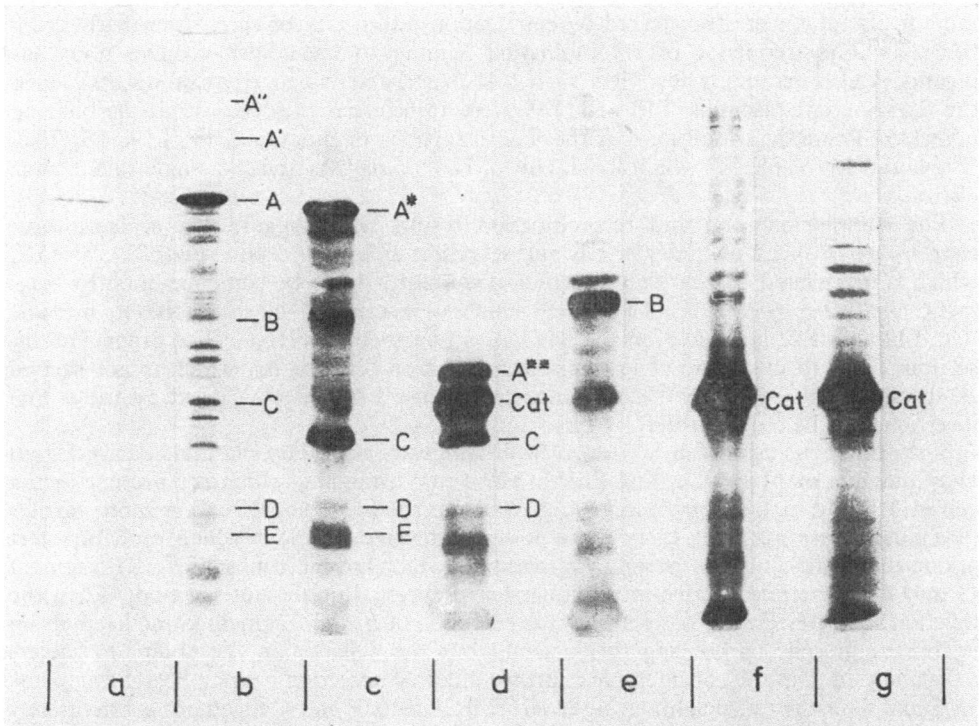

Fig. 4. Gene products identified in minicells of E. coli which are determined by the whole haemolytic plasmid pHly152, and the recombinant plasmids carrying either EcoRI- or HindIII-fragments from the Hly-region of pHly152. Minicells of E. coli P678-54 harbouring the indicated plasmids were isolated as described([32]), preincubated in methionine assay medium (Difco) for 25 min at 37 °C, which was then supplemented with ^{35}S-methionine (40 μCi/ml – spec activity 594.6 Ci/mmol) and further incubated at 37 °C for 30 min. An excess of unlabelled L-methionine was added to the reaction mixture, which was further incubated for 25 min at 37 °C. Cells were lysed as described([32]) and the proteins separated on SDS-containing polyacrylamide gradient gels (10 to 20%). The labelled protein bands were visualized by autoradiography([33]) (a) purified ^{35}S-labelled extracellular haemolysin, proteins determined by (b) pHly 152, (c) pAN104, (d) pAN202, (e) pAN250, (f) pAN215 and (g) vector plasmid pACYC184 as control. (cat) is chloramphenicol acetylase.

that the haemolysis-positive phenotype is caused by recombination between both plasmids rather than by complementation. The frequency of recombination between pAN215 and these Tn3 mutant Hly plasmids (T5, T10, T21 and T12) is only slightly reduced in a recA background suggesting some kind of illegitimate recombination between these plasmids. The recombined large plasmids appear to be present in the cell in a higher copy number than the original Tn3 insertion mutant plasmid. The restriction fragment pattern of the recombined plasmid DNA indicates that the entire recombinant plasmid pAN215 is integrated in the Tn3 mutant plasmids. The integration occurs between sites on the EcoRI fragment L. In addition mating between strains carrying the recombined plasmids and a suitable K12 recipient strain leads to haemolytic transconjugants that are resistant to ampicillin and chloramphenicol. Since chloramphenicol resistance is determined by pAN215 this further indicates recombination of the entire pAN215 plasmid with the pHly152::Tn3 insertion mutants. Only very few Hly$^+$ transconjugants ($<1\%$) carrying recombined plasmids were obtained when these Tn3 mutant plasmids were transferred by conjugation into a rec$^+$ or recA strain harbouring pAN250. The frequency of recombination leading to haemolysis-positive transconjugants is also extremely low (less than 0.1% of the obtained transconjugants) when the Tn3 mutant plasmids T18 and T19 were transferred in a rec$^+$ strain harbouring pAN215. These data indicate that the Tn3 insertions of mutants T18, T19, T5, T10, T21 and T12 (Table II.) which are all Hly$^-_{in}$/Hly$^-_{ex}$ may inactivate a single rather large cistron.

The complementation and recombination results thus suggest that at least three cistrons are involved in the synthesis and secretion of α-haemolysin (Fig. 3). Cistron C which is inactivated by the Tn3 insertion of mutant T4 can be complemented by fragment HindIII-E (pAN202). Cistron B which is inactivated in the secretion mutants T8, T14 and T22 is located on the left half of fragment EcoRI-G. The other Tn3 insertions seem to inactivate at least one other cistron (cistron A), which is not carried as an entire unit by either of the cloned fragments. The defects caused by these Tn3 insertions can be only restored by recombination.

The same conclusion can be drawn from studies of mini-Hly plasmids derived from copy mutants of pHly152::Tn3 (T15). Most copy mutants, which are present in the cell in 3–5 fold higher copy numbers, give rise to the formation of one or more smaller plasmids. These plasmids carry more or less extensive deletions which probably start at one end of the Tn3 transposon. Miniplasmids which have lost most of EcoRI fragment G may still determine active intracellular haemolysin, which is not secreted. When the deletion abolishes EcoRI fragment F, or the left end of EcoRI fragment G, no haemolysin is formed in cells harbouring this plasmid but the cells retain the ability to secrete α-haemolysin. Despite considerable effort we did not succeed in cloning EcoRI fragments F, L and G or the two neighbouring HindIII fragments E and C together on a multicopy vector plasmid. These two combinations should yield haemolytic clones capable of synthesizing either intracellular and extracellular haemolysin (by combination of the three EcoRI fragments F, L and G) or only intracellular haemolysin (by combination of the HindIII fragments E and C) as they comprise the complete DNA stretch of cistrons C and A, both of which are required for active haemolysin as stated above. The reason for this failure is presently unknown but certainly represents more than a technical problem.

Evidence that the Hly cistrons represent a transcriptional unit

The expression of the genetic information in a cloned fragment is dependent on its orientation relative to a promotor of the vector plasmid if no promotor is carried by the cloned fragment. Fragments without promotors cloned on pACYC184 will therefore depend on a suitable promotor of the vector plasmid. The genetic information of such fragments when inserted into the vector plasmid pACYC184 will therefore become expressed only in one orientation, whereas that of fragments carrying promotor will be independent of its orientation within the vector plasmid. The above described fragments were cloned in both orientations except for *Hind*III-C, where only one orientation could be obtained for yet unclear reasons (Noegel, to be published). Both types of hybrid plasmids were tested for the ability to complement the mutations as described above. The results obtained (Table III. and IV.) clearly indicated that only *Eco*RI fragment F and Hind*III* fragment E are independent of the orientation, whereas *Eco*RI fragment G is orientation-dependent. This suggests that probably only one promotor, located on the *Hind*III-fragment E controls the transcription of all genes required for the synthesis of biologically active α-haemolysin and its secretion. As Tn3 insertions located on *Eco*RI-F left of the *Hind*III-site 1 do not affect the α-haemolytic property of this strain (Noegel, to be published), it can be concluded that the stretch of DNA on plasmid pHly152 starting behind the *Hind*III-site 1 and running to *Eco*RI site 4 (Fig. 2) represents a transcriptional unit comprising the three cistrons required for the synthesis and excretion of α-haemolysin.

Identification of gene products involved in the production of biologically active α-haemolysin and its secretion

To obtain information on the nature of the gene products required for the production and secretion of active α-haemolysin, we analysed the proteins expressed in minicells of *E.coli*[24] harbouring the complete haemolysin plasmid pHly152 and the recombinant plasmids carrying the *Eco*RI fragments F and G and the *Hind*III fragments C and E. Fig. 4b shows the complete set of ^{35}S-labeled proteins coded by pHly152. Protein A (mw 58.000 dalton) is presumably α-haemolysin, as demonstrated by comparison with purified active extracellular haemolysin (Fig. 4a) (Springer, Rdest and Goebel, in preparation). The larger proteins A' (mw 75 000 d) and A'' (mw 90,000 d) also carry the α-haemolysin antigen determinant, as demonstrated with specific antibody against Hly$_{ex}$ (data not shown) and may represent precursors of α-haemolysin, which are processed to protein A. Hybrid plasmid pAN104 (Fig 4c) expresses four major proteins, all of which are also determined by the complete haemolysin plasmid pHly152. (1) Protein A$^+$ (mw 56,000 d) reacts with α-haemolysin antibody indicating that it may represent α-haemolysin, which is however smaller by about 2,000–3,000 dalton than protein A. This suggests that *Eco*RI site 2 cleaves off part of the C-terminal sequence of the haemolysin protein. (2) Protein C (mw 18,000 d), which is also expressed by pAN202 (*Hind*III fragment E inserted into pACYC184), is required for the activity of α-haemolysin as it is missing in the Tn3 mutant T4 and can be complemented by the recombinant plasmids pAN104 and pAN202. (3) Protein D is expressed by minicells carrying pAN104 or pAN202, indicating that it may also be involved in a step leading to active haemolysin. (4) Protein E (mw 10,000 d), which is expressed by pAN104 and the complete hemolytic plasmid pHly152 but not by the hybrid plasmid pAN202 (Fig. 4d), is determined by a gene located left of the *Hind*III site 1 and is therefore not involved in the synthesis of active haemolysin. The recombinant plasmid pAN202 expresses, in addition to proteins C and D, another protein, which bands above *cat* (Fig. 4e). This protein seems to represent a N-terminal fragment of haemolysin (protein A) (data not shown) which indicates that the start of the gene for haemolysin is located the *Hind*III fragment E. The recombinant plasmid pAN215, carrying the *Hind*III fragment C inserted into pACYC184, does not seem to express specific proteins besides those determined by the

187

vector (Fig. 4 f and g), which further suggests that this fragment carries the C-terminal part of the haemolysin gene (cistron A) but no additional cistron. The recombinant plasmid pAN250 (carrying the EcoRI fragment G) expresses one major protein in minicells (Fig. 4e), which is termed protein B (mw 33,000 d). It is determined by pAN250 but not by pAN215, indicating that the cistron for this product is located on the right half of EcoRI fragment G (Fig. 2). This protein seems to be absent or altered in mutants which are unable to secrete haemolysin (Hly^+_{in}/Hly^-ex).

Protein B and protein C were found in the membrane fraction, when the minicell preparations harbouring the recombinant plasmids pAN104 and pAN250, respectively, were separated into cytoplasmic and membrane fractions. Protein A of pHly152 and A^+ of pAN104 are distributed among both fractions, suggesting that these proteins which represent α-haemolysin and a derivate of it are either located in part in the cytoplasm or are loosely bound to the cytoplasmic or the outer membrane.

DISCUSSION

The data described indicate that haemolysis exhibited by some *E. coli* strains is a rather complex process requiring at least three clustered cistrons which comprise about 3.4×10^6 dalton of a transmissible plasmid in the *E. coli* strain PM152. Similar to the plasmid-determined heat stable (ST) and heat labile (LT) enterotoxins of *E. coli*[25-27], α-haemolysin represents an exotoxin. The genetic determinants for ST and LT which were recently isolated by the gene cloning technique, seem however to be considerably smaller[28, 29]. There is no evidence for a specific excretion function for either ST or LT enterotoxins, although more than one polypeptide seems to be coded for by the LT determinant.

The production of active α-haemolysin requires at least two cistrons (A and C). One of them (cistron A) determines a gene product which appears to be the precursor of the protein identified as extracellular α-haemolysin. The gene product of the other cistron C is required for the production of active haemolysin and may be involved in the conversion of an inactive haemolysin into its active form. The gene product of cistron C may function as a processing protease converting the precursor protein into the active form. The active intracellular haemolysin appears to accumulate first in the periplasmic space (Springer, unpublished). Its secretion through the outer membrane is controlled by at least one cistron (cistron B). It seems to determine a protein which is located in the outer membrane of haemolytic *E. coli* cells.

REFERENCES

1. WISEMAN, G. M. and CAIRD, J. D. (1967): Can. J. Microbiol. **13**, 369.
2. HAUSCHILD, A. H., LECROISEY, A. and ALOUF, J. E., (1973): Can. J. Microbiol. **19**, 881.
3. LOCHMANN, O., VÝMOLA, F., and CHALOUPECKÝ, V. (1975): J. Hyg. Epidemiol. Microbiol. Immunol. (Praha) **19**, 61.
4. SAKURAI, J., MATSUZAKI, A., TAKEDA, Y., and MIWATANI, Z. (1974): Infect. Immun. **9**, 777.
5. FUJITA, M., KOSHIMURA, S., J. Exp. Med. (JPN) (1975): **45**, 457.
6. SEEGER, R., BURKHARDT, M., HAUPT, M. and FEULNER, L. (1976): Arch. Pharmacol. **293**, 163.
7. FREER, J. H., ARBUTHNOTT, J. P. and BERNHEIMER, A. W. (1968): J. Bact. **95**, 1153.
8. RENNIE, R. P., FREER, J. H. and ARBUTHNOTT, J. P. (1974): J. Med. Microbiol. **7**, 189.
9. FACKRELL, H. B. and WISEMAN, G. M., J. Gen. Microbiol. (1976): **92**, 1.

10. BERNHEIMER, A. W. (1974): Biochim. Biophys. Acta **344**, 27.
11. RATINER, I. U. A., KANARE'IKINA, S. K., BONDARENKO, V. M. and GOLUBERA, I. V. (1976): ZH Mikrobiol. Epidemiol. Immunobiol., 117–21.
12. SMITH, H. W. (1969): In Bacterial Episomes and Plasmids, 213–226, J. and A. Churchill Ltd. London.
13. LOWELL, R. and REES, T. A. (1960): Nature **188**, 755.
14. SNYDER, J. S. and KOCH, N. A. (1966): J. Bact. **91**, 763.
15. WALTON, J. B. and SMITH, D. H. (1969): J. Bact. **98**, 304.
16. SMITH, H. W. and HALLS, S., J. Gen. Microbiol. **47**, 153 (1967).
17. GOEBEL, W., ROYER-POKORA, B., LINDENMAIER, W. and BUJARD, H. (1974): J. Bact. **18**, 964.
18. ROYER-POKORA, B. and GOEBEL, W. (1976): Molec. gen. Genet. **144**, 177
19. MONTI-BRAGADIN, L., SAMER, L., ROTTINI, G. D. and PANI, B. (1975): J. Gen. Microbiol. **86**, 367.
20. LE MINOR, S. and LE COUEFFIC, E. (1975): Ann. Microbiol. (Paris) **126**, 313.
21. SHORT, E. C. J. R. and KURTZ, H. J. (1971): Infect. Immun. **3**, 678.
22. GOEBEL, W., LINDENMAIER, W., PFEIFER, F., SCHREMPF, H. and SCHELLE, B. (1978): Molec. gen. Genet. **157**, 119.
23. CHANG, A. C. Y. and COHEN, S. N. (1978): J. Bact. **134**, 1141.
24. ADLER, H. J., FISCHER, W. D., COHEN, A. and HARDIGREE, A. A. (1967): Proc. natn. Acad. Sci. **57**, 321.
25. SMITH, H. W., and HALLS, S. (1968): J. Gen. Microbiol. **52**, 319.
26. GYLES, C. L., So, M. and FALKOW, S. (1974): J. Infect. Dis. **130**, 40.
27. GYLES, C. L., PALCHAUDHURI, S. and MAAS, W. K. (1977): Science **198**, 198.
28. So, M., BOYER, H. W., BETLACH, M. and FALKOW, S. (1976): J. Bact. **128**, 463.
29. DALLAS, W. S., DOUGAN, D. and FALKOW, S. (1978): Genetic Engineering, H. W. Boyerand S. Nicosia, eds., Elsevier/North Holland, Biomedical Press.
30. BLOHM, D. and GOEBEL, W. (1978): Molec. gen. Genet. **167**, 119.
31. So, M., GILL, R. and FALKOW, S. (1975): Molec. gen. Genet. **142**, 239.
32. LEVY, S. B. (1974): J. Bact. **120**, 1451.
33. BONNER, W. M. and LASKEY, R. A. (1974): Eur. J. Biochem. **46**, 83.

W. G., Inst. für Genetik und Mikrobiologie,
Röntgenring 11,
D-8700 Würzburg, B. R. D.

MOLECULAR FINE STRUCTURE OF ENTEROTOXIN PLASMID OBTAINED FROM *ESCHERICHIA COLI* AND INSERTED WITH DRUG-RESISTANCE TRANSPOSONS

T. YOKOTA, T. YAMAMOTO and R. SEKIGUCHI

Department of Bacteriology, School of Medicine, Juntendo University Tokyo, Japan

INTRODUCTION

Two types of enterotoxins, one heat-labile (LT) and the other heat-stable (ST) have been identified in *Escherichia coli* strains isolated from patients with acute diarrhea (Smith, 1968). The genes encoding the production of toxins are located on transferable plasmids called ENT plasmids (Gyles, So, and Falkow, 1974).

On the other hand, genes on R plasmids specifying antibiotic resistance in bacteria are often found as components of transposable genetic elements named by transposons (Hedges and Jacob, 1974). Since clinical isolates of gram-negative bacteria often carry more than two kinds of transferable or non-transferable plasmids, it is likely that such transposons would lead to the formation of new plasmids bearing the both genes for enterotoxin-production and drug-resistance, which may become an important biohazard for public health. In fact, several papers appeared on naturally occurring recombinant plasmids that carry both enterotoxin-producing and drug-resistance genes (Gyles, Palchaudhuri, and Mass, 1977), on the frequency of association between R and ENT plasmids in clinical isolates of *E. coli* in south-east Asian countries (Echeveria, Verheart, Uylango, Komalarini, Ho, Ørskov, and Ørskov, 1977) and on the method for genetic labelling of ENT plasmids with drug-resistance transposons (So, Heffron, and Falkow, 1978). The authors propose the name ENT-R for such plasmids.

ENT PLASMIDS LABELLED EXPERIMENTALLY WITH DRUG-RESISTANCE TRANSPOSONS

Throughout the experiments, the production of LT and ST in bacteria was assayed by the tissue culture method with CHO-K1 cells (Nozawa, Yokota, and Kuwahara, 1978) and intrastomachic injection method with suckling mice, respectively.

Two different methods were employed for labelling ENT plasmids with drug-resistance transposons. One, an R plasmid, RP4, *amp, tet, kan*, was transferred to an enterotoxigenic *E. coli*, H10407 producing both LT and ST, and the ampicillin(ABPC)-resistance of transconjugants was further transferred to the plasmid-free strain of *E. coli* 20SO. The subclones of second transconjugants resistant to only ABPC but not to tetracycline (TC) and kanamycin (KM) were selected, and their enterotoxigenicities were assayed. The second, a plasmid sensitive to high temperature for the replication, R402, *amp, str, spc*, was transferred to *E. coli* TD218C1 producing only LT, and the transconjugants were successively subcultured in L broth containing 100 μg/ml of ABPC at 42°C for 10 days. The ABCP-resistance of transconjugants was transferred to *E. coli* 20SO at 42°C and the second transconjugats were confirmed to be sensitive to streptomycin (SM) and spectinomycin (SPC) and enterotoxigenic. Retransfer of ENT-R plasmids thus obtained was performed with *E. coli* CSH2 and cyclic AMP-less mutants of *E. coli*, YE100 and CA7902, as the recipients.

Although the cotransfer of enterotoxigenicity and ABPC-resistance in the ENT-R plasmid obtained from *E. coli* TD218C1 by the second method was always 100 per cent (pJY13), that of ENT-R obtained from *E. coli* H10407 by the first method was found to be unstable and the enterotoxigenicity was easily lost by conjugative transfer and transformation. It is, however, unlikely that the ABPC-resistance is harbored apart from the ENT plasmid in the latter case, since other R plasmids belonging to the P group to which the source of ABPC-transposon also belongs are compatible with the ENT-R plasmid.

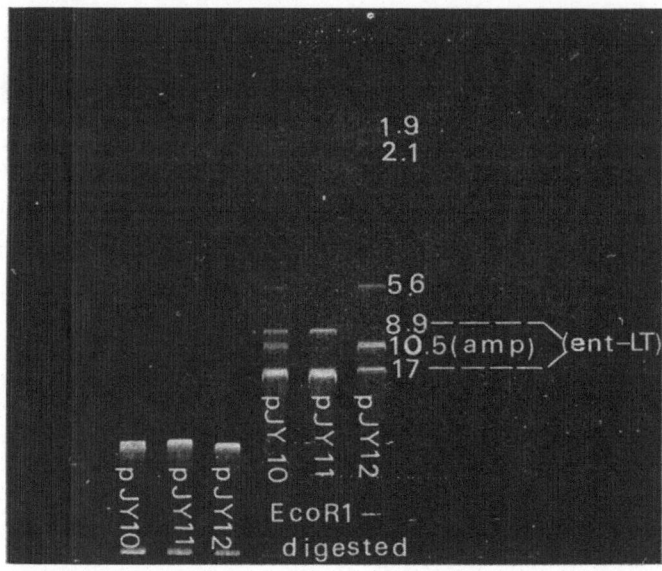

Fig. 1. Electrophoretic patterns of cccDNAs of pJY10(*ent-amp*), pJY11(*ent*) and pJY12(*amp*) treated and not treated with *EcoR1* restriction enzyme. Numbers indicate approximate MW of each fragment.

Furthermore, it was confirmed that *E. coli* 20SO bearing the latter ENT-R plasmid (pJY10) or its segregants from which either enterotoxigenicity or ABPC-resistance was eliminated spontaneously (pJY12 and pJY11) harbored only one size each of covalently closed circular (ccc) DNA with molecular weights (MW) of a little larger than 40 megadaltons, as shown in Fig. 1. No cccDNAs with MW 36 or less corresponding to RP4 and its segregants were found in any host bacteria. The figure indicates also the electrophoretic patterns of DNA fragments of pJY10, pJY11 and pJY12 digested with the *EcoR1* restriction enzyme, showing that pJY10 (ENT-*amp*) consists of 9 *EcoR1* fragments with MW 17, 10.5, 8.9, 5.6, 2.1, 1.9, 0.5, 0.45, and 0.4 megadaltons. One segregant plasmid missing the *amp* gene (pJY11) lacks 6 fragments, 2nd, 4th, 5th, 6th, 8th, and 9th, although a new fragment with MW 18 megadaltons appears. The other segregant missing the *ent* gene (pJY12) is deficient in only the 3rd fragment but the 2nd fragments is in double. Since it has been well known that the RP4 plasmid possesses only one *EcoR1* site (DePicker, Montagu, and Schell, 1977), the result indicates again the *amp* gene translocated from RP4 to other plasmids (pressumably ENT).

Alkaline sucrose gradient analysis of cccDNAs of *E. coli* 20SO carrying pJY10, pJY11 or PJY12 showed also only one peak indicating that the bacteria harbor one size of plasmid.

Molecular weights of pJY10, pJY11 and pJY12 calculated from the sums of *Eco*R1 fragments and by the alkaline sucrose gradient analysis are 47, 44 and 49 megadaltons, respectively, agreeing with those measured from electron micrograms.

CLEAVAGE MAPPING AND CLONING OF *AMP*, *ENT(LT)* AND *ENT(ST)* GENES OF THE UNSTABLE ENT-R PLASMID

The cloning of *amp*, *ent*(LT) and *ent*(ST) genes in pJY10 was attempted to clarify the locations on the plasmid DNA. For the first step of cloning, pMK1 that is ColE1 harboring a KM-transposon, was employed as the vector. The DNAs of vector and donor

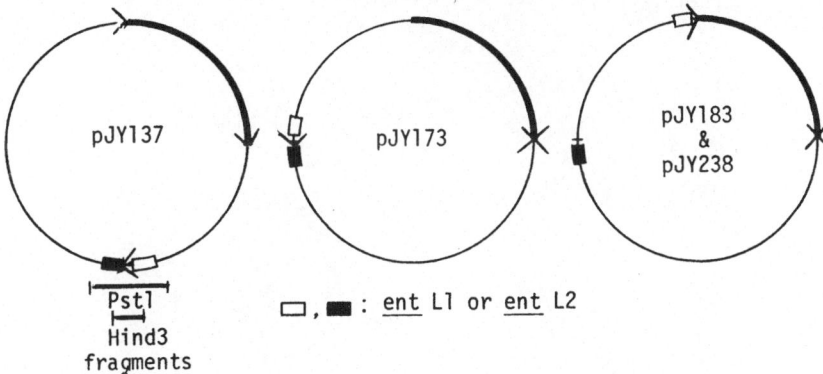

Fig. 2. Representative illustration of orientations of *EcoR1* fragments in the chimeric DNAs between pJY10(*ent-amp*) and pMK1(*ColE1-Tn, kan*) and location of the *ent(LT)* genes.

(pJY10) were digested with *Eco*R1 enzyme and ligated with *T4*-ligase after mixing. The transformatts of recombinant DNAs were selected by KM-resistance, and the enterotoxigenicity or ABPC-resistance was assayed by the CHO-K1 cell culture method or replica plating. The cccDNAs of enterotoxigenic or ABPC-resistant transformants were digested with *Eco*R1 enzyme and compared with that of pJY10 by agarose slab gel electrophoresis. It was confirmed that the *amp* gene located in the 2nd *Rco*R1 fragment with MW 10.5 megadaltons, whereas, the *ent*(LT) gene was found to be extending over the first and third fragments with MW 17 and 8.9 megadaltons (Fig. 1). The chimeric DNAs thus obtained showed two interesting characteristics. First, one of four recombinant plasmids, pJY137, confers higher and the other two, pJY183 and pJY238, confer lower LT-productions on their host bacteria than the parent ENT plasmid does. The remaining recombinant plasmid, pJ1173, was confirmed to confer the same level of LT-production as the parent ENT plasmid. The reason why chimeric plasmids confer different levels of LT-production was analyzed by cleavage mapping with various restriction enzymes such as *Hind*3, *Pst*1, *Bam*H1, *Sma*1 and *Sal*1, resulting in that the orientations of two *Eco*R1 fragments against the vector DNA are different each others. As shown in Fig. 2, if one can assume the orientations of *Eco*R1-fragments of pJY137, the highest LT-producer, as clockwise against that of vector DNA, those of pJY173, the moderate LT-producer, are counterclockwise, and the orientations of *Eco*R1-fragments in pJY183 and pJY238, the lowest LT-producers, are inverted, i.e. one fragment is clockwise and the other is counterclockwise. The results may suggest that the genetic determinant for the LT-toxin on the ENT-R plasmid are multigenes that can be separated as the *Eco*R1-sensitive site. The second interesting characteristic of first the chimeric plasmid is its DNA structure. The electronmicrograms of pJY137 and pJY173 DNAs revealed three inverted repeat structures as shown in Fig. 3, indicating the KM-transposon of vector DNA with rather large

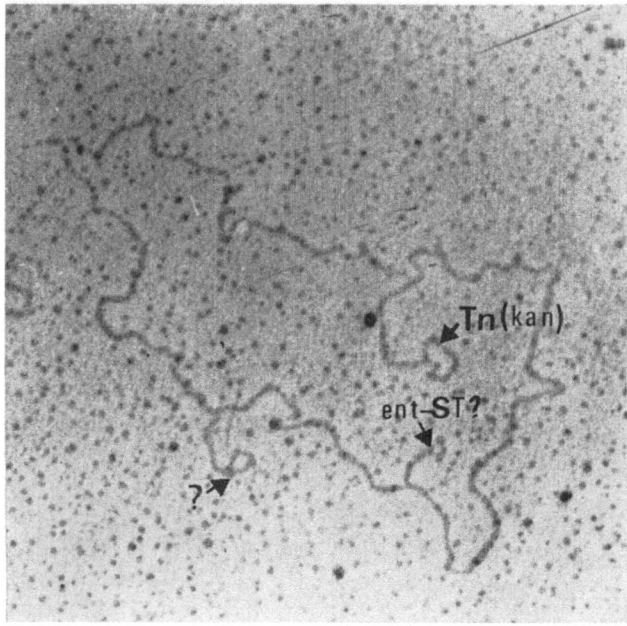

Fig. 3. Electronmicrogram of self-annealed pJY137 DNA possessing 3 inverted repeat structures.

head loop and two others that are assumed to be closely related to the *ent*(LT) and *ent*(ST) genes by cleavage mapping. Although it has been confirmed that the *ent*(ST) gene in another ENT plasmid is a transposon flanked by inverted repeats of *IS*1 (So, Heffron, and McCarthy, 1979), it is not clear yet whether or not the *ent*(LT) and *ent*(ST) genes in this ENT-R plasmid are transposons.

Since the *Pst*1- and *Hind*3-sensitive sites of pJY137 and pJY173 were found to locate on either side of *Eco*R1-sensitive site in the *ent*(LT) genes, the central 4.5 kilobase *Pst*1 fragment was cloned in the *Pst*1 site of pBR322. The single *Pst*1 site within pBR322 occurs in the ABPC-resistance gene. Transformants were selected by culture plates containing 10 μg/ml of TC and enterotoxigenicity of ABPC-sensitive transformants was checked with CHO-K1 cells. The secondary chimeric plasmids thus obtained, pJY1212 and pJY2299, were found to consist of only vector DNA and the 4.5 kilobase insert encoding the LT (Fig. 4). Amount of LT produced by *E. coli* 20SO carrying pJY1212 or PJY2299 was as high as that specified by pJY137, the highest LT-producer in the first chimeric plasmids. It is assumed that higher LT-production encoded by pJY137, pJY1212 and pJY2299 is based upon the larger copy number regulated by ColE1 replicator gene.

194

The cloning of ent(ST) gene was also performed by the same manner. The *Pst*1 fragments of pJY137 were inserted into the *Pst*1 site of pBR322, and the ST-production of TC-resistant but ABPC-sensitive transformants was checked. Although digestion with the *Pst*1 enzyme of pJY007 and pJY255 thus obtained resulted in 3 fragments other than vector DNA, the two of 1.5 and 3.0 kirobase fragments were found to be common

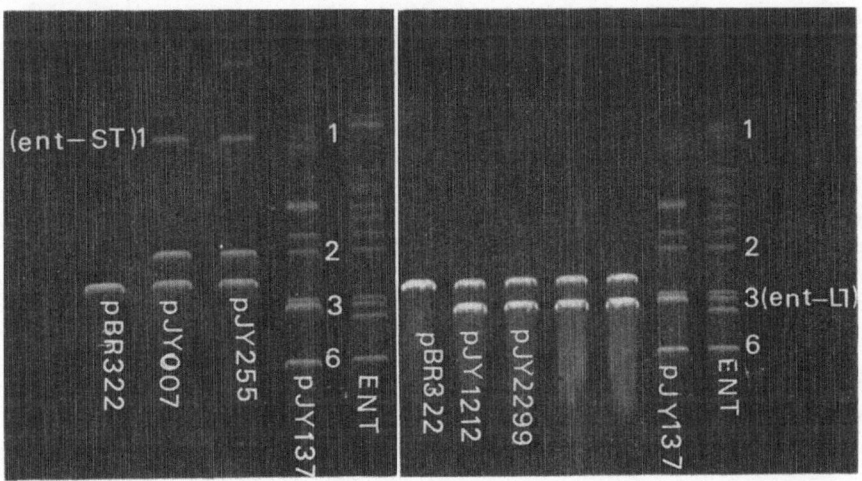

Fig. 4. Electrophoretic patterns of *Pst1*-digested DNAs of the parent ENT plasmid, pJY137 [first cloning of *ent(LT)* and *ent(ST)* genes], yJY1212 and pJY2299 [final cloning of *ent(LT)* genes], and of pJY255 and pJY007 [intermediate cloning of the *ent(ST)* gene].
(FOOT NOTE Numbers on the figure indicate approximate MW of each fragment.

in the plasmids. The DNAs of common fragments were extracted from the agarose gel and cloned again with the same vector. All the transformants by the new recombinant DNA possessing the 1.5 kilobase fragment were toxigenic, whereas none of the transformants by the DNA containing the 3.0 kilobase fragment produced ST. The result indicates that the ent(ST) gene locates in the 1.5 kilobase *Pst*1 fragment of the ENT-R plasmid.

CYCLIC ADENYLIC ACID-INDEPENDENT IN VIVO AND IN VITRO PRODUCTION OF LT IN *ESCHERICHIA COLI*

Requirement of adenosine 3'.5'-cyclic monophospahte (cyclic AMP) for the production of choleratoxin was reported by the authors (Yokota and Kuwahara, 1974). Since LT of *E. coli* has been known to be related to choleratoxin enzymologically and immunologically, the role of cyclic AMP in the LT-production was studied. An ENT-R plasmid, pJY10, was conjugatively transferred to adenylate cyclase-less mutants, YE100 and CA7902, of *E. coli*, and amounts of LT produced in the presence and absence of exogenous cyclic AMP were measured. Contrary to the case of *Vibrio cholerae*, LT of *E. coli* was produced regardless of the presence of cyclic AMP. The chemical rather suppressed the LT-production.

In vitro synthesis of LT was also attempted by the modified Zubay's method with purified DNAs of ENT plasmid and cloned ent(LT) gene as the template. The protein fractions incorporated with ^{14}C-leucine were identified by the fluorographic method of acrylamide slab gel electrophoresis (Bonner and Laskey, 1974) of reaction mixtures. One protein fraction with molecular weight of about 40,000 that is produced by pJY10 (ENT-*amp*) DNA but not by pJY12 missing the *ent* gene was found to be successively concentrated by stepwise cloning of *ent*(LT) gene described previously. The nature of fraction, however, has not been characterized yet.

DISCUSSION

The authors presented here ENT-R plasmids stable or unstable genetically. Although the reason why some ENT-R plasmids are unstable has not been clarified, a hypothesis may be postulated suggesting that ENT-R plasmids carrying multitransposons are unstable by changing easily the arrangement of cccDNA. In fact, pJY10 carries at least 3 inverted repeats, i.e. the ABPC-transposon (*Tn*1) from RP4 and two others assumed to be related with the *ent*(LT) and *ent*(ST) genes, whereas a stable ENT-R plasmid obtained from *E. coli* TD218C1 produces only LT. To confirm the hypothesis, it is nescessary to determine whether or not the *ent*(LT) and *ent*(ST) genes in pJY10, an unstable ENT-R, are transposons. Since an *ent*(ST) gene has been already confirmed to be a transposon (So, Heffron, and AcCarthy, 1979), it is likely that the *ent*(ST) gene of pJY10 is also a transposon. Papers showing that *ent*(LT) genes are transposons have not appeared yet. Details of transposons in unstable ENT-R plasmids are now under investigation.

REFERENCES

BONNER, W. M. and R. A. LASKEY (1974): A film detection method for tritium-labelled proteins and nucleic acid in polyacrylamide gel. Eur. J. Biochem. 46, 83–88.
DEPICKER, A., M. V. MONTAGU, and J. SHELL (1977): Physical map of RP4. *In* DNA Insertion, Elements, Plasmids and Episomes, Bukhari, A. I., J. A. Shapiro, and S. L. Adhya, *Eds.* p678, Cold Spring Harbor Laboratory, New York, N. Y.
ECHEVERRIA, P., L. VERHEART, C. V. UYLANGO, S. KOMALARINI, M. T. HO, S. ØRSKOV, and I. ØRSKOV (1977): Plasmid-mediated antimicrobial resistance and enterotoxin production among isolates of enterotoxigenic *Escherichia coli* in the Far East. Proc. 13th Joint Conf. Cholera, 81–102, US Dept. Hlth., Educ., Welf., Washington DC.
GYLES, C. L., M. SO, and S. FALKOW (1974): The enterotoxin plasmids of *Escherichia coli*. J. Infect. Dis. 130, 40–49.
GYLES, C. L., S. PALCHAUDHURI, and W. K. MAAS (1977): Naturally occurring plasmid carrying genes for enterotoxin production and drug resistance. Science, 198, 198–199.
HEDGES, R. W. and A. E. JACOB (1974): Transposition of ampicillin resistance from RP4 to other replicons. Molc. Gen. Genet. 132, 31–40.
NOZAWA, R. T., T. YOKOTA, and S. KUWAHARA (1978): Assay method for *Vibrio cholerae* and *Escherichia coli* enterotoxins by automated counting of floating Chinese hamster ovary cells in culture medium. J. Clin. Microbiol. 7, 479–485.
SMITH, H. W. and S. HALLS (1968): The transmissible nature of the genetic factor in *Escherichia coli* that controls enterotoxin production. J. Gen. Microbiol. 32, 319–334.
SO, M., F. HEFFRON and S. FALKOW (1978): Method for the genetic labelling of cryptic plasmids. J. Bacteriol. 133, 1520–1523.
SO, M., F. HEFFRON, and B. J. MACCARTHY (1979): The *E. coli* gene encoding heat stable toxin is a bacterial transposon flanked by inverted repeats of *ISl*. Nature, 277, 453–456.
YOKOTA, T. and S. KUWAHARA (1974): Adenosine 3′, 5′-cyclic monophosphate deficient mutants of *vibrio cholerae*. J. Bacteriol. 120, 106–113.

T. Y., Juntendo University,
2-1-1 Hongo, Bunkyoku,
Tokyo, Japan

CONDITIONS FOR THE EXISTENCE OF R-PLASMIDS
IN BACTERIAL POPULATIONS

B. R. LEVIN

Department of Zoology, University of Massachusetts, Amherst
Massachusetts, U.S.A.

INTRODUCTION

It is generally assumed that a relaxation of the selection pressure that will result from the more prudent use of antibiotics will be followed by a concomitant decline in the frequency of bacteria with chromosomal or plasmid determined antibiotic resistance. This optimistic view is consistant with traditional wisdom and observations about regressive evolution (Dobzhansky, 1970). Characters that are not directly favored by selection are eventually lost. The observed changes in the bacterial flora following the termination of antibiotic treatment supports the proposition that a relaxation of the intensity of antibiotic selection will result in a decline in the frequency of antibiotic resistant bacteria (see the review by, Richmond, 1974). However, at this juncture it is not at all clear whether these observed replacements of antibiotic resistant organisms by sensitive ones are due to selection against the resistant cell types or due to the replacement of these antibiotic resistant clones by sensitive ones for other reasons. In fact, there are a variety of observations that can be interpreted to suggest that in cases of relaxed and/or limited antibiotic selection the rate of loss of resistant organisms is low and that antibiotic resistant bacteria could maintain substantial frequencies: 1) cell carrying R-factors represent a significant portion of the coliform bacteria isolated from domestic sewage (Linton et al., 1974), 2) R-factors were sufficiently common prior to the onset of this antibiotic era as to be detected in *E. coli* culures lyophilized during the period, (Smith, 1967), and 3) bacteria resistant to antibiotics produced by soil dwelling organisms are sufficiently common to be detected in the normal enteric flora of mammals from "pre-antibiotic" environments.

In part to examine the above problem of the fate of antibiotic resistant bacteria in situations of relaxed antibiotic selection, we have developed mathematical models of the population biology of plasmids and have been studying R-factors in experimental populations of *E. coli*. In the following, I summarize the results of these *a priori* studies of the existence conditions for plasmids and briefly discuss their implications to the prognosis for the maintenance of plasmid mediated resistance in the absence of antibiotic selection.

PARAMETERS OF CONCERN

The fate of a conjugationally transmitted plasmid in a bacterial population would depend on 1) the relative growth and survival rates of bacteria carrying that plasmid and bacteria free of that factor (their relative fitness), 2) the rate at which the plasmid is lost by vegetative segregation, and 3) the rate at which the plasmid is infectiously transmitted by conjugation. It seems reasonable to assume that the magnitude of the former two parameters would, in general, be independent of population density. On the other hand, the rate of infectious transmission of plasmids would depend on the absolute density of donor and recipient cells.

The relative roles of these parameters in the determination of the fate of a conjugative plasmid are illustrated in Figure 1. The trajectories in this figure were generated by numerical solutions to the differential equations used in our model of the population biology of conjugative plasmids in bacteria maintained in an "equable" (continuous flow) habitat (Stewart and Levin, 1977). When the relative growth and survival rates of plasmid-

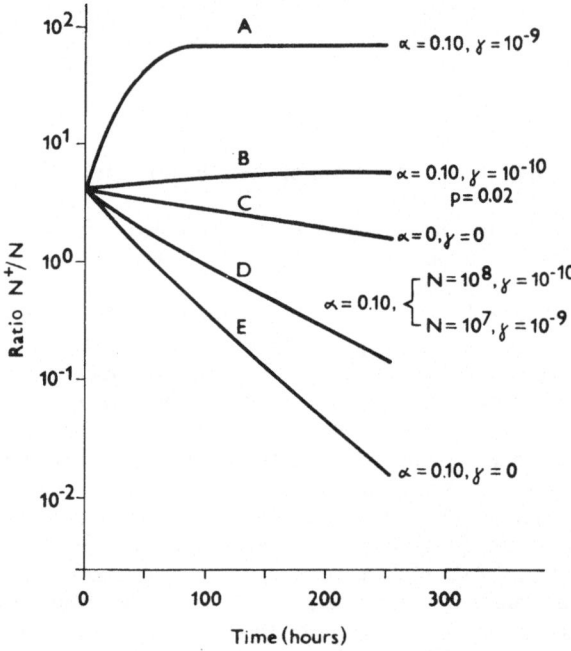

Fig. 1. Ratio of plasmid-bearing to plasmid-free cells as a function of time for different sets of parameter values. Numerical solution to the differential equations of the "equable" habitat model of the population biology of conjugative plasmids in Stewart and Levin (1977). In all runs the segregation rate, $\tau = 10^{-3}$ per cell per hour. Unless otherwise stated, the dilution rate $\rho = 0.20$ hr^{-1} and the equilibrium density of recipients in a plasmid-free culture $N = 10^8$ cells per ml. All runs were intiated with the total number of bacteria near the density (chemostat) equilibrium.

-bearing and plasmid-free cells are equal, (the selection coefficient, $\alpha = 0$), then in the absence of plasmid transfer (the transfer rate constant, $\gamma = 0$ ml per cell per hour) and with a finite rate of segregation (the segregation rate $\tau = 10^{-3}$ per cell per hour) the frequency of cells carrying the plasmid would continually decline (line C). When the carriage of the plasmid confers a 10% growth disadvantage on the cells carrying it ($\alpha = 0.10$), the rate of decline in the frequency of plasmid-bearing cells is considerably augmented (line E). With some plasmid transfer ($\alpha > 0$) the rate of decline in the ratio of plasmid-bearing cells due to segregation and negative selection is retarded (line D). As can be seen from the parameter values specified for the latter line, the rate of plasmid transmission is proportional to the product of the population density, N, and the transfer rate parameter, γ. With sufficiently high rates of transfer the effects of segregation and negative selection can be overcome and a stable equilibrium, with both plasmid-

-bearing and plasmid-free cells present, obtains when the rate of plasmid loss due to selection and segregation is equal to its rate of infectious transmission (line A). In this continuous flow model the response to selection is dependent upon the rate of flow through the habitat (the dilution rate ρ per hour) and as a result the fate of the plasmid would depend on the turnover rate.

From the formal, mathematical, analysis of the properties of the model considered above we can conclude that the plasmid could become established and would be maintained in previously plasmid-free population of density N (cells per ml) as long as the following inequality is met,

$$\gamma > \frac{\alpha\rho + \tau}{N} \tag{1}$$

where the parameters α, τ, ρ, and γ, are defined above (Stewart and Levin, 1977). In that theoretical investigation we also found a situation where conjugative plasmids could become established and would be maintained by infectious transfer overcoming selection and segregation in seasonal (serial transfer) types of habitats where the population is maintained by sequential transfers to fresh environments. More recently we have developed a model for the population biology of nonconjugative plasmids that can be mobilized by conjugative factors (Levin and Stewart, 1979). Our analysis of the properties of this model suggest that, as a consequence of infectious transmission via mobilization, there are situations where a nonconjugative plasmid could become established and would be maintained in a bacterial population even when the cells carrying that factor are not the most fit members of that host clone. However, the conditions under which this would obtain without some selection favoring the nonconjugative element are highly restrictive and therefore not anticipated to occur.

PARAMETER VALUES IN POPULATIONS OF E. COLI

1) *Selection coefficient*: Due to the cost of DNA and protein synthesis, it seems intuitively reasonable to expect that, in the absence of selection favoring the carriage of plasmid borne genes, bacteria carrying those factors would have a selective disadvantage relative to plasmid-free members of the same host clone. In fact, the results of a feeding experiment with R-plasmid-bearing and R-plasmid-free bacteria are consistent with hypothesis that in the absence of the appropriate antibiotic (s) cells carrying these factors are at a selective disadvantage (Anderson, 1974). In an effort to obtain some idea of the magnitude of this "cost" of carrying "unselected" plasmids, parameter α, we have been performing competition experiments with strains of *E. coli* K-12 carrying the plasmids R1 (Cm-Km-Ap) and R1-drd-19 (Km-Cm-Ap) in glucose limited minimal medium chemostats maintained at 37°C.

For these experiments the plasmids were transferred from their original, J53 host to CSH 7 (*lac*Y str) and a *lac*+ spontaneous revertant of the latter clone. Stationary phase cultures of plasmid-bearing and plasmid-free cells of different lactose phenotypes were mixed in approximately equal amounts in chemostats of the type described in the appendix to Chao *et al.* (1977). At daily intervals samples were taken and the relative numbers of *lac*+ and *lac*− phenotypes estimated by plating on a tetrazolium lactose indicator medium. The numbers of transconjugants produced were monitored by plating appropriate dilutions on selective media. At periodic intervals donor and transconjugant colonies from the R1-drd-19 experiments were tested with the male specific phage MS2 to ascertain whether these permanently derepressed plasmids were replaced by ones coding for repressible conjugative pili synthesis (Meynell, 1973).

The results of a set of competion-selection experiments of this type are presented in Figure 2. During the period depicted, the frequency of transconjugants remained less than 10^{-2} that of the clone with the same lactose phenotype and therefore did not contribute to the changes in the lac^+/lac^- ratio. As can be seen in these figures, when the plasmids are carried by the lac^+ cells, those cells have a selective disadvantage and when the plasmid is carried by the lac^- cells, the competing plasmid-free cell types have a selective advantage. The apparent change in slope of the regression line in the R1-drd-19 experiment with a lac^- plasmid bearing host is, at least in part, due to the evolution of MS2 resistant (repressed transfer) plasmids.

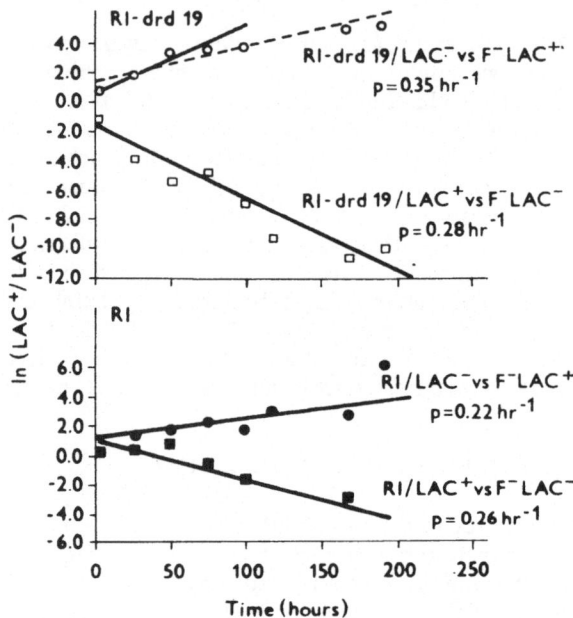

Fig. 2. Competition between plasmid-bearing and plasmid-free cells in glucose limited chemostat culture. Natural log of the ratio of lac^+ to lac^- cells. R1 and R1-drd 19 plasmids in *E. coli* K-12 hosts.

For reasons that we are currently trying to understand, the magnitude of the selection coefficients estimated in experiments of this type varies between experiments, and some of our long-term chemostat results suggest that the absolute magnitude of this "cost" of plasmid carriage can change in time. However, taken in total, results of these *in vitro* selection studies suggest that the carriage of these "unselected" conjugative plasmids confer a significant selective disadvantage on their host bacteria with the selection coefficients, α, of the order of 0.10 for R1 and of the order of 0.15 for R1-drd-19. At this juncture, we don't know why these extra chromosomal elements impose this seemingly high cost on their hosts.

ii) *Segregation rate*: Although with sufficient searching it is possible to pick up segregants for R-factors in the absence of some selective mechanisms, it is difficult to obtain estimates of the actual rates of vegetative segregation for plasmids of this type. However, from counts of the fermentation negative colonies on tetrazolium indicator plates from a fluctuation test experiment where transfer is precluded by low densities, it is possible

TABLE I.

Minimum value of the transfer rate constant, necessary for the establishment and maintenance of conjugative plasmids in the absence of positive selection*

α	N = 10^6		N = 10^7		N = 10^8	
	ρ = 0.2	ρ = 0.02	ρ = 0.2	ρ = 0.02	ρ = 0.2	ρ = 0.02
0.0	1.0×10^{-9}	1.0×10^{-9}	1.0×10^{-10}	1.0×10^{-10}	1.0×10^{-11}	1.0×10^{-11}
0.05	1.1×10^{-8}	2.0×10^{-9}	1.1×10^{-9}	2.0×10^{-10}	1.1×10^{-10}	2.0×10^{-11}
0.10	2.1×10^{-8}	3.0×10^{-9}	2.1×10^{-9}	3.0×10^{-10}	1.1×10^{-10}	3.0×10^{-11}
0.15	3.1×10^{-8}	4.0×10^{-9}	3.1×10^{-9}	4.0×10^{-10}	3.1×10^{-10}	4.0×10^{-11}

* calculated from formula (1), $\gamma > \dfrac{\alpha\rho + \tau}{N}$

N = equilibrium density of recipients in plasmid-free population (cells per ml).
α — The selection coefficient for plasmid-bearing cells.
ρ — Ahe dilution rate (hr^{-1}).
γ — The transfer rate constant (ml. per cell per hour).
In all cases the segregation rate, $\tau = 10^{-3}$ (hr^{-1}).

to obtain estimates of segregation rates for plasmids carrying fermentation markers. Using this type of procedure, we estimated a segregation rate, τ, of approximately 10^{-3} per cell generation for an F-lac-pro plasmid on a CSH 50 (Δ lac pro) host. This experiment was done at 37°C in glucose limited minimal medium with the bacteria growing exponentially with about a one hour generation time.

iii) *Transfer rate constant*: In an earlier paper, Levin, Stewart and Rice (1979) presented a method for estimating the transfer rate parameter γ in exponential and equilibrium chemostat cultures of bacteria. For strains of *E. coli* K-12 growing exponentially in glucose limited minimal medium at 37°C, the rate constants for R1 and R1-drd-19 were respectively 1.5×10^{-12} and 1.7×10^{-9} ml per cell per hour. For the more slowing dividing bacteria in equilibrium chemostats the corresponding values of γ were 2.6×10^{-14} and 1.8×10^{-11}.

iv). *Population densities and dilution rates*: At this time we have not obtained direct estimates of the densities and turnover rates of *E. coli* in their natural habitats nor am I aware of existing studies specifically estimating these parameters. Consequently, for this present consideration, I shall assume the seemingly realistic ranges of 10^6 to 10^8 bacteria per ml for N and of 0.20 to 0.02 hr^{-1} for ρ.

DISCUSSION AND CONCLUSIONS

In Levin *et al.* (1979) we concluded that for *E. coli* in steady-state (exponential or equilibrium chemostat) liquid cultures the kinetics of conjugative plasmid transmission can be reasonably well approximated by mass-action models of the type considered here. Nevertheless, I do not consider these simple models to be quantitatively precise analogues of the population dynamics of plasmids in biologically and physically complex natural habitats. These models were constructed for a general consideration of the conditions for the existence of plasmids and the previously described experiments were performed to get a rough idea of the magnitude of the parameters of this model in real populations. It is with this in mind that the results of this type of *a priori* investigation have to be interpreted.

The necessary conditions for the existence of conjugative plasmids as calculated from formula (1) are presented in Table I. for the parameter values in the range considered in the previous section. When the transfer rate constant is below the indicated value, the plasmid could not become established and if it were present in a population it would not be maintained.

Since estimated of the transfer rate constant in excess of 10^{-9} ml per cell per hour do obtain, it is necessary to conclude that there are "realistic" situations where conjugative R-plasmids could become established and would be maintained even in the face of segregational loss and the relatively intense negative selection against R-plasmid-bearing cells anticipated to occur in the absence of antibiotic selection. However, transfer rate constants of this high magnitude obtain only for plasmids with permanently derepressed conjugative pili synthesis and therefore would only be anticipated for wild-type, repressed transfer, plasmids in newly formed transconjugants rather than for the population of plasmid-bearing cells at large. Furthermore, transfer rate parameters of this high magnitude obtain only under the lush conditions of unrestricted exponential population growth. In the conditions of relatively low rates of cell division that are likely to exist in natural habitats, the rate constants of plasmid transfer even for the permanently derepressed factors appear too low to allow for the establishment and maintenence of the plasmid in the absence of positive selection. This view of low rates of plasmid transfer in natural habitats is also suggested by the generally negative results obtained from feeding experiments studying plasmid transfer "in vivo" (see for example, Smith 1969).

Based on the above considerations it seems most reasonable to interpret the results of these theoretical and experimental studies as support for the hypothesis that the absence of antibiotic selection, the frequency of bacteria carrying R-factors would continually decline and that these R-plasmids would eventually be lost.

REFERENCES

ANDERSON, J. D. (1974): The effect of R-factor carriage on the survival of *Escherichia coli* in the human intestine. J. Med. Microbiol. 7: 85–90.

CHAO, L, B. R. LEVIN and F. M. STEWART (1977): A complex community in a simple habitat. An experimental study with bacteria and phage. Ecology. 58: 369–378.

DOBZHANSKY, T. (1970): *Genetics of the Evolutionary Process*. Columbia Univ. Press. New York, 5050 pp.

LEVIN, B. R. and F. M. STEWART, (1979): The population biology of bacterial plasmids: *A priori* conditions for the existence of mobilizable nonconjugative factors. Genetics (in press).

LEVIN, B. R., F. M. STEWART and V. A. RICE (1979): The kinetics of conjugative plasmid transmission: Fit of simple mass action models. Plasmid 2: 247–260.

LINTON, K. B., M. H. RICHMOND, R. BEVAN and W. A. GILLESPIE (1974): Antibiotic resistance and R-factors in coliform bacilli isolated from hospital and domestic sewage. J. Med. Microbiol. 7: 91–103.

MEYNELL, G. G. (1973): *Bacterial Plasmids* M. I. T. Press Cambridge. 164 pp.

RICHMOND, M. H., (1974): R factors in man and his environment. *Microbiology 1974*, D. Schlessinger ed. 27–35 Amer. Soc. of Micro. Wash.

SMITH, D. H. (1967): R factor infection of *Escherichia coli* lyophilized in 1946. J. Bact. 94: 2071–2072.

SMITH, H. W. (1969): Transfer of antibiotic resistance from farm animals and human strains of *Escherichia coli* to resident *E. coli* in the alimentary tract in man. Lancet i: 1174–1176.

STEWART, F. M. and B. R. LEVIN, (1977): The population biology of bacterial plasmids: *A priori* conditions for the existence of conjugationally transmitted factors. Genetics 87: 209–228.

B. L., Dept. of Zoology,
Univ. of Massachusetts, Amherst,
Massachusetts 01003, U.S.A.

R-FACTOR TRANSFER UNDER ENVIRONMENTAL CONDITIONS

MARYLYN D. COOKE

Cawthron Institute, Nelson, New Zealand

INTRODUCTION

The majority of studies on R-factor prevalence, ecology and transferability have been heavily biased at 35–37° because of the public health implications of these transmissible elements in human and veterinary medicine. Furthermore, the majority of wild-type R-factors isolated from clinical material are normally repressed with respect to donor phenotype, so that only about 1 in 10^5 cells of the donor population are capable of plasmid transfer at 35° (Meynell, Meynell, and Datta, 1968). Certain plasmids may be essentially non-transferable at lower temperatures because of non-expression of donor phenotype (Curtiss, 1976).

Among a random collection of R^+ *Escherichia coli* strains isolated from environmental sources (Cooke, 1976a, b and unpublished), I have found a spectrum of R-factor transfer frequencies at environmental and higher temperatures, and which may include examples of naturally occurring derepressed R-factors.

METHODS

Apart from strains 58.161 R1, 58.161 R1*drd*19, 58.161 F⁻ and J6-2 *nal*ʳ which were obtained from E. Meynell, all other *E. coli* strains were isolated from samples of drinking water, streams, seawater and effluents (Cooke, 1976a, b and unpublished). Oxoid Nutrient broth No. 2 and Blood agar base were used throughout with 0.1% peptone as diluent. For selection of R^+ transconjugants, mercuric chloride 10 μg/ml as Hg, streptomycin 15 or 20 μg/ml or tetracycline 20 μg/ml were added to Blood agar base plus 50 μg/ml nalidixic acid. Selection plates were incubated at 35°.

Modifications of the method used to determine the R-factor transfer rate previously reported (Cooke, 1976a) are outlined in table and figure legends. All transfer frequencies are expressed relative to the viable count of the donor culture at the time of mixing.

RESULTS AND DISCUSSION

Transferability at different temperatures

Results presented in Table I. are representative of the transfer behaviour of 95 of the 193 resistant *E. coli* strains tested initially in a 2 hr mating at 35°, and which can be regarded as R^+ on the basis of demonstrable transferability of one or more R-determinants. The 2 hr mating period was initially chosen in order to concentrate on isolates presumed to carry a derepressed R-factor, and therefore more likely to transfer in an unfavourable environmental situation. However, depending upon the temperature and time of mating, there was a wide spectrum of transfer frequencies among these isolates. The majority, in a 2 hr mating at 35°, appeared to carry typical repressed R-factors, the R-factor transferring with a frequency of 10^{-4} or less per donor cell, similar to the classical wild-type R-factor, R1. 67 of the 95 isolates showed some transfer within this 2 hr period, and of these, 23 carried R-factors which transferred with frequencies greater than 1%, approaching that of the control derepressed R-factor, R1drd19.

Sixteen isolates which did not show any transfer in the 2 hr mating, however did transfer their R-factor in 24 hr stationary mixed culture at 35°. This prompted a more detailed investigation of the effect of environmental temperatures on transfer.

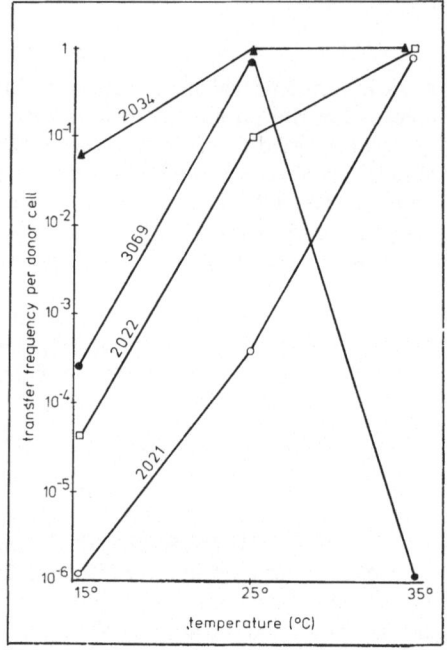

Fig. 1. Effect of temperature on R-factor transfer frequency.
0.1 ml overnight donor: 5 ml overnight recipient: 5 ml nutrient broth incubated 24 hr in stationary culture at temperature shown.

At 25°, again with a 24 hr mating period, 88/95 (92.6%) isolates transferred one or more R-determinants, and of these, 39 showed high frequency transfer ($> 10^{-2}$ per donor cell). At 15°, a temperature often attainable in the environment, the frequency of transfer of any one R-factor was always less than at 25°. However, 65/95 (68.4%) isolates did transfer their R-factor at 15°, and of these 8 showed high frequency transfer, similar to that of R1drd19.

The transfer frequencies exhibited by these R-factors at different temperatures can be roughly grouped into four types of response (Fig. 1):

a; Group I — represented by strains 2033, 15001, 15013, 16420 (Table I.) and 2021 (Fig. 1) show a similar response to the classical wild-type R-factor, R1, with optimum transfer at 35°, a sharp decrease in transfer frequency with decreasing temperature, and essentially little or no transfer at 15°.

b; Group II — represented by 15010, 15217 (Table I.) and 2022 (Fig. 1) give high frequency transfer at 35° and at 25°, and some transfer at 15°.

c; Group III — represented by 3079, 5056 (Table I.) and 3069 (Fig. 1) are probably similar to other temperature sensitive plasmids recently described (Ishiguro, Oka, and Sato, 1978, Smith, Parsell, and Green, 1978). However, none of my isolates are citrate-positive, nor are they resistant to chloramphenicol.

TABLE I.

Transfer frequency of R plasmids from wild-type E coli					
R$^+$ strain	Resistances	Transfer frequency per donor cell			
		2hr 35°^{a}	24hr 35°^{b}	24hr 25°^{c}	24hr 15°^{d}
5056	Ap Tc Hg	–	–	1	3×10^{-6}
2033	Sm Su Tc Hg	–	1×10^{-3}	–	–
3079	Ap Tc	4×10^{-7}	4×10^{-5}	1	5×10^{-3}
15010	Ap Sm Su Tc	3×10^{-6}	1×10^{-2}	2×10^{2}	1×10^{-4}
15001	Tc	7×10^{-5}	1	4×10^{-2}	4×10^{-6}
15013	Tc	2×10^{-4}	2×10^{-1}	7×10^{-5}	–
14606	Ap Sm	3×10^{-4}	1	1	2×10^{2}
5059	Sm Su Tc Hg	1×10^{-3}	1×10^{-3}	1	2×10^{2}
16420	Su Tc	2×10^{-2}	2×10^{-1}	2×10^{-4}	3×10^{-7}
15217	Tc	1×10^{-1}	1	5×10^{-1}	3×10^{-5}
58.161-R1	Ap Cm Km Sm Su	3×10^{-4}	1	3×10^{-3}	6×10^{-6}
58.161 R1drd19	Ap Cm Km Sm Su	1	1	1	3×10^{-2}

— No transfer detectable ($<10^{-7}$ per donor cell).

Transfer method and selection of transconjugants (Cooke, 1976a) with following modifications: (a) 0.5 ml logarithmic phase donor: 4.5 ml overnight recipient J6-2 nal^r: 5 ml nutrient broth, rotated 2 hr at 35°; (b, c, d) 0.1 ml overnight donor: 5 ml overnight recipient: 5 ml nutrient broth, incubated in stationary culture 24 hr (b) at 35°, (c) at 25°, (d) at 15°.

Ap, ampicillin; Cm, chloramphenicol; Km, kanamycin; Sm, streptomycin; Su, sulphafurazole; Tc, tetracycline; Hg, mercuric chloride.

d; Group IV — represented by strains 5059, 14606 (Table I.) and 2034 (Fig. 1), is composed of those isolates which transfer their R-factor with high frequency at 35°, 25° and 15°. This group may include examples of naturally occurring derepressed R-factors. There have been two reports of derepressed isolates from clinical material but one reverted at high frequency to the wild-type repressed state.

Temperature shift experiments

Further investigation of two isolates, representative of Group II (15010) and of Group IV (14606), showed that pregrowth of the donor culture at 35° did not specifically restrict transfer, unlike some temperature-sensitive R-factors (Yoshida, Terawaki, and Nakaya, 1974, Rodriguez-Lemoine, Jacob, Hedges, and Datta, 1975). The R-factor was also transferred from the wild-type isolate to 58.161 F⁻, so that the behaviour of the new plasmid could be studied against the same genetic background as that of R1 (Table II.).

The transconjugant 58.161 p14606 shows a higher frequency of transfer than the environmental isolate under all combinations of donor pregrowth and mating. However, transconjugant 58.161 p14606b, selected after two transfers at 15°, is more variable in transfer frequency compared with both the wild-type isolate and with 58.161 p14606a.

These results indicate that the temperature at which one transfer occurs might affect the frequency of future transfer at the same or different temperatures. This has public health implications in situations where R^+ bacteria can pass from the intestine to the environment and back again.

TABLE II.

Transfer frequency of R plasmids					
Culture temperature (°C)		Transfer frequency			
Donor	Mixture	WT 14606	58.161 p14606a	58.161 p14606b	58.161 R 1
15		5×10^{-7}	3×10^{-5}	1×10^{-6}	—
25	15	6×10^{-6}	2×10^{-5}	3×10^{-6}	—
35		1×10^{-7}	2×10^{-6}	1×10^{-7}	2×10^{-7}
15		7×10^{-5}	1×10^{-3}	4×10^{-6}	—
25	25	9×10^{-5}	5×10^{-3}	1×10^{-3}	1×10^{-6}
35		4×10^{-6}	7×10^{-5}	4×10^{-6}	6×10^{-4}
15		3×10^{-4}	8×10^{-4}	8×10^{-4}	5×10^{-5}
25	35	1×10^{-3}	3×10^{-2}	4×10^{-6}	5×10^{-5}
35		1×10^{-4}	3×10^{-3}	1×10^{-4}	3×10^{-4}

— No transfer detectable ($<10^{-7}$ per donor cell).
The R-factor was transferred from WT14606 to J6-2 nal^r at 15°, then from J6-2 R^+ to 58.161 F^-; 58.161 R^+ transconjugants were selected after mating at 35° (a) and 15° (b), and used as donors. Each donor was grown 24 hr at temperature shown and mixed in proportion of 0.5 ml donor: 4.5 ml overnight recipient: 5 ml nutrient broth. The mixtures were incubated for 2 hr in stationary culture at 15°, 25° or 35°.

Further investigation of this aspect, using strain 15010 and a 2 or 24 hr mating period, showed that passage at 15° gave a frequency of 5×10^{-3} for 5/5 58.161 R^+ transconjugants, 10-fold higher than that of the original environmental isolate (5×10^{-4}), whereas passage at 35° increased the transfer frequency of 7/7 transconjugants 100-fold, from 5×10^{-6} to 4×10^{-4} per donor cell. With a temperature shift from 35° to 15° at successive transfers, as might happen when an R^+ bacterium is passed from the intestine to the environment, the frequency of transfer at the final 15° mating was increased further, to 2×10^{-2} per donor cell. This high frequency was shown by 14/14 transconjugants, and was stable on repeated testing at 15°.

R-factor transfer in river and seawater

However, mating in nutrient broth, even at environmental temperatures, remains an artificial environment. Nutrient levels in oxidation ponds, sewage and sediments may be of similar enrichment to broth, but transferability in a nutrient-poor environment, such as river or seawater might still be regarded as of low probability although it may occur at low frequency (Grabow, Prozesky, and Burger, 1975; Fontaine and Hoadley, 1976; Curtiss, 1976).

Fig. 2 shows the behaviour of two environmental strains (14606, 15010) compared with R1. In nutrient broth matings, both environmental strains exhibit a lower frequency of

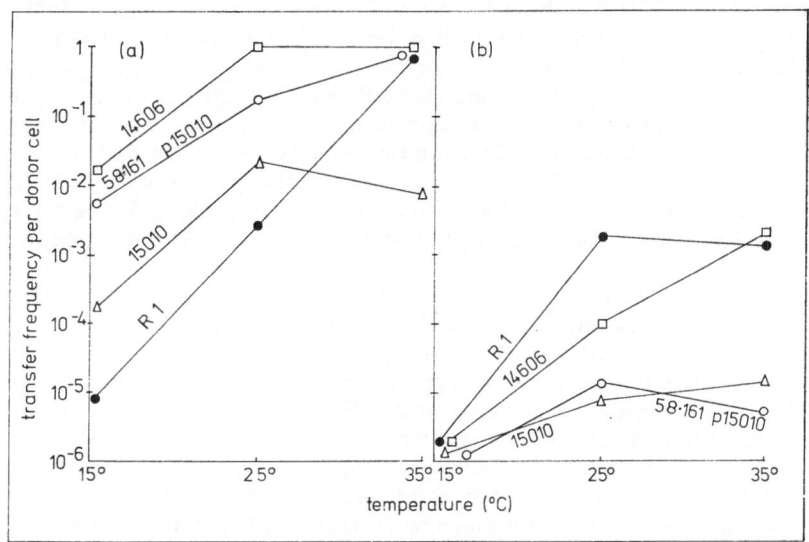

Fig. 2. Effect of temperature on R-factor transfer frequency in nutrient broth and riverwater
a: 0.1 ml overnight donor: 5 ml overnight recipient J6-2 *nal*$_r$: 5 ml nutrient broth.
b: cells resuspended in unfiltered, nonsterile riverwater and mixed in the same proportions using riverwater instead of nutrient broth. Mixed cultures were incubated in stationary culture for 24 hr at temperature shown.

transfer than R1 at 35°, whereas at 25° and 15° the frequency of transfer of these R-factors is considerably higher. Transconjugant 58.161 p14606a was identical to the original 14606 isolate, and is not shown in Fig. 2, whereas transconjugant 58.161 p15010 exhibits a higher frequency of transfer than the environmental isolate at all temperatures. Parallel with these experiments, matings were done with the bacteria resuspended in unfiltered, non--sterile riverwater. Under these conditions, absolute frequencies of transfer are lower at all temperatures than in nutrient broth matings, although the ability of R1 to transfer in riverwater at 25° (1.8 × 10^{-3}) was only slightly less than in nutrient broth (2.8 × 10^{-3}). However, the considerably reduced transfer frequency of R-factors from 14606 and 15010 in riverwater matings, indicates that the ability to transfer at high frequency at environmental temperatures in nutrient broth cannot be used to predict the ability of a certain R-factor to transfer under conditions of nutrient limitation. In addition, the ability of an R-factor to transfer at environmental temperatures, and under environmental conditions, is a function of the R-factor and of the host bacterial strain.

In unfiltered, non-sterile seawater, transfer frequencies are lower, and showed more experimental fluctuation, than in riverwater. All strains showed peak transfer at 25°.

CONCLUSIONS

The results presented show that non-clinical *E. coli* strains isolated from drinking water supplies and polluted water sources carry R-factors which differ in their optimum temperature for transfer. They can be roughly grouped on the basis of their response to temperature, and include some which appear to be examples of naturally occurring derepressed R-factors.

Nearly 50% of the isolates would transfer their R-factors at temperatures attainable in the environment. The fact that 68% of the R-factors would transfer at 15°, and 93% would transfer at 25°, in some cases at high frequency, and furthermore that some were capable of transfer when mated under conditions of nutrient limitation, suggest that it may be advisable to re-evaluate standards for effluent discharges into natural waters.

Water quality standards in many countries are based on the concept that coliform bacteria may be regarded as harmless indicators of faecal pollution. No limits are at present set on discharges of R$^+$ bacteria and conventional treatment methods do not completely remove such bacteria. Yet, in the natural environment, R$^+$ bacteria may in fact have a selective advantage compared with antibiotic-sensitive coliforms (Cooke, 1976b).

The possibility of spread of genetic information from R$^+$ bacteria under environmental conditions has been regarded as an unlikely possibility because of constraints put upon such transfer by temperature optima. The public health aspects of dissemination of R-factors into the environment, and the current concern about the 'escape' of recombinant DNA become of increasing importance since I have shown that plasmid-bearing components of the intestinal flora of humans and other animals, isolated from the environment outside their usual ecological niche, are capable of transferring genetic information at environmental temperatures. Since more than half the genetic complement of an R-factor specifies information as yet unknown, this also will be transferred under environmental conditions along with the marker resistance determinants.

REFERENCES

COOKE, M. D. (1976a): Antibiotic resistance among coliform and faecal coliform bacteria isolated from sewage, seawater, and marine shellfish. Antimicrob. Ag. Chemother. 9, 879–884.
COOKE, M. D. (1976b): Antibiotic resistance in coliform and faecal coliform bacteria from natural waters and effluents. N. Z. J. Mar. Freshw. Res. 10, 391–397.
CURTISS, R. (1976): Genetic manipulation of micro-organisms: potential benefits and biohazards. Ann. Rev. Microbiol. 30, 507–533.
FONTAINE, T. D. and A. W. HOADLEY (1976): Transferable drug resistance associated with coliforms isolated from hospital and domestic sewage. Health Lab. Sci. 13, 238–245.
GRABOW, W. O. K., O. W. PROZESKY and J. S. BURGER (1975): Behaviour in a river and dam of coliform bacteria with transferable or non-transferable drug resistance. Water Res. 9, 777–782.
ISHIGURO, N., C. OKA and G. SATO (1978): Isolation of citrate-positive variants of *Escherichia coli* from domestic pigeons, pigs, cattle, and horses Appl Environ. Microbiol. 36, 217–222.
MEYNELL, E., G. G. MEYNELL and N. DATTA (1968): Phylogenetic relationships of drug-resistance factors and other transmissible bacterial plasmids. Bacteriol. Rev. 32, 55–83.
RODRIGUEZ-LEMOINE, V., A. E. JACOB, R. W. HEDGES and N. DATTA (1975): Thermosensitive production of their transfer systems by Group 5 plasmids. J. Gen. Microbiol. 86, 11–114.
SMITH, H. W., Z. PARSELL and P. GREEN (1978): Thermosensitive antibiotic resistance plasmids in *Enterobacteria*. J. Gen. Microbiol. 109, 37–47.
YOSHIDA, Y., Y. TERAWAKI and R. NAKAYA (1974): Temperature sensitive R plasmid originated from *Salmonella typhimurium*. Biochem. Biophys. Res. Commun. 59, 361–369.

M. D. C., Cawthron Inst. P.O. Box 175,
Nelson, New Zealand.

INCOMPATIBILITY TESTING OF FI+ R-FACTORS
FROM THE AREA OF BERNE
(PROPOSITION FOR A TESTING METHOD)

G. LEBEK and L. PETRI

Institute of Hygiene and Medical Microbiology, University of Berne, Switzerland

R-factors are widely spread in nosocomial infections with *Enterobacteriaceae*. Therefore a classification of R-factors could be of great importance. It is not sufficient to follow the resistance patterns which they confer on their bacterial hosts, as the same drug resistance patterns may be determined by unrelated plasmids. For instance, the analysis of a hospital outbreak in Munich showed that the outbreak was only uniform in the R-factors, whereas strains of *Klebsiella*, *Enterobacter* and *Serratia* were involved. Observations of hospital outbreaks permit us to suppose that R-factors influence epidemics to a remarkable degree.

TABLE I.

F-indicatorplasmids

FI :	F'*lac*/R386
FII :	R1/R6
FIV:	R124/pGL 109
FV :	F$_0$*lac*
FVI:	pGL 625/pGL 611

A classification can first be made by the analysis of the sexpilus with sexphages. Then incompatibility testing can be made using of indicator plasmids.* The incompatibility effect can be hidden by the recombination of related plasmids. Therefore substrains of *E. coli* K$_{12}$ with defective recombination systems were examined in order to find out which of them show incompatibility phenomena more clearly. We subsequently tested the incompatibility effect in strains with intact and defective rec systems. The following were found: *rec* A$^-$ (*rec* B$^-$) *rec* C$^-$) *rec* B$^-$C$^-$. We obtained the following results: the incompatibility effect in the *rec* A$^-$ strain turned out to be most distinct. Thus we carried out our tests on *E. coli* K$_{12}$ C600 *rec* A$^-$.

Incompatibility could also be shown in *Klebsiella* and *Salmonella* strains. Thus a classification can also be made out in these strains by transferring the indicator plasmid. Non-transferable R-factors can also be classified by mobilizing them with a helper plasmid or by transforming them in the usual way.

It was very striking that we could not definitely identify any representative of Inc FIII. In agreement with the assumption of Datta, we examined the indicator-plasmid for Inc FIII (ColB-K98), to find whether it showed incompatibility with other plasmids of the F groups. Our results showed that the ColB-K98 plasmid was incompatible with 5 representatives of Inc FII, but compatible with the plasmids of the other F groups.

Therefore we propose to classify this plasmid in the Inc FII-group.

---

* compare with Table I

TABLE II.

Effect of sexphages on plasmid-determined pili repressed and derepressed state of plasmid pGL 108

Strain	Plasmid	Inc Group	Plaques		Multiplied	
			IKe	MS$_2$	Ike	MS$_2$
C600 S^{res}	N$_3$	fi$^-$ Inc N	+	—	+	—
921 S^{res}	F$'lac$	fi$^+$ Inc FI	—	+	—	+
921 S^{res}	pGL 108	fi$^-$ Inc FI$_N$	+	+	+	+
C600 S^{res}	F$'lac$. pGL 108		+	+	+	+
C600 S^{res}	pGL 108$\star$ [1]		+	+	+	+

[1] F$'lac$. pGL 108 → pGL 108$\star$
after being transferred into *E. coli* K12 C600
recA$^-$ harboring plasmid F$'lac$

TABLE III.

Frequency of transfer into E. coli K$_{12}$ 921

Plasmid	without phages	with 10 PFU MS$_2$/Donor	with 10 PFU IKE/Donor
pGL 108 (Inc FI$_N$)	2×10^{-5}	$<10^{-8}$	$<10^{-8}$
N3 (Inc N)	1×10^{-5}	1×10^{-5}	$<10^{-8}$
R1 (Inc FII)	1×10^{-3}	$<10^{-8}$	1×10^{-3}

During the testings of Inc FI-group we found plasmids which make their host strains sensitive to MS$_2$ and IKe phages. These plasmids show a very strong incompatibility to the Inc FI indicator-plasmid. When such a plasmid is transferred to a recipient harboring the indicator-plasmid, the latter is lost with high frequency on the selective medium. In a colony that contains both plasmids cultivated in drug-free broth, both plasmids can coexist stably. Some of these qualitites are demonstrated on R-factor pGL 108 (Table II.): the host strain which contains pGL 108 forms plaques with MS$_2$ as well as with IKe phages, but not with If$_1$ and PRR1 phages. Thus the host cell forms sexpili, which are different from F-pili. They have receptors for MS$_2$ as well as for IKe phages. These results also show that the R-factor is derepreseed. For a freshly isolated R-factor this is rather unusual. So far all the R-factors isolated (several 100 R-factors) were repressed. After 4–5 years of storage in stab culture at 4°C, the R-factors lose their repressor to a degree of approx. 35%.

The doubles showed corresponding results in the plaque-test and in the phage--multiplying-test. Furthermore, it should be noted that pGL 108 was compatible with Inc N. Table III. shows that both sexphages can supress conjugation. The next step will be to investigate whether the R-factor contains one or two different tra$^+$ genes. We would like to classify these R-factors in a subgroup of Inc FI, termed Inc FI^N, to indicate the ability of the sexpili to function as receptors for IKe phages.

In the hospital outbreak in Munich mentioned above, we found that the plasmids

TABLE IV.

Distribution of tested fi+ R-factors

Incompatibility group	Total No.	No. of E. coli	No. of other organisms
FI	80	76	4
FI$_N$	7	7	0
FII	8	8	0
FIV	7	7	0
FVI	23	0	23 (Klebs.)
no Inc	2	2	0
Total	127	100	27

belonged to the fi+-class and that they were compatible with all known indicator plasmids. As they were incompatible with each other, we postulated a new group, Inc FVI. It is interesting that we found R-factors from another hospital outbreak also to belong to the same group.

We could find no plasmid which was incompatible with F_0lac (belonging to Inc FV). Therefore we presume that this plasmid possibly expresses no incompatibility. During our investigations, we also found two wild-R-factors which were fi+ and compatible with all existing indicator-plasmids. This suggests that these plasmids also belong to the group without incompatibility or to a new Inc F group.

Our experience suggested the following testing method: the fi+-R-factors to be tested are transferred into E. coli K_{12} C600 *rec* A⁻, which contains the indicator-plasmid. In some cases the incompatibility for *Inc* FI can be seen on the selective plate in a heavy loss of *lac+* colonies when using the indicator plasmid F'*lac*. If only a few colonies become *lac⁻*, one *lac+* colony is picked up and incubated in drug-free broth. When plated out again on McConkey Agar, incompatibility becomes visible in a loss of *lac+* greater than 30%. The incompatibility of the other groups was tested by the replica technique after incubating the doubles in drug-free broth as before. In case of incompatibility, loss of one or both plasmids was 30–90%.

Tested R-factors of bacterial isolates of patients and of animal feces could be classified as in Table IV.

REFERENCES

DATTA, N. (1974): Epidemiology and classification of plasmids. Schlessinger (ed.) Microbiology, p. 9–15.

GRINDLEY, N. D. F., G. O. HUMPHREYS and E. S. ANDERSON (1973): Molecular studies of R-factor compatibility groups. J. Bact. 115, 378–398.

HEDGES, R. W. and N. DATTA (1972): R124, an fi+ R-factor of a new compatibility class. J. Gen. Microbiol. 71, 403–405.

HOLLOWAY, G. W. and M. H. RICHMOND (1973): R-factors used for genetic studies in strains of *Pseudomonas aeruginosa* and their origin. Genet. Res. 21, 103.

ZUEND, P. and G. LEBEK (1978): Classification of fi+ R-factors from enterobacteriaceae isolated from medical material. Int. Soc. Chemoth. (ed.). Current Chemotherapy 1, 451–453.

G. L., Hygiene-Institut,
Friedbrühlstrasse 51,
CH-3008 Bern, Schweiz

II. MEDICAL PART

A) NEW DRUGS AGAINST RESISTANT BACTERIA

Chairman: S. M. NAVASHIN

IN VITRO AND *IN VIVO* ANTIBACTERIAL ACTIVITY OF
T-1551, A NEW SEMISYNTHETIC CEPHALOSPORIN

N. MATSUBARA, S. MINAMI, T. MURAOKA, I. SAIKAWA
and S. MITSUHASHI

Department of Microbiology, School of Medicine, Gunma University,
Maebashi, and Research Laboratory, Toyama Chemical Co., Ltd., Toyama, Japan

INTRODUCTION

Many kinds of β-lactam antibiotics have been used for infections with gram-positive and gram-negative bacteria. β-lactam antibiotics are characteristic for the low virulence in men. However, it is a fault that β-lactam antibiotics are ineffective against some species of gram-negative bacteria because of hydrolysis by β-lactamase and intrinsic resistance. Therefore, infection with gram-negative bacteria including *Pseudomonas aeruginosa* has recently become an issue (Altermier, Hummel, Hill, and Lewis, 1973, Finland, 1970). As counterplan, some broad spectrum penicillins, carbenicillin (CB-PC), sulbenicillin (SB-PC), piperacillin (PI-PC), apalcillin (AP-PC), etc., have been developed up to the

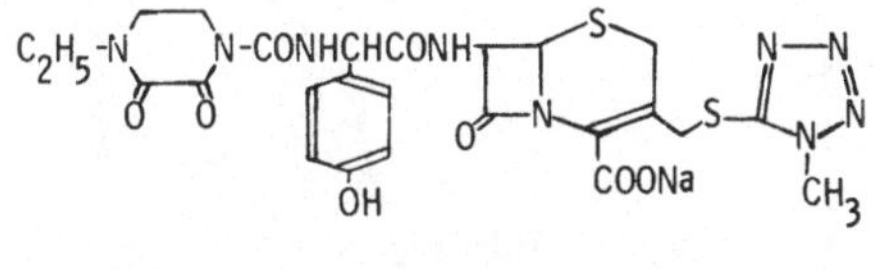

Fig. 1. Chemical structure of T-1551.

present (Acred, Brown, Knudsen, Rolinson, and Sutherland, 1967; Ueo, Fukuoka, Hayashi, Yasuda, Taki, Tai, Watanabe, Saikawa, and Mitsuhashi, 1977). T-1551, a new semisynthetic cephalosporin, shows a broader spectrum of antibacterial activity than other cephalosporins, and is effective against *P. aeruginosa*, *Serratia marcescens*, and *Enterobacter cloacae*. This paper presents the *in vitro* and *in vivo* evaluation of T-1551.

The spectrum of antibacterial activity against standard strains of gram-positive and gram-negative bacteria was examined. T-1551 was effective against gram-positive bacteria susceptible to cefazolin (CEZ). MIC values of T-1551 against most gram-positive bacteria were on the same level as those of CEZ. T-1551 was also highly effective against gram-negative bacteria, and T-1551 inhibited the growth of all the tested strains at concentrations lower than those of CEZ. It is a remarkable characteristic that T-1551 shows high antibacterial activity against *S. marcescens*, indole-positive *Proteus* species, *E. cloacae* and *P. aeruginosa* strains which are resistant to CEZ. Thus, it can be concluded that T-1551 shows a broad spectrum of antibacterial activity against gram-positive and gram-negative bacteria.

The antibacterial activity of T-1551 against clinical isolates was compared with those of CEZ and CB-PC (Nishida, Matsubara, Murakawa, Mine, Yokota, Kuwahara, and Goto, 1970, Reller, Karney, Beaty, Holmes, and Turck, 1973). T-1551 was less active against *Staphylococcus aureus* than CEZ, and the peaks of the MIC distributions of these drugs were found to be 1.56 and 0.39 µg/ml, respectively. But T-1551 was more active than CEZ against *Escherichia coli*, *Klebsiella pneumoniae*, and *Proteus mirabilis* strains, most of which were susceptible to CEZ. The concentrations of the drugs required to inhibit the growth of 70 % of the total number of tested strains, MIC$_{70}$ against *P. mor-*

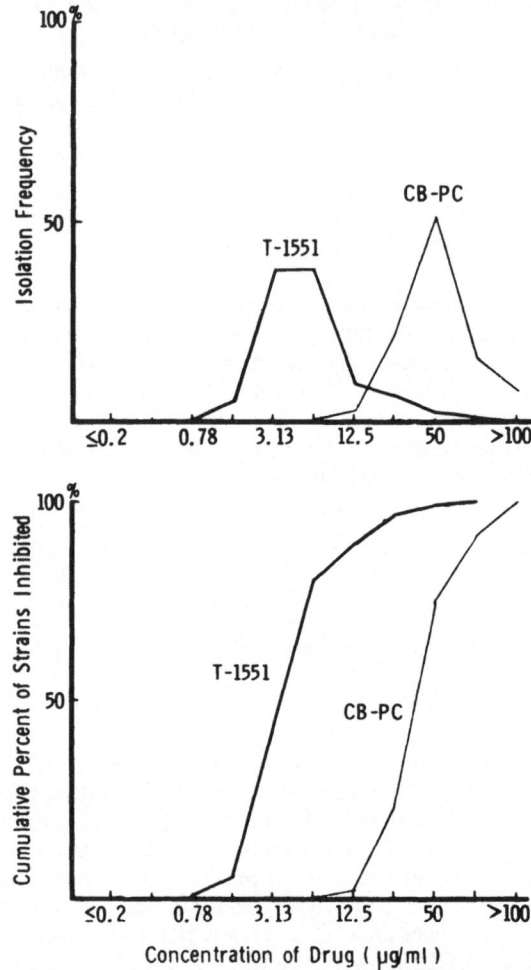

Fig. 2. Antibacterial activity of T-1551 against 200 *P. aeruginosa* strains.

ganii, *P. vulgaris*, *P. rettgeri*, *P. inconstans* and *Citrobacter freundii* were found to be 2.05, 4.50, 3.13, 2.40 and 2.00 µg/ml, respectively, while those of CEZ were more than 100 µg/ml. T-1551 was also highly active against *P. aeruginosa*, *S. marcescens* and *E. cloacae* resistant to CEZ, and the peaks of the MIC distributions of T-1551 to these species were found to be 3.13 to 6.25, 3.13 and 0.2 µg/ml, respectively. It is a notable characteristic that it shows a potent activity in particular against *P. aeruginosa* on which cephalosporins used for chemotherapy are now hardly effective.

T-1551 was effective against ampicillin (AB-PC)-resistant *E. coli* strains CB-PC- and

TABLE I.

Stability of T-1551 against cephalosporinases

Enzyme source	Specific[b] activity	Relative rate of hydrolysis[a]					
		CER	T-1551	CEZ	CET	CEX	PC-G
E. coli GN5482	0.24	100	0.04	135	691	55.5	28.7
P. aeruginosa GN918	0.24	100	0.04	160	480	62.9	24.8
P. vulgaris GN76	0.40	100	7.00	375	204	52.0	21.0
E. cloacae GN7471	3.68	100	0.80	50	402	54.0	83.1
C. freundii GN346	3.27	100	0.01	120	127	81.1	7.00

[a] Hydrolysis of each substrate by CSase is expressed as relative rate of hydrolysis, taking the absolute rate of hydrolysis of CER as 100.
[b] Units per mg of protein.

gentamicin (GM)-resistant *P. aeruginosa* strains. MIC_{70} values of T-1551 and CEZ against AB-PC-resistant *E. coli* were 6.25 and 12.5 μg/ml, respectively. T-1551 also showed high antibacterial activity against CB-PC-resistant *P. aeruginosa*, and the MIC_{70} value of T-1551 was 37.5 μg/ml. A cross-resistance was not observed between the activity of T-1551 and GM against GM-resistant *P. aeruginosa* strains.

Inoculum sizes greatly affected the antibacterial activity of T-1551 against *E. coli*, *Klebsiella pneumoniae* and *P. aeruginosa*. A great difference was observed in the antibacterial activity of T-1551, particularly between the inoculation of one loopful of 10^6 and 10^8 cells/ml of culture on an agar plate.

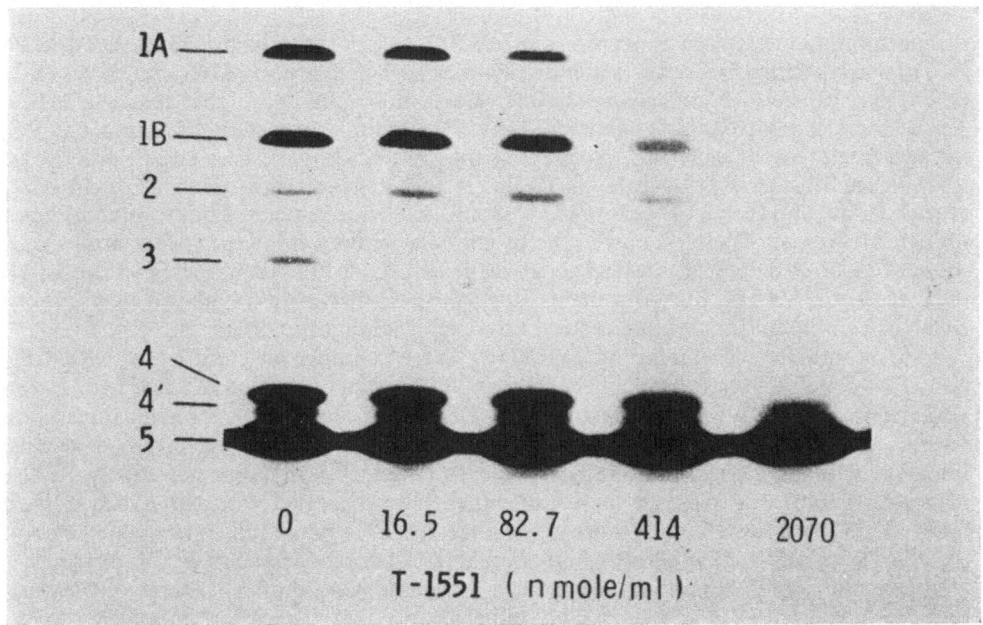

Fig. 3. Competition of T-1551 for ^{14}C-penicillin G binding to *P. aeruginosa* penicillin-binding proteins (PBPs). The concentration of ^{14}C-penicillin G is 82.7 n mole/ml.

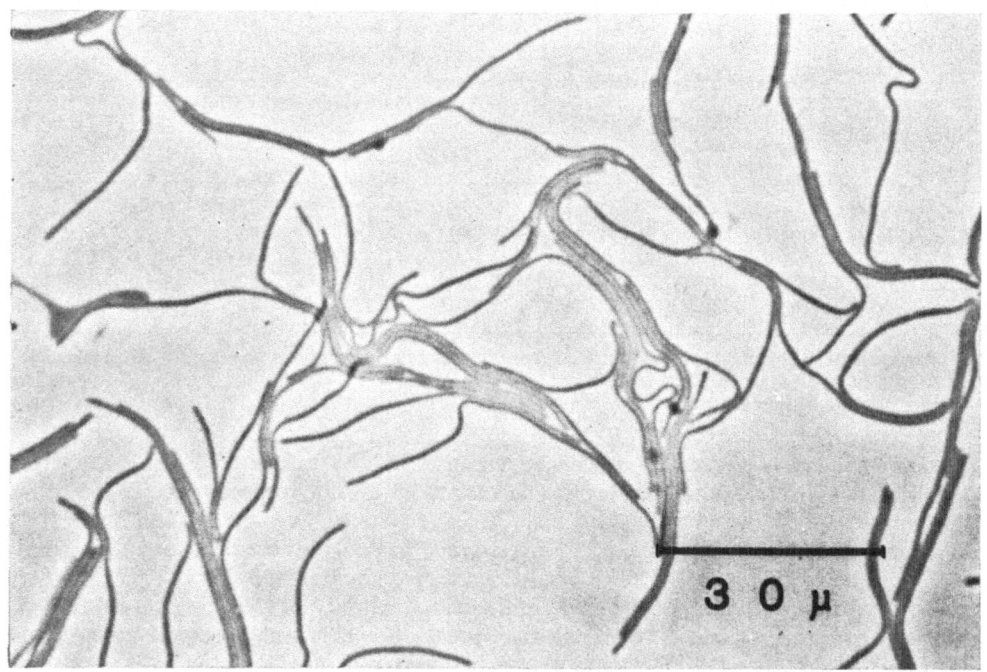

Fig. 4. Filamentous cells of *P. aeruginosa* IFO3445 treated with 12.5 μg/ml of T-1551 for 4.5 hours.

The minimum bacterial concentrations (MBCs) of T-1551 against 25 *E. coli* and 25 *K. pneumoniae* strains were the same or two-fold higher than the MICs, and the MBCs of T-1551 against 25 *P. aeruginosa* strains were at most four-fold higher than the MICs. To evaluate the bactericidal activity of T-1551 in comparison with CEZ and CB-PC, the number of viable cells was counted at intervals of 2 hours after adding the drugs. T-1551 acted bactericidally on *E. coli* GN6307 and *K. pneumoniae* GN5558 at the MIC of each strain, and the viable cells of these strains were not detected 4 hours after addition of T-1551. Namely, T-1551 showed the bactericidal activity at lower concentrations than those of CEZ. A bacteristatic effect on *P. aeruginosa* GN6736 was observed at the MICs of T-1551 and CB-PC, but both drugs killed the cells at two-fold higher concentrations than MICs of each drug within about 4 hours after adding the drugs.

In vivo antibacterial activity of T-1551 was also examined in comparison with CEZ and CB-PC. T-1551 was more effective than CEZ against infection of mice with *E. coli* and *K.pneumoniae*. A fifty % effective dose (ED$_{50}$) of T-1551 and CEZ against infection with *E. coli* ML4707 was 1.40 and 16.7 mg/kg, respectively, and ED$_{50}$ values against infection with *K. pneumoniae* GN6445 were 2.1 and 14.3 mg/kg, respectively. ED$_{50}$ values of T-1551 and CEZ in these experiments were parallel with the MICs of each drug. T-1551 was also 5 times more active than CB-PC against infection with *P. aeruginosa* NC-5. Thus, T-1551 showed high antibacterial activity *in vivo* as well as *in vitro*.

The stability of β-lactam antibiotics to β-lactamase is a significant factor of the antibacterial activity. T-1551 was highly stable against β-lactamase produced by various species of gram-negative bacteria. T-1551 was 100 to 10,000 times more stable than cephaloridine (CER) against cephalosporinases (CSases) produced by *E. coli, P. aerugi-*

nosa, E. cloacae, C. freundii and *P. morganii,* as shown in Table 1, giving also the relative rate of T-1551 hydrolysis by the CSase from *P. vulgaris,* that has a broad substrate specificity (Yaginuma, Sawai, Ono, Yamagishi, and Mitsuhashi, 1973). T-1551 was also stable against type I, II and IV penicillinases (PCases) mediated by R plasmids and the PCase produced by *K. pneumoniae,* that is T-1551 was 10 times more stable than penicillin G (PC-G) against type I and *K. pneumoniae* PCases, and T-1551 was not hydrolyzed by type II and IV PCases (Sawada, Yaginuma, Tai, Iyobe, and Mitsuhashi, 1976, Sawai, Mitsuhashi, and Yamagishi, 1968, Yamagishi, Ohara, Sawai, and Mitsuhashi, 1969). It is supposed that the stability of T-1551 against β-lactamase greatly contributes to its high antibacterial activity against β-lactamase-producing strains.

B. G. Spratt (1975) demonstrated the penicillin-binding proteins (PBPs) in *E. coli* by SDS polyacrylamide slab gel electrophoresis. Similarly, the PBPs of *P. aeruginosa* were detected by this laboratory (Noguchi, Matsuhashi, Takaoka, and Mitsuhashi, 1978). It seems that these PBPs are the target sites of β-lactam antibiotics. According to the competitive experiment of T-1551 for the binding of ^{14}C-PC-G to the PBPs, T-1551 showed high affinity for *E. coli* PBP-1A, -1Bs, -2 and -3, and *P. aeruginosa* PBP-1A and -3. B. G. Spratt reported that *E. coli* PBP-1A and -1Bs participate in cell elongation, PBP-2, in maintenance of rod form, and PBP-3, in cell division, respectively. It is presumed that *P. aeruginosa* PBP-1A, -1B, -2 and -3 correspond to *E. coli* PBP-1Bs, -1A, -2 and -3, respectively. As shown in Fig. 4, *E. coli* and *P. aeruginosa* cells showed a filamentous shape because of the high affinity of T-1551 particularly for PBP-3. On the other hand, T-1551 had a weak affinity for *E. coli* PBP-4 and -5 which are *D-alanine* carboxypeptidase IB and IA, respectively. T-1551 inhibited *in vitro* the activity of *D*-alanine carboxypeptidase IA and IB as well as did PC-G, and T-1551 inhibited a crosslinking reaction in cell wall synthesis *in vitro* as powerfully as PC-G.

REFERENCES

ACRED, P., D. M. BROWN, E. T. KNUDSEN, G. N. ROLINSON, and R. SUTHERLAND (1967): New semisynthetic penicillin active against *Pseudomonas pyocyanea.* Nature (London) New Biol. 215, 25–30.

ALTERMIER, W. A., R. P. HUMMEL, E. O. HILL, and S. LEWIS (1973): Changing patterns in surgical infections. Ann, Surg. 178, 436–445.

FINLAND, M. (1970): Changing ecology of bacterial infections as related to antibacterial therapy. J. Infect. Dis. 122, 419–431.

NISHIDA, M., T. MATSUBARA, T. MURAKAWA, Y. MINE, Y. YOKOTA, S. KUWAHARA, and S. GOTO (1970): *In vitro* and *in vivo* evaluation of cefazolin, a new cephalosporin C derivative. 236–243. Antimicrob. Agents Chemother.

NOGUCHI, H., M. MATSUHASHI, M. TAKAOKA, and S. MITSUHASHI (1978): New antipseudomonal penicillin, PC-904: affinity to penicillin-binding proteins and inhibition of the enzyme crosslinking peptidoglycan. Antimicrob. Agents Chemother. 14(4), 617–624.

RELLER, L. B., W. W. KARNEY, H. M. BEATY, K. K. HOLMES, and N. TURCK (1973): Evaluation of cefazolin, a new cephalosporin antibiotic. Antimicrob. Agents Chemother. 3, 488–497.

SAWADA, Y., S. YAGINUMA, M. TAI, S. IYOBE, and S. MITSUHASHI (1976): Plasmid-mediated penicillin beta-lactamase in *Pseudomonas aeruginosa.* Antimicrob. Agents Chemother. 9, 55–60.

SAWAI, T., S. MITSUHASHI, and S. YAMAGISHI (1968): Drug resistance of enteric bacteria. XIV. Comparison of β-lactamases in gram-negative rod bacteria resistant to α-aminobenzylpenicillin. Jpn. J. Microbiol. 12, 423–434.

UEO, K., Y. FUKUOKA, T. HAYASHI, T. YASUDA, H. TAKI, M. TAI, Y. WATANABE, I. SAIKAWA, and S. MITSUHASHI (1977): In vitro and in vivo antibacterial acitivity of T-1220, a new semisynthetic penicillin. Antimicrob. Agents Chemother. 12(4), 455–460.

YAGINUMA, S., T. SAWAI, H. ONO, S. YAMAGISHI, and S. MITSUHASHI (1973): Biochemical properties of a cephalosporin β-lactamase from *Pseudomonas aeruginosa*. Jpn. J. Microbiol. 17, 141–149.

YAMAGISHI, S., K. OHARA, T. SAWAI and S. MITSUHASHI (1969): The purification and properties of penicillins β-lactamases mediated by transmissible R factors in *Escherichia coli*. J. Biochem. 66, 11–20.

N. M., *Gunma University,*
Maebashi, Gunma, Japan

IN *VITRO* SUSCEPTIBILITY TO CEFAMANDOLE OF STAPHYLOCOCCI, ENTEROCOCCI, ENTEROBACTERIACEAE, AND PSEUDOMONAS SPECIES

E. STRAUBE and G. NAUMANN

Institute of Medical Microbiology and Epidemiology,
Department of Medicine,
Wilhelm-Pieck-University Rostock, G.D.R.

Cefamandole exhibits in vitro an antibacterial activity against a wide range of gram-positive and gram-negative bacteria including all cephalothin-susceptible gram-positive cocci, some methicillin-resistant staphylococci, and strains of most genera within the *Enterobacteriacaeae* family (Lode, Baruch, Koeppe, and Lehmann-Brauns, 1978, Eickhoff and Ehret, 1976, and Hall, Gerding, and Schierl, 1977).

Most enterococci are inhibited by concentrations greater than 16 µg/ml.

Also most *Pseudomonas aeruginosa* strains are resistant against therapeutical concentrations.

A total of 245 strains of staphylococci, enterococci, *Enterobacteriacaeae*, and *Pseudomonas species* isolated in 1977 and 1978 from clinical specimens and clinical background, especially from an urologic department, were tested for susceptibility to Cefamandole in a tube dilution test.

Organisms susceptible to a Cefamandole concentration of 32 µg/ml were defined as being resistant. (Plaue, Müller, Fabricius, and Bethke, 1978, Höffler, Moecke, and Sassmann, 1978).

TABLE I.

In vitro susceptibility to cefamandole determined by tube dilution test
with Mueller-Hinton broth

Organism	No. of strains	0,06	0,125	0,25	0,5	1	2	4	8	16	32	64 and more µg/ml
Staph. aureus	52	4	4	6	18	9	8	3				
Staph. epidermidis	8				4	3	1					
Enterococci	36									1	14	21
E. coli	21				2	7	3	2	1	1	3	2
Klebsiella	24					2	3	2	1	1	2	13
Enterobacter	11										3	8
Citrobacter	7										4	3
Proteus mirabilis	18				2	3	4	2				7
Proteus indole-positive	16			1	1	3		1	2		1	7
Proteus inconstans	20					6		4	5	3		2
Pseudomonas	32									2		30

The 52 strains of *Staphylococcus aureus* were sensitive to 4 μg/ml Cefamandole and the geometric mean of inhibitory concentration (MIC) was determined as 0.56 μg/ml and the geometric mean of bactericidal concentration (MBC) as 1.62 μg/ml. The patterns of susceptibility of *Staphylococcus epidermidis* to Cefamandole are about the same.

The 36 strains of enterococci show no susceptibility to Cefamandole except one strain. The MIC of 21 strains of *Escherichia coli* tested was 3.62 μg/ml and the MBC 7.25 μg/ml. 76% of these strains are sensitive to a concentration of 16 μg/ml. In contrast to other authors (Bodey and Weaver, 1976, Eickhoff and Ehret, 1976, and Hall, Gerding, and Schierl, 1977) we found the 24 klebsiella strains tested to be more resistant. About 60% of these strains are resistant to 32 μg/ml Cefamandole or even more.

The susceptibility of 11 strains of *Enterobacter* and 7 strains of *Citrobacter* tested was very similar.

61% of Proteus mirabilis strains tested were sensitive to 4 μg/ml. Cefamandole. A group of 7 strains was resistant to more than 32 μg/ml. The MIC for *Proteus mirabilis* was within the range of 6.9 μg/ml and the MBC within 14.8 μg/ml.

The Indole-positive proteus strains seem to be a bit more resistant to Cefamandole. From 16 strains tested 8 were sensible to 8 μg/ml and 8 strains were resistant to more than 32 μg/ml Cefamandole. The MIC was within the range of 11.3 μg/ml and the MBC within 30.6 μg/ml.

More sensitive than expected were the 20 *Proteus inconstans* (Providencia) tested. Only two of them were resistant to more than 32 μg/ml and the MIC was determined as 5.1 μg/ml and the MBC as 12.1 μg/ml.

All pseudomonas strains — a total of 32 was tested — are resistant to more than 32 μg/ml Cefamandole except two strains.

REFERENCES

BODEY, G. P., and S. WEAVER (1976): *In Vitro* Studies of Cefamandole. Antimicrob. Agents Chemother. 9, 452–456.

EICKHOFF, T. C., and J. M. EHRET (1976): *In Vitro* Comparison of Cefoxitin, Cefamandole, Cephalexin, and Cephalothin. Antimicrob. Agents Chemother. 9, 994–999.

HALL, W. H., D. N. GERDING, and E. A. SCHIERL (1977): Antibacterial Activity of Cefamandole, Eight Other Cephalosporins, Cefoxitin and Ampicillin. Curr. Ther. Res. 21, 374–388.

HÖFFLER, D., D. MOECKE und M. SASSMANN (1978): Cefamandol: Pharmakokinetik bei normaler und eingeschränkter Nierenfunktion. Dtsch. med. Wschr. 103, 1334–1338.

LODE, H., B. BARUCH, P. KOEPPE und S. LEHMANN-BRAUNS (1978): Neue Entwicklungen bei den Cephalosporin-Antibiotika. Infection 6, S197–202.

PLAUE, R., O. MÜLLER, K. FABRICIUS und R. O. BETHKE (1978): Untersuchungen über die Diffusionsrate von Cefamandol in verschiedene menschliche Gewebe. Arzneim.-Forsch.: Drug Res. 28, 2343–2349.

E. S., Wilhelm-Pieck-Univerzity Dept.
Med. Inst. Med. Microbiol. Epidemiol.,
Leninallee 70, 25 Rostock, G.D.R.

YM09330, A NEW BROAD-SPECTRUM SEMISYNTHETIC CEPHAMYCIN ANTIBIOTIC
ANTIBACTERIAL ACTIVITIES
AND SOME BIOCHEMICAL APPROACHES

T. SAITO, M. TODA, M. INOUE, M. SAITO, K. SUZAKI,
K. YANO, T. OSONO and S. MITSUHASHI

*Laboratory of Microbial Resistance, School of Medicine, Gunma University,
Maebashi, Japan, Yamanouchi Pharmaceutical Co., Ltd.
Tokyo, Japan*

INTRODUCTION

We have been screening new semisynthetic cephalosporins which will be stable against β-lactamases of bacterial sources of wide range, will show higher permeability through the bacterial cell wall and thus will have broader antibacterial spectra and a stronger activity especially for stubborn gram-negative pathogens.

YM09330, 7β-[4-(Carbamoylcarboxymethylene)-1,3-dithietan-2-yl] carboxamido-7α--methoxy-3-(1-methyltetrazol-5-yl)thiomethyl-$\triangle^3$-cephem-4-carboxylic acid, is a new parenteral semisynthetic cephamycin derivative thus selected, and the chemical properties of YM09330 were presented at the Amer. Chem. Soc.-Chem. Soc. Jap. Chemical Congress, April, 1979, Honolulu. This paper deals with the *in vitro* and *in vivo* antibacterial activities of this drug.

RESULTS and DISCUSSION

The spectrum of antibacterial activity of YM09330 against gram-positive and gram-negative organisms was compared with cefoxitin (CFX), cefamethazole (CMZ) and cefazolin (CEZ). The MIC values of YM09330 against *Staphylococcus aureus* were larger than that of CEZ, inhibiting the growth at concentrations ranging from 3.13 to 12.5 μg/ml using one loopful of inoculation of 10^8 colony forming units (CFU) per ml. Against gram-positive bacteria, YM09330 was about 2 to 4 times less active than CFX and CEZ. YM09330 showed a high antibacterial activity against gram-negative bacteria except for *Pseudomonas aeruginosa* and *Acinetobacter calcoaceticus*. Against standard strains of *Escherichia coli, Klebsiella pneumoniae, Citrobacter freundii, Salmonella* spp., *Shigella* spp., *Enterobacter aerogenes, Proteus mirabilis,* indole-positive *Proteus* spp., *Seratia marcescens* and *Haemophilus influenzae,* it showed growth inhibition at 0.05 to 0.78 μg/ml, and for *Enterobacter cloacae* at 3.13 μg/ml. Against these strains, YM09330 was about two to over 100 times more potent than CFX and CEZ. But like other references, YM09330 did not inhibit the growth of *Pseudomonas aeruginosa* and *Acinetobacter calcoaceticus*. YM09330 was as potent as CFX against *Bacteroides* spp., giving MICs of 6.25 to 100 μg/ml using one loopful of inoculation of 10^6 CFU per ml. Against other anaerobes, YM09330 was as active as the reference used.

The susceptibility distribution of YM09330 was examined against about 1500 clinical isolates of gram-positive and gram-negative bacteria at an inoculum size of 10^6 CFU/ml

TABLE I.

Susceptibility of clinical isolates to YM09330, cefoxitin, cefametazole and cefazolin

Organisms	No. of Strain	YM09330 MIC for		CFX MIC for		CMZ MIC for		CEZ MIC for	
		50%	70%	50%	70%	50%	70%	50%	70%
S. aureus	200	3.84	5.10	1.28	1.54	0.56	0.68	0.19	0.28
Str. pyogenes	175	0.96	1.28	0.32	0.48	0.22	0.32	0.01	0.15
E. coli	160	0.32	0.38	5.32	6.88	1.21	1.44	1.87	2.81
K. pneumoniae	200	0.13	0.17	4.38	5.79	1.17	1.40	3.60	5.16
H. influenzae	58	0.52	0.66	1.28	1.48	1.71	2.50	7.04	10.20
P. mirabilis	100	0.18	0.25	5.01	5.95	2.65	3.75	5.95	9.40
P. vulgaris	54	0.21	0.30	4.22	5.47	2.18	2.70	>100	>100
P. rettgeri	31	0.18	0.86	5.63	30.00	3.91	21.25	80.00	>100
P. morganii	54	2.96	4.22	10.31	12.50	7.18	10.60	>100	>100
Ent. cloacae	81	3.13	30.00	84.50	100	62.50	100	>100	>100
C. freundii	65	0.34	17.50	45.00	95.00	20.00	45.00	11.50	>100
Serr. marcescens	241	1.36	4.53	25.00	65.00	13.80	45.00	>100	>100
Ps. cepacia	79	5.90	12.50	60.00	80.00	35.00	50.00	>100	>100

One loopful of 10^6 CFU/ml.
CFX, cefoxitin; CMZ, cefametazole; CEZ, cefazolin. MIC, μg/ml.

(Table I.). Against both S. aureus and Streptococcus pyogenes, YM09330 was slighty less active than CFX and CEZ. Against gram-negative bacteria, YM09330 showed the highest degree of activity, being 3.5 to over 100-fold more potent than CEZ. Seventy per cent of clinical isolates of E. coli, K. pneumoniae, H. influensae, P. mirabilis, indole-positive Proteus spp., and S. marcescens were inhibited at concentrations of 0.17 to 4.5 μg/ml of YM09330, indicating that YM09330 is one of the most active cephamycin derivatives. YM09330 was shown to be more active than any of the above cephalosporins against Pseudomonas cepacia strains which were shown to produce a strong β-lactamase hydrolyzing cefuroxime (CXM) and cefamandol (CMD). MIC_{73} of YM09330 against P. cepacia strains was found to be 12.5 μg/ml. Further it was more active than the above two cephalosporins against E. cloacae and C. freundii, but its activity was rather weak, showing MIC_{70} against clinical isolates at 30 and 17.5 μg/ml, respectively.

The MICs of YM09330 against ten gram-negative organisms and a S. aureus strain were compared under various conditions. Changes in inoculum size had some effect on MICs for some bacterial strains, i.e. S. aureus Terajima, P. aeruginosa NCTC 10490 and E. cloacae 963, which became more susceptible to YM09330 by reducing the size of inoculum from 10^9 to 10^5 CFU/ml, giving a change of MICs from 6.25 to 1.56 μg/ml, from more than 100 to 25 μg/ml and from 25 to 3.13 μg/ml, respectively and indicating no significant effect by the changing inoculum size. No significant difference in MICs of YM09330 was observed on change of growth media: Heart infusion agar, Mueler-Hinton (MH) agar, Nutrient agar, Brain heart infusion agar and Trypticase soy agar in combination of pH changes ranging from 5.5 to 7.5. The MBC and MIC values of YM09330 were completely identical at 10^4 CFU per ml inoculum against E. coli and K. pneumoniae, but against S. marcescens the MBC value was about twice that of MIC.

Bactericidal activity of YM09330 against E. coli GN6370 was examined in Antibiotic medium 3 (Bacto-Penassay Broth) containing serial twofold dilutions of the antibiotic. A clear decrease of viable bacteria was observed at 0.1 μg/ml of YM09330 which was the same concentration as the MIC determined by the agar dilution method. At a concen-

224

TABLE II.

Substrate profiles of β-lactamase produced by various gram-negative bacteria

CSase producer	Substrate	Relative rate of hydrolysis							
		CER	CEZ	CET	CEX	CTM	CMD	CXM	YM09330
E. coli	GN 5482	100%	70	370	54	42	0	0	0
Ent. cloacae	GN 7471	100	52	420	56	76	0	0	0
C. freundii	GN 346	100	116	125	80	8	0	0	0
Serr. marcescens	GN 48	100	457	389	30	60	0	0	0
P. vulgaris	GN 7919	100	555	190	48	150	225	219	0
P. morganii	GN 5407	100	68	232	28	36	4	0	0
P. rettgeri	GN 4425	100	36	44	0	22	4	0	0
Ps. aeruginosa	GN 98	100	131	496	33	33	0	0	0
Ps. cepacia	GN 11164	100	115	176	60	69	147	85	0

β-Lactamase activity was assayed according to Samuni.
CTM, cefotiam; CMD, cefamandole; CXM, cefuroxime.

tration of 0.1 μ/ml, regrowth was not observed during an incubation time of twenty-four hours.

The dose-response curve of YM09330 showed steep gradient and gave ID_{50} values (Kato et al., 1978) of 0.023 μg/ml against *E. coli* NY-17, 0.041 μ/ml against *K. pneumoniae* ATCC 10031, 0.074 μg/ml against *P. mirabilis* 1278, 0.044 μg/ml against *P. vulgaris* IID 874, 0.016 μg/ml against *P. morganii* Kono and 0.045 μg/ml against *S. marcescens* IID 620, respectively.

Enzymatic studies using spectrophotometric assay (Samuni, 1975) were carried out on the resistance of YM09330 to β-lactamases obtained from various bacteria which were shown to produce cephalosporinase. Results are given in Table II indicating that YM09330 has strong resistance to all types of β-lactamase. It was not hydrolyzed by any of the β-lactamases tested, while CMD and CXM were hydrolyzed by *P. vulgaris* type and *P. cepacia* type β-lactamases.

According to Spratt, B. G. (1977), inner membrane protein fractions were prepared from *E. coli* JE 1011, *S. marcescens* GN7403 and *P. aeruginosa* NCTC 10490. Affinity of YM09330 for PBPs was detected by measuring competition for binding of [^{14}C]-PC-G to the PBPs and by direct binding of [^{14}C]-YM09330 to them. The fluorograms thus obtained are given as Fig. 1. As seen from the figure, YM09330 showed high affinity for almost all PBPs, with stronger affinity for PBP 3, with strong affinity for PBPs 1A, 1Bs and 5/6, with moderate affinity for PBP 4 and with little affinity for PBPs 2 and 4'.

In the course of our experiments, new binding proteins were found in the inner membrane fraction of *E. coli* JE 1011.

By direct binding of [^{14}C]-YM09330, there were detected four new binding proteins: Y-1 (MW = 75,000), Y-2 (MW = 68,000), Y-3 (MW = 57,000) and Y-4 (MW = 53,000). But their functions are not yet known. In addition to the above four proteins, several bands of proteins were observed at about MW 30,000.

The 50 per cent effective dose (ED_{50}) of YM09330 was 1.18 mg/kg in mice infections with *E. coli* NY-17. Mice were challenged intraperitoneally, with 2.5×10^6 cells per mouse ($25LD_{50}$), and the drug was given subcutaneously 1 and 4 hr after the infection. Thus YM09330 was shown to have about 30 times greater protective effect than CEZ and CFX.

On the other hand, the 50 per cent effective dose (ED_{50}) of YM09330 was 13.0 mg/kg in mice infections with *S. marcescens* GN 7577. Mice were intraperitoneally challenged with 1.0×10^5 cells per mouse (90LD$_{50}$), and the drug was given subcutaneously 1 and 4 hr after the infection. Thus YM09330 was shown to have about 35 times greater protective effect than CFX.

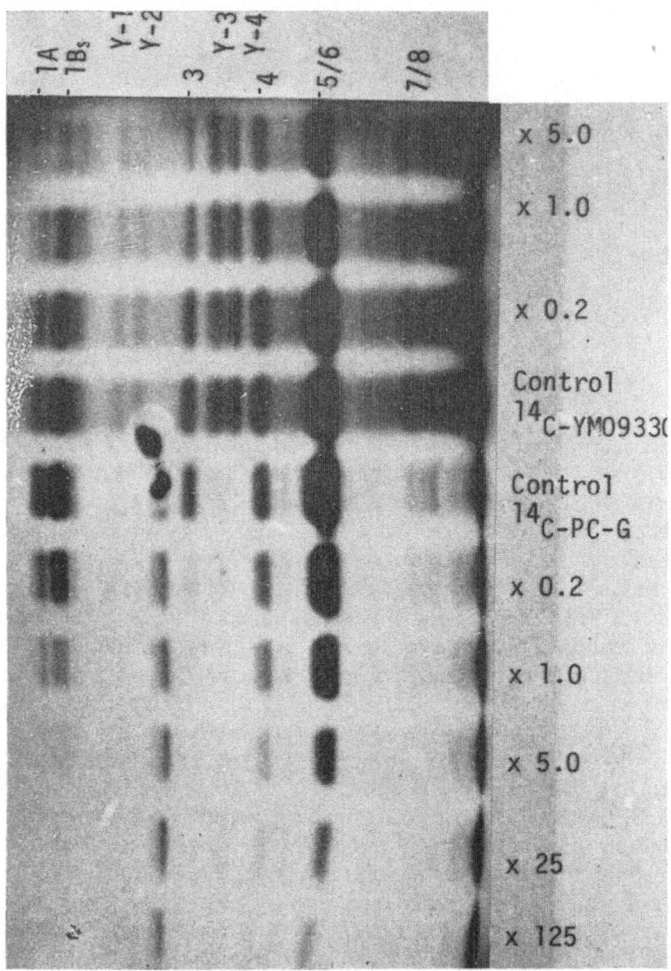

Fig. 1. Fluorogram showing competition of YM09330 for [14C]-benzylpenicillin (PC-G) and for [14C]-YM09330 binding to envelopes of *E. coli* JE1011. Washed cell envelopes were preincubated with water (each control). Concentrations of antibiotics are expressed as molar ratios to [14C]-PC-G and [14C]-YM-09330, respectively.

The effect of YM09330 on the morphology of *E. coli* NY-17 and *S. marcescens* IID 620 was examined by phase contrast and scanning electron microscopies. On exposing to serial dilution of YM09330 in MH broth, both strains showed long filamentous shape at concentrations near MIC after 2 hr of incubation, and then lysis proceeded in the long filaments down to the MIC to give only elongated filaments attached with spheroplast-like structures at the MIC within 18 hours. (cf. Fig. 2.) But CFX and CEZ did not show

226

Fig. 2. Morphological changes of the cells of *E. coli* NY-17 (left) and *Serratia marcescens* IID 620 (right) induced by YM09330. Both organisms were exposed at the concentration of 0.78 μg/ml for 5 hours in MH broth at 37 °C. The samples for the scanning electron microscope were prepared after doublefixing with glutaraldehyde and OsO₄. (Scale bar = 5 μm)

such a lytic change under the corresponding conditions. O'Callaghan, C. H. et al. (1976) recently suggested that stability against β-lactamases produced by gram-negative organisms is important in determing the range of the spectrum and the antibacterial effectiveness of a compound. It is also advantageous for the compound to be lytic and to be able to kill bacterial cells rapidly at low concentrations. The lytic character revealed by YM09330 is considered to have a definite meaning for its much stronger activity against gram-negatives than CFX which was resistant to most of the β-lactamases tested.

REFERENCES

KATO, T., S. KURASHIGE, Y. A. CHABBERT and S. MITSUHASHI (1978): Determination of the ID_{50} value of antibacterial agents in agar. J. Antibiot. 31, 1299–1303.

O'CALLAGHAN, C. H., R. B. SYKES, A. GRIFITHS and J. E. THORNTON (1976): Cefuroxime, a new cephalosporin antibiotic; activity in vitro. Antimicrob. Ag. Chemother. 9, 511–9.

SAMUNI, A. (1975); A direct spectrophotometric assay and determination of Michaelis constants for the β-lactamase reaction. Anal. Biochem. 63, 17–26.

SPRATT, B. G. (1977): Properties of the penicillin binding proteins of *Escherichia coli* K 12. Eur. J. Biochem. 72, 341–352.

T. S., School of Medicine,
Gunma University, Maebashi, Japan

IMMUNOCHEMICAL PROPERTIES OF THE SURFACE STRUCTURES OF *SALMONELLA TYPHIMURIUM* STRAINS RESISTANT TO CEPHALORIDINE

C. PLOCZEKOVÁ, J. KAROLČEK and I. ČIŽNÁR

*Research Institute of Preventive Medicine, Bratislava,
Czechoslovakia*

INTRODUCTION

The resistance of gramnegative bacteria to β-lactam antibiotics seems to depend on three factors: first, on the ability of the antibiotic to inhibit peptidoglycan synthesis in the cell wall; second, on the presence of β-lactamases produced after acquisition of the specific plasmid, and third, on the ability of the antibiotic to penetrate through the outer layer of the cell envelope (Richmond and Wotton, 1976). The third factor often called nonspecific bacterial antibiotic resistance is poorly understood.

Variations in composition of proteins located in the cell wall (Wu, 1972), as well as lipoproteins (Nikaido, Bavoil and Hirota, 1977), phospholipids (Suling and O'Leary, 1977) and lipopolysaccharides (Roantree, Kuo and Mac Phee, 1977) have been correlated to the nonspecific resistance to antibiotics, detergents and dyes in bacteria.

Lipopolysaccharide represents the dominant structure of the cell wall of gramnegative bacteria. It has been shown that in many cases the penetration of some substances to the inner space of the cell is limited by the molecular weight of 700 daltons. The structure responsible for this limit was identified to be lipopolysaccharide (Nakae and Nikaido, 1977). For characterisation of the substances which resist the penetration of the antibiotics to the target structures, chemical analysis has often been used (Schlecht and Westphal, 1970). This has shown that rough R-mutants sensitivity to antibiotics increased progressively with degradation of the R-core of lipopolysaccharide. It has been found that the lipopolysaccharide of the same bacterial strain can contain a population of macromolecules with different lengths of side chains (Galanos, Lüderitz, Rietschel and Westphal, 1977). This stimulates to the assumption whether differences in the structure of the polysaccharide moiety are related to nonspecific resistance to cephalosporin antibiotics.

We have analysed *Salmonella typhimurium* lipopolysaccharide of the strain resistant and sensitive to cephaloridin.

MATERIAL AND METHODS

Salmonella typhimurium was isolated from the patient and was resistant to cephaloridin and cephalotin when tested by the disc diffusion method. The strain was maintained on DST agar plates containing cephaloridin at 32 µg/ml and was further used in analysis. Two strains of *Salmonella typhimurium* isolated from the patients were used as control strains. Both were sensitive to cephaloridin.

Lipopolysaccharides of the tested strains were extracted from the cell wall by phenol-water procedure and purified by ultracentrifugation (Westphal and Jann, 1965).

The complete antigenic extract of the cell walls was prepared by osmolysis and ultrasonic desintegration of the cells and further by extraction with buffered physiological solution pH 7.2. The extract was dialyzed and concentrated by lyophilization.

Rabbit antiserum was prepared by imunization of rabbits with *Salmonella typhimurium* vaccine as described previously (Čižnár and Shands, 1970).

Double diffusion was performed in agar gel.

Immunoelectrophoresis in agar gel was performed by standard procedure in veronal-citrate buffer, pH 8.6, 6 V/cm, 1.5 hr.

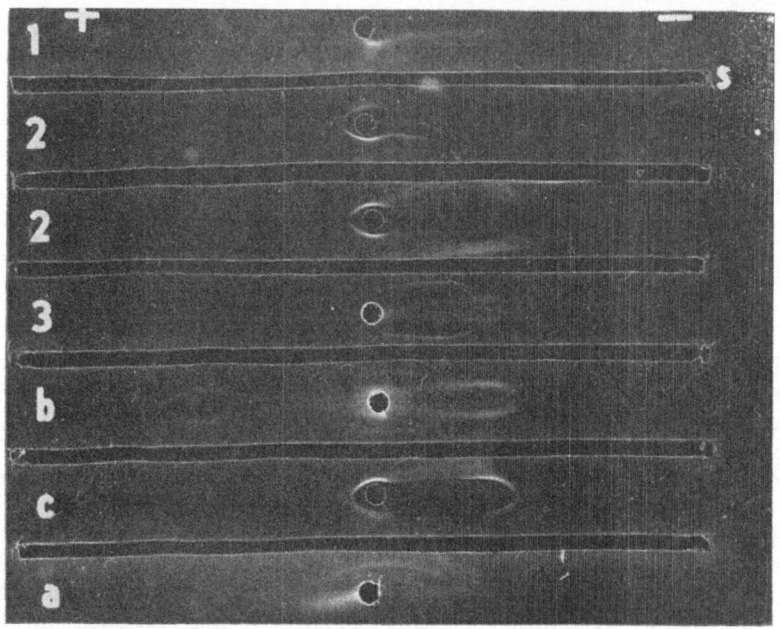

Fig. 1. Agar gel precipitin test.

s = rabbit antiserum
1,3 = lipopolysaccharides from sensitive strains
2 = lipopolysaccharide from the resistant strain
a,c = antigenic extract from sensitive strains
b = antigenic extract from the resistant strain

RESULTS AND DISCUSSION

Fig. 1. shows the reactivity of lipopolysaccharide from sensitive strains (1 and 3) with specific rabbit antiserum. Number two represents lipopolysaccharide isolated from the resistant strain. This lipopolysaccharide produces a different pattern of precipitin bands with rabbit antiserum as compared with lipopolysaccharide of the sensitive strains. It forms one strong and one weak precipitin band. The lipopolysaccharide of sensitive strains produces more bands, however, their intensity is lower. Contrary to these reactions, the complete antigenic extract of resistant strain (b) produces more precipitin bands than the extracts from sensitive strains (a, c).

Similar results were obtained by the immunoelectrophoretic analysis of the preparations. This was done by the standard procedure in agar gel. (Fig. 2.). Again, the number of precipitin bands as well as their position was different. The lipopolysaccharide from the resistant strain (2) formed precipitin bands on the cathodic side of the immuno-electrophoreogram, similarly lipopolysaccharide from the sensitive strains (1, 3). However, the dense line formed by lipopolysaccharide from the sensitive strains was missing in the reaction of lipopolysaccharide from the resistant strain.

The immunoelectrophoretic analysis of ultrasonic antigenic extract showed that there are more precipitin bands produced by the extract from the resistant strain (b) than from

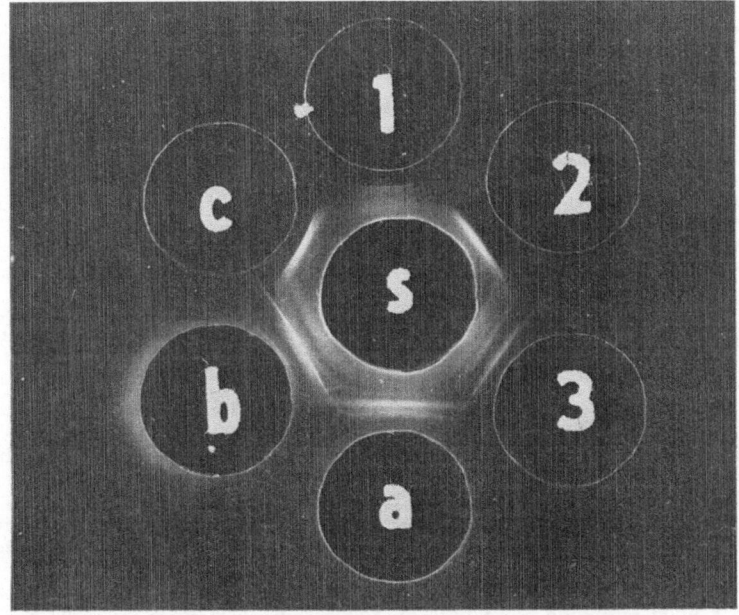

Fig. 2. Immunoelectrophoresis of cell wall substances.

s   = rabbit antiserum
1,3 = lipopolysaccharide from sensitive strains
2   = lipopolysaccharide from the resistant strain
a,c = antigenic extract from sensitive strains
b   = antigenic extract from the resistant strain

Fig. 1. Determination of the ID_{50} of AM-715 against *E. coli* ML4707 and *P. aeruginosa* NC-5.

the sensitive strains (a, c). Similarly, when these extracts were analysed in polyacrylamide electrophoresis, more protein fractions were detected in the extract from the resistant strain.

These results would indicate that there is a relation between resistance to cephaloridin and the structures of the outer layer of the cell wall in *Salmonella typhimurium*. Comparison of the immunodiffusion pattern and the immunoelectrophoretic behaviour showed that the lipopolysaccharide of the resistant strain was more homogeneous. The homogenicity in this type of analysis is expressed by the number of precipitin bands and by the mobility and the position of bands in immunoelectrophoresis. Lipopolysaccharide is a polyanion which due to its amphipathic character can form aggregates up to a molecular weight of several millions. In double diffusion analysis it always forms several precipitin bands and in immunoelectrophoresis it migrates towards the cathode forming

231

a line spread from the start. Whether the alteration of this pattern of lipopolysaccharide from the resistant strain is related to the permeability function of this layer is difficult to say.

Since more fractions or more protein bands were detected in the cell wall of the resistant strain, this would indicate that the permeability function of the outer layer may be changed. Similar results were obtained with Neisseria gonorrhoeae antibiotic sensitive and resistant strains (Guymon, Walstad and Sparling, 1978).

The significance of these observations and the possibility that the permeability function of the outer layer of the cell wall is altered in the *Salmonella typhimurium* strain resistant to cephaloridin is being further investigated.

REFERENCES

ČIŽNÁR, I. and J. W. SHANDS (1970): Effect of alkali on the Immunological reactivity of Lipopolysaccharide from *Salmonella typhimurium*. Infect. Immunity. 2, 549–555.

GALANOS, C., O. LÜDERITZ, E. T. RIETSCHEL and O. WESTPHAL (1977): Newer aspects of the chemistry and biology of bacterial lipopolysaccharides, with special reference to their Lipid A component. Inter. Rev. Biochem. 14, 240–335.

GUYMON, L. F., D. L., WALSTAD and P. F. SPARLING (1978): Cell Envelope Alterations in Antibiotic – Sensitive and – Resistant Strains of *Neisseria gonorrhoeae*. J. Bacteriol. 136, 391–401.

LELVE, L. (1974): The barrier function of the gramnegative envelope. Ann. N. Y. Acad. Sci. 235, 109–129.

NAKAET, T. and NIKAIDO, M. (1975): Outer membrane as a diffusion barrier in *Salmonella typhimurium*. J. Biol. Chem. 250, 7359–7365.

NIKAIDO, H., P. BAVOIL and Y. HIROTA (1977): Outer membranes of gramnegative bacteria XV. Transmembrane diffusion rates in Lipoprotein – deficient mutants of *Escherichia coli*. J. Bacteriol. 132, 1045–1047.

RICHMOND, M. H. and S. WOTTO (1976): Comparative study of seven cephalosporins: susceptibility to beta – lactamases and ability to penetrate the surface layers of *Escherichia coli*. Antimicrob. Agents. Chem. 10, 219–222.

ROANTREE, R. J., T. T. KUO and D. G. MAC PHEE (1977): The effect of different Lipopolysaccharide core defects upon antibiotic resistances of *Salmonella typhimurium*. J. Gen. Microbiol. 103, 223–234.

SCHLECHT, S. and O. WESTPHAL (1970): Untersuchungen zur Typisierung von *Salmonella* – R – Formen IV. Typisierung von s. minnesota – R – Mutanten mittels Antibiotica. Zentralbl. Bacteriol. Parasitenkr. Infektionskr. Hyg. Abt. I Orig. 213, 356–381.

SULING, W. J. and O. LEARY, W. M. (1977): Lipids of antibiotic – resistant and – susceptible members of the *Enteribacteriaceae*. Canad. J. Microbiol. 23, 1045–1051.

WESTPHAL, O. and K. JANN (1965): Methods in carbohydrate chemistry. 5, 63.

WU, H. C. (1972): Isolation and characterisation of an *Escherichia coli* mutant with alteration in the outer membrane proteins of the cell envelope. Biochim. Biphys. Acta. 290, 274–289.

C. P.,
Research Institute of Preventive Medicine
Sasinkova 13, 809 58 Bratislava
Czechoslovakia

IN VITRO AND *IN VIVO* ANTIBACTERIAL ACTIVITY, β-LACTAMASE STABILITY AND BINDING AFFINITY TO PENICILLIN-BINDING PROTEINS OF CEFOTAXIME (HR756), A NEW CEPHALOSPORIN DERIVATIVE

S. MASUYOSHI, S. ARAI and M. MIYAMOTO

Department of Microbiology, School of Medicine, Gunma University, Gunma and Development Laboratories, Hoechst Japan Limited, Saitama, Japan

INTRODUCTION

Many cephalosporins with broad antibacterial spectrum have been in clinical use for the treatment of a wide range of bacterial infections. Recently, however, the effectiveness of the existing cephalosporins has been shown to be limited, there has been a steady increase in clinical infections caused by gram-negative bacteria resistant to β-lactam antibiotics, *e.g., Pseudomonas* spp., *Enterobacter* spp., *Serratia marcescens* and indole-positive *Proteus* spp., which have been reported to produce β-lactamase (Mitsuhashi, Yamagishi, Sawai and Kawabe 1977). Much efforts have then been expended in developing broad-spectrum antibiotics with resistance to the enzyme. As a result, cefoxitin (CFX) and cefuroxime (CXM) have been so far semisynthesized but they have been shown to be still ineffective against some strains of *Enterobacter* spp. and *Pseudomonas* spp. *in vitro* (Une and Mitsuhashi, 1977, Neu and Fu, 1978).

HR 756, a new cephalosporin, has been reported to have a broad antibacteria spectrum (Drasar, Farrell, Howard, Hince, Leung and Williams, 1978, Hamilton-Miller, Brumfitt and Reynolds, 1978, Heymes, Lutz and Schrinner, 1978) and to be a potent inhibitor of β-lactamase that is stable to the enzyme (Fu and Neu, 1978). The present study investigated the *in vitro* and *in vivo* antibacterial activity, β-lactamase stability and binding affinity to penicillin-binding proteins of the compound provided by Hoechst Japan Limited using strains isolated clinically in Japan.

RESULTS AND DISCUSSION

Antibiotics. Cefotaxime (HR 756) is a white powder soluble in water and its chemical name is sodium 7-[2-(2-amino-4-thiazolyl)-2-methoxyimino]-acetamido cephalosporanate. Cefoxitin, cefuroxime and SCE 963 were gifts from Daiichi Seiyaku Co., Ltd., Shinnihon Jitsugyo Co., Ltd., and Takeda Chemical Industries, Ltd., respectively. The following antibiotics were commercially available products: cefazolin (CEZ), cephaloridine (CER), cephalothin (CET), cephalexin (CEX), carbenicillin (CBPC), penicillin G (PC-G), gentamicin (GM), and kanamycin (KM).

Determination of MICs. Organisms tested were standard strains obtained from stock cultures of the Laboratory of Microbial Resistance, Gunma University, and strains isolated recently from patients at several hospitals in Japan.

Minimal inhibitory concentrations (MICs) were determined by an agar dilution method. Twofold serial antibiotic dilutions were added to heart infusion (HI) agar, brain heart infusion (BHI) agar or BHI agar containing hemin and nicotinamide adenine dinucleotide

TABLE I.

Antibacterial activity of HR 756 and other antibiotics against
gram-positive and gram-negative clinical isolates

Species	No. of strains	Drug concentration (μg/ml) MIC$_{70}$				
		HR 756	Cefazolin	Cefoxitin	Carbenicillin	Gentamicin
E. coli	110	0.086	2.24	4.80	—	—
K. pneumoniae	110	0.066	3.00	3.40	—	—
P. aeruginosa	139	14.0	>100	—	50.0	1.25
P. cepacia	32	9.40	—	—	>100	>100
E. cloaceae	129	0.560	100	—	90.0	—
S. marcescens	135	0.380	>100	—	>100	0.570
P. mirabilis	62	0.044	1.56	1.62	—	—
P. vulgaris	61	0.360	>100	10.2	—	—
P. morganii	60	0.170	>100	13.8	—	—
P. rettgeri	37	0.105	>100	8.00	—	—
P. inconstance	32	0.280	>100	4.70	—	—
Salmonella spp.	107	0.120	1.50	2.80	—	—
H. influenzae	21	0.040	40.0	3.60	—	—
S. pyogenes	32	0.014	0.079	0.35	—	—
S. aureus	112	2.25	0.420	2.70	—	—

(NAD), and the agar plates were inoculated with one loopful of approximately 10^6 colony-forming units (CFU) using an inoculum-replicating apparatus. The test organisms were grown at 37°C overnight in different media; growth media used for the preculture were: peptone water containig 0.3% KNO_3 for *Pseudomonas* strains, BHI broth for *S. pyogenes*, BHI broth containing 10 μg/ml of hemin and 2 μg/ml of NAD for *H. influenzae*, and peptone water for the rest strains. The inoculum was prepared by diluting the cultures in buffer saline gelatine (BSG) solution. End points were read following 18 hr of incubation at 37°C. The *in vitro* antibacterial activity of HR 756 against gram-positive and gram-negative organisms was compared with that of CEZ, CFX, CBPC and GM by cumulative percentages of strains inhibited by various concentration of the drugs.

Table I. shows concentrations of the antibiotics required to inhibit the growth of 50 and 70% of the total number of strains examined (MIC$_{50}$ and MIC$_{70}$). Against *S. aureus*, HR 756 was slightly less active than CEZ but similarly active or a little more active than CFX. HR 756, however, showed a higher degree of activity than CEZ and CFX against gram-negative bacteria, *E. coli, K. pneumoniae, Salmonella* spp. and *P. mirabilis*, reported to be sensitive to the cephalosporins. HR 756 was 25- to 60-fold more active than CEZ and CFX against *E. coli, K. pneumoniae* and *P. mirabilis*, and 10 to 25 times as active as the two drugs against *Salmonella* spp. Furthermore, for species usually reported as being "cephalosporin-resistant" (indole-positive *Proteus* spp., *Enterobacter* spp., *S. marcescens* and *Pseudomonas* spp.), HR 756 was just as effective as for the cephalosporin-sensitive strains. The activity of HR 756 was greater than that of CFX against indole-positive strains; the activity of the compound was approximately 100-fold higher than that of CFX against *P. morganii* and *P. rettgeri* and 20- to 50-fold higher against *P. vulgaris* and *P. inconstance* than that of CFX.

In addition, HR 756, unlike the existing β-lactam antibiotics, exhibited a high degree of activity also to *P. aeruginosa, P. cepacia, E. cloacae* and *S. marcescens*. Toward *P. aeruginosa*, the MIC$_{50}$ value of HR 756 was 10.0 μg/ml compared with 39.0 μg/ml of CBPC.

TABLE II.

Effect of inoculum size on antibacterial activities of HR 756 cefazolin, SCE 963 and carbenicillin

Organism	Inoculum size (CFU)	HR 756		Cefazolin		SCE 963		Carbenicillin	
		MIC (µg/ml)	MBC (µg/ml)	MIC (µg/ml)	MBC (µg/ml)	MIC (µg/ml)	MBC (µg/ml)	MIC (µg/ml)	MBC (µg/ml)
	1.2×10^8	1.56	1.56	6.25	12.5	12.5	25.0		
	1.2×10^7	0.78	0.78	0.78	12.5	1.56	12.5		
	1.2×10^6	0.013	0.05	0.78	1.56	0.10	0.39		
E. coli	1.2×10^5	<0.006	<0.006	0.39	1.56	0.10	0.10		
NIHJ JC-2	1.2×10^4	<0.006	<0.006	0.39	0.78	0.05	0.10		
	1.2×10^3	<0.006	<0.006	0.39	0.78	0.05	0.05		
	1.2×10^2	<0.006	<0.006	0.39	0.78	0.05	0.05		
	2.9×10^7	12.5	100					12.5	200
	2.9×10^6	6.25	25					12.5	100
S. marcescens	2.9×10^5	0.10	1.56					3.13	12.5
IID620	2.9×10^4	0.05	0.39					1.56	12.5
	2.9×10^3	0.05	0.10					0.78	6.25
	2.9×10^2	0.025	0.05					0.78	1.56

Fifty percent of *P. cepacia* strains were inhibited by 7.80 µg/ml of HR 756 and by 92.0 µg/ml of GM. Against *E. cloacae*, the MIC_{50} and MIC_{70} values of HR 756 were, respectively, 0.23 and 0.56 µg/ml while those of CBPC were 8.20 and 90.0 µg/ml, respectively. HR 756 was also active to the same degree as GM and more active than CBPC against *S. marcescens* that has recently caused clinical infections increasingly, the MIC_{50} and MIC_{70} values of HR 756 were, respectively, 0.36 and 0.38 µg/ml compared with 0.52 and 0.57 µg/ml of GM and 70 and > 100 µg/ml of CBPC, respectively. Further, HR 756 was 6- to 100-fold more active than CEZ, and 25- to 65-fold more active than CFX toward *Salmonella* spp., *H. influenzae* and *S. pyogenes*.

Determination of bactericidal activity. *E. coli* ML4707 was grown in antibiotic medium 3 (ABM 3) at 37°C overnight, and inoculated at the density of 10^5 cells/ml into freshly prepared ABM 3. Antibiotics at concentrations ranging from 1/4 of the MIC to 4 times the value were added to the culture incubated in a shaker at 37°C for 2 hr. Samples collected in the course of time and applied to drug-free bromothymol blue (BTB) agar, and colonies were counted after 18 hr of incubation at 37°C. Against this organism HR 756 displayed a much higher degree of bactericidal activity than did CEZ and CFX, they apparently decreased viable bacterial cells in number at the concentration of 0.05, 1.56, and 3.13 µg/ml, respectively, and all of the drugs inhibited regrowth of the cells even after 20 hr of incubation.

Determination of MBCs. Each organism was grown overnight in ABM 3 and inoculated at the final concentration of 10^4 cells/ml into ABM 3 containing twofold serial antibiotic dilutions. The MICs were recorded, after 18 hr of incubation at 37°C, as the lowest concentration of antibiotic that inhibited the development of visible growth in the broth. Minimum bactericidal concentrations (MBCs) were measured by spotting 0.005 ml from culture broth tubes onto antibiotic-free BTB agar, and incubating the culture at 37°C for 18 hr. They were defined as the lowest concentration of drug that produced no colonies on the agar. Comparison between the MICs and MBCs of HR 756 and other 5 antibiotics was made using 25 strains of *E. coli*, 23 *P. aeruginosa* and 25 *S. marcescens*.

TABLE III.

β-lactamase hydrolysis of HR 756 compared with other known cephalosporins[a]

Organism	Type of β-lactamases	Relative rate of hydrolysis[b]						
		CER	HR 756	CXM	CEZ	CET	CEX	PC-G
E. coli W3630 (Rms212)[+]	PCase type I	18.2	<0.1	<0.1	7.2	7.3	<1.3	100
E. coli W3630 (Rms213)[+]	PCase type II	3.9	<0.1	<0.1	4.6	9.2	<2.6	100
P. aeruginosa MI4259 (Rms139)[+]	PCase type VI	8.6	<0.1	<0.1	<0.5	<0.5	<0.6	100
K. pneumoniae GN69	PCase	15.1	<0.1	<0.1	2.7	2.8	<0.5	100
E. coli GN5482	CSase	100	<0.1	<0.1	135	691	55.5	28.7
P. aeruginosa GN918	CSase	100	<0.1	<0.1	160	480	62.9	24.8
P. vulgar s GN76	CSase	100	28.0	148	357	204	52.0	21 0
E. cloacae GN7471	CSase	100	<0.1	<0.1	50	402	54.0	83.1
C. freundii GN346	CSase	100	0.1	<0.1	120	127	81.1	7.0
P. morganii GN5406	CSase	100	0.1	<0.1	74	242	31.0	121.0

[a] β-lactamase activity was determined by the spectrophotometric method.
[b] Hydrolysis of each substrate by PCase and CSase is expressed as a relative rate of hydrolysis taking the absolute rate of PC-G or CER hydrolysis as 100.

The MBC of HR 756, like CEZ, CFX, KM and CBPC, seemed to be almost the same as the MIC against *E. coli* and *P. aeruginosa*, but the MBC was four times higher against *S. marcescens*.

Effects of inoculum size on MICs and MBCs. Each test organism was grown in ABM 3 at 37°C overnight. Tenfold serial dilutions of the culture in ABM 3 were inoculated into ABM 3 containing twofold serial antibiotic dilutions, and the MICs and MBCs were measured. The antibacterial activity of β-lactam antibiotics is known to be influenced by inoculum size (Noguchi, Kubo, Kurashige and Mitsuhashi, 1978). Therefore, its effects on the MICs and MBCs for *E. coli* NIHJ-JC-2 and *S. marcescens* IID620 were investigated using 6 or 7 different inoculum sizes ranging from 10^2 to 10^7 or 10^8 CFU. The results are given in Table II. With both the test species, there was great influence on the MICs and MBCs of all the antibiotics examined. Large difference in the values occurred between 10^7 adn 10^6 CFU in *E. coli* and between 10^6 and 10^5 CFU in *S. marcescens*, and the concentrations became lower as the inoculum size decreased. The MBCs of HR 756, CEZ and the SCE 963 were almost identical to the MICs against *E. coli*, but the MBC values of HR 756 and CBPC were 2- to 16-fold higher against *S. marcescens*.

β-Lactamase preparation. *E. coli* W3630 (Rms212)[+], *E. coli* W3630 (Rms213)[+], *P. aeruginosa* ML4259 (Rms 139)[+] and *K. pneumoniae* GN69 were employed as strains producing penicillinase (PCase) (Mitsuhashi, Yamagishi, Sawai and Kawabe, 1977). Six strains were used as producers of cephalosporinase (CSase). These comprised *E. coli* GN5482, *P. aeruginosa* GN918, *P. vulgaris* GN76, *E. cloacae* GN7471, *C. freundii* GN346 and *P. morganii* GN5406. Overnight BHI broth cultures were diluted 10-fold in 250 ml od Medium B. Cultures of the PCase producers were harvested by centrifugation after incubated in a shaker at 37°C for 5 hr. With regard to cultures of the CSase producers, they were incubated in a shaker at 37°C for 3 hr, PC-G was then added as an inducer, and the bacteria were harvested after grown for another 2 hr. All the harvested organisms were washed twice with 0.05 M phosphate buffer (pH 7.0), and were suspended in 5 ml of phosphate buffer. The cells were disrupted by sonic treatment by an ultrasonicator at 30-s intervals for 10 min in an ice bath. The extracts were centrifuged for 30 min at 4°C at 9,000 × g, and the supernatant was used as the source of β-lactamase.

Susceptibility to β-lactamase. β-Lactamase activity was determined by the spectro-photometric method (Samuni, 1975) in a spectrophotometer controlled at 30°C. The reaction mixture was 3 ml of 50 mM phosphate buffer at pH 7.0 containing 100 μM antibiotic. The decerase in its optical density was recorded after addition of 50 μl of crude β-lactamase. A decomposition product of HR 756 hydrolyzed by the enzyme was obtained at 264 nm. PCase and CSase activities were expressed as units per milligram of

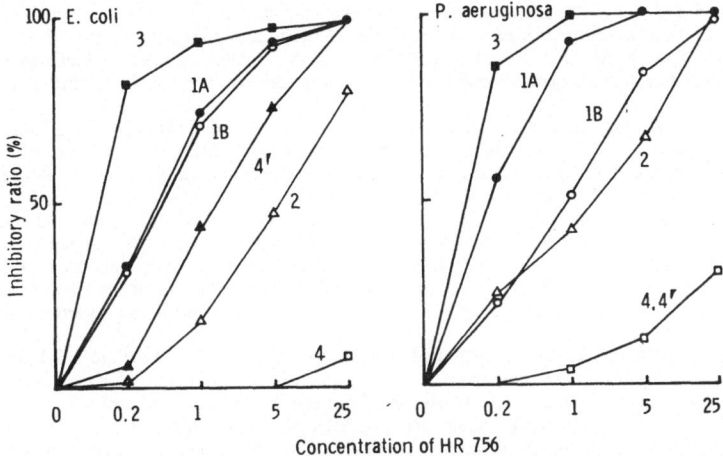

Fig.1. Competition of HR 756 with ^{14}C-labeled penicillin G for binding to PBPs in cytoplasmic membranes of *E. coli* and *P. aeruginosa in vitro.* Concentration of HR 756 is expressed as molar ratios to labelled penicillin G.

protein when PC-G and CER were used as substrates, respectively. Substrate specificity was expressed as the relative rate of hydrolysis of the 6 antibiotics, taking the absolute rate of hydrolisis of PC-G or CER as 100. The amount of protein in the samples was estimated by the method of Lowry.

Rates of hydrolysis of the 6 cephalosporins relative to those of PC-G and CER are presented in Table III. HR 756, like CXM, was stable to PCase produced by *K. pneumoniae*, R-mediated PCase, and CSase produced by *E. coli* GN5482, *P. aeruginosa* GN918, *E. cloacae* GN7471, *C. freundii* GN346 *P. morganii* GN5406., To CSase produced by *P. vulgaris* GN76, HR 756 was the most resistant of all the tested antibiotics, it was 3.5-fold more stable than CER and 2- to 13-fold more resistant than CXM, CEZ, CET and CEX.

Affinity for penicillin-binding proteins. Affinity of HR 756 for penicillin-binding proteins in the cytoplasmic membrane of *E. coli* JE1011 and *P. aeruginosa* ATCC13048 was measured in accordance with the method of Spratt and Noguchi *et al.* (Spratt, 1975, Noguchi, Matsuhashi, Nikaido, Itoh, Matsubara, and Mitsuhashi, 1979). HR 756 showed high affinity for *E. coli* proteins-1A, -1Bs and -3 and for *P. aeruginosa* proteins-1A, 1B and -3.

In vivo antibacterial activity. Curative effects of HR 756 against bacterial infections were studied in comparison with those of SCE 963 and CEZ. Groups of 20 four-week-old male mice weighing 18–20 g were intraperitoneally infected with 1 × 10[7] cells of *E. coli* ML4707 per mouse and of *K. pneumoniae* GN6445. The animals were given the anti-

237

biotics 1 and 4 hr after infection and observed for 6 days. HR 756 was shown to be 10–20 times and 10–25 times more effective against *E. coli* and *K. pneumoniae*, respectively, than SCE 963 and CEZ.

REFERENCES

DRASAR, F. A., W. FARRELL, A. J. HOWARD, C. HINCE, T. LEUNG and J. D. WILLIAMS (1979): Activity of HR 756 against *Haemophilus influenzae, Bacteroides fragilis* and gram-negative rods. J. Antimicrob. Chemother. 4, 445–450.

FU, K. P. and H. C. NEU (1978): Beta-lactamase stability of HR 756, a novel cephalosporin, compared to that of cefuroxime and cefoxitin. Antimicrob. Agents Chemother. 14, 322–326.

HAMILTON-MILLER, J. M. T., W. BRUMFITT and A. V. REYNOLDS (1978): Cefotaxime (HR 756) a new cephalosporin with exceptional broadspectrum activity *in vitro*. J. Antimicrob. Chemother. 4, 437–444.

HEYMES, R., A. LUTZ and E. SCHRINNER (1978): Experimental evaluation of HR 756, a new cephalosporin derivative, P. 823–824. *In* W. Siegenthaler and R. Lüthy (ed.), Current Chemotherapy. Proceedings of the 10th International Congress of Chemotherapy, vol. II. American Society for Microbiology, Washington, D. C.

MITSUHASHI, S., S. YAMAGISHI, T. SAWAI and H. KAWABE (1977): Biochemical mechanisms of plasmid-mediated resistance, P. 195–254. *In* S. Mitsuhashi (ed.), R. factor-Drug Resistance Plasmid. University of Tokyo Press; University Park Press, Tokyo, Baltimore and London.

NEU, H. C. and K. P. FU (1978): Cefuroxime, a beta-lactamase-resistant cephalosporin with a broad spectrum of gram-positive and -negative activity. Antimicrob. Agents Chemother. 13, 657–664.

NOGUCHI, H., M. KUBO, S. KURASHIGE and S. MITSUHASHI (1978): Antibacterial activity of apalcillin (PC-904) against gram-negative bacilli, especially ampicillin-, carbenicillin-, and gentamicin- resistant clinical isolates. Antimicrob. Agents Chemother. 13, 745–752.

NOGUCHI, H., M. MATSUHASHI, T. NIKAIDO, J. ITOH, N. MATSUBARA, M. TAKEDA and S. MITSUHASHI (1979): Affinities of beta-lactam antibiotics to bind to penicillin-binding proteins in *E. coli* and *Pseudomonas aeruginosa* in relation to their antibacterial potencies. *In* S. Mitsuhashi (ed.), Microbial drug resistance, vol. 2: Microbial drug resistance and related plasmids. University of Tokyo Press, Tokyo, in press.

SAMUNI, A. (1975): A direct spectrophotometric assay and determination of michaelis constants for the β-lactamase reaction. Anal. Biochem. 63, 17–26.

SPRATT, B. G. (1977): Properties of the penicillin-binding proteins of *Escherichia coli* K12. Eur. J. Biochem. 72, 341–352.

UNE, T. and S. MITSUHASHI (1977): Antimicrobial evaluation of cefoxitin: a new semisynthetic cephamycin. Comparative study with cefazolin and cephalothin. Arzneim.-Forsch. 27, 89–93.

S. M., Gunma University,
Maebashi, Gunma, Japan

IN VITRO AND *IN VIVO* ANTIBACTERIAL ACTIVITY OF AM-715
— A NEW ANALOG OF NALIDIXIC ACID

A. ITO, K. HIRAI, S. SUZUE, T. IRIKURA, M. INOUE,
and S. MITSUHASHI

*Department of Microbiology, School of Medicine, Gunma University, Maebashi,
Japan and Central Research Laboratory, Kyorin Pharmaceutical Co., Ltd., Tochigi, Japan*

INTRODUCTION

Nalidixic acid (NA) has been proved to be highly effective against the infection with gram-negative bacteria except for most strains of gram-positive bacteria, *Pseudomonas aeruginosa* and *Serratia marcescens* (Lesher et al., 1962, Deitz et al., 1964).

The incidence of infections with *P. aeruginosa* and *S. marcescens* has progressively increased, along with infections due to other gram-negative bacteria (Altermier et al., 1973, Finland, 1970), and these infections are known to be extremely refractory to chemotherapy.

Pipemidic acid, a NA analog, was a first effective drug against *P. aeruginosa* (Shimizu et al., 1975; 1976). In our laboratory, we synthesized series of quinolinecarboxylicacid derivative. These are structurally related to nalidixic acid and pipemidic acid. A quinolinecarboxylic acid derivative, 1-ethyl-6-fluoro-1,4-dihydro-4-oxo-7(1-piperazinyl)-3-quinolinecarboxylic acid (Fig. 1), showed more potent antipseudomonal activity, in addition to high activity against other gram-negative and gram-positive bacteria. This paper presents *in vitro* and *in vivo* antibacterial activity of this compound, AM-715.

RESULTS

Antibacterial spectrum. The spectrum of antibacterial activity of AM-715 against standard strains of gram-positive and gram-negative bacteria is shown in Table 1, compared with that of nalidixic acid (NA), pipemidic acid (PPA), miloxacin (AB-206), (Agui et al., 1977), gentamicin (GM) and carbenicillin (CBPC).

Minimal inhibitory concentrations (MICs) were determined by a standard twofold serial dilution method. Plates were inoculated with the loopful of 10-fold diluted suspension (about 10^8 cells/ml) of overnight culture or a bacterial suspension (about 10^8 cells/ml) in brain heart infusion (Eiken), which has been adjusted to an optical density of 0.3 at 560 nm.

Most gram-positive bacteria were susceptible to AM-715 but were less susceptible or resistant to NA, PPA and AB-206. AM-715 inhibited the growth of *Streptococcus pyogenes*, *Streptococcus faecelis* and *Corynebacterium pyogenes* resistant to NA, PPA and AB-206 at concentrations between 1.56 and 3.13 µg/ml. AM-715 proved to be remarkably active against gram-negative bacteria including *P. aeruginosa*, and had more potent antibacterial activity than NA, PPA, AB-206, GM and CBPC. AM-715 inhibited the growth of gram-negative bacteria tested at low concentrations less than 1.56 µg/ml. Thus, it was found that AM-715 had a broader spectrum and more potent antibacterial activity than NA, PPA and AB-206.

TABLE I.

Antibacterial spectrum of AM-715

Organism	MIC (μg/ml)					
	AM-715	NA	PPA	AB-206	GM	CBPC
S. aureus 209P	0.78	100	25	12.5	0.10	0.39
S. aureus ATCC14775	1.56	>100	100	12.5	0.10	6.25
S. pyogenes S-8	1.56	>100	>100	>100	6.25	0.10
S. faecalis IID682	3.13	>100	>100	100	50	100
C. pyogenes IID548	3.13	>100	100	>100	0.78	0.20
M. phlei IFO3142	0.39	100	12.5	6.25	6.25	>100
E. coli NIHJ JC-2	0.05	3.13	1.56	0.39	0.39	1.56
E. coli ATCC10536	0.05	3.13	1.56	0.39	0.39	1.56
K. pneumoniae IFO3512	0.05	1.56	1.56	0.39	0.20	>100
S. marcescens IID620	0.05	6.25	3.13	0.78	1.56	50
P. vulgaris IFO3167	0.05	3.13	3.13	0.39	0.20	6.25
S. enteritidis IID604	0.20	12.5	12.5	6.25	1.56	12.5
S. sonnei IID969	0.10	1.56	1.56	0.39	0.78	1.56
H. influenzae IID986	0.05	1.56	3.13	0.39	1.56	0.39
N. perflava IID856	0.05	1.56	1.56	0.20	0.78	1.56
P. aeruginosa V-1	0.39	100	12.5	50	0.78	12.5
P. aeruginosa IID1130	0.78	>100	25	50	3.13	>100

Inoculum size one loopful of bacterial suspension (10^8 cells/ml).

Antibacterial activity against clinical isolates. *In vivo* antibacterial activity of AM-715 against clinical isolates of gram-negative and gram-positive bacteria is shown in Table II., compared with that of NA, PPA and AB-206. Plates were inoculated with one loopful of 10^3-fold diluted suspension (about 10^6 cells/ml) of overnight culture. Table II. shows the concentrations required to inhibit the growth of 70% of the total number of strains tested (MIC_{70}).

MIC_{70} values of AM-715 were 0.032 μg/ml against 219 *Escherichia coli* strains, 0.36

TABLE II.

In vivo antibacterial activity of AM-715

	Organism	Challenge dose (cells/body)	Drug[a]	ED_{50}[b] (mg/kg)	95% confidence limit (mg/kg)	
Systematic infection[c]	E. coli ML4707	1.3×10^7 (in saline)	AM-715	5.0	4.5	— 5.6
			NA	34.8	32.1	— 37.5
			PPA	35.4	32.5	— 38.3
			AB-206	19.8	17.9	— 21.5
	K. pneumoniae GN6445	1.4×10^7 (in saline)	AM-715	4.6	3.5	— 5.5
			NA	28.4	25.0	— 32.1
			PPA	29.1	25.2	— 34.5
			AB-206	14.1	10.6	— 16.1
	P. aeruginosa GN11189	4.8×10^4 (in BHI containing 5% mucin)[d]	AM-715	27.1	25.9	— 29.1
			NA	364	307	— 546
			PPA	200	176	— 224
			AB-206	107	76.1	— 146

[a] Oral administration of AM-715, NA, PPA and AB-206.   [b] 50% effective dose.
[c] Intraperitoneal infection.   [d] Brain heat infusion.

240

μg/ml against 175 *P. aeruginosa* strains, 0.094 μg/ml against 248 *Klebsiella pneumoniae* strains, 0.40 μg/ml against 160 *S. marcescens* strains, 0.058 μg/ml against 130 *Enterobacter* sp., 0.071 μg/ml against 152 indole-positive *Proteus* sp., 0.047 μg/ml against 50 indole-negative *Proteus* sp., 1.00 μg/ml against 297 *Staphylococcus aureus* strains and 0.59 μg/ml against 116 NA-resistant strains of gram-negative bacteria. MIC_{70} values of NA, PPA and AB-206 against the above gram-negative bacteria were more than 4.0, 1.70 and 0.58 μg/ml, respectively, those of NA, PPA and AB-206 against gram-positive bacteria were 66, 19.5 and 18.0 μg/ml, respectively, and those of PPA and AB-206 against NA-resistant strains of gram-negative bacteria were 13.0 and 20.2 μg/ml, respectively.

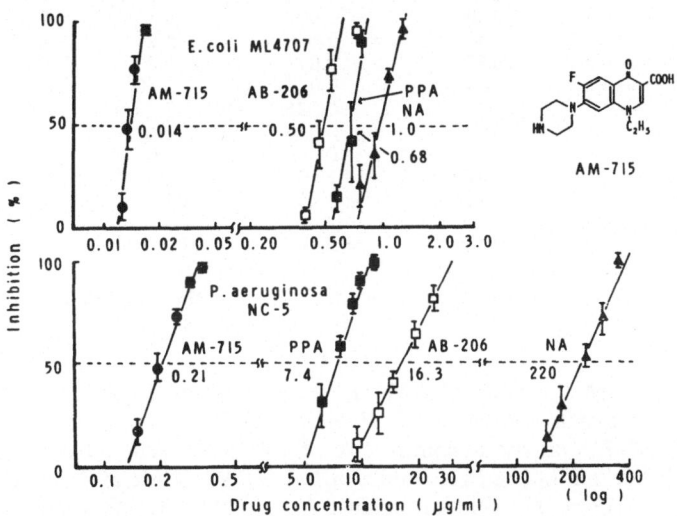

Fig. 1. Determination of the ID-- of AM-715 against *E. coli* ML4707 and *P. aeruginosa* NC-5.

Bactericidal activity. The bactericidal activity of AM-715 was determined by counting the number of viable cells at appropriate time intervals after addition of AM-715. AM-715 showed bactericidal action against *E. coli* ML4707, *K. pneumoniae* GN6445 and *P. aeruginosa* NC-5 at MICs 0.10, 0.10 and 0.78 μg/ml, respectively. The viable cells of *E. coli* ML4707 and *K. pneumoniae* GN6445 were reduced to 1/100 after 2 hr and were not detected after 6 hr at MICs. The viable cells of *P. aeruginosa* NC-5 were reduced to 1/100 after 2 hr and were not detected after 6 hr at MICs and the viable cells of this strain were not detected even after 24 hr at a twofold higher concentration than the MIC (1.56 μg/ml).

Effect of inoculum size on MIC and MBC. MICs and minimal bactericidal concentrations (MBCs) of AM-715 against 3 strains of gram-negative bacteria were determined by using different inoculum sizes and were compared with those of NA, PPA and AB-206 (*E. coli* ML4707, 2×10^2 to 2×10^8 cells/ml, *P. aeruginosa* NC-5 and *S. marcescens* GN6486, 1×10^2 to 1×10^7 cells/ml). The MIC and the MBC were determined by a standard twofold serial dilution method with Antibiotic Medium 3 (Difco). The MIC was determined after incubation at 37°C for 18 hr, and one loopful of each culture tube in the MIC test series was spotted onto drug-free heart infusion (HI) agar (Nissui) plates. After incubation at 37°C for 18 hr, the MBC of a drug was determined as the lowest concentration of the drug that had prevented visible growth on HI agar plates.

TABLE III.

Antibacterial activity of AM-715 against clinical isolates

Organism	No. of strains	Drug concentration (μg/ml)[a]			
		AM-715 MIC$_{70}$	NA MIC$_{70}$	PPA MIC$_{70}$	AB-206 MIC$_{70}$
E. coli	219	0.032	4.00	1.70	0.41
P. aeruginosa	175	0.36	>100	14.9	20.5
K. pneumoniae	248	0.094	6.20	2.70	2.40
S. marcescens	160	0.40	5.30	9.30	1.85
Enterobacter sp.	130	0.058	5.20	4.15	1.85
indole-positive Proteus sp.	152	0.071	5.20	2.65	0.70
indole-negative Proteus sp.	50	0.047	5.10	2.70	0.64
Salmonella sp.	199	0.046	3.40	—[b]	—
Shigella sp.	150	0.020	1.40	—	—
H. influenzae	21	0.034	1.00	—	—
S. aureus	297	1.00	66.0	19.5	18.0
S. pyogenes	80	1.45	>100	61.0	100
NA-resistant gram-negative bacteria	116	0.59	>100	13.0	20.2

[a] The concentrations required to inhibit the growth of 70% of the total number of strains used.
[b] — not tested.

The MIC and the MBC values of AM-715 were the same on the inoculum size of 10^2 to 10^5 cells/ml against *E. coli* ML4707, *P. aeruginosa* NC-5 and *S. marcescens* GN6486, the values of AM-715 on the inoculum size of 10^6 to 10^8 cells/ml were twofold higher than those of 10^5 cells/ml. The MIC values of NA, PPA and AB-206 were the same on the inoculum size of 10^2 to 10^5 cells/ml against *E. coli* ML4707, *P. aeruginosa* NC-5 and *S. marcescens* GN6486 as well as AM-715, and the values of NA, PPA and AB-206 on the inoculum size of 10^6 and 10^7 cells/ml were twofold or fourfold higher than those of 10^4 cells/ml, but the MBC values of NA, PPA and AB-206 were influenced by inoculum size. The effects of inoculum size on the MIC and the MBC of AM-715 against *E. coli*, *P. aeruginosa* and *S. marcescens* were smaller than those of NA, PPA and AB-206.

Determination of ID$_{50}$. The dose response curve was determined at a bacterial concentration of 200 to 500 cells per plate, and the concentration of a drug (ID$_{50}$) required to inhibit the growth of 50% of the total number of bacterial cells was determined (Kato et al., 1978). As shown in Fig. 1, AM-715 showed a steep gradient and gave ID$_{50}$ values of 0.014 μg/ml against *E. coli* ML4707 and 0.21 μg/ml against *P. aeruginosa* NC-5. The dose response curve of NA, PPA and AB-206 showed a steep gradient as well as AM-715. DI$_{50}$ values of NA, PPA and AB-206 were 1.0, 0.68 and 0.50 μg/ml, respectively, against *E. coli*, and 220, 7.4 and 16.3 μg/ml, respectively, against *P. aeruginosa*, indicating much higher values than those of AM-715.

In vivo antibacterial activity. The *in vivo* antibacterial activity of AM-715 was determined in systemic and ascending kidney infections, and the results are shown in Table III, compared with those of NA, PPA and AB-206. Mice were infected intraperitoneally with *E. coli* ML4707, *K. pneumoniae* GN6445 and P. aeruginosa GN11189. AM-715, NA, PPA and AB-206 were administered orally to the mice twice a day, immediately after and 4 hr after infection, and the therapeutic effect of drugs was judged from the survival rate. A comparison of *in vivo* antibacterial activity was made on the basis of the mean effective dose (ED$_{50}$) calculated by probit analysis (Miller et al., 1944) and a 95 % confidence limit that was calculated by the method of Litchfield and Wilcoxon (Litchfield et al., 1948). ED_{50} value of AM-715 was 5.0 mg/kg against *E. coli* ML4707, 4.6 mg/kg against *K. pneumoniae* GN6445 and 27.1 mg/kg against *P. aeruginosa* GN11189. ED_{50} values of NA, PPA and AB-206 were 34.8, 35.4 and 19.8 mg/kg respectively, against *E. coli*, 28.4, 29.1 and 14.2 mg/kg, respectively, against *K. pneumoiae* and 364, 200 and 107 mg/kg, respectively, against *P. aerugiosa*.

Ascending kidney infection was induced in female mice by instilling *E. coli* NIHJ JC-2 suspended in 0.04 ml tryptosoya broth (Nissui) into the urinary tract of the anesthetized mice and holding them with clamped outlet of the urinary tract for 6 hr in plastic tubes. AM-715, NA, PPA and AB-206 were orally administered 3 hr after infection. Kidneys were removed aseptically at 48 hr after infection and bisected, and the two halves were streaked on an agar plate containing desoxycholate. The absence of bacterial growth on the agar surface after incubation at 37°C for 20 hr was the sole end point of the therapeutic achievement. ED_{53} values of AM-715, NA, PPA and AB-206 were 15.0, 86.0 74.0 and 135 mg/kg respectively.

DISCUSSION

AM-715 is a new synthetic antibacterial drug structurally related to NA and PPA. It has been known that these drugs are primarily active against gram-negative bacteria. AM-715 inhibited the growth of most gram-positive and gram-negative bacteria at lower concentrations than 1.56 μg/ml, and was more effective than NA, PPA, AB-206, GM and CBPC against gram-negative bacteria including *P. aeruginosa*. Thus, NA-715 possessed a broader spectrum and more potent antibacterial activity than other NA-analogs. Most strains of NA-resistant gram-negative bacteria were susceptible to AM-715, probably due to the high penetrability of AM-715 into bacterial cells and a to different mode of action (unpublished.) Cross-resistance was not observed between AM-715 and various antibiotics.

The antibacterial activity of most β-lactam antibiotics is known to be influenced by inoculum size (Sabath et al., 1975). By contrast, the effect of inoculum sizes on the MIC and the MBC of AM-715, NA, PPA and AB-206 was found to be smaller, and that of AM-715 was smaller than that of other NA analogs. The *in vivo* evaluation of antibacterial activity of AM-715 was clearly more effective than that of NA, PPA and AB-206 against systemic infections of mice with *E. coli*, *K. pneumoiae* and *P. aeruginosa*, and ascending kidney infection of mice with *E. coli*.

REFERENCES

AGUI, H., T. MITANI, A. IZAWA, T. KOMATSU and T. NAKAGOME (1977): Studies on quinoline derivatives related compounds. 5. Synthesis and antimicrobial activity of novel 1-alkoxy-1,4-dihydro-4-oxo-3-quinolinecarboxylic acids. J. Med. Chem. 20, 791–796.

ALTERMIER, W. A., R. P. HUMMEL, E. O. HILL, and S. LEWIS (1973): Changing patterns in surgical infections. Ann. Surg. 178, 436–445.

DEITS, W. H., J. H. BAILEY and E. J. FROELICH (1964): *In vitro* antibacterial properties of nalidixic acid, a new drug active against gram-negative organisms. Antimicrob. Agents Chemother. 583–587. 1963.

FINLAND, M. (1970): Changing ecology of bacterial infections as related to antibacterial therapy. J. Infect. Dis. 122, 419–431.

KATO, T., S. KURASHIGE, Y. A. CHABBERT and S. MITSUHASHI (1978): Determination of the ID_{50} values of antibacterial agents in agar. J. Antibiotics 31, 1299–1303.

Lesher, G. Y., E. J. Froelish, M. D. Gruett, J. M. Bailey and R. P. Brundage (1962): 1,8.-Naphtyridine derivatives: a new class of chemotherapeutic agents. J. Med. Pharm. Chem. 5, 1063–1065.

Litchfield, J. T. and F. Wilcoxon (1948): A simplified method of evaluating dose-effect. J. Pharmacol. 92, 99–113.

Miller, L. C. and M. L. Tainter (1944): Estimation of the ED_{50} and its error by means of logarithmic-probit graph paper. Proc. Soc. Exp. Biol. Med. 57, 261–264.

Sabath, L. D., C. Grner, C. Wilcox and M. Finiand (1975): Effect of inoculum and beta-lactamase on the anti-staphylococcal activity of thirteen penicillins and cephalosporins. Antimicrob. Agents Chemother. 8, 344–349.

Shimizu, M., Y. Takase, S. Nakamura, H. Katae, A. Minami, K. Nakata, S. Inoue, M. Ishiyama and Y. Kubo (1975): Pipemidic acid, a new antibacterial agent active against *Pseudomonas aeruginosa*: In vitro properties. Antimicrob. Agents Chemother. 8, 132–138.

Shimizu, M., Y. Takase, S. Nakamura, H. Takae, A. Minami, K. Nakata and N. Kurobe (1976): Pipemidic acid, Its activities against various experimental infections. Antimicrob. Agents Chemother. 9, 569–574.

A. I., Dept. of Microbiology,
Gunma University, Maebashi,
Japan

INTERACTIONS OF AZLOCILLIN AND MEZLOCILLIN WITH AMINOGLYCOSIDES AGAINST CARBENICILLIN RESISTANT AND SENSITIVE *P. AERUGINOSA*

E. J. PEREA, M. C. GARCÍA IGLESIAS, E. MARTIN, J. C. PALOMARES
and M. C. NOGALES

Department of Microbiology and Medicine, Preventive.
University Hospital. Scholl of Medicine. University of Seville.
Seville, Spain

The *"in vitro"* activity of the new semisynthetic ureidopenicillins Azlocillin (AZ) and Mezlocillin (MZ) was determined against 25 carbenicillin susceptible (MIC's < < 128 µg/ml; CAR$_S$) and 25 carbecillin resistant (MIC's ≥ 128 µg/ml; CAR$_R$) *Pseudomonas aeruginosa* clinical isolates.

In both groups of strains, AZ was more active than MZ with MICs two-fold lower dilution.

Interaction of these Beta-lactams and five aminoglycosides: Gentamicin (GM), Tobramicin (TM), Dibekacin (DKB), Sisomicin (SSM) and Amikacin (AMK) was studied against the same strains using a full checquer board titration and an agar (MH) dilution technique. The results obtained were expressed in terms of FIC's (Fractionary Inhibitory Concentrations).

No significant differences were encountered in the results obtained in the CAR$_R$ and CAR$_S$ strains groups, nor with the eleven aminoglycoside resistant strains.

MZ showed more favourable interaction (synergism or addition) than AZ with all five aminoglycosides. TM with AZ showed synergism in 32%, and with MZ in 48% of the strains. GM with AZ showed 52% and with MZ 78% synergism, respectively. With SSM the percentages were 68 and 90%; with DKB 48 and 66%; and with AMK 42% and 64%, respectively.

The combinations AZ + TM and AZ + GM showed the highest percentages of antagonism, 14 and 18, respectively.

MZ + SSM showed 96% favourable interactions, the highest percentage of all combinations.

INTRODUCTION

After the introduction of carbenicillin in clinical usage, reports of resistant strains of *Ps. aeruginosa* appeared, and in some centres such strains constituted a significant proportion of isolates. (Perea et al., 1979). With aminoglycosides we have similar problems but to a lesser extent. (Kluge et al., 1974).

The combination of gentamicin and carbenicillin is synergistic *"in vitro"* and *"in vivo"* against isolates of *Ps. aeruginosa* which are susceptible to both agents. But some reports indicate problems when the isolate is resistant to one or both antibiotics (Kluge et al., 1974).

Two new β-lactam antibiotic with anti-*Pseudomonas* activity, the ureido penicillin derivatives Azlocillin (AZ) and Mezlocillin (MZ), have been described (Stewart and Bodey, 1977).

TABLE I.

Antimicrobial activity (MIC) of AZ, MZ, GM, TM, SSM, AMK and DKB against 50 Ps. aeruginosa, 25 car^r and 25 car^s strains

Test agent	Strains group	Varying percentage of inhibited test strain at MIC (µg/ml) of:													
		>0.25	0.25	0.5	1	2	4	8	16	32	64	128	256	512	>512
AZ	CAR^r*						20	52	64	84	96	100			
	CAR^s**				4	12	24	60	84	88	96	100			
MZ	CAR^r						8		12	40	88	96	100		
	CAR^s								8	56	100				
GM	CAR^r	12		16		64	88	92							100
	CAR^s				24	48	64	68	84	88	96	100			
TM	CAR^r	12		60	88		92					100			
	CAR^s	8		60	64	76		80			96	100			
SSM	CAR^r	4		12	24	60	92								100
	CAR^s			4	24	44	60	72		80	84	96	100		
AMK	CAR^r	8				20	64	84	88	100					
	CAR^s				4	36	72	84	92	96			100		
DKB	CAR^r	8		12	32	76	92							100	
	CAR^s			8	44	52	56	68		72	76	80	96	100	

* CAR^r – Carbenicillin resistant (MIC —128 µg/ml)
** CAR^s – Carbenicillin sensitive (MIC ≥ 128 µg/ml)

TABLE II.

Interactions of AZ and five aminoglycosides against 25 CAR^s and 25 CAR^r Ps. aeruginosa strains

Test agent combination	Strains group	Synergy <0.75*	Addition 0.75–1	Indifference 1	Antagonism >1
AZ + GM	CAR^r	12	3	6	4
	CAR^s	14	1	5	5
AZ + TM	CAR^r	7	1	13	4
	CAR^s	9	2	11	3
AZ + SSM	CAR^r	19	0	4	2
	CAR^s	15	2	7	1
AZ + AMK	CAR^r	11	0	11	3
	CAR^s	10	1	14	0
AZ + DKB	CAR^r	13	1	9	2
	CAR^s	11	3	9	2

* FICs index

The present report presents an in vitro evaluation of these antibiotics alone and in combination with five aminoglycosides, Gentamicin (GM), Tobramicin (TM), Dibekacin (DKB), Sisomicin (SSM) and Amikacin (AMK), against 50 Ps. aeruginosa (25 Cerbenicillin resistant (CAR^r) and 25 carbenicillin sensitive (CAR^s) strains, eleven of which were aminoglycoside resistant.

MATERIAL AND METHODS

Bacteria: 50 bacterial strains were isolated from clinical material and identified in the Bacteriology Laboratory of the University Hospital of Seville. (Perea et al., 1979).

The *Ps. aeruginosa* clinical isolates were classified in to two groups as 25 CAR^r strains with MICs $\geq$ 128 µg/ml and 25 CAR^s strains with MIC < 128 µg/ml. There were eleven *Ps. aeruginosa* (resistant to two or more aminoglycosides) with MICs of: GM 8 µg/ml, Tm 8 µg/ml, SSM 8 µg/ml, AMK 20 µg/ml and DKB 8 µg/ml.

Antibiotics: AZ, MZ and SSM were generously provided by Bayer, GM by Schering, TM by Lilly, DKB by Almirall and AMK by Bristol Labs.

Media: Mueller-Hinton (MH) agar or broth (Difco Laboratories) adjusted to pH 7.4 was used for cultures and serial dilutions.

Minimal Inhibitory Concentrations: These values were determined by the two-fold agar dilution method described previously (Perea et al., 1978).

TABLE III.

Interactions of MZ and five aminoglycosides against 25 CAR^s and 25 CAR^r Ps. aeruginosa strain

Test agent combination	Strains group	Synergy <0.75*	Addition 0.75–1	Indifference 1	Antagonism >1
MZ + GM	CAR^r	20	5	0	0
	CAR^s	19	0	3	3
MZ + TM	CAR^r	10	8	6	1
	CAR^s	14	1	10	0
MZ + SSM	CAR^r	23	2	0	0
	CAR^s	22	1	1	1
MZ + AMK	CAR^r	15	0	10	0
	CAR^s	17	0	5	3
MZ + DKB	CAR^r	16	2	7	0
	CAR^s	17	0	6	2

* FICs index

TABLE IV.

Interactions of AZ and MZ with five aminoglycosides against eleven aminoglycoside resistant strains

Test agent combination	Synergy 0.75*	Addition 0.75–1	Indifference 1	Antagonism 1
GM + AZ	5	3	0	3
MZ	6	3	2	0
TM + AZ	4	1	1	5
MZ	5	2	2	2
SSM + AZ	8	2	0	1
MZ	9	0	2	0
AMK + AZ	4	2	0	5
MZ	4	2	0	5
DKB + AZ	8	1	0	2
MZ	6	4	0	1

* FICs index

Studies of combined antimicrobial activity: So-called chequerboard MIC studies with two drug combinations were performed simultaneously on MH agar with AZ, MZ and GM, TM, SSM, DKB, and AMK by the agar plate dilution method using a Steer replicator (Perea et al., 1978). Results of the interaction are expressed by the fractional inhibitory concentration index (FIC index) (Perea et al., 1978).

RESULTS

The activities of these antibiotics against 50 *Ps. aeruginosa* are shown in Table I. AZ was more active than MZ which has a narrow range of MIC (between 16 and 64 µg/ml). No significant differences were encountered in the MICs obtained with the groups of CAR[s] and CAR[r] strains. At low concentrations AZ showed more activity in the CAR[s] than in the CAR[r] group.

Among the aminoglycoside antibiotics, TM was the most active. All the aminoglycosides showed significantly more activity on the CAR[r] strains than on the CAR[s] strains. The only exception was AMK which showed the same MICs on both groups.

There were eleven strains resistant to the aminoglycosides.

Interactions of AZ and MZ with the aminoglycoside antibiotics.

The results of the interactions of these aminoglycosides and AZ against the 25 CAR[r] and 25 CAR[s] *Ps. aeruginosa* strains are shown in Table II. There were no significant differences between the CAR[r] and CAR[s] strains. The most active association was that of AZ + SSM which showed the highest degrees of favourable interaction (synergism or addition).

The interactions of the aminoglycosides with MZ are shown in Table III. CAR[r] and CAR[s] strain groups behaved similarly, with no significant differences between them. The most favourable interaction was MZ + SSM.

The eleven aminoglycoside resistant strains exhibited results similar to those of the aminoglycoside susceptible strain with all of the various combinations. (Table IV.).

DISCUSSION

AZ showed more activity than MZ. At a concentration of 32 µg/ml AZ inhibited 88% of strains and MZ only 50%. AZ and MZ had the same activity on CAR[r] and CAR[s] groups. Similar results with a low number of strains were found by Bywater et al., 1977, but these findings disagree with those of Stewart and Bodey, 1977 who found 9 *Ps. aeruginosa* carbenicillin resistant strains to be AZ resistant.

Except for AMK all aminoglycosides showed more activity on CAR[r] strains. Schassan et al., 1978 found MZ to be less active against GM-resistant than against GM-sensitive strains.

AZ + SSM was the most active combination against *Ps. aeruginosa*, both resistant and susceptible strains, with better results in the CAR[r] group. We think that this finding could be of great interest for the clinical use of both antibiotics.

With combinations of the other aminoglycosides we obtained results similar to those of Farrel et al. 1977, who used AZ in combination with GM, TM and AMK, with the exception of AZ + TM which showed less synergy in our results.

Interaction of MZ with the five aminoglycosides also exhibited good activity. The MZ + SSM combination showed the highest number of favourable interactions fo all the combinations tested. Similar results were obtained in the aminoglycoside resistant group.

These results are in disagreement with those of Schassan et al., 1978 who found better activity of MZ + TM and MZ + AMK, against *Ps. aeruginosa*, but these authors used only sub-inhibitory concentrations of the aminoglycosides in their tests for interactions and did not use all the combinations used here.

The interaction MZ + SSM proved to have the best inhibitory activity against *Ps. aeruginosa*.

REFERENCES

BYWATER, M. J., HOLT, H. A., BROUGHALL, J. M., and REEVES, D. S. (1977): Azlocillin, Mezlocillin, Pirmecillinam, Mecillinam and PC-904 in Current Chemotherapy. Proceeding of the 10th International Congress of Chemotherapy. A.S.M. Washington.

FARRELL, W., WILKS, M., and DRASAR, F. A. (1977): Sinergy between Aminoglycosides and Semisynthetic Penicillins Against Gentamicin-Resistant Gram-Negative Rods in Current Chemotherapy. Proceeding of the 10th International Congress of Chemotherapy. A.S.M. Washington.

KLUGE, R. M., STANDIFORD, H. C., TATEM, B., YOUNG, V. M., GREENE, W. H., SCHIMPFF, S. C., CALIA, F. M. and HORNICK, R. B. (1974): Comparative Activity of Tobramycin, Amikacin, and Gentamicin Alone and with Carbenicillin Against *Pseudomonas Aeruginosa*. Antimicrob. Agents and Chemother. 6:442-446.

PEREA, E. J., TORRES, M. A., and BOROBIO, M. V. (1978): Synergism of Fosfomycin-Ampicillin and Fosfomycin-Chloramphenicol against *Salmonella* and *Shigella*. Antimicrob. Agents and Chemother. 13:705–709.

PEREA, E. J., PALOMARES, J. C., and GARCIA-IGLESIAS, M. C. (1979): *In Vitro* Evaluation of New Penicillins and Cephalosporins upon *Ps. aeruginosa* and their Interaction with Mecillinam. Chemotherapy (In press).

SCHASSAN, H. H., KOPERSKI, K., and SHEERF, H. (1978): Mezlocillin: A new Acyl Ureido-penicillin. Antimicrobial Activity and Combination Effects with Four Aminoglycoside Antibiotics. Chemother. 24(3), 134–142.

STEWART, D., and BODEY, G. P. (1977): Azlocillin: In vitro Studies of a New Semisynthetic Penicillin. Antimicrob. Agents Chemother. 11:865–870.

E.J.P., Dept. of Microbiology,
University Hospital. Univ. of Seville,
Seville, Spain

II. MEDICAL PART
B) NATURAL SPREAD OF ANTIBIOTIC RESISTANCE
Chairman: R. GOMEZ-LUS

NATIONAL INSTITUTES
OF HEALTH PROGRAMS IN ANTIBIOTICS RESISTANCE, RECOMBINANT DNA, HOSPITAL ASSOCIATED INFECTIONS AND SEXUALLY TRANSMITTED DISEASES

I. P. DELAPPE

Molecular Microbiology and Parasitology Branch
Microbiology and Infectious Diseases Program
National Institute of Allergy and Infectious Diseases
National Institutes of Health, Bethesda, U.S.A.

Two public health problems of major concern in the United States are hospital associated infections and sexually transmitted diseases; gonorrhea, in the latter category, is the most common infection reportable to the Center for Disease Control in Atlanta, Georgia. Substantial evidence indicates that both of these medically important problems have a common origin in the phenomenon of bacterial resistance to antimicrobial agents. Basic research on the mechanisms of resistance to antimicrobial agents is an NIH program directed to the solution of these problems. I will discuss this program and a growing new research program, Recombinant DNA, which had its origin in studies of antibiotic resistance.

Both of these programs are supported by the Molecular Microbiology and Parasitology Branch which is located in the Microbiology and Infectious Diseases Program of the National Institute of Allergy and Infectious Diseases, one of the National Institutes of Health.

The first of these two programs is Mechanisms of Resistance ot Antimicrobial Agents.

The principal goal of this program is to elucidate fundamental biological mechanisms involved in the development of drug resistance in microorganisms and to increase our basic understanding of this phenomenon. More specific goals involve investigation of the origin, development, evolution, expression and mechanisms of drug resistance in a variety of specific microorganisms. Of particular interest to the program are the *Enterobacteriaceae*, *Pseudomonas*, *Neisseria*, staphylococci, streptococci, mycobacteria, mycoplasmas, and pathogenic fungi.

Research of special interest to this program is included in one or more of the following categories: (1) genetic and structural studies of R factors and related plasmids; (2) origin, development, and evolution of drug resistance in microorganisms; (3) replication and conjugal transfer of plasmids; (4) biochemistry and genetics of plasmid determined functions, especially resistance to antimicrobial agents; (5) correlated epidemiological and microbiological studies of naturally occurring plasmids with special reference to R factors.

The branch currently has approximately 3.3 million dollars invested in this program.

In addition, the branch has supported the Stanford Plasmid Reference Center for approximately two and one-half years. This serves as the sole collection and coordination center of its type in the United States, and, as such, is an important establishment that is very useful to workers in this rapidly expanding area of research.

We have budgeted approximately $400,000 for this center over a four year period.

Research included in this program is carried out by a high quality group of workers who, while conducting basic studies, are surprisingly goal-oriented.

Contributions and potential contributions of this relatively small number of grantees to the general area of biomedical science have been many. Some examples of these follow:

(1) Work in plasmid-determined enterotoxins which may offer a method of dealing with diseases of a diarrheal nature.

(2) The mechanisms of sulfonamide resistance in pneumococci have been elucidated at the molecular level.

(3) Demonstration that bacterial mutation can increase the number of resistance factor copies; an explanation of the high resistance of certain strains to the antibiotic, ampicillin. This could lead to the identification of mutants which would specifically inhibit resistance factor replication.

(4) Incompatibility typing of plasmids that can be used to identify the origin and evolution of plasmids and may be important for epidemiological purposes.

(5) Mapping of resistance determinants in *Salmonella*, utilizing resistance factors. The structure of *Salmonella*, utilizing resistance factors. The structure of *Salmonella* antigens is highly important for diagnostic work; aberrations in these antigens are often attributable to plasmids.

(6) The entire molecular basis of bacterial resistance to aminoglycoside antibiotics (gentamycin and neomycin) has been determined by another grantee. This could culminate in a "tailoring" of antibiotics.

(7) The mechanism of evolution of plasmids has been explored and may lead to future control of the spread of plasmids.

(8) Rapid and accurate enzymetic assays for antibiotics in patient sera have been produced.

(9) An ultimate contribution may be the use of plasmids as models of choice for the study of replication in general, including that of bacteria and viruses.

Much of this information has been transferred already to the clinical and epidemiological areas, but additional and more efficient methods must be explored by which the transfer of the remainder can be accomplished.

The second program, Recombinant DNA, had its origin in the first. Dr. Hamilton Smith, who pioneered in the area of restriction enzymes, was supported by our Mechanisms of Resistance to Antimicrobial Agents Program. Another grantee, Dr. Stanley Cohen, supported by the same program, was successful in introducing DNA from an unrelated bacterium, *Staphylococcus aureus*, into a plasmid in *E. coli*. This specific piece of DNA was one that conferred resistance to penicillin. He was able to prove that it was working in the new host because the *E. coli* became resistant to penicillin. He next took DNA from a South African toad and inserted it into an *E. coli* plasmid and the *E. coli* plasmid, now part toad, was able to replicate.

Our most important goal in this program is the utilization of the recombinant DNA technology to provide us with a greater knowledge of the molecular basis of pathogenicity. Once obtained, this information may lead to improved prevention, diagnosis and treatment of infectious diseases.

Another goal is the productions of a variety of biologicaly useful substances through the construction of bacterial cells containing functional DNA of animal origin. Currently some of our people are attempting to isolate the messenger RNA produced by the gene which codes for interferon and then artificially construct the complementary DNA. This will then be cloned either with a bacterial episome or lambda phage and then clones producing interferon will be identified and subsequently propagated.

254

An equally important goal is the identification, assessment and elimination of any and all potential biohazards encountered in the exploitation of this technology.

Research projects of special interest to this program are included in one or more of the following categories: (1) Construction *in vitro* of biologically functional DNA or RNA molecules which combine genetic information from two different sources; (2) Utilization of plasmids, bacteriophages, viruses and organellar DNA (chlorplast or mitochondria) as *cloning* vehicles or vectors for introducing this genetic information into hosts (bacteria, algae, fungi, protozoa, helminths or insects); (3) Studies which involve breaking DNA or RNA molecules, utilizing restriction enzymes (endonucleases) or other methods and the joining of short naturally occurring or synthetic sequences of nucleotides into longer segments of DNA or RNA; (4) Research involving molecular biology of transformation of microorganisms (pathogens and non-pathogens) with foreign DNA and development of methods for the introduction of DNA into clones of cells of higher organisms (protozoa, helminths and insects), (5) Use of the technology to elucidate the mechanisms utilized by plasmids and bacterial viruses to contribute to the ability of bacteria to colonize and overcome normal defense mechanisms of hosts, produce toxins and/or specify resistance to antimicrobial agents, (6) Research employing recombinant DNA techniqeus to elucidate the molecular mechanism of viral, bacterial, mycotic or parasitic infection and pathogenesis, (7) Epxloitation of the technology culminating in the design of more effective vaccines, (8) Studies on the mechanism of infectious agent transmission by species which act as vectors (e.g., insects, snails, rodents, *et al.*); (9) Experiments utilizing this technology to study the role of intermediate or reservoir hosts in the life cycle of infectious disease agents; (10) Design, construction and testing of safer host-vector systems for use in this research.

Apropos the last category (10) mentioned above, Drs. Wallace Rowe and Malcolm Martin of the National Institute of Allergy and Infectious Diseases have recently concluded studies involving risk assessment of a cloned DNA virus, mouse polyoma. They inserted genetic material from this virus into *Escherichia coli* which they subsequently introduced into mice. This experiment confirmed their prediction that this form of recombinant DNA research would be completely safe. The microbes proved to be either non-infectious or far less infectious (by a factor of 10^9) than the mouse polyoma virus itself. However, these results must not be extrapolated to the cloning of RNA tumor viruses which are considered more hazardous.

Currently the branch has approximately 4.6 million dollars in this program.

About 95% of the grants supported by both of these programs are devoted to basic research. Our original support of basic research in drug resistance mediated by plasmids opened a veritable "Pandora's Box" that made possible a massive technological breakthrough known as recombinant DNA. Proper utilization of this technology should increase our knowledge of the mechanisms of pathogenicity (at the molecular level) of viral, bacterial, mycotic and parasitic agents. This information, in turn, will lead to improved prevention, diagnosis and treatment of infectious diseases.

Having concluded my remarks concerning our programs in mechanisms of resistance to antimicrobial agents and recombinant DNA, I would like to devote the remainder of my time to a discussion of programs involving two highly relevant major public health problems that were mentioned in my opening statement: hospital associated infections and sexually transmitted diseases. Programs in both of these critical areas are supported by the Bacteriology and Virology Branch headed by Dr. Milton Puziss and also are located in the Microbiology and Infectious Diseases Program of the National Institute of Allergy and Infectious Diseases.

The first of these two programs is Hospital Associated Infections. These infections constitute a problem of the utmost gravity, not confined solely to the United States where they make an important contribution to spiraling increases in hospitalization costs. Of greater significance is the annual death rate attributable to these infections. Although precise figures are difficult to ascertain, current estimates by expert epidemiologists are in the range of 100,000 deaths per year in the United States. In addition, these same epidemiologists conservatively estimate the annual costs of hospital associated infections in the above country at one billion dollars. During the past 15 years the endogenous (resident) population of gram negative microbial flora has shown a dramatic increase in antibiotic resistance, especially in immunosuppressed patients. This has resulted in a high incidence of bacteremia accompanied by a high mortality rate in these patients.

The goal of this program is to enlarge our store of fundamental knowledge in order to understand the multiple factors involved in the causation of these infections. The long range goal will be to apply this knowledge to the development of methods for the prevention and treatment of these infections.

Research of special interest to this program is included in one or more of the following major categories: (1) host resistance studies embracing the full range of normal resistance processes and immune mechanisms, both cellular and humoral, of healthy people against the microbial flora commonly encountered, the organisms involved do not usually threaten a healthy individual; (2) disturbances in resistance mechanisms in hospitalized, debilitated, or immunosuppressed patients are to be investigated, including colonization immunity (the ability to ward off an abnormal or potentially invasive microbial flora), natural antimicrobial properties (as found on body surfaces and within body fluids and their components), and cellular immune defenses (polymorphonuclear leukocytes, macrophages, and lymphocytes, as related to the problem of hospital associated infections); (3) interaction of microbial products (e.g., endotoxins) from gram-negative pathogenic species with serum components that alter the blood-clotting mechanism, vascular permeability, and smooth muscle reactivity; (4) influence of direct interbacterial interference (e.g. by lactobacillus) on prevention of colonization or disease, together with induction of cross-immunity or specific immunity (shared or common antigens); (5) in addition to various aspects of host resistance, environmental studies are being carried out on the value of the intensive care unit concept in the immunosuppressed or debilitated patient.

The branch currently has approximately 3 million dollars invested in this program.

Some examples of research performed by grantees in this program follow:

Dr. A. I. Braude (University of California, San Diego) has developed a therapeutic antiserum of great potential for treating patients with life-threatening gram negative bacterial infections. In earlier reports he described the J5 antiserum produced in humans vaccinated with the core glycolipid from an *E. coli* strain. In a double blind study of patients in profound bacteremic shock, he found that recovery from shock in the J5 serum treated group was nearly three times that of the control group. Tests in rabbits given the J5 antiserum, to study optimum dosage against fatal *Pseudomonas* sepsis, indicated that protection was not dose dependent, and was undiminished after two weeks. The J5 antiserum administered prophylactically to animals prior to inoculation with *Pseudomonas* provided as strong protection (75%) as antiserum given for therapy against an established bacteremia. The feasibility was therefore established of using 75 human antiserum prophylactically in patients with a high risk of developing gram negative bacteremia (Ziegler, McCutchan and Braude, 1979; Ziegler, McCutchan and Braude, 1979; McCutchan, Ziegler and Braude, 1979).

Dr. P. F. Bartell (New Jersey College of Medicine) has investigated severe infections with *Pseudomonas aeruginosa* which are one of the major problems in some hospitalized patients. The pathogenic role of the slime layer of the organism is still undertermined. Dr. Bartell has isolated and purified a glycolipoprotein (GLP) from this slime, and has shown that it acts as a toxin and as a protective antigen. In a rabbit model, immunization with GLP resulted in hemagglutinating and bacterial agglutinating activity. These paralleled mouse passive protection activity against lethal challenge with the *Ps.* organism; the protection was dependent on activity levels in the plasma. The activity was found in the IgM serum antibody fraction. This suggests that the GLP may offer a potential means of passive protection against the toxic and lethal effects of *Ps. aeruginosa* infections in hospital patients (Sensakovic and Bartell, 1975; Sensakovic and Bartell, 1977; Lynn, Sensakovic and Bartell, 1977; Dimitracopoulos and Bartell, 1979).

Dr. J. W. Alexander (University of Cincinnati) has investigated nosocomial infections in surgical and burn patients for immunologic variables of host resistance to infection. Infected burn patients had significantly worse neutrophil bactericidal activity, this was consistently observed in association with infection. In the bacteremic episodes, immunologic measurements in the burn patients showed significant predisposing abnormalities, primarily of neutrophil function and opsonization activity. Kidney transplant patients are in a similar category, with transient defects in the primary defense mechanism. In a guinea pig model, malnourished individuals may not maintain adequate levels of C3 complement factor. Patients are possibly at much greater risk when malnourished because of the failure of synthesis of the C3 complement factor, although this could be related to cellular defects. Nutritional support may therefore be a vital factor in host response to infection in these patients (Ogle, Ogle and Alexander, 1978).

Dr. W. R. McCabe (Boston City Hospital) has demonstrated in previous studies (McCabe, 1972) that both active and passive immunization with the Re mutant of *Salmonella minnesota* protected mice against lethal infections with *Klebsiella pneumoniae* and *Proteus morganii*. Antibody to the Re mutant protects against both initial bloodstream invasion and lethality of the bacteremia. Studies in volunteers were recently initiated after completion of extensive animal testing and Food and Drug Administration approval of an Investigational New Drug (IND) was obtained. Intradermal injection of 10 mcg of Re lipopolysaccharide showed a significant increase in antibody titers to the Re mutant, with no adverse systemic reactions observed. Subcutaneous immunization with the Re mutant was also initiated. No systemic reactions were seen and local arm reaction was equivalent to that seen after typhoid vaccination, these volunteer studies are continuing (Johns, Bruins and McCabe, 1977).

The second of these two programs is Sexually Transmitted Diseases. This program is devoted to the etiological agents of diseases that are still considered major public health problems, with genorrhea classed as a national epidemic. Most recent statistics indicate, however, that there has been a slight decline in the rate of increase of reported cases of gonorrhea and syphilis, for unknown reasons. Cases of herpes type 2 virus seem to be increasing; this is also true for nongonococcal urethritis cases, although definitive data are lacking. There is widespread interest in this research program, although the number of people already in specialized training or seeking such training for research careers in this area is not large. M. D. candidates, especially, seem to be reluctant to undertake postdoctoral training in infectious diseases.

The immediate goal of this program is to narrow the existing gap areas in our fundamental knowledge of the basic biology and immunology of venereal diseases. This is urgently needed to check the alarming spread of these diseases.

Research of special interest to the program is included in the following categories: (1) basic biology — growth, nutrition and subcellular structures of the organisms; (2) cellular antigens — cell walls, capsules , and their characterization and purification;

(3) host defense mechanisms – specific antibody development, cellular and/or humoral defenses, pathogenesis of disease, reinfection mechanisms; (4) experimental models — *in vivo* and *in vitro* systems; (5) drug resistance and susceptibility; (6) therapeutic and clinical studies.

Including contracts, grants and training, we have a current investment of almost 4 million dollars in this program.

A few examples of research by grantees in this program follow:

The pathogenesis of gonorrhea has been difficult to study for lack of a suitable non-human model. Dr. Z. A. McGee (Vanderbilt University) is studying the mechanisms by which gonococci cause disease in the human genital tract by means of an organ culture model system using human fallopian tube sections. The rate of damage to the mucosa can be quantitated and the interaction with the gonococci studied microscopically. Data obtained show that attachment of gonococci may enhance mucosal damage by a toxic factor. This toxic factor is present during the first 24 hours of infection, and remains at a high level thereafter. The toxic factor appears to be an endotoxin, and is responsible for at least part of the damage to the recipient cells. There is also a strong indication that the toxin(s) is specific for human genital mucosa and not for that of other species. Of interest also is the finding that type 4 gonococci, which lack pili, nevertheless do attach to fallopian tube mucosal cells, albeit more slowly than the piliated type 1 organisms (Johnson, Taylor-Robinson, and McGee, 1977).

Chlamydia trachomatis is a causal agent of nongonococcal urethritis (NGU), the second most prevalent of the sexually transmitted diseases. In the male it causes NGU, while in the female the infection is primarily a cervicitis. Newborns of infected mothers frequently develop inclusion conjunctivitis and the recently described infant pneumonia syndrome. Dr. A. L. Barron (University of Arkansas) is studying the immune responses of chlamydial infections in a guinea pig model infected with a chlamydial inclusion agent (gp-ic). The effect of the immunosuppressant drug cyclophosphamide (cy) on both humoral and cellular immunity was investigated. When both B and T cells were depressed by cy, the infection persisted for 35–45 days, the T cell function, depending on regimen, returned by day 15, but antibody to gp-ic was not detected until day 21 when it remained at a high level. Recovery from infection was coincident with antibody development. Treatment with higher doses of cy resulted in selective suppression of B cells only, and animals remained infected until death at day 30. No antibody response was detected in these animals. In cy-treated animals as ascending genital infection, including endo-metritis and cystic salpingitis was seen. The data indicate that in the presence of a functional cell-mediated system animals fail to recover in the face of a suppressed humoral antibody response. In other infection experiments, results showed that involvement of guinea pig vaginal tissue was superficial, but an intense inflammatory response of the exocervix was observed, no evidence of endometrial infection was seen in untreated animals (Rank, White and Barron, 1978).

In a prospective evaluation of antibiotic prophylaxis against gonorrhea, 1080 men were given 200 mg of oral minocycline or placebo after sexual intercourse with prostitutes in a Far Eastern port. Later, at sea, gonococcal infection was detected in 57 of 565 men given placebo and 24 of 515 men given minocycline (P < 0.001). Minocycline prophylaxis completely prevented infection by gonococci susceptible to 0.75 µg or less of tetracycline per milliliter, reduced the risk of infection or prolonged the incubation period in men exposed to gonococci susceptible to 1.0 to 2.0 µg per milliliter, but did not prevent infection or prolong incubation in men exposed to gonococci resistant to 2.0 µg. Mino-cycline did not increase the proportion of asymptomatic infections. Minocycline pro-phylaxis would probably have limited effectiveness as a public health measure because of the tendency to select resistant gonococci (Harrison, Hooper, Wiesner, Campbell, Karney, Reynolds, Jones, and Holmes, 1979).

In conclusion, an attempt has been made to give you an administrator's view of some of the research conducted by our grantees. It is our hope that you are convinced of the vital interrelationships existing among four of the many programs supported by the National Institute of Allergy and Infectious Diseases. Our current investment in these programs approximates 15.6 million dollars of the money provided so generously by the United States taxpayer.

REFERENCES

DIMITRACOUPOLOS, G. and P. F. BARTELL (1979): Phage-related surface modifications of *Pseudomonas aeruginosa*: effects on the biological activity of the viable cells. Infection and Immunity 23, 87.

HARRISON, W. O., R. R. HOOPER, P. J. WIESNER, A. F. CAMPBELL, W. W. KARNEY, G. H. REYNOLDS, O. G. JONES and K. K. HOLMES (1979): A trial of minocyckline given after exposure to prevent gonorrhea. N. Engl. J. Med. 300, 1074–1078.

JOHNS, M. A., S. C. BRUINS and W. R. McCABE (1977): Immunization with R mutants of *Salmonella minnesota*. II. Serological response to lipid A and the lipopolysaccharide of Re mutants. Infection and Immunity 17, 9–15.

JOHNSON, A. P., D. TAYLOR-ROBINSON and Z. A. McGEE (1977): Species specificity of attachment and damage to oviduct mucosa by *Neisseria gonorrhoeae*. Infection and Immunity 18, 833–839.

LYNN, M., J. W. SENSAKOVIC and P. F. BARTELL (1977): *In vivo* distribution of *Pseudomonas aeruginosa* slime glycolipoprotein: association with leukocytes. Infection and Immunity 15, 109–114.

McCABE, W. R. (1972): Immunization with R mutants of *S. minnesota*. I. Protection against challenge with heterologous gram-negative bacteria. J. Immun. 108, 601–610.

McCUTCHAN, J. A., E. J. ZIEGLER and A. I. BRAUDE (1979): Treatment of gram-negative bacteremia with antiserum to core glycolipid. II. A controlled trial of antiserum in patients with bacteremia. Europ. J. Cancer. In press.

OGLE, C. K., J. D. OGLE and J. W. ALEXANDER (1978): Comparison of C3 levels in patients' sera measuring C3 levels with both anti C3 (β1c/β1a) and anti C3 (B). J. Lab. and Clin. Med. (Submitted).

RANK, R. G., H. J. WHITE and A. L. BARRON (1978): Altered pathogenesis by immunosuppression with cyclophosphamide of genital infection in female guinea pigs infected with the chlamydial agent of guinea pig inclusion conjuctivitis (gp-ic). 18th Interscience Conf. on Antimicrob. Agents and Chemotherapy. 1–4 Oct., 1978, Atlanta, Ga.

SENSAKOVIC, J. W. and P. F. BARTELL (1975): Biological activity of fragments derived from the extracellular slime glycolipoprotein of *Pseudomonas aeruginosa*. Infection and Immunity 12, 808–812.

SENSAKOVIC, J. W. and P. F. BARTELL (1977): Glycolipoprotein from *Pseudomonas aeruginosa* as a protective antigen against *P. aeruginosa* infection in mice. Infection and Immunity 18, 304–309.

ZIEGLER, E. J., J. A. McCUTCHAN and A. I. BRAUDE (1979): Clinical trial of core glycolipid antibody in gram-negative bacteremia. Trans. Assoc. Am. Physicians. In press.

ZIEGLER, E. J., J. A. McCUTCHAN and A. I. BRAUDE (1979): Treatment of gram-negative bacteremia with antiserum to core glycolipid. I. The experimental basis of immunity to endotoxin. Europ. J. Cancer. In press.

I.P.D., National Institute of Health,
Bethesda, Maryland,
U.S.A.

THE SPREAD OF TRANSPOSON 7 AMONG BACTERIA
OF MEDICAL AND VETERINARY IMPORTANCE

H. RICHARDS

*Department of Bacteriology, Royal Postgraduate
Medical School, London, U.K.*

INTRODUCTION

Trimethoprim is an antibacterial drug which has been heavily used in both clinical and veterinary medicine over the last 10 years. Its mode of action is to interfere with the enzyme dihydrofolic acid reductase which converts dihydrofolic to tetrahydrofolic acid, an essential step in bacterial DNA synthesis. This enzyme acts at a stage which immediately follows the enzyme conversion of para-aminobenzoic acid to dihydrofolic acid, which can be competitively blocked by sulphonamides. This sequential action of sulphonamides and trimethoprim explains the synergistic action of this combination against some bacteria. In treatment trimethoprim is always used in combination with a sulphonamide, sulphamethoxazole, as cotrimoxazole. However, sulphonamide resistance is common, in 1976 Grüneberg found that greater than 33% of urinary pathogens were resistant to sulphonamides. Therefore, the spread of a transposon carrying genes determining trimethoprim resistance among bacteria of medical and veterinary importance has serious implications on the future therapeutic use of trimethoprim. This paper describes the incidence and spread of transposon 7 among such bacteria. The findings illustrate that the distribution of this transposon has contributed to the increase in trimethoprim resistance which is now found in clinical and veterinary medicine.

TRANSPOSON 7

Transposon 7 was originally identified on R483 in 1976 by Barth *et al*. Tn7 encodes resistance to trimethoprim and to streptomycin and spectinomycin and has a molecular weight of 9 Md. It transposes well in the laboratory with a frequency of 5×10^{-4} which suggests that under selective pressure Tn7 will spread rapidly among wild bacteria.

TRIMETHOPRIM RESISTANCE IN HAMMERSMITH HOSPITAL

In Hammersmith Hospital we have been collecting bacteria which are resistant to the newer antibacterial drugs, including trimethoprim. Like ampicillin, trimethoprim is a broad spectrum antibiotic active against a wide range of bacteria. Because of this and its lack of toxicity it is much used both in and out of hospitals in the treatment of a variety of infections. It has, up to now, been particularly effective against urinary infections. However, Table I. demonstrates how the percentage of trimethoprim resistant urinary pathogens has increased dramatically over the last few years.

Trimethoprim resistance plasmids have been isolated from the resistant bacteria and they are listed together with their properties in Table II. This table illustrates that

261

TABLE I.

Trimethoprim resistance in urine infections in Hammersmith Hospital

Year	Total tested	% Resistant
1970	429	5.4
1975	421	13.5
1977	883	17.0
1978	806	20.6

plasmids carrying trimethoprim resistance have spread among a variety of bacteria isolated from different wards in different parts of the hospital.

Trimethoprim R plasmids were first identified in London in 1971 (Fleming *et al.* 1972). In 1972 Datta and Hedges tested trimethoprim resistant Enterobacteria isolated in hospitals in the U.K. They found a klebsiella strain carrying a sulphonamide-trimetroprim (SuTp) resistance plasmid of incompatibility group W in two London hospitals. In 1974 Jobanputra and Datta found that this klebsiella R$^+$ strain had spread to Hammersmith Hospital. This strain is no longer detected but IncW plasmids determining SuTp and, in addition, ampicillin, mercuric chlordie and gentamicin resistance are now found suggesting that the original plasmid has acquired additional resistance genes, probably by transposition (Hughes and Datta, this volume).

Table III. shows the range of incompatibility groups in which plasmid-borne trimethoprim resistance has so far been identified. The trimethoprim resistance W plasmids isolated in 1974, of which R388 is an example do not carry Tn7 and neither do the W

TABLE II.

Properties of trimethoprim R plasmids in Hammersmith Hospital

Bacterial Species	Ward	Plasmid	R Pattern[2]	Group[1]
E. coli	C6	pHH1169	Cm Su Hg Sm Tp	FII
Coliform	B2	pHH1170	Su Tp	N
E. coli	C6	pHH1172	Cm Su Hg Sm Tp	FII
Enterobacter	B2	pHH1173	Su Hg Tp	P
Klebsiella sp.	C6	pHH1183	Ap Tc Cm Su Hg Sm Tp	
E. coli	B6	pHH1187	Su Sm Tp	X
E. coli	B2	pHH1188	Ap Su Hg Km Gm Tp	W
E. coli	B1	pHH1189	Sm Tp	Iδ
E. coli	D2	pHH1190	Ap Cm Su Hg Km Gm Tb Sm Tp	C
E. coli	I3	pHH1191	Su Km Tp	W
Citrobacter	B1	pHH1192	Ap Su Hg Gm Tp	W
Citrobacter	C6	pHH1193	Ap Su Hg Km Gm Tp	W
Citrobacter	B1	pHH1302b	Ap Su Hg Gm Tp	W
Enterobacter	B2	pHH1303	Ap Su Hg Km Gm Tp	W
E. coli	B6	pHH1305	Sm Tp	Iδ
E. coli	C6	pHH1306b	Ap Su Hg Gm Tp	W
E. coli	I3	pHH1307	Ap Su Hg Km Gm Tp	W
Enterobacter	D5	pHH1308	Ap Tc Cm Su Hg Sm Tp	

[1] These are the unpublished results of Naomi Datta and Victoria Hughes. pHH1183 and PHH1308 have not yet been classified.

[2] Abbreviations: Ap, ampicillin; Cm, chloramphenicol; Gm, gentamicin; Hg, mercuric chloride; Km, kanamycin; Sm, streptomycin; Su, sulphonamides; Tc, tetracycline; Tb, tobramycin; Tp, trimethoprim.

plasmids isolated in 1977. However, Tn7 has been identified on an IncC plasmid (pHH1190), on an Iδ plasmid (pHH1189), on an FII plasmid (pHH1169) and on an X plasmid (pHH1187) see Table II. (S. Dacey, personal communication). Experiments are in progress to determine if plasmids of the other incompatibility groups also carry Tn7.

TABLE III.

Trimethoprim resistant plasmids in Hammersmith Hospital

Range of Incompatibility groups:

1974	W						
1977	W	C	I	FII	N	P	X

The evidence does suggest that Tn7 is spreading amongst plasmids of different incompatibility groups and that this is contributing to the increasing incidence of trimethoprim resistance in hospital pathogens. Investigation of plasmids from bacteria isolated during an outbreak of infection from another London hospital showed that the epidemic strain of *Klebsiella aerogenes* contained a multiple resistance plasmid carrying Tn7. This plasmid has also spread to environmentally related *Escherichia coli* and to *Citrobacter koseri* and the transposon had entered the chromosome of all the wild bacteria (Datta *et al.* 1979; Richards and Nugent, 1979).

TRANSPOSON 7 IN *SALMONELLAE*

Trimethoprim is widely used in the treatment of respiratory and enteric infections in animals. Table IV. shows the incidence of trimethoprim R plasmids in *Salmonellae* tested at the Central Veterinary Laboratory, Weybridge. Table V. shows the salmonella species from which these plasmids were derived and the origin of infection. These tables together illustrate the increasing number of trimethoprim R plasmids found in *Salmonellae*. Trimethoprim R plasmids first appeared in *S. typhimurium* but are now found in a wide variety of salmonellae species (Table V.). Of particular importance is their appearance in *S. hadar* which infects poultry especially chickens and turkeys.

Phage type 204 of *S. typhimurium* is still the predominent type in bovine and human infections in the British Isles (Threlfall *et al.* 1978a). It first appeared in 1974 and the most common resistance pattern was to sulphonamides and tetracycline. In 1975 one strain was found to carry a trimethoprim R plasmid (Table V.) and this plasmid carried

TABLE IV.

Salmonellae with trimethoprim R plasmids, Central Veterinary Laboratory[1]

Year	Total Tested	Number (%) with Tp plasmids
1975	4,000	2 (0.05)
1976	1,333	4 (0.3)
1977	3,333	10 (0.3)
1978	4,417	53 (1.2)

[1] The information in this table and in Table V was kindly provided by Dr. C. Wray Ph. D. MRCVS, Central Veterinary Laboratory, Weybridge.

TABLE V.
Origin of plasmid determined trimethoprim resistance among salmonellae

Year	Serotype	Origin
1975	S. typhimurium 204	Cattle
1976	S. typhimurium 49	Cattle
		Horses
	S. typhimurium 204	Cattle
1977	S. typhimurium 204	
	S. typhimurium 104	Cattle
	S. typhimurium 12	
	S. typhimurium 145	
	S. poona	Seagull
	S. tennessee	Poultry
	S. heidelberg	Pig
	S. krefeld	Water
	Salmonella 4, 12:d	Feather meal
1978		Cattle
	S. typhimurium 204	Cat
		Sheep
	S. typhimurium 145	
	S. typhimurium 56	Cattle
	S. typhimurium 49	
	S. typhimurium 193	
	S. typhimurium 193	Cattle
		Horses
	S. typhimurium	Cattle
	S. typhimurium 168	Pig
	S. typhimurium 18	
	S. give	River
	S. hadar	Poultry
	S. newport	Pig

Tn7 (Richards *et al.* 1978). Threlfall *et al.* 1978b, showed that phage type 204 can be changed to phage type 193 by the acquisition of a particular Iα plasmid. They also report that some bovine isolations of the type 193 strain have recently been found to carry trimethoprim R plasmids.

Tn7 has also been identified in another plasmid isolated from a phage 49 strain of *S. typhimurium* (Richards *et al.* 1978). Again the evidence suggests that the spread of Tn7 is contributing to the increase in trimethoprim resistance in *Salmonellae*. It is now ten years since the Joint Committee on the Use of Antibiotics in Animal Husbandry and Veterinary Medicine published their report (Report, 1969), and yet resistant salmonellas in food animals are not even under control, much less eliminated (Editorial, 1979).

CONCLUSIONS

Trimethoprim resistance is increasing in pathogens causing disease in man and animals. Presumably this is because trimethoprim-sulphonamide combination therapy is much used, since it is non-toxic and effective against a wide spectrum of disease.

Because sulphonamide resistance is common, the spread of trimethoprim R plasmids threatens the effectiveness of this treatment. A transposon carrying trimethoprim resistance genes increases the chances of dissemination of resistance even more. This paper presents results which illustrate that such a transposon, Tn7, is spreading among pathogenic bacteria. Another transposon determining ampicillin resistance has spread throughout the world among plasmids of different incompatibility groups and bacteria of many genera presumably as a result of very heavy worldwide use of ampicillin. Because of this ampicillin is no longer so often the drug of choice, and other antibacterials, including trimethoprim, have been used instead. However, the excessive therapeutic use of trimethoprim could lead to worldwide distribution of the trimethoprim-resistant transposon. Trimethoprim will then no longer be such an effective antibacterial drug.

ACKNOWLEDGEMENTS

This work was supported by a Medical Research Council grant to Naomi Datta.

REFERENCES

BARTH, P. T., DATTA, N., HEDGES, R. W. and GRINTER, N. J. (1976): Transposition of a deoxyribonucleic acid sequence encoding trimethoprim and streptomycin resistances from R483 to other replicons. J. Bacteriol. 125, 800–810.
DATTA, N. and HEDGES, R. W. (1972): Trimethoprim resistance conferred by W plasmids in Enterobacteriaceae. J. Gen. Microbiol. 72, 349–355.
DATTA, N., HUGHES, V. M., NUGENT, M. E. and RICHARDS, H. (1979): Plasmids and transposons, their stability and mutability in bacteria isolated during an outbreak of hospital infection. Plasmid (in press).
EDITORIAL (1979): Salmonellosis – an unhappy turn of events. Lancet i, 1009–1010.
FLEMING, M. P., DATTA, N. and GRÜNEBERG, R. N. (1972): Trimethoprim resistance determined by R factors. Brit. Med. J. i, 726–728.
GRÜNEBERG, R. N. (1976): Susceptibility of urinary pathogens to various antimicrobial substances: a four year study. J. Clin. Path. 29, 292–295.
HUGHES, V. M. and DATTA, N. (1979): Gentamicin resistance plasmids in Hospital Infections. This volume p.
JOBANPUTRA, R. and DATTA, N. (1974): Trimethoprim R factors in Enterobacteria from clinical specimens. J. Med. Microbiol. 7, 169–177.
Report of the JOINT COMMITTEE ON THE USE OF ANTIBIOTICS IN ANIMAL HUSBANDRY AND VETERINARY MEDICINE (1969): H. M. Stationery Office, London.
RICHARDS, H., DATTA, N., SOJKA, W. J. and WRAY, C. (1978): Trimethoprim-resistance plasmids and transposons in Salmonellae. Lancet ii, 1194–1195.
RICHARDS, H. and NUGENT, M. E. (1979): The incidence and spread of transposon 7. Plasmids of Medical, Environmental and Commercial importance. Ed. K. Timmis and A. Pühler (in press).
THRELFALL, E. J., WARD, L. R. and ROWE, B. (1978a): Epidemic spread of a chloramphenicol-resistant strain of Salmonella typhimurium phage type 204 in bovine animals in Britain. Veterinary Record 103, 438–440.
THRELFALL, E. J., WARD, L. R. and ROWE, B. (1978b): Spread of multiresistant strains of Salmonella typhimurium phage types 204 and 193 in Britain. Brit. Med. J. ii, 997.

H.R., Royal Postgraduate
Medical School,
London W12 OHS, U.K.

TRIMETHOPRIM RESISTANCE PLASMIDS
FROM ENTEROBACTERIACEAE ISOLATED IN GREECE

G. SAROGLOU, P. PARASKEVOPOULOU, O. PANIARA
and P. KONTOMICHALOU*
*Department of Clinical Therapeutics, School of Medicine,
University of Athens, Greece*

INTRODUCTION

Trimethoprim resistance among Enterobacteriaceae has slightly increased despite the wide use of the combination Trimethoprim-Sulfamethoxasole against a variety of infections for over a decade. Resistance to Trimethoprim has been reported ranging from 4.3% (Brumfitt, 1977) to 11.6% (Amyes 1978). Reports from Italy showed increased resistance up to 30% and positive transfer of Trimethoprim resistance in 10% of the nosocomial isolates during 1973 and 20% during 1975 (Romero 1977).

Plasmids carrying resistance to Trimethoprim belong to different compatibility groups and carry also resistance to different combinations of other antibiotics (Acar 1977).

We decided to study the problem of Trimethoprim resistance in our hospital because since 1975 we noticed that Enterobacteriaceae isolates were resistant to Trimethoprim in a proportion much higher than that reported in the literature (Marks, 1977, Wong 1975).

INFECTIONS DUE TO PATHOGENS RESISTANT TO TRIMETHOPRIM

In our hospital we found during 1978 a high percentage of resistance to Trimethoprim among *Enterobacteriaceae*. We tested 1460 clinical isolates and 42% resistance was found, while it was 30% for 1975. Specifically blood isolates showed 42% resistance to Trimethoprim, sputum isolates 36% and urine isolates 51%.

Because of the increased resistance to Trimethoprim we decided to survey the infections that were caused by pathogens resistant to Trimethoprim for a period of one month (April 1978). We found 25 infections with pathogens *Enterobacteriaceae* resistant to Trimethoprim. Epidemiologic follow of these infections showed that 19 of them were hospital aquired while 6 were community acquired infections. Among the 19 nosocomial infections 13 were urinary tract infections, two pneumonias, two septicemias with unknown primary focus and two peritonitis cases in patients undergoing peritoneal dialysis. Among the six community aquired infections five were urinary tract and one septicemia.

The duration of hospitalization before the development of a nosocomial infection varied. Nine patients developed infection during their first two weeks of hospitalization while four other patients five weeks after hospitalization.

Thirty two Enterobacteriaceae isolated from the above 25 patients were incriminated as pathogens of the infections. For all isolates conjugation by broth mating experiments

---

* This work has been supported by Grant from the Hellenic Ministry of Hygiene.

TABLE I.

Resistance pattern from the 11 transfers

Donors	Number of clones	Resistance pattern in transconjugants
Klebsiella pneumoniae 70	3/5 1/5 1/5	Tp Tp Su Tp Su Cm
E. coli 71	2/5 3/5	Tp Tp Su Cm
E. coli 72	5/5	Tp Su bla
Klebsiella pneumoniae 73	3/3	Tp Su bla
E. coli 74	3/5 2/5	Tp Su bla Cm Km Tp Su bla Cm Km Sm
Proteus mirabilis 75	3/3	Tp Su bla Cm Km
Proteus vulgaris 76	5/5 2/5	Tp Su bla Cm Km Tp Su bla Cm Km Sm Nm
Enterobacter cloacae 77	2/5 1/5	Tp Su bla Cm Km Sm Nm Tc Tp Su bla Cm Km Nm Tc
E. coli 78	1/5 1/5 1/5 1/5 1/5	Tp Tp Su Tp Su bla Cm Km Tp Su bla Cm Nm Tp Su bla Cm Km Nm
Proteus morgani 79	1/1	Tp Su bla Cm Tc Km Co
Serratia marcesens 80	2/5 2/5 1/5	Tp Su bla Cm Tp Su bla Cm Km Nm Tp Su bla Cm Km

were performed using *E. coli* RC85 R⁻ as a recipient host. Conjugal transferability for Trimethoprim resistance was found positive in 11 out of 32 isolates (transfer rate 34%).

From each donor five clones of transconjugants were tested for Trimethoprim resistance. The transformed resistance pattern by selection with Trimethoprim (Table I.) was found identical in all five clones for 3 donors only, while it varied for the rest. We found clones with resistance pattern to Trimethoprim alone, to Trimethoprim-Sulfonamide, to Trimethoprim-Sulfonamide-β-lactamase. We also found clones with different combinations of Trimethoprim, Sulfonamide, β-lactamase, Chloramphenicol, Streptomycin, Tetracycline, Kanamycin, Neomycin.

Trimethoprim resistance phenotype in all donor and recipient cultures was high (Table II.). Particularly, from nine strains of wild type *E. coli* used as donors, four had MIC 1000 μg/ml, one 5000 μg/ml and two 10 000 μg/ml. Highly resistant strains were found also among all the other *Enterobacteriaceae* species tested. The Trimethoprim resistance from all 6 donor strains with MIC 10 000 μg/ml was succesfully transferred to the recipient host.

The phenotype of Trimethoprim resistance in all recipient cultures was found over 1000 μg/ml.

Dihydrofolate reductase activity assay was performed on representative clones from each transconjugant. In all clones examined two-fold to eight-fold increase enzyme activity compared to R⁻ host culture was found.

TABLE II.

Minimal inhibitory concentrations (MIC) of trimethoprim for the 32 donors and 11 recipients

Donor Cultures	Total No. of strains		MICγ/ml															
			125		250		500		1 000		2 000		4 000		5 000		10 000	
	D	DS	D	DS	D	DS	D	DS	D	DS	D	DS	D	DS	D	DS	D	DS
E. coli	9	5	1	–	–	–	1	1	4	1					1	1	2	2
Serratia marcesens	1						1											
Proteus mirabilis	8	1					2		2		1		1		1		1	1
Proteus morgani	1	1															1	1
Proteus vulgaris	2	1									1						1	1
Klebsiella pneumoniae	7	2			1				2	1			2		1		1	1
Enterobacter cloacae	3	1	1												2	1		
Acinetobacter	1						1											
Reciepient Cultures K^{12} R^{+}	11												9		2			

D = Total number of Donors for each Enterobacteriaceae species.
DS = Number of Donors with successful transfer.

The eleven donor strains with positive Trimethoprim transfer originated from nine patients. We analyse the predisposing factors associated with the nine infections. Eight infections were nosocomial. Common underlying factor in seven infections was the prolonged use of broad spectrum antibiotics, five to ten weeks before the appearence of the hospital acquired infection. All patients had chronic debilitating underlying disease and prolonged hospitalization. In three cases predisposing factor was the insertion of a permanent Foley catheter.

Only one patient was an out-patient case, with the diagnosis of chronic pyelonephritis. She was on Co-Trimoxazol prophylaxis for over six months before an exacerbation of her pyelonephritis. An *E. coli* resistant to Trimethoprim was isolated from the urine (S-72).

An example of epidemiologic follow of a nosocomial infection with acquired Trimethoprim resistance is as follows. A patient suffering from chronic respiratory insufficiency was admitted for bronchitis. A *Klebsiella pneumoniae* strain was isolated from sputum culture, resistant to Sulfonamide, but sensitive to Chloramphenicol-Tetracycline and Trimethoprim. The patient improved on antibiotics and supportive therapy. He remained in the hospital for his respiratory problem and five days later we isolated from his sputum a *Klebsiella pneumoniae* with the same resistance pattern as the previous one.

TABLE III.

Agarose gel electrophoresis of plasmids coding for Tp resistance

	Culture	Resistance patterns	No. of plasmids detected	Molecular weight Md	
Recipient	*E. coli* K$_{12}$ RC85+ pPK70-3	Tp	1		63
Recipient	*E. coli* K$_{12}$ RC85+ pPK71-2	Tp	1		63
Donor	S-70	Tp, Su, bla, Cm, Tc, Gm, Km, Co, Nm	1		94
Recipient	*E. coli* RC85+ pPK70-2	Tr, Su	1		87
Donor	S-78	Tp, Su, bla, Cm, Tc, Km, Nm	1		87
Recipient	*E. coli* RC85+ pPK78-5	Tp, Su	1		87
Donor	S-72	Tp, Su, bla, Sm	2	a) 7.2 b) 52.5	
Recipient	*E. coli* RC85+ pPK72-1	Tp, Su, bla	1		52.5
Donor	S-73	Tp, Su, bla, Sm, Gm	2	a) 91 b) 108	
Recipient	*E. coli* RC85+ pPK 73-1	Tp, Su, bla	1		108

A *Pseudomonas aeruginosa* strain was also isolated resistant to Chloramphenicol-Tetracycline and Trimethoprim. After one month hospitalization the patient developed septicemia and bronchopneumonia. Both sputum and blood cultures grew a *Klebsiella pneumoniae* strain resistant to Trimethoprim, Tetracycline, Sulfonamide and Chloramphenicol. We speculate from the antibiograms that the resistance to Trimethoprim was transferred within the hospital from the *Pseudomonas* strain that colonized the upper respiratory tract of the patient to the *Klebsiella pneumoniae* strain that caused the septicemia. To support the above hypothesis we were able to transfer the resistance to Chloramphenicol, Sulfonamide, Tetracycline and Trimethoprim from the *Klebsiella* strain that caused the septicemia to our recipient host.

AGAROSE GEL ELECTROPHORESIS OF PLASMID DNA

Agarose gel Electrophoresis was performed on single cell lysates of selected clones in order to examine their plasmid content (Table III.). Two plasmids PK 70-3 and PK 71-2, coding for Trimethoprim resistance only had mol. weight of 63 Md. The plasmids PK 70-2 and PK 78-5 coding for Trimethoprim and Sulfonamide resistance had molecular weight 87 Md. In the wild strain *E. coli* 78, from where the plasmid PK 78-5 originated, we detected a plasmid DNA band with the same molecular weight 87 Md. In the other donor *Klebsiella pneumoniae* 70 a different plasmid DNA band was formed with molecular weight 94 Md. The community acquired pPK 72-1 coding for Tp, Su and β-lactamase resistance had molecular weight of 52.5 Md. From the donor strain where this plasmid originated two bands were detected one of 7.2 Md. and the other 52.5 Md.

The nosocomial plasmid PK 73-1 also coding for Trimethoprim, Sulfonamide, β-lactamase resistance had molecular weight 108 Md.

In the donor *Klebsiella pneumoniae* from where this plasmid originated two bands were found one of 91 Md and the other 108 Md.

In recipient cultures, carrying Trimethoprim resistance plasmids, only one plasmid DNA band was found. The molecular weight of the nosocomial plasmids is increasing as they code resistance to increased number of antibiotics.

CONCLUSIONS

After one month surveillance of isolates resistant to Trimethoprim in our hospital we found the majority of them to cause nosocomial infections. Transferability of Trimethoprim resistance occurred in 34%. Prolonged hospitalization of debilitated patients and use of wide spectrum antibiotics were the principal predisposing factors for the development of nosocomial infections with pathogenic *Enterobacteriaceae* carrying plasmid-mediated resistance to Trimethoprim.

In agarose gel electrophoresis only one plasmid DNA band was seen for plasmids carrying Trimethoprim resistance. Our findings indicate that Trimethopriim resistance is carried on different plasmids with molecular weights varrying from 67 to 108 Md.

Plasmids from hospital acquired infections were found with the molecular weight increasing progressively from 63 to 108 Md. as they code resistance to increased number of antibiotics. This may indicate the evolution mechanism of nosocomial multiple resistant plasmids.

REFERENCES

ACAR, J. F., GOLDSTEIN, F. W., GERBAUD, G. R., CHABBERT, Y. A. (1977): Plasmides de resistance en Trimethoprime, transferabilité et groups d'incompatibilité. Ann Microb. (I.P.) 128A:41–47.
AMYES, S. G. B., EMMERSON A. M., SMITH, J. T. (1978): R-factor mediated Trimethoprim resistance. Result of two three month clinical survey s. J. Clin. Path. 31:850–854.
BRUMFITT, W., HAMILTON MILLER, J. M. T., GREY, D. (1977): Trimethoprim-resistant coliforms. Lancet II : 926.
MARKS, P. J., BRUTEN, D. M., SPELLER, D. C. E., (1977): Trimethoprim-resistant coliforms. Lancet I : 774.
ROMERO, E. and PERDUCA, M. (1977): Compatibility group of R-factors for Trimethoprim resistance isolated in Italy. J. Antimicrob. Chemoth. 3 (suppl. C) : 35–38.
WONG, C. K., HARDING, G. K. M., RONALD, A. R., HOBAN, S. (1975): Trimethoprim-resistant *Enterobacteriaceae* in urinary tract infections CMA. Journal, 112, Suppl. p. 54–58.

G.S., Dept. of Clinical Therapeutics,
School of Medicine, Univ. of Athens,
Athens, Greece

GENTAMICIN RESISTANCE PLASMIDS IN HOSPITAL INFECTIONS

VICTORIA HUGHES and N. DATTA

*Department of Bacteriology, Royal Postgraduate Medical School,
London U.K.*

INTRODUCTION

Gentamicin, introduced to medicine in 1968, and tobramycin (1973) are two of the newer aminoglycoside drugs of value in the treatment of hospital infections. They are active against the often multiply resistant coliform bacteria and pseudomonads that cause "opportunistic" infections. This value has been exploited, sometimes for life-threatening infections, by commencing antibacterial therapy before identification and sensitivity pattern of the causative organism has been established. Increasing usage has therefore led to an increase in resistance and to reports of outbreaks of infection by gentamicin/tobramycin resistant (GmTm^R) bacteria. The role of plasmids in this increase in resistance has been well documented. However, the incidence and levels of resistance are variable between regions and hospitals, and indeed vary chronologically within hospitals.

This paper describes a survey in our own hospital (Hammersmith Hospital) of GmTm^R amongst bacteria and their plasmids, and in a second hospital of an outbreak of infection localised to one ward. Both studies illustrate the association of major incompatibility (Inc) groups with particular genes for the enzymic inactivation of aminoglycoside drugs. Examples of these associations have been reported several years previously in various places.

HOSPITAL A - HAMMERSMITH HOSPITAL

Our study covered the first half of 1977. Bacteria identified as causing infections and found to be Gm or Tm resistant in the diagnostic laboratory were collected and tested for transfer of plasmids to *Escherichia coli* K12. Hammersmith Hospital is a general hospital of about 600 beds, and amongst these are special units for renal dialysis and

TABLE I.

Gentamicin resistance of coliform bacteria from Hammersmith Hospital

Year	Total tested	% Resistant
Jul-Dec 1975[1]	646	2.32
1976	1703	2.17
1977	1636	3.91
1978	1752	4.79

[1] Prior to 1975.

Anderson, Datta and Shaw, 1972. Coliform infections in Hammersmith Hospital 524, none Gm^R.

Harkness, Anderson and Datta, 1975. Coliform infections in Hammersmith Hospital and West Middlesex Hospital 186, 0.5% Gm^R.

273

TABLE II.

Gentamicin and tobramycin R plasmids in Hammersmith Hospital
(Jan-June 1977)

Group	R pattern[1]	Bacteria Species and Serotype	No.[2]
IncC	Tc Cm Su Km Gm Tm	*Citrobacter*	1
	Ap Tc Cm Su Km Gm Tm	*E. coli*	2
	Ap Cm Su Km Gm Tm Sm Tp Hg	*Klebsiella aerogenes* K10	1
	Ap Su Km Gm Tm Sm Hg	*Proteus mirabilis*	1
		Providencia	1
IncW	Ap Su Tp Gm Hg	*Citrobacter*	2
	Ap Su Tp Gm Hg Km	*Enterobacter*	1
	Ap Su Gm Hg	*E. coli*	3
IncFII	Cm Su Hg Km Gm Tm	*Citrobacter*	1
	Ap Cm Su Hg Km Gm Tm	*Enterobacter*	2
		E. coli	1
		Klebsiella aerogens K10	8
		Klebsiella aerogens K21	2
		Klebsiella aerogenes NT	1

1. Abbreviations: Ap, ampicillin; Cm, chloramphenicol; Gm, gentamicin; Hg, mercuric chloride; Km, kanamycin; Sm, streptomycin; Su, sulphonamides; Tc, tetracycline; Tm, tobramycin; Tp, trimethoprim.

2. Where repeated isolations were made from one patient of the same bacterium and its plasmid, only one example has been given.

transplantation, intensive care and the treatment of anaemia and leukaemia. As a consequence of their illness, patients in these units are at special risk of infection.

Resistance to gentamicin of coliform bacteria in the hospital is shown in Table I. *Klebsiella* sp. and *E. coli* are the most frequently isolated resistant bacteria. Resistance is steadily increasing, but has not paralleled the rate of the rise in the incidence of resistance to trimethoprim (Richards, this volume).

After transfer to *E. coli* K12 and determination of resistance pattern (R pattern), the plasmids were classified into Inc groups. Table II. summarises this and the bacteria from which plasmids came. Infected patients were located throughout the hospital in both medical and surgical wards and the special units. No GmTm[R] bacteria from childrens or geriatric wards were found. Resistance to gentamicin and/or tobramycin was determined by plasmids of three major Inc groups.

IncC GmTm[R] plasmids

A variety of bacterial genera harboured plasmids of IncC with varying R patterns. All inactivated aminoglycosides by production of aminoglycoside 2″-0 adenylyltransferase [AAD(2″)]. In Paris in 1969 (Witchitz and Chabbert, 1972) and subsequently, an "epidemic" plasmid (pIP55) of IncC, determining resistance to ApCmSuGm was identified in numerous bacterial species. It too determined the inactivation of gentamicin and tobramycin by adenylylation. It is possible that our IncC plasmids were derived from this earlier GmTm[R] plasmid, acquiring further resistances; indeed one example, pHH1190, includes the trimethoprim-streptomycin transposon Tn7 (Richards, this vol.). A random sample of Gm[R] bacteria from our diagnostic laboratory, collected in March this year, included a *Klebsiella* and *E. coli* from the peritoneal dialysis fluid of one patient. Both bacteria harboured an IncC plasmid determining Tc Cm Su Km Gm Tm[R]. The patient was in a ward where, in the 1977 survey an *E. coli* with apparently this same IncC plasmid had been found. It appears that the same host bacterium and its plasmid have remained unchanged in an ecological niche for two years.

IncFII Gm Tm[R] plasmids

In March 1976, two very sick men (suffering from leukaemia and sickle cell anaemia) occupying adjacent beds, became infected with *Klebsiella aerogenes* capsular serotype K10 that was GmTm[R]. A plasmid of IncFII (determining CmSuHgKmGmTm[R]) was transferred to *E. coli* K12. No other patients were then infected, and no environ-

TABLE III.

Plasmids of IncW in Hammersmith Hospital

Year	R pattern
1974	Su Tp
1977	Su Tp
	Ap Su Tp Hg Gm
	Ap Km Su Tp Hg Gm
	Ap Su Hg Gm

mental source was detected. During the 1977 survey, this plasmid was repeatedly found in *K. aerogenes* type K10 and in other bacteria. The incidence of infection never reached epidemic proportions, but for over a year the same host bacterium with its plasmid was endemic. Two main locations were noticed, a medical block including some dialysis patients (*Inc*FII in *Enterobacter cloacae* and *K. aerogenes* K10) and renal transplant unit (RTU) (IncFII in *E. coli*, *K. aerogenes* types K10 and K21). In the RTU one patient was infected with both *E. coli* and *K. aerogenes* K21; both harboured the *Inc*FII plasmid. The plasmid-determined resistance pattern was constant except for the addition (or loss) of Ap[R]. The plasmid determined the modification of gentamicin and tobramycin by AAD(2″); in this and its Inc group it resembled previously-reported GmTm[R] plasmids identified in N. America viz. JR66b (Benveniste and Davies 1971; Datta and Hedges, 1973) and plasmids from an outbreak of infection in a Toronto hospital (Rennie and Duncan 1977; Datta, Hughes and Nugent, 1979).

IncW Gm[R] plasmids

In surveys in the early 1970's at Hammersmith Hospital (Jobanputra and Datta, 1974) of trimethoprim resistant bacteria, a *Klebsiella* strain carried a sulphonamide trimetroprim (SuTp) resistance plasmid of IncW. This host plasmid combination is no longer found, but from 1977, IncW plasmids were found with ApGmHg[R] in addition to SuTp[R] (Table III.). Thus R388, the original 1974 SuTp[R] plasmid could have acquired these resistances, possibly by transposition. IncW plasmids inactivate gentamicin by production of aminoglycoside 3-N acetyltransferase I or II [AAC(3)I or II] so cannot have acquired this resistance from plasmids of IncFII or C.

The Hammersmith Hospital survey has shown examples of the establishment of a bacterium with its plasmid in an ecological niche and its successful maintenance there for over a year. The second London hospital we have studied provides an example of a limited, and contained, outbreak of infection with one plasmid-carrying bacterial strain.

HOSPITAL B

On a male urological ward, over a five month period a multiply resistant *Klebsiella aerogenes* was isolated from 17/237 patients (Casewell, Dalton, Webster and Phillips,1977). Infection was associated with catheterisation of patients and with antibiotic therapy· The strain was finally eradicated by prompt isolation of infected patients and strict ward hygiene.

The epidemic strain was *K. aerogenes* capsular type K16 and its characteristics are shown in Table IV. For several months the bacterium and its plasmid remained unchanged in the ward during spread between patients. From the index case *Citrobacter koseri* and *E. coli* were isolated in addition to the epidemic *Klebsiella*. Both bacteria harboured the same 65 Md IncM plasmid, but in each the plasmid had undergone slight changes. The *Eco*RI restriction fragments of pTH2, from *C. koseri*, were not quite identical to those from pTH1, suggesting a minor DNA rearrangement, and pTH3b conferred no Cm[R]. As pTH3b did not differ in molecular weight or *Eco*RI restriction fragment pattern from pTH1 a small rearrangement of DNA in the Cm gene region is a likely explanation.

Gentamicin resistance conferred by these plasmids was due to production of AAC(3)1. Again a similar plasmid, RIP135 of IncM, and determining the production of AAC(3)1 has previously been described (Witchitz and Gerbaud 1972), and it may be that this plasmid has acquired further antibiotic resistance genes (including in this case the Sm Tp transposon Tn7) in the intervening years. Similar plasmids in incompatibility and gentamicin modifying enzymes to pTH1 and RIP135, have been reported from several hospitals in Melbourne, Australia (Davey and Pittard, 1977; Richards and Datta, 1979). The Melbourne plasmids had all the antibiotic resistance determinants of pTH1 except for the transposon-determined trimethoprim resistance.

CONCLUSIONS

We have illustrated by experiences in our own hospital and others, the ability of R plasmids to form a stable relationship with a host bacterial species, and to become a successful infective partnership. Examples of plasmids determining gentamicin or tobramycin resistance in a medical block, renal transplant unit and a urological ward serve to illustrate this stability.

The plasmids we have encountered of *Inc*FII, C and M appear to be examples of

TABLE IV.

Bacteria and Plasmids from a Urological Ward — Hospital B[1]

INDEX CASE

Bacterial species	Plasmid	R pattern	Inc	M.wt ($\times 10^6$)
K. aerogenes K16	pTH1	Ap Tc Cm Km Su Gm Hg *Sm Tp*	M	65
C. koseri	pHT2	Ap Tc Cm Km Su Gm Hg *Sm Tp*	M	65
E. coli	pHT3a	Ap Tc Cm		58
	pTH3b	Ap Tc Km Su Gm Hg *Sm Tp*	M	65

EPIDEMIC STRAIN

K. aerogenes K16	Ap Tc Cm Km Su Gm Hg *Sm Tp*

Sm Tp a transposable sequence — indistinguishable from Tn7.

[1] Datta, Hughes, Nugent and Richards, 1979.

276

a plasmid acquiring genes for specific aminoglycoside modifying enzymes, and over a period of years maintaining this association and acquiring further resistances — possibly by transposition.

Plasmids of IncW at Hammersmith Hospital have shown broader resistance patterns over 3 years, from Su Tp to Ap Su Tp Hg Gm. Gentamicin resistance in some cases is due to AAC(3)1. This enzyme has been associated for many years and in different places with plasmids of IncM, yet IncM plasmids were not found in Hammersmith Hospital, so the origin of the Gm^R genes of the IncW plasmids is unknown.

The small but steady increase in gentamicin resistance over several years, may have resulted from dissemination of the plasmids of three major Inc groups that early acquired particular Gm/Tm modifying genes — in response to increased aminoglycoside usage. The possibility of a transposon determining Gm or Tm^R is of some concern in the light of experience with trimethoprim.

ACKNOWLEDGEMENTS

We are grateful to Dr. K. P. Shannon for identifying the aminoglycoside modifying enzymes, and to Dr. M. Casewell for serotyping our *Klebsiellae*.

REFERENCES

ANDERSON, F. M., DATTA, N. and SHAW, E. J. (1972): R factors in hospital infection. Brit. Med. J. 3, 82–85.
BENVENISTE, R. and DAVIES, J. (1971): R-factor mediated gentamicin resistance: a new enzyme which modified aminoglycoside antibiotics. FEBS Letters 14, 293–296.
CASEWELL, M. W., DALTON, M. T., WEBSTER, M. and PHILLIPS, I. (1977): Gentamicin-resistant *Klebsiella aerogenes* in a urological ward. Lancet 2, 444–446.
DATTA, N. and HEDGES, R. W. (1973): An I pilus-determining R factor with anomalous compatibility properties, mobilising a gentamicin-resistance plasmid. J. Gen. Microbiol. 77, 11–17.
DATTA, N., HUGHES, V. and NUGENT, M. (1979): Gentamicin resistance plasmids. *Plasmids of Medical Environmental and Commercial Importance.* Ed. K. Timmis and A. Pühler (in press).
DATTA, N., HUGHES, V. M., NUGENT M. E. and RICHARDS H. (1979): Plasmids and transposons and their stability and mutability in bacteria isolated during an outbreak of hospital infection. Plasmid 2, in press.
DAVEY, R. B. and PITTARD, J. (1977): Plasmids mediating resistance to gentamicin and other antibiotics in *Enterobacteriaceae* from four hospitals in Melbourne. Austral. J. Exp. Biol. Med. Sci. 55, 299–307.
HARKNESS, J. L., ANDERSON, F. M. and DATTA, N. (1975): R-factors in urinary tract infection. Kidney International 8, S130–S133.
JOBANPUTRA, R. S. and DATTA, N. (1974): Trimethoprim R factors in *Enterobacteria* from clinical specimens. J. Med. Microbiol. 7, 169–177.
RENNIE, R. P. and DUNCAN, I. B. R. (1977): Emergence of gentamicin-resistant *Klebsiella* in a general hospital. Antimicrob. Ag. Chemother. 11, 179–184.
RICHARDS, H. (1979): The spread of transposon 7 among bacteria of medical and veterinary importance. This volume p.
RICHARDS, H. and DATTA, N. (1979): Reclassification of Incompatibility group L (IncL) plasmids. Plasmid 2, in press.
WITCHITZ, J. L. and CHABBERT, Y. A. (1972): Résistance transférable à la gentamicine II – Transmission et liaisons du caractère de résistance. Ann Inst. Pasteur 122, 367–378.
WITCHITZ, J. L. and GERBAUD, G. R. (1972): Classification de plasmides conférant la résistance à la gentamicine. Ann. Inst. Pasteur 123, 333–339.

V.H. and N.D.
Dept. of Bacteriology,
Royal Postgraduate Medical School,
London W12 U.K.

DRUG RESISTANCE OF *BORDETELLA BRONCHISEPTICA* ISOLATED FROM PIGS

N. TERAKADO, Y. ISAYAMA and S. MITSUHASHI

National Institute of Animal Health, Ibaraki, Japan and School of Medicine, Gunma University, Maebashi, Japan

Until now, drug resistance plasmids (R plasmids) have been demonstrated in various species of bacteria, and the genetic and molecular nature of R plasmids has been extensively studied (Mitsuhashi, 1977). We have also demonstrated R plasmids from *Bordetella bronchiseptica* which is thought to be a causative agent of infectious atrophic rhinitis in young pigs (Terakado, Azechi, Ninomiya and Shimizu, 1973). Epidemiological surveys have revealed that R plasmids play a major role in the drug resistance of *B. bronchiseptica* (Terakado, Azechi, Ninomiya, Fukuyasu and Shimizu, 1974). The genetic and biochemical properties of the representative R plasmid (Rte16) from *B. bronchiseptica* have been described previously (Terakado and Mitsuhashi, 1974; Yaginuma, Terakado and Mitsuhashi, 1975). This paper deals with further observation on drug resistance of *B. bronchiseptica*, especially on demonstration of non-conjugative R plasmids from this organisms.

Epidemiology of drug resistance of B. bronchiseptica

As examined so far, a total of 5 kinds of drug resistance patterns have been detected in naturally occurring *B. bronchiseptica* (Table I.). Strains of quadruple resistance to streptomycin (Sm), sulfonamide (Su), ampicillin (Ap) and mercury (Hg) and triple (Sm.Su.Ap)-resistance were isolated most frequently, followed by single (Sm or Su)- and double (Sm.Su)-resistance in that order. Among them, strains of quadruple (Sm.Su.Ap.Hg)- and triple (Sm.Su.Ap)-resistance were identified as conjugative types. In contrast, double (Sm.Su)- and single (Sm or Su)-resistance transfer could not be demonstrated, suggesting that these strains are non-conjugative types.

TABLE I.

Drug resistance patterns of Bordetella bronchiseptica[a] and isolation frequency of R plasmids[b]

Type of resistance pattern[c]	Number of strains (%)	Number of R+ strains	Resistance pattern of R plasmids
Quadruple (Sm.Su.Ap.Hg)	94 (64)	94	Sm.Su.Ap.Hg
Triple (Sm.Su.Ap)	38 (26)	38	Sm.Su.Ap
Double (Sm.Su)	2 (1)	0	
Single (Sm)	3 (2)	0	
(Su)	10 (7)	0	
Total	147 (100)		

[a] Strains tested were isolated from 1969 to 1977.
[b] R plasmids were detected by a mixed cultivation method using *E. coli* ML1410 as a recipient strain.
[c] Abbreviations: Sm, streptomycin; Su, sulfonamide; Ap, ampicillin; Hg, mercury.

Genetic properties and beta-lactamase mediated by conjugative R plasmid, Rte16

Rte16 is a representative R plasmid derived from *B. bronchiseptica* strain O-16 and is considered to be the same as R906 reported by Hedges and Jacob (1974). This R plasmid can confer resistance to Sm, Su, Ap and Hg on the host bacteria, belongs to incompatibility group P and produces oxacillin hydrolyzing β-lactamase. The molecular weight of the plasmid DNA was estimated to be about 35 megadaltons.

Demonstration of non-conjugative R plasmids from B. bronchiseptica

1) *Mobilization of non-conjugative drug resistance markers of B. bronchiseptica*

As described above, all strains of double (Sm.Su)- and single (Sm or Su)-resistance were found to be non-selftransmissible. From the resistance patterns of these strains, however, it was suspected that their resistance might be mediated by non-conjugative R plasmids. There has been no report, however, on non-conjugative R plasmids from *B. bronchiseptica*. So, we attempted to examine the presence of non-conjugative R plasmids in double (Sm.Su)- and single (Sm or Su)-resistant *B. bronchiseptica* strains. First, we examined whether this resistance can be mobilized in the sensitive recipient. For the transfer experiments, an RP4 (Datta, Hedges, Shaw, Sykes and Richmond, 1971) was employed as mobilizer. Bordetella strains tested had markers of resistance to rifampicin (Rfp) being obtained spontaneously on the selective plate containing this drug. Substrains of *Escherichia coli* K-12 which were resistant to nalidixic acid (Na) or Rfp were also employed. Table II. shows the results of conjugal transfer of RP4 into the Bordetella recipients. It should be noted that Su-resistance of Bordetella recipients 004 and 006 was eliminated at a very high frequency by introducing RP4. On the contrary, such loss of resistance markers was not seen when strains of double (Sm.Su)- and single (Sm)-resistant Bordetella were used as recipients. Table III. shows the co-transfer of double (Sm.Su)-resistance markers with RP4. In transfer experiments, (Sm.Su)-resistant Bordetella transconjugants carrying RP4, NM60RP4 and 065RP4, were used as donors, and *E. coli* ML1410 was used as recipient. As shown in this table, double (Sm.Su)-resistance of Bordetella was transferred at a frequency of 10^{-4} to 10^{-5}. Furthermore, it was found that transferred (Sm.Su)-resistance in the transconjugant ML1410

TABLE II.

Conjugal transfer of an RP4 to Bordetella bronchiseptica carrying nontransferable drug resistance markers

Donor (E. coli)	Recipient[a] (B. bronchiseptica)	Selector	Transfer frequency[b]	Resistance pattern of trnasconjugants[c]
W3630 (Km.Tc.Ap)	MM60 (Sm.Su)	Km+Rfp	2.0×10^{-3}	50/50 Km.Tc.Ap.Sm.Su
	065 (Sm.Su)	Km+Rfp	5.0×10^{-4}	50/50 Km.Tc.Ap.Sm.Su
	071 (Sm)	Km+Rfp	5.0×10^{-3}	50/50 Km.Tc.Ap.Sm
	069 (Sm)	Km+Rfp	5.6×10^{-3}	50/50 Km.Tc.Ap.Sm
	004 (Su)	Km+Rfp	5.4×10^{-3}	15/50 Km.Tc.Ap.Su
				35/50 Km.Tc.Ap
	006 (Su)	Km+Rfp	5.4×10^{-3}	1/50 Km.Tc.Ap.Su
				49/50 Km.Tc.Ap

[a] All of the recipient strains had the marker of rifampicin (Rfp)-resistance.
[b] Expressed as the number of transconjugants in relation to donors after mixed culture at 37 C for 1 hr.
[c] Abbreviations: Km, kanamycin; Tc, tetracycline; Ap, ampicillin; Sm, streptomycin; Su, sulfonamide.

TABLE III.

Co-transfer of (Sm.Su)-resistance markers of B. bronchiseptica with an RP4

Donor (B. bronchiseptica)	Recipient (E. coli)	Selector	Transfer frequency[a]	Resistance pattern of transconjugants[b]
MM60 RP4 (Km.Tc.Ap.Sm.Su)	ML1410	Km + Na	2.0×10^{-4}	10/50 Km.Tc.Ap.Sm.Su 40/50 Km.Tc.Ap
		Sm + Na	2.5×10^{-4}	7/50 Km.Tc.Ap.Sm.Su 43/50 Sm.Su
065 RP4 (Km.Tc.Ap.Sm.Su)	ML1410	Km + Na	3.9×10^{-5}	21/50 Km.Tc.Ap.Sm.Su 29/50 Km.Tc.Ap
		Sm + Na	5.3×10^{-5}	15/50 Km.Tc.Ap.Sm.Su 35/50 Sm.Su

[a] Expressed as the number of transconjugants in relation to that of donors after mixed culture at 37 C for 1 hr.
[b] Abbreviations: Km, kanamycin; Tc, tetracycline; Ap, ampicillin; Sm, streptomycin; Su, sulfonamide; Na, nalidixic acid.

could be transferred to another recipient when RP4 was co-existent in the same cells. In contrast, however, such transmission was not seen when RP4 was not co-existent indicating that (Sm.Su)-resistance markers of Bordetella origin are mobilized with RP4. It was also found that single (Sm or Su)-resistance markers were not mobilizable with RP4 at least under the conditions we employed.

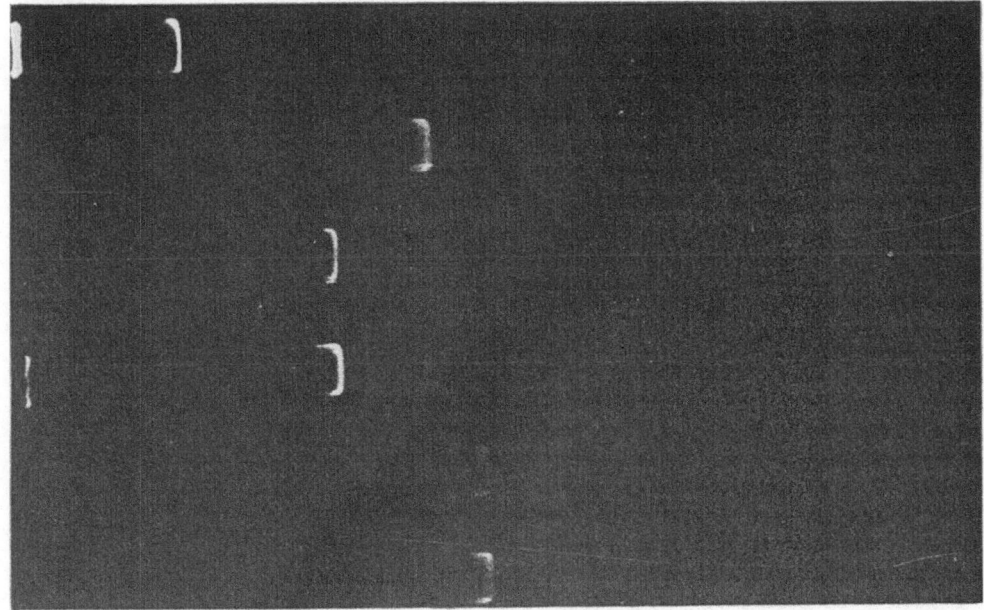

Fig. 1. Agarose gel electrophoresis of plasmid DNAs. The plasmid DNA was purified by dye-CsCl centrifugation. Electrophoresis was carried out for 3 hrs at 100 V. Agarose concentration was 0.7 %, the gel thickness was 0.3 cm. From left to right the samples are RNM60 (molecular weight, 15×10^6), R065 (molecular weight, 15×10^6), Rte16 (molecular weight, 35×10^6), RP4 (molecular weight, 36×10^6), R6K (molecular weight, 26×10^6), R100 (molecular weight, 70×10^6).

2) *Transduction of (Sm.Su)-resistance markers of B. bronchiseptica*

Pl*vir* lysates were prepared by propagation on the transconjugant ML1410 in which (Sm.Su)-resistance markers had been mobilized with RP4 and were employed for transduction experiments with the Rfp-resistant mutant of ML1410. After phage absorption, cells were plated on selective plates containing either kanamycin (Km) or Sm. When selection was performed with Km, transductants carrying triple (Km.Tc.Ap)-resistance markers were obtained at a frequency of 10^{-7}. All of these transductants examined were susceptible to both Sm and Su, suggesting that (Sm.Su)-resistance markers were not transducible under the conditions we employed. In contrast, when selection was performed with Sm, all transductants obtained were resistant to both Sm and Su, but were susceptible to Km, Tc and Ap, suggesting that RP4 was not transduced. Considering that transducing phage particles carry only a single plasmid type (Anderson and Natkin, 1972), we conclude that (Sm.Su)-resistance markers detected in Bordetella strains NM60 and 065 might be mediated by non-conjugative R plasmids.

3) *Isolation of non-conjugative R plasmid DNAs from B. bronchiseptica*

Isolation and purification of plasmid DNAs were performed by cleared lysate and the ethidium bromide-cesium chloride density gradient centrifugation method (Guerry, Leblanc and Falkow, 1974; Meyers, Sanchez, Elwell and Falkow, 1976). By ultracentrifugation, precipitation satellite bands were obtained from both double (Sm.Su)-resistant Bordetella and transconjugant *E. coli* strains. It was also found that *E. coli* transformed with DNA acquired resistance to both Sm and Su. From single (Sm or Su)-resistant Bordetella strains, however, no plasmid DNA satellite bands were obtained. The molecular weight of plasmid DNA was estimated by comparing their respective migrations rates in 0.7% agarose gel electrophoresis with known plasmid standards. Fig. 1 show agarose gel electrophoresis of plasmid DNAs isolated from Bordetella strains. From the migration rates, the molecular weights of non-conjugative R plasmids RNM60 and R065 were estimated to be about 15 megadaltons, respectively.

CONCLUSION

So far, a total of 5 kinds of drug resistant strains have been isolated from naturally occurring *B. bronchiseptica* of pig origin. The results reported here indicate that all quadruple (Sm.Su.Ap.Hg)- and triple (Sm.Su.Ap)-resistance detected in *B. bronchiseptica* is mediated by conjugative R plasmids. Furthermore, it was also found that double (Sm.Su)-resistance markers are mediated by non-conjugative R plasmids. Although we failed to reveal a plasmid DNA from single (Sm or Su)-resistant strains of *B. bronchiseptica*, it is suggested that at least Su-resistance may be mediated by non-conjugative R plasmids, because this resistance marker can be eliminated at a very high frequency from host bacteria by introducing RP4. From these results, we conclude that nearly all drug resistance detected in *B. bronchiseptica* is plasmid borne.

REFERENCES

ANDERSON, E. S. and NATKIN, E. (1972): Transduction of resistance determinants and R-factors of the transfer system by phage Plkc. Mol. Gen. Genet. 114, 261–265.

DATTA, N., HEDGES, R. W., SHAW, E. J., SYKES, R. B. and RICHMOND, M. H. (1971): Properties of an R factor from *Pseudomonas aeruginosa*. J. Bacteriol. 108, 1244–1249.

GUERRY, P., LEBLANC, D. J. and FALKOW, S. (1973): General method for the isolation of plasmid deoxyribonucleic acid. J. Bacteriol. 116, 1064–1066.

HEDGES, R. W., and JACOB, A. E. (1974): Properties of an R factor from *Bordetella bronchiseptica*. J. Gen. Microbiol. 84, 199–204.

MEYERS, J. A., SANCHEZ, D., ELWELL, P. L. and FALKOW, S. (1976): Simple agarose gel electrophoretic method for the identification and characterization of plasmid deoxynucleic acid. J. Bacteriol. 127, 1529–1537.

MITSUHASHI, S. (1977): R factor, drug resistance plasmid. University of Tokyo Press.

TERAKADO, N. AZECHI, H., NINOMIYA, K. and SHIMIZU, T. (1973): Demonstration of the R factor in *Bordetella bronchiseptica* isolated from pigs. Antimicrob. Ag. Chemother. 3, 555–558.

TERAKADO, N., AZECHI, H., NINOMIYA, K. FUKUYASU, T. and SHIMIZU, T. (1974): The incidence of the R factor in *Bordetella bronchiseptica* isolated from pigs. Jap. J. Microbiol. 18, 45–48.

TERAKADO, N. and MITSUHASHI, S. (1974): Properties of R factors from *Bordetella bronchiseptica*. Antimicrob. Ag. Chemother. 6, 836–840.

YAGINUMA, S., TERAKADO, N. and MITSUHASHI, S. (1975): Biochemical properties of a penicillin beta-lactamase mediated by the R factor from *Bordetella bronchiseptica*. Antimicrob. Ag. Chemother. 8, 238–242.

Dr.N.T.
National Institute of Animal Health,
Kodaira, Tokyo, Japan

283

ANTIBIOTIC RESISTANCE AND R PLASMIDS IN
SERRATIA MARCESCENS ISOLATED FROM CLINICAL SPECIMENS
IN JAPAN

T. EDA, T. IKEDA, M. KIMURA, K. KAWAHARA, Y. KANDA
and S. KIMURA

*Department of Bacteriology, Teikyo University School of Medicine,
Tokyo, Japan*

In recent years, *Serratia marcescens* has been recognized as one of the causative micro-organisms in hospital infections, especially in urinary tract and septic infections. Investigations of antibiotic resistance in this species have shown that *Serratia* is resistant to a broader range of antibiotics than other Enterobacteriaceae. It has been also reported that the multiple drug-resistance in this species is determined by R plasmid prevalent widely among Enterobacteriaceae (Medeiros and O'Brien, 1968; Schaefler, Winter, Catelli, Greene and Toharski, 1971; Hedges, Rodriguez-Lemoine and Datta, 1975; Knothe, Krčméry and Wiedemann, 1975, Cooksey, Thorne and Farrar Jr., 1976; Olexy, Bird, Grieble and Farrand, 1979). This paper presents annual surveys on the antibiotic resistance and R plasmids in the strains of *Serratia marcescens* isolated from clinical specimens in Japan during the four-year period from 1975 to 1978.

I. *Serratia* Strains Used

A total of 1,029 strains of *Serratia* was obtained from eight hospitals. Among the 1,029 strains examined, 110 were obtained from two hospitals in 1975, 210 from three hospitals in 1976, 286 from five hospitals in 1977 and 423 from seven hospitals in 1978 (Table 1). In addition, the 1,029 strains studied were isolated from the following clinical specimens: 414 (40.2%) from urines, 210 (20.4%) from sputa, 65 (6.3%) from puses and 259 (25.2%) from unknown origin. Duplicate strains from the same patient were excluded from this study.

II. Antibiotic Resistance in *Serratia*

In this study resistance to antibiotics was represented with maximum allowing concentration of growth (MAC). MAC of the antibiotics against all *Serratia* strains was determined by the agar dilution method using the multiple inoculator. Antibiotics used were streptomycin (Sm or S), chloramphenicol (Cm or C), tetracycline (Tc or T), kanamycin (Km or K), gentamicin (Gm or G), ampicillin (Am or A) and nalidixic acid (Nx or N). The strains allowing the growth at the following concentrations of antibiotics were regarded as resistant to each of the antibiotics: Sm, Cm, Tc, Km and Nx, 25 µg/ml; Gm, 12.5 µg/ml; Am, 100 µg/ml.

1. Antibiotic Resistance Rate

Table I. shows the yearly change in the resistance rate and the isolation frequency of the strains resistant to each of the antibiotics in *Serratia* isolates. Among the 1,029 strains isolated from 1975 to 1978, 913 (88.7%) were found to be resistant to one or more

TABLE I.

Change in resistance rate and isolation frequency of strains resistant to each antibiotic in Serratia per year of isolation

	Year of isolation				
	1975	1976	1977	1978	Total
Total No. of strains	110	210	286	423	1029
No. of resistant strains	104	181	238	390	913
Resistance Rate (%)	94.6	86.2	83.2	92.2	88.7

Antibiotic	No. of resistant strains (Isolation frequency %)				
Sm	45 (40.9)	80 (38.1)	179 (62.6)	235 (55.6)	539 (52.4)
Cm	29 (26.4)	28 (13.3)	97 (33.9)	157 (37.1)	311 (30.2)
Tc	77 (70.0)	136 (64.8)	145 (50.7)	252 (59.6)	610 (59.3)
Km	34 (30.9)	80 (38.1)	151 (52.8)	219 (51.8)	484 (47.0)
Gm	3 (2.7)	16 (7.6)	64 (22.4)	100 (23.6)	183 (17.8)
Am	81 (73.6)	131 (62.4)	192 (67.1)	345 (81.6)	749 (72.8)
Nx	40 (36.4)	19 (9.0)	94 (32.9)	184 (44.0)	339 (32.9)

of the antibiotics. Considerable fluctuation in the resistance rate was not observed during the period studied.

2. Resistance to Each Entibiotic

Among the 1,029 strains, the strains with Am-resistance (72.8%) were isolated most frequently, followed by those with Tc(59.3%)-, Sm(52.4%)-, Km(47.0%)-, Nx(32.9%)-, Cm(30.2%)- and Gm(17.8%)-resistance (Table I.). It was observed that the isolation of the strains resistant to Gm or Km was increasing markedly during four years (Table I.).

Compared with the resistance rate reported by other workers (Medeiros and O'Brien, 1968); Schaefler, Winter, Catelli, Greene and Toharski, 1971), the resistance rates to each antibiotic in *Serratia* studied were shown to be lower.

3. Antibiotic Resistance Pattern

In relation to resistance to seven antibiotics examined, the isolation frequency of the strains with one drug-resistance was highest (27.6%), followed by those with five drug (15.8%)-, six drug (14.7%)-, two drug (13.7%)-, four drug (12.3%)-, seven drug (8.4%)- and three drug (7.4%)-resistance (Table II.). Of the 913 resistant strains, however, 661 (72.4%) were found to be resistant to two or more antibiotics, whereas the remaining 252 (27.6%) were resistant to only one antibiotic. In addition, the isolation of the strains with multiple antibiotic-resistance to more than four antibiotics tended to increase year by year. Detailed antibiotic resistance patterns in *Serratia* strains are listed in Table II.

TABLE II.

Change in drug resistance patterns and detection of R plasmids in Serratia strains per year of isolation

Resistance pattern	Year of isolation				Total
	1975	1976	1977	1978	
	No. of strains (No. of strains with R plasmids)				
S C T K G A N			16 (13)	62 (41)	78 (54)
S C T K G A	3 (3)	6 (6)	26 (26)	3 (3)	38 (38)
S C T K A N	4 (0)	2 (1)	27 (6)	44 (15)	77 (22)
S T K G A N			2 (2)	16 (3)	18 (5)
S C T K G		1 (1)	1 (1)		2 (2)
S C T K A	8 (3)	9 (4)	5 (5)	18 (10)	40 (22)
S C T A N		2 (0)	1 (0)	1 (1)	4 (1)
S C K A N			12 (0)	3 (2)	15 (2)
S T K G A		8 (7)	12 (12)	5 (4)	25 (23)
S T K A N	7 (3)	2 (2)	15 (1)	17 (0)	41 (6)
S K G A N		1 (1)	1 (1)	3 (2)	5 (4)
C T K A N	5 (0)	3 (1)	1 (0)	1 (1)	10 (2)
S C T A	1 (0)			2 (1)	3 (1)
S C K A		1 (0)	3 (3)	2 (1)	6 (4)
S T K A		28 (1)	11 (1)	7 (6)	46 (8)
S T A N	9 (0)	8 (0)	3 (0)	3 (0)	23 (0)
S K G A			6 (2)	4 (4)	10 (6)
S K A N			5 (0)	3 (1)	8 (1)
C T K A	2 (0)		1 (0)	6 (0)	9 (0)
T K A N	2 (0)			2 (1)	4 (1)
S C T	3 (0)			1 (0)	4 (0)
S C A	1 (0)			1 (1)	2 (1)
S T A	1 (0)	1 (0)	2 (0)	6 (2)	10 (2)
S K G				1 (1)	1 (1)
S K A	1 (0)	2 (0)	1 (0)	1 (1)	5 (1)
S A N			1 (0)	4 (0)	5 (0)
C T A	2 (0)	3 (0)			5 (0)
C A N				4 (0)	4 (0)
T K G				3 (1)	3 (1)
T K A		13 (5)	1 (0)	1 (0)	15 (5)
T A N	1 (0)			6 (1)	7 (1)
K A N	2 (0)		1 (0)		3 (0)
S T	7 (0)	2 (0)	5 (0)	12 (0)	26 (0)
S A		2 (0)	3 (0)	7 (1)	12 (1)
S N			5 (0)		5 (0)
T A	10 (0)	5 (0)	4 (1)	15 (1)	34 (2)
K A		1 (0)	2 (1)	9 (0)	12 (1)
A N	9 (0)	1 (0)	2 (0)	13 (5)	25 (5)
S		3 (0)	14 (0)	4 (0)	21 (0)
T	12 (0)	41 (0)	8 (0)	16 (0)	77 (0)
A	13 (0)	33 (0)	34 (0)	69 (3)	149 (3)
The others	1 (0)	3 (0)	7 (0)	15 (0)	26 (0)
Total No. of resistant strains	104 (9)	181 (29)	238 (75)	390 (113)	913 (226)

III. Detection of R Plasmids from *Serratia* Strains

To examine the presence of conjugally transferable R plasmids in *Serratia* isolates, mating experiments were carried out with a total of 913 resistant strains as donors. *Escherichia coli* K12 W3104 *rif* and J53 *nal*, which are rifampicin (Rf)-resistant and Nx-resistant mutant respectively, were employed as recipient bacteria. 0.1 ml of donor culture grown overnight in Penassay broth (PB, Difco) was mixed with 0.9 ml of a recipient culture similarly prepared. After adding 4 ml of fresh PB to this, the mixture was incubated at 37°C for 24hr, Transconjugants were selected on nutrient agar plate (Eiken Kagaku, Tokyo) containing the following concentrations of the antibiotics: Sm, Cm, Km and Nx, 25 μg/ml, Tc and Gm, 12.5 μg/ml, Am and Rf, 100 μg/ml. Transconjugants obtained were tested for resistance to seven antibiotics as described above.

1. Detection of R Plasmids

The detection of transmissible R plasmids from the resistant *Serratia* strains per year of isolation is shown in Table II. Of the 913 resistant strains, 226 (24.8%) carried R plasmids. A considerable increase in the detection rate of R plasmid-harboring strains was observed during the period studied. R plasmids were detected in 9 (8.7%) of the 104 resistant strains in 1975, in 29 (16.0%) of 181 in 1976, in 75 (31.5%) of 238 in 1977 and in 113 (29.0%) of 390 in 1978 (Table II.).

Detection of R plasmids from *Serratia marcescens* has been reported by Medeiros and O'Brien (1968), Cooksey *et al.* (1976) and Hedges *et al.* (1975). The detection rate in this study was lower than that in the studies of Medeiros and O'Brien (1968) and Cooksey *et al.* (1976), but was higher than that in the observation of Hedges *et al.* (1975). Moreover, it was indicated that the detection rate of R plasmids was lower in *Serratia* than in *Shigella, Salmonella* or other enteric bacteria (Mitsuhashi, 1977).

2. Detection of R Plasmids and Resistance to Each Antibiotic

Among the resistances to each of the antibiotics examined during four years, resistance to Sm, Cm, Tc, Km, Gm and Am mediated by R plasmids was 31.2%, 30.9%, 12.6%, 37.2%, 47.0%, and 21.8%, respectively. On the other hand, in the strains carrying R plasmids, Km-(86.1%) resistance transferred most frequently, followed by Sm-(82.0%), Am-(73.4%), Cm-(65.3%), Gm-(64.2%) and Tc-(39.5%) resistance. Resistance to Nx and Tc alone was never observed to transfer.

3. Detection of R Plasmids and Antibiotic Resistance Patterns

The detection of R plasmids was related to the drug resistance patterns of the resistant strains. Of the 252 strains with one drug-resistance, only 3 (1.2%) carried R plasmids. R plasmids were detected in 9 (7.2%) of 125 strains with two drug-resistances, in 12 (17.6%) of 68 with three ones, in 21 (18.8%) of 112 with four ones, in 62 (43.1%) of 144 with five ones, in 65 (48.5%) of 134 with six ones and in 54 (69.2%) of 78 with seven ones (Table II.). It was noted that approximately 50% of the strains resistant to five or more antibiotics harbored R plasmids. In addition, among the 226 R plasmids detected, 223 (98.7%) were derived from the strains resistant to two or more antibiotics, although only 3 (1.3%) were derived from the strains with one drug resistance. Detection of R plasmids and antibiotic resistance patterns are shown in Table II. in detail.

IV. Some Characters of R Plasmids Derived from *Serratia* strains

1. Antibiotic Resistance Patterns of R Plasmids

Three hundred and seventy five of R plasmids were isolated from the 226 strains carrying R plasmids during the period studied. Among the R plasmids isolated, the R plasmids with one resistance marker (25.1%) were found most frequently, followed by those with four resistance ones (24.8%), three resistance ones (15.9%), two resistance ones (15.7%), five resistance ones (11.7%) and six resistance ones (3.2%). R plasmid with Nx resistance marker was never observed to isolate in the R plasmids derived from

TABLE III.

Change in drug resistance patterns of R plasmids derived from Serratia strains per year of isolation

| Resistance pattern | Year of isolation | | | | Total |
| | 1975 | 1976 | 1977 | 1978 | |
	No. of plasmids (%)				
S C T K G A			11 (7.0)	1 (0.6)	12 (3.2)
S C T K G			1 (0.6)		1 (0.3)
S C T K A	3 (33.3)	4 (12.1)	10 (6.4)	10 (5.7)	27 (7.2)
S C K G A			2 (1.3)	8 (4.5)	10 (2.7)
S T K G A			6 (3.8)		6 (1.6)
S C T K		2 (6.1)	2 (1.3)	4 (2.3)	8 (2.1)
S C K G			2 (1.3)	5 (2.8)	7 (1.9)
S C K A			22 (14.0)	15 (8.5)	37 (9.9)
S T K G			16 (10.2)		16 (4.3)
S T K A		2 (6.1)	2 (1.3)	3 (1.7)	7 (1.9)
S K G A			8 (5.1)	7 (4.0)	15 (4.0)
C T K A			1 (0.6)	1 (0.6)	2 (0.5)
C K G A				1 (0.6)	1 (0.3)
S C K				3 (1.7)	3 (0.8)
S C A			1 (0.6)	9 (5.1)	10 (2.7)
S T A				2 (1.1)	2 (0.5)
S K G	3 (33.3)	9 (27.3)	10 (6.4)	11 (6.3)	33 (8.8)
S K A			3 (1.9)	1 (0.6)	4 (1.1)
C T K		2 (6.1)	1 (0.6)	1 (0.6)	4 (1.1)
C K A		1 (3.0)	9 (5.7)	1 (0.6)	11 (2.9)
T K G				3 (1.7)	3 (0.8)
T K A			1 (0.6)		1 (0.3)
K G A				2 (1.1)	2 (0.5)
S C			4 (2.5)		4 (1.1)
S K				7 (4.0)	7 (1.9)
S A			1 (0.6)	2 (1.1)	3 (0.8)
C K		1 (3.0)			1 (0.3)
C A			3 (1.9)		3 (0.8)
T A				1 (0.6)	1 (0.3)
K A	3 (33.3)	7 (21.2)	13 (8.3)	17 (9.7)	40 (10.7)
S			3 (1.9)	21 (11.9)	24 (6.4)
K			1 (0.6)	7 (4.0)	8 (2.1)
A		5 (15.2)	24 (15.3)	33 (18.8)	62 (16.5)
Total	9 (100)	33 (100)	157 (100)	176 (100)	375 (100)

Serratia strains. However, 281 (74.9%) of the 375 plasmids isolated had markers determining resistance to two or more antibiotics. Table III. shows antibiotic resistance patterns of R plasmids derived from *Serratia* strains in detail.

2. *Fi* Type of R Plasmids

Seventeen R plasmids isolated from *Serratia* strains were examined for *fi* type. *Fi* type was determined with the method observing lysis by *f*2 phage described by Watanabe *et al*. (1962). All of the R plasmids tested were shown to be *fi⁻* type.

3. Incompatibility Group of R Plasmids

Seventeen R plasmids derived from *Serratia* strains were examined for incompatibility group. Classification of R plasmids by incompatibility was performed by the method described by Coetzee *et al*. (1972) using standard plasmids listed by Hedges (1974). Among the 17 R plasmids tested, 5 belonged to group L, 2 to group A-C, 7 to group M and 3 to group H2(S). It was observed that R plasmids isolated from a hospital tended to belong to one incompatibility group. Hedges *et al*. (1975) reported that R plasmids isolated from *Serratia* belonged to any of incompatibility groups C, FII, L, M, P and S, and group L could be found only in R plasmids derived from *Serratia*. Recently, it is said that groups L and M belong to the same incompatibility group, and group S is designated as group H2. Up to date, it was shown that R plasmids derived from *Serratia* isolated in Japan belonged to any of groups A-C, L or M and H2(S).

4. Molecular Weight of R Plasmids

Some R plasmids belonging to incompatibility groups A-C, L and M were examined for determination of molecular weight. DNA in R plasmid was isolated by the modified method described by Cohen and Miller (1969) using ethidium bromide-cesium chloride density gradient centrifugation. The DNAs isolated were examined in the electron microscope by the method reported by Follet and Crawford (1967). All of DNAs isolated consisted of covalently closed circular DNA fibers. DNA derived from incompatibility group A-C of R plasmids measured 105.2 megadaltons in molecular weight. Molecular weights of DNAs derived from incompatibility groups L and M were determined as 48.9 to 49.0 megadaltons and 41.8 to 43.7 megadaltons, respectively. Table 4 shows resistance marker, incompatibility group, *fi* type and molecular weight of R plasmids derived from *Serratia* in detail.

ACKNOWLEDGEMENT

This study was supported in part by a grant provided by the Ministry of Education, the Japanese Government (Grant No. 311201).

REFERENCES

COETZEE, J. N., N. DATTA and R. W. HEDGES (1972): R factors from *Proteus rettgeri*. J. Gen. Microbiol., 72, 543–552.
COHEN, S. N. and C. A. MILLER (1969): Multiple molecular species of circular R-factor DNA isolated from *Escherichia coli*. Nature, 224, 1273–1277.
COOKSEY, R. C., G. M. THORNE and W. E. FARRAR Jr. (1976): R factor-mediated antibiotic resistance in *Serratia marcescens*. Antimicrob. Agents Chemother., 10, 123–127.

Follet, E. A. C. and L. V. Crawford (1967): Electron microscope study of the denaturation of human papilloma virus DNA. 1. Loss and reversal of supercoiling turns. J. Mol. Biol., 28, 455–459.

Hedges, R. W. (1974): R factors from *Providence*. J. Gen. Microbiol, 81, 171–181.

Hedges, R. W., V. Rodriguez-Lemoine and N. Datta (1975): R factors from *Serratia marcescens*. J. Gen. Microbiol., 86, 88–92.

Knothe, H., V. Krčméry and B. Wiedemann (1975): A new type of R plasmid in *Serratia marcescens* transferring gentamicin, tobramycin, DKB, lividomycin and kanamycin resistance. In Microbial Drug Resistance. Edited by S. Mitsuhashi and H. Hashimoto, University of Tokyo Press, Tokyo, pp. 217–222.

Medeiros, A. A. and T. F. O'Brien (1968): Contribution of R factors to the antibiotic resistance of hospital isolates of *Serratia*. Antimicrob. Agents Chemother., 1968, 30–35.

Mitsuhashi, S. (1977): Epidemiology of R factors. In R-Factor: Drug Resistance Plasmid, Edited by S. Mitsuhashi, University of Tokyo Press, Tokyo, pp. 25–45.

Olexy, V. M., T. J. Bird, H. G. Grieble and S. K. Farrand (1979): Hospital isolates of *Serratia marcescens* transferring ampicillin, carbenicillin and gentamicin resistance to other gram-negative bacteria including *Pseudomonas aeruginosa*. Antimicrob. Agents Chemother,. 15, 93–100.

Schaefler, S., J. Winter, A. Catelli, J. Greene and B. Toharski (1971): Specific distribution of R factors in *Serratia marcescens* strains isolated from hospital infections. Appl. Microbiol., 22, 339–343.

Watanabe, T., T. Fukasawa and T. Takano (1962): Conversion of male bacteria of *Escherichia coli* K-12 to resistance to *f* phages by infection with the episome "resistance transfer factor". Virol., 17, 218–219.

T.E., Dept. of Bacteriology,
Teikyo University School of Medicine,
Tokyo, Japan

II. MEDICAL PART
C) ENZYMES INACTIVATING ANTIBIOTICS
Chairman: R. GOMEZ-LUS

AAC (3) AND AAC (6') ENZYMES PRODUCED BY R PLASMIDS ISOLATED IN GENERAL HOSPITAL

R. GOMEZ-LUS, L. LARRAD, M. C. RUBIO-CALVO, M. NAVARRO
and M. P. LASIERRA

Department of Microbiology, School of Medicine,
University of Zaragoza, Spain

INTRODUCTION

The aminoglycoside acetylating enzymes (AAC) catalyze the transfer of acetate from acetyl-coenzyme A to an aminogroup on the antibiotic. The AAC can be classed into three groups: AAC (6'), AAC (2'), and AAC (3).

Aminoglycoside-6'-acetyltransferases AAC (6')

These substances catalyze the acetylation of the 6'amino group of deoxystreptamine antibiotics; all compounds with a free $6'-NH_2$ group are susceptible to this modification (neomycin B, kanamycin A and B, tobramycin, gentamycin C_1a, sisomicin, netilmicin, dibekacin, butirosin, amikacin and ribostamycin), depending on the class of AAC (6'). 6'-N-acetylase [AAC(6')-I] was first described by Umezawa et al. (18) and characterized by Benveniste and Davies (1). Le Goffic et al. (14) described a new-6'-N-acetylating enzyme from a Moraxella strain, and named this enzyme AAC (6')-II. Haas et al. (11) described a third type of 6'-acetyltransferase, isolated from *P. aeruginosa*, which shows a novel substrate profile characterized by markedly reduced activity towards butirosin and amikacin. Mitsuhashi (16) has classified 6'-N-acetyltransferases encoded by R plasmids into four categories on the basis of in vitro substrate profiles. Class I enzymes modify kanamycin A and B and neomycin. The enzyme AAC (6')-II inactivates kanamycin A and B, neomycin, and gentamicin C_1a and C_2. Class III contains the substrates of class II plus dibekacin. Finally, AAC (6')-IV has all the resistances of class III, plus amikacin.

Aminoglycoside 2'-acetyltransferases AAC (2')

AAC (2') catalyzes the acetylation of the 2-amino group on the amino glucose II moiety of kanamycin C, gentamicins, sisomicin, netilmicin and tobramycin. Butirosin and neomycin are poorer substrates. The strains producing this enzyme are resistant to lividomycin A and gentamicin C but sensitive to kanamycin A, because this aminoglycoside lacks the 2-amino group. AAC (2') has been found in clinical isolates of *Proteus* and *Providencia* strains.

Aminoglycoside 3-acetyltransferases AAC (3)

The 3-N-acetyltransferases AAC (3) are widely distributed in clinical isolates of *E. coli*, *Klebsiella pneumoniae* and *Pseudomonas aeruginosa*. AAC (3)-I was originally isolated from a P. aeruginosa strain by Brzezinska et al. (3), this type of acetylating enzyme confers resistance to gentamicin and sisomicin. AAC (3)-II was isolated from *Klebsiella* harboring an R plasmid (R 176) by Le Goffic et al. (13). This enzyme catalyzes the acetylation of the 3-NH_2group on the deoxystreptamine ring, conferring resistance to gentamicin, sisomicin, tobramycin and kanamycin. Biddlecome et al. (2) described a new 3-N-aminoglycoside acetyltransferase, AAC (3)-III, which inactivates gentamicins, sisomicin, tobramycin, kanamycins, netilmicin, neomycin and paromomycin. We have previously isolated several strains of *Enterobacteriaceae* resistant to gentamicin, sisomicin, tobramycin, netilmicin and dibekacin but sensitive to kanamycin and amikacin, and able to synthetize an acetyltransferase (7, 8). Comparison of the substrate activities of this enzyme with the resistance spectrum of the R^+ strains, suggests that it may contain a 3-N-acetyltransferase, probably an ACC (3)-IV. Davies and O'Connor (5) described a plasmid-mediated 3-N-acetylating enzyme, AAC (3)-IV, with broad substrate range that includes all of the substrates of AAC (3)-III and, in addition, the novel mono-substituted antibiotic apramycin.

RESULTS

In earlier publications (7, 8) we communicated our primary results on the control of plasmids conducted in the University Hospital of Zaragoza; this hospital was inaugurated in 1975 and had an in-patient capacity of approximately 900 beds.

TABLE I.

General hospital. University of Zaragoza, Spain
April 1th, 1976 April 1th, 1977
Number of antibiotic-resistant strains of gram-negative bacilli 1,360

	Number	%
Strains carrying R plasmid	428	31.47
Strains resistant to gentamicin	42	3.08

Percentage related to R^+ strains 9.81%

TABLE II.

General hospital. University of Zaragoza. Spain
April 1th, 1977   May 22th, 1979
Number of antibiotic-resistant strains of gram-negative bacilli 2,160

	Number	%
Strains carrying R plasmids	804	37.22
Strains resistant to gentamicin	335	15.50

Percentage related to R^+ strains 41.66

Beginning April 1, 1976 to April 1, 1977, a total of 1360 resistant strains of gram-negative bacilli were isolated; of these, we demonstrated the transference of resistance by conjugation in 428 isolated strains (31,47%). From these R plasmids harboring strains, 42 (3,08%) conferred resistance to gentamicin and other antibiotics. In our continuing surveillance of resistance patterns during the period from April 1, 1977 to May 22, 1979, we analyzed 2,160 resistant gram-negative bacilli in which the number of isolates gentamicin-resistant increased from the earlier valued 3.08% to 15.5%, a five fold increase. (Table I., II.).

TABLE III.
Properties of seven R+ aminoglycoside-resistant strains

Organism	Strain no.	Plasmid	Incompatibility group	Transferred resistance pattern	Aminoglycoside modifying enzymes
P. aeruginosa	CRT	pUZ1	P1	Ap Tc Gm Ss Km Cm Sm Su	AAC(3)I+APH(3')I+ +AAD(3'')
E. coli	15159	pUZ2	M	Ap Tc Gm Ss Tm	AAC(3)-IV
P. vulgaris	18182	pUZ3a	M	Ap Tc Gm Ss Tm	AAC(3)-IV
C. Freundii	19162	pUZ4	M	Ap Tc Gm Ss Tm	AAC(3)-IV
K. Pneumoniae	17434	pUZ5	M	Ap Tc Gm Ss Tm	AAC(3)-IV
E. coli	20092	pUZ6	M	Ap Tc Gm Ss Tm	AAC(3)-IV
E. coli	20013	pUZ11	P1	Ap Tc Gm Ss Tm Km Cm Sm Su	AAC(3)-IV+APH(3')- -I+AAD(3'')

TABLE IV.
Properties of eleven R+ aminoglycoside-resistant strains

Organism	Strain no.	Plasmid	Incompatibility group	Transferred resistance pattern	Aminoglycoside modifying enzymes
P. vulgaris	18182	pUZ3b	P1	Ap Tc Gm Ss Km Cm Sm Su	AAC(3)I+APH(3')+ +AAD(3'')
S. marcescens	965	pUZ7	P1	Ap Tc Gm Ss Km Cm Sm Su	AAC(3)I+APH(3')I+ +AAD(3'')
P. aeruginosa	15875	pUZ8	P1	Tc Km	APH(3')-I
K. pneumoniae	21938	pUZ9	P1	Ap Tc Gm Ss Km Cm Sm Su	AAC(3)I+APH(3')I+ +AAD(3'')
E. coli	16306	pUZ10	P1	Ap Tc Gm Ss Km Cm Sm Su	AAC(3)I+APH(3')I+ +AAD(3'')
A. calcoaceticus	1976	pUZ12	P1	Ap Km Sm Su	APH(3')I+AAD(3'')
A. calcoaceticus	21493	pUZ13	P1	Ap Tc Gm Ss Km Cm Sm Su	AAC(3)I+APH(3')+ +AAD(3'')
K. Pneumoniae	19932	pUZ14	P1	Ap Tc Gm Ss Km Cm Sm Su	AAC(3)I+APH(3')+ +AAD(3'')
K. Pneumoniae	19990	pUZ15	P1	Ap Tc Gm Ss Km Cm Sm Su	AAC(3)I+APH(3')+ +AAD(3'')
P. vulgaris	21314	pUZ16	P1	Ap Tc Gm Ss Km Cm Sm Su	AAC(3)I+APH(3')+ +AAD(3'')
K. pneumoniae	19560	pUZ17	P1	Ap Tc Gm Ss Km Cm Sm Su	AAC(3)I+APH(3')+ +AAD(3'')

Acetylating Enzymes Produced by Group M Plasmid

The five plasmids pUZ*2, pUZ3a, pUZ4, pUZ5 and pUZ6, have in common the characteristics, not only of belonging to the same compatibility group (12) possessing identical determinants of resistance (Ap, Tc, Gm, Ss, Tm) and synthesizing a beta-

pUZ*2 (plasmids University Zaragoza): denomination proposed by Drs. Naomi Datta and R.W. Hedges of the Royal Postgraduate School of Medicine, London, England.

297

lactamase type TEM-1, but also of producing an acetylase with affinity for gentamicin, C, sisomicin, tobramycin, dibekacin and netilmicin (Table III.). Using gentamicin C_1 as reference (100%), the affinity of the five cell-free extracts for sisomicin varied from 106 to 111%, while tobramycin showed an average affinity of 64.80%, having a maximum of 67% and a minimum of 62%; the acetylases also exhibited affinity for netilmicin ($\bar{X} = 41\%$) and dibekacin ($\bar{X} = 73\%$). The determination of the minimum inhibitory

TABLE V.

58 AAC(3)-I producer strains (Gm Ss-resistant)

Organism	No. of strains	Percentage
K. pneumoniae	13	22.41
S. marcescens	12	20.68
P. aeruginosa	6	10.34
P. mirabilis	4	6.89
E. cloacae	3	5.17
P. vulgaris	3	5.17
A. calcoaceticus	2	3.44
E. aerogenes	2	3.44
H. alvei	1	1.72
P. morganii	1	1.72
S. enteritidis	1	1.72

TABLE VI.

44 AAC(3)-IV producer strains (Gm Ss Tm Nt DKB-resistant)

Organism	No. of strains	Percentage
E. coli	14	31.81
K. pneumoniae	10	22.72
E. cloacae	9	20.45
P. mirabilis	3	6.81
S. marcescens	3	6.81
P. morganii	2	4.54
C. freundii	1	2.27
P. vulgaris	1	2.27
S. enteritidis	1	2.27

concentration (MIC) with respect to these five aminoglycosides shows that the strains harboring plasmids pUZ2, pUZ4, pUZ5, and pUZ6, are resistant to gentamicin, sisomicin, tobramycin, netilmicin and dibekacin, and thus we may conclude that the acetylating enzymes for which they code have identical profiles and that their affinity also implies inactivation. The behavior of these enzymes with respect to kanamycin B shows affinity ($\bar{X} = 59\%$) without inactivation (MIC = 2 mcg/ml.). Comparison of the substrate specifities of this enzyme with the resistance spectrum of the 5 strains suggests that they contain a 3-N-acetyltransferase, provisionally classified as AAC (3)-IV.

Acetylating Enzymes Produced by Plasmids of the P1 Group

Nine plasmids of the P1 group (pUZ3b, pUZ7, pUZ9, pUZ10, pUZ13, pUZ14, pUZ15, pUZ16 and pUZ17) produce a completely homogenous set of enzymes, coding for one beta-lactamase type TEM-1 (15) and three aminoglycoside inactivating enzymes (Table IV.):

TABLE VII.

118 APH producer strains

Organisms	No. of strains	Enzyme	Resistance pattern
E. coli	67	APH(3')-I	Km Nm Lv Rm Pm
	9	APH(3')-II	Km Nm Bt Rm Pm
K. pneumoniae	12	APH(3')-I	Km Nm Lv Rm Pm
	2	APH(3')-II	Km Nm Bt Rm Pm
Enterobacter sp.	4	APH(3')-I	Km Nm Lv Rm Pm
	1	APH(3')-II	Km Nm Bt Rm Pm
S. enteritidis	4	APH(3')-I	Km Nm Lv Rm Pm
	1	APH(5'')	Rm
Citrobacter sp.	2	APH(3')-I	Km Nm Lv Rm Pm
	1	APH(3')-II	Km Nm Bt Rm Pm
P. aeruginosa	9	APH(3')-I	Km Nm Lv Rm Pm
P. mirabilis	4	APH(3')-I	Km Nm Lv Rm Pm
P. vulgaris	1	APH(3')-I	Km Nm Lv Rm Pm

1. Aminoglycoside 3''-adenylyltransferase [AAD(3'')] with affinity for streptomycin and spectinomycin.

2. Aminoglycoside 3'-phosphotransferase I [APH(3')-I] with affinity for kanamycins, neomycin, ribostamycin and lividomycin, but not for butirosin.

3. Aminoglycoside 3-acetyltransferase I [AAC(3)-I] with affinity for gentamicin and sisomicin. An acetylase, coded by plasmid pUZ1 (formerly R1033), with an identical substrate profile range was found in *P. aeruginosa* (17).

Aminoglycoside Inactivating Enzymes produced by 222 R$^+$ strains:

We reviewed the data obtained from 222 R$^+$ strains isolated in the period from 1976 to 1978 regarding those producing aminoglycoside inactivating enzymes and found the following (10):

1. *Strains producing AAC (3)-I.* Fifty-eight strains of various genera that are capable of producing AAC (3)-I, and are resistant to gentamicin and sisomicin. We have not encountered any AAC(3)-I producing strains that at the same time were capable of demonstrating resistance to gentamicin, sisomicin and netilmicin. (Table V.).

2. *Strains producing AAC (3)-IV.* Forty-four strains that produced an AAC(3)-IV and were resistant to gentamicin, sisomicin, tobramycin, dibekavin and netilmicin. (Table VI.).

3. Strains producing AAC (6'). One AAC (6') producing strain of K. pneumoniae was shown to be resistant to kanamycin, neomycin, ribostamycin, butirosin, tobramycin, sisomycin, netilmicin, dibekacin and amikacin, and sensitive to the gentamicin complex (MIC = 2 mcg/ml).

4. Strains producing 3'-O-phosphotransferases [(APH(3')-I and II] and 5''-O-phosphotransferase [APH(5'')] (Table VI.).

   A. One hundred and three strains, belonging to the genera *Escherichia, Citrobacter, Klebsiella, Enterobacter, Proteus,* and *Pseudomonas,* produced APH(3)-I.

   B. Thirteen strains belonging to the genera *Escherichia, Klebsiella, Citrobacter* and *Enterobacter,* coding for an APH(3')-II.

   C. One strain of *Salmonella* enteritidis that produced an APH(5'') which determined the resistance to ribostamycin.

TABLE VIII.

Resistance spectra of E. coli/pUZ 1, E. coli/R 176, P. aeruginosa PST 1, E. coli/JR 225, and E. coli/pUZ 25 expressed AS MIC (mcg/ml)

Antibiotic	AAC (3)-I E. coli/ pUZ 1	AAC (3)-II E. coli/ R176	AAC (3)-III P. aeruginosa PST 1	AAC (3)-IV E. coli/ JR 225	AAC (3)-V E. coli/ pUZ 25
GENTAMICIN C_1	32	32	32	32	32
SISOMICIN	32	32	32	16	32
TOBRAMYCIN	1	16	32	32	8
KANAMYCIN A	2	16	32	16	4
NEOMYCIN B	2	2	32	8	2
PAROMOMYCIN	2	4	32	32	2
APRAMYCIN	2	2	8	32	2
5-EPI-SISOMICIN	0.5	1	64	4	0.5

TABLE IX.

Enzymatic activity of five AAC (3) classes*

SUBSTRATE 4.6 DOS	AAC (3)-I E. coli/ pUZ 1	AAC (3)-II E. coli/ R 176	AAC (3)-III P. aeruginosa PST 1	AAC (3)-IV E. coli/ J 225	AAC (3)-V E. coli/ pUZ 25
GENTAMICIN C_1	100	100	100	100	100
SISOMICIN	147	89	190	118	106
TOBRAMYCIN	32	49	115	120	72
KANAMYCIN A	6	15	71	33	10
KANAMYCIN B	37	50	92	78	37
DIBEKACIN	29	69		126	79
NETILMICIN	38	68		124	63
AMIKACIN	0	0	0	4	0

* Expressed as percentage relative to gentamicin C,
** Data taken from reference, 2.

TABLE X.

Enzymatic activity of five AAC (3) classes*

	AAC (3)-I E. coli/ pUZ 1	AAC (3)-II E. coli/ R 176	AAC (3)-III P. aeruginosa PST 1	AAC (3)-IV E. coli J 225	AAC (3)-V E. coli pUZ 25
NEOMYCIN B	0	0	158	94	0
PAROMOMYCIN	0	0	105	53	0
RIBOSTAMYCIN	0	0		89	0
BUTIROSIN	0	0		3	0
LIVIDOMYCIN	0	0			0
APRAMYCIN	0	0		135	0

* Expressed as percentage relative to gentamicin C_1.
** Data taken from reference, 2.

5. *Strains producing 2″-O-nucleotidyltransferase ANT(2″)*

Two strains of E. coli and one of K. pneumonia producing ANT (2″) were found to be resistant to kanamycins, gentamicins, sisomicin, tobramycin, and dibekacin, while they were sensitive to netimicin and amikacin.

Comparison of five classes of AAC(3)

In an attempt to establish a relationship between our provisionally denominated AAC(3)-IV (Henceforth referred to as AAC(3)-V and the other classes of 3-N-acetyl-transferases, we decided to compare their substrate ranges and their resistance pheno-types. Thus we have compared the following clases of AAC(3) enzymes under the same laboratory conditions: I, II, III, IV (Davies et al. 5, 6) and V. First of all we applied the following criteria to assign our AAC(3)-V to the 3-N-acetyltransferase group:
1. We determined that they did not belong to the AAC(2') group, since this enzyme group has affinity not only for gentamicin, sisomicin, tobramycin and netilmicin but also for lividomycin A, neomycin and butirosin;
2. The enzymes are not of the AAC(6') group as this enzyme, among other distinguishing characteristics, inactivates neomycin B, kanamycin A and B, and exhibits no affinity for the gentamicin C_1 substrate;
3. The affinity profile, which includes 4.6 deoxystreptamine aminoglycosides, is that of a 3-N-acetyltransferase.

Of the five enzymes studied we may initially classify them into two groups, the first consists of AAC(3)-I, AAC(3)-II and AAC(3)-V, with affinity only for 4.6 deoxystrept-amine antibiotics; the second group consists of AAC(3)-III and AAC(3)-IV, which exhibit affinity for the 4.6 and 4.5 deoxystreptamine aminoglycosides (Tables VIII., IX., X.), it should be noted that with respect to the first three enzymes they possess the following differential characteristics: AAC(3)-I has affinity for gentamicin, and sisomicin, inactivating them, with only moderate affinity for tobramycin, kanamycin B, dibekacin and netilmicin, without inactivation. AAC(3)-V displays an elevated affinity for genta-micin (100%) and sisomicin (106%), and slightly less for tobramycin, dibekacin and netilmicin, but confers resistance to all. The affinity for kanamycin B is of the order of 37 per cent, and for kanamycin A approximately 10 per cent, without inactivation. For the AAC(3)-II class there is slightly less affinity for sisomicin (89%) than for gentamicin (100%) as compared to the slightly greater affinity for sisomicin (106%) than for genta-micin (100%), as displayed by AAC(4)-V, and, as could be expected, both enzymes confer resistance to gentamicin and sisomicin. The affinity for tobramycin (49%), dibe-kacin (69%) and netilmicin (68%) is expressed as resistance to these antibiotics. In spite of the low degree of affinity displayed by the enzyme for kanamycin A (15%), the AAC(3)-II producing *E. coli*/R176 conditions resistance to kanamycin A (16 mcg/ml). Paradoxically, AAC(3)-II displays much greater affinity for kanamycin B but without inactivation (MIC = 2 mcg/ml).

TABLE XI.

Effect of 3-N-acetylating enzymes on antibacterial activity of aminoglycosides

Antibiotic	AAC (3)-I	AAC (3)-V	AAC (3)-II	AAC (3)-III	AAC (3)-IV
GENTAMICIN C*	+	+	+	+	+
SISOMICIN	+	+	+	+	+
TOBRAMYCIN	—	+	+	+	+
KANAMYCIN A	—	—	+	+	+
NEOMYCIN B	—	—	—	+	+
PAROMOMYCIN	—	—	—	+	+
APRAMYCIN	—	—	—	—	+

* Gentamicin complex

301

TABLE XII.

Substrate ranges of 6'-acetyltransferases from E. coli R5/W677
S. marcescens 1830 and P. aeruginosa 3796

Antibiotic	E. coli/ R5	S. marcescens 1830	P. aeruginosa** 3726
KANAMYCIN A	100	100	100
KANAMYCIN B	53	47	29
KANAMYCIN C	0	0	0
GENTAMICIN C_{1a}	72	52	35
GENTAMICIN C_1	2	1	7
SISOMICIN	49	96	44
TOBRAMYCIN	79	80	24
DIBEKACIN	85	93	
NETILMICIN	81	52	
NEOMYCIN B	53	74	106
BUTIROSIN	59	62	7
AMIKACIN	46	54	2
RIBOSTAMYCIN	38	58	
LIVIDOMYCIN	0	0	0
PAROMOMYCIN	0	0	0

* Expressed as percentage relative to Kanamycin A
** Data taken from reference, 11

The separation of the enzymes of the second group, AAC(3)-III and AAC(3)-IV is based primarily on the degree of affinity displayed for some of the 4.5 deoxystreptamine antibiotics. Basically, AAC(3)-III demonstrates a much greater affinity for neomycin B(158%) and for paromomycin (105%) than AAC(3)-IV (94% and 53%, respectively), both enzymes determining resistance to neomycin B and paromomycin. As described previously (5), AAC(3)-IV exhibits affinity for and inactivates apramycin, a valuable distinguishing characteristic.

As a practical consequence the five classes of AAC(3) enzymes can be easily distinguished by their antibacterial activity, as can be seen in Table XI. We have grouped the enzymes accordingly from that with the narrowest spectrum [AAC(3)-I] to that with the broadest [AAC(3)-IV]. Our AAC(3)-V can be seen to occupy an intermediate position between AAC(3)-I and AAC(4)-II.

Currently, the prevalent acetylating enzymes in ·our hospital are AAC(3)-I and AAC(3)-V, only one AAC(6') producing strain has been isolated earlier from an ambulatory patient. Very recently we have isolated an AAC(6') producing strain of S. marcescenses 1830 (resistant to kanamycin A and B, ribostamycin, butirosin, tobramycin, sisomicin, netilmicin, dibekacin and amikacin) for the first time from a hospitalized patient, whose substrate profile is shown in Table XII.

ACKNOWLEDGEMENT

We deeply appreciate the valuable help of Drs. Julian Davies, Naomi Datta, Robert W. Hedges and Margaret Matthew, as well as the assistance of P. E. Davis and the secretarial aid of M. Soledad Pardo.

REFERENCES

1. BENVENISTE, R. and J. DAVIES (1971): Enzymatic acetylation of aminoglycoside antibiotics by *Escherichia coli* carrying an R factor. Biochemistry. 10: 1787–1796
2. BIDDLECOME, S., M. HAAS, J. DAVIES, G. H. MILLER, D. F. RANE, and D. J. L. DANIELS, (1976): Enzymatic Modification of Aminoglycoside Antibiotics: a New 3-N-Acetylating Enzyme from a *Pseudomonas aeruginosa* Isolate. Antimicrob. Agents Chemother, 9:951–955.
3. BRZEZINSKA, M., R. BENVENISTE, J. DAVIES, D. J. L. DANIELS, and J. WEINSTEIN (1972): Gentamicin resistance in strains of *Pseudomonas aeruginosa* mediated by enzymatic N-acetylation of the deoxystreptamine moiety. Biochemistry 11:761–766.
4. DAVIES, J. (1975): Some aspects of antibiotic resistance in bacteria. Plasmids. Medical and Theoretical aspects. Ed. by S. Mitsuhashi, L. Rosíval, V. Krčméry. 121–128. Avicenum, Czechoslovak Medical Press, Prague.
5. DAVIES, J. and S. O'CONNOR (1978): Enzymatic Modification of Aminoglycoside Antibiotics: 3-N-Acetyltransferase with Broad Specificity that determines Resistance to the novel Aminoglycoside Apramycin. Antimicrob. Agents Chemother. 14:69–72.
6. DAVIES, J. and D. I. SMITH (1978): Plasmid-determined resistance to antimicrobial agents. Am. Rev. Microbiol. 34:669–518
7. GÓMEZ-LUS, R. and M. C. RUBIO-CALVO (1977): Bacterial susceptibility to sisomicin in a general hospital. Isolation of two R plasmid conferring sisomicin-resistance. Drugs Exptl. Clin. Res. 3 (1) 161–165.
8. GÓMEZ-LUS, R., M. C. RUBIO-CALVO and L. LARRAD MUR (1977): Aminoglycoside inactivating enzymes produced by R plasmids of *Escherichia coli*, *Citrobacter freundii*, *Klebsiella pneumoniae*, *Proteus vulgaris*, *Providencia stuartii* and *Serratia marcescens*. J. of Antimicrob. Chemother. 3 (Suppl. C) 39–41.
9. GÓMEZ-LUS, R.: Enzimas acetilantes de aminoglicosidos producídos por plasmidos R de los grupos M y P1. Proceedings of the Congress on Microbiology 1977. Santiago de Compostela. Spain (In press).
10. GÓMES-LUS, R., M. C. RUBIO-CALVO, L. LARRAD, M. P. CHOCARRO, M. NAVARRO, A. VILLARROYA and VITORIA (1979): *In vitro* activity of netilmicin against R⁻ and R⁺ gram-negative bacilli. Drugs Exptl. Clin. Res. (In Press).
11. HAAS, M., S. BIDDLECOME, J. DAVIES, C. E. LUCE and P. J. L. DANIELS (1976): Enzymatic Modification of Aminoglycoside Antibiotics: a new 6'-N-Acetylating Enzyme from a *Pseudomonas aeruginosa* isolate. Antimicrob. Agent. Chemother. 9:945–950.
12. HEDGES, R. W.: Personal communication.
13. LE GOFFIC, F., and A. MARTEL (1974): La resistance aux aminosides provoquée par une isoenzyme de la kanamycin acetyltransferase. Biochemie. 56:893–897.
14. LE GOFFIC, F., A. MARTEL (1974) and J. WITCHITZ: 3-N Enzymatic Acetylation of Gentamicin, Tobramycin, and Kanamycin by *Escherichia coli* carrying an R factor. Antimicrob. Agents Chemother. 6:680–684.
15. MATTHEW, M.: Personal communication.
16. MITSUHASHI, S. (1975): Drug-inactivating Enzymes and Antibiotic Resistance Plasmids. Medical and Theoretical aspects. Ed. by S. Mitsuhashi, L. Rosíval and V. Krčmery, 157–163. Avicenum, Czechoslovak Medical Press, Prague.
17. SMITH, D. I., R. GÓMEZ-LUS, M. C. RUBIO-CALVO, NAOMI DATTA, A. E. JACOB, and R. W. HEDGES (1975): Third Type of Plasmid Conferring gentamicin resistance in *Pseudomonas aeruginosa*. Antimicrob. Agents Chemother. 8: 227–230.
18. UMEZAWA, H., M. OKANISHI, R. UTAHARA, K. MAEDA and S. KONDO (1967): Isolation and Structure of Kanamycin Inactivated by a cell-free system of kanamycin resistant *E. coli* J. Antibiot. 20: 136–141.

R.G.L., Facultad de Medicina,
Dept. de Microbiologia,
Zaragoza, Spain

INACTIVATION OF AMINOGLYCOSIDES BY ENZYMES: BIOLOGICAL PROPERTIES OF INACTIVATION PRODUCTS

S. M. NAVASHIN, YU. O. SAZYKIN, I. P. FOMINA, L. G. VINOGRADOVA,
R. M. PETYUSHENKO, V. L. GANELIN, M. K. KUKHANOVA
and L. V. NIKOLAEVA

*National Research Institute of Antibiotics, Moscow, Institute of Molecular
Biology, Academy of Sciences of the U.S.S.R., Moscow,, U.S.S.R.*

The paper deals with the study on the mechanism of aminoglycoside resistance in clinical strains of *E. coli* and *Ps. aeruginosa*. It is known that one of the most frequent mechanisms of aminoglycoside resistance is associated with enzymatic inactivation of the antibiotics.

Aminoglycoside resistant strains of *E. coli* and *Ps. aeruginosa* were isolated from clinical sources, and enzymes inactivating kanamycin, neomycin, streptomycin, paromycin and some other aminoglycosides were found. Aminoglycoside phosphotransferases were isolated from the cells and purified to a homogenous state by affinity chromatography. The following enzymes were detected: aminoglycoside-3′-phosphotransferases I, aminoglycoside-3′-phosphotransferases II and streptomycin-3″-phosphotransferases. Phosphorylated antibiotics in preparation amounts were obtained with the use of the purified enzymes. The structure of these products of fermentative inactivation was determined by the methods of NMR ^{1}H and ^{13}C, as well as by chemical analysis (Ganelin, Razlogova, Petyushenko, Chernyshov, Esipov, Sazykin and Navashin, 1978). All the phosphorylated aminoglycosides were homogenous according to the data of paper chromatography with the use of various systems and completely lost their activity when determined by microbiological methods.

Considering the problem of the causes of the activity lacking in such compounds, it should be noted that resistance to phosphorylated aminoglycoside antibiotics might be connected either with the loss of the capacity for penetration of the modified antibiotics into the cells of the aminoglycoside resistant microorganisms or with inability of the modified antibiotics to interact with the cell targets, .ie. 30S subunits of the bacterial ribosome.

There is evidence that adenylated or phosphorylated streptomycins, as well as lividomycin-5″-phosphate have lower activity as inhibitors of polypeptide synthesis compared to streptomycin and lividomycin (Yamaguchi, Kobayashi and Mitsuhashi, 1973). At the same time it was recently shown that adenylated streptomycin penetrated into the cells of *E. coli* capable of synthesizing streptomycin adenylyltransferase (Dickie, Bryan and Pickard, 1978).

It is known that binding of aminoglycosides with 30S subunits of the bacterial ribosomes resulted in conformation changes in the latter affecting many manifestations of their functional activity. Under various experimental conditions it is possible to observe different effects of aminoglycosides on the ribosomes, such as inhibition of protein synthesis, misreading, sticking effect, formation of abnormal initiation complexes, etc. We tested the phosphorylated aminoglycosides for their ability to inhibit protein synthesis in the presence of endogenic matrix and polyphenylalanine in the presence of polyuridilate. The system of protein synthesis contained a fraction of *E. coli* polysomes, a set of ^{14}C-amino acids, and GTP and ATP-generating systems. The system of ^{14}C-poly-phenylalanine synthesis contained 70S ribosomes, a fermentative fraction of yeast amino-acyl-tRNA-synthetases, tRNA, poly-U, ^{14}C-phenylalanine, GTP and ATP-generating system.

The capacity of the phosphorylated derivatives for induction of the sticking effect or prevention of dissociation of the ribosomal subunits at decreased concentrations of the magnesium ions in the medium was also tested. The studies revealing the sticking effect

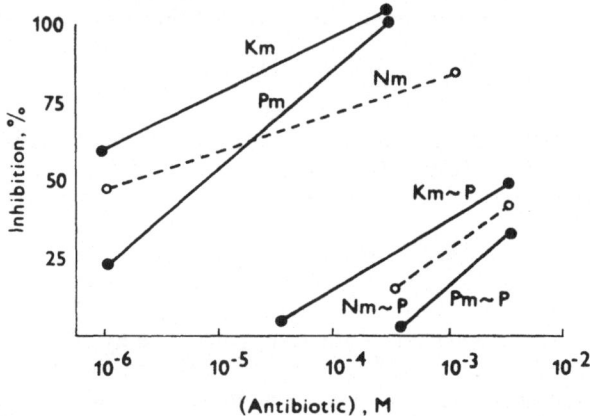

Fig.1. Effect of aminoglycosides and phophorilated aminoglycosides on the poly-U directed synthesis of polyphenylalanine. Km – kanamycin, Nm – neomycin, Pm – paromycin.

of the antibiotics and their derivatives were performed with using analytical ultracentri-fugation of 70S ribosomes in 0.02 M tris-HCl buffer, pH 7.8 containing 1 mM of MgCl$_2$.

Testing of phosphorylated kanamycin, neomycin and paromomycin as well as strepto-mycin for their ability to inhibit the protein synthesis showed the lack of activity in the phosphorylated derivatives at concentrations up to 1000 µg/ml. Nonphosphorylated antibiotics markedly inhibited the protein synthesis at a concentration of 1–5 µg/ml.

Fig. 1 shows the effect of the phosphorylated antibiotics on the synthesis of poly-phenylalanine. In this case not polyribosomes but ribosomes subjected to treatment eliminating the endogenic matrix and nascent peptide chains were used. The data show that phosphorylated kanamycin had a marked inhibitory effect on polyphenylalanine synthesis in low concentrations, i.e. about 0.5 µg/ml. However, it should be noted that the nonphosphorylated antibiotic was effective in concentrations many times lower, as is evident from the same figure. Similar data were also obtained on comparison of the activity of phosphorylated and nonphosphorylated neomycin and paromomycin as inhibitors of polypeptide synthesis.

On the basis of these data it is possible to conclude that enzymatic inactivation of aminoglycosides results in marked decreasing of their capacity for affecting the ribosome function, especially under conditions close to those *in vivo*, i.e. in the presence of the endogenic matrix and polyribosomes. It should be of interest to study the ability of phosphorylated aminoglycosides to penetrate through the bacterial cell membrane to more detailed data on the cause of resistance to aminoglycoside antibiotics as a result of their modification.

It is expedient to compare the above data with those of streptomycin dinitrophenyl hydrazone activity in the cell-free system of polyphenylalanine synthesis.

This streptomycin derivative was obtained from dinitrophenyl hydrazine and streptomycin. The derivative did not inhibit the growth of a streptomycin sensitive strain of *Bacillus mycoides* but inhibited the synthesis of polyphenylalanine in the cell-free system

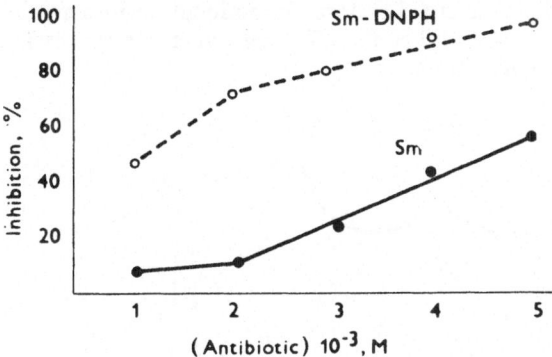

Fig. 2. Effect of streptomycin and streptomycin dinitrophenyl hydrazone on the poly-U directed synthesis of polyphenylalanine.

to a greater extent than streptomycin (Fig. 2). According to these data it is possible to suppose that the lack of antibacterial activity in streptomycin dinitrophenyl hydrazone is due to its inability to penetrate into the bacterial cell.

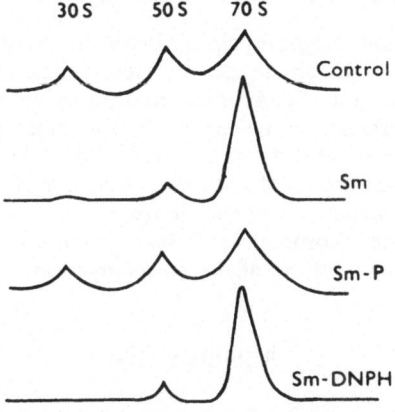

Fig. 3. Sticking effect of kanamycin-3'-phosphate and kanamycin on the ribosomes of *E. coli* (Mg^{2+} 1 mM)

It is known that some analogues of natural aminoglycoside antibiotics produce a sticking effect but have no antibacterial activity (β-anomer RV 23468) (Cousin, Lando, Oyasoo and Raynand, 1977). We tested the ability of phosphorylated aminoglycosides to induce the sticking effect.

Fig. 3 presents sedimentograms obtained as a result of analytical ultracentrifugation.

As is evident from the control data at a low concentration of the magnesium ions, 70S ribosomes of *E. coli* dissociated into 30S and 50S subunits. In the presence of kanamycin in a concentration of 50 µg/ml there is practically no dissociation, which is evidence of the presence of the sticking effect of this antibiotic. In this case the peak representing the content of 70S units is the main one. At the same time in the presence of phosphorylated kanamycin (100 µg/ml) the sedimentogram practically corresponded to the control one, i.e. there was no sticking effect. Therefore, the aminoglycosides modified by the phosphorylating enzymes lost their ability for affecting either the synthesis of proteins or dissociation of the bacterial ribosomes.

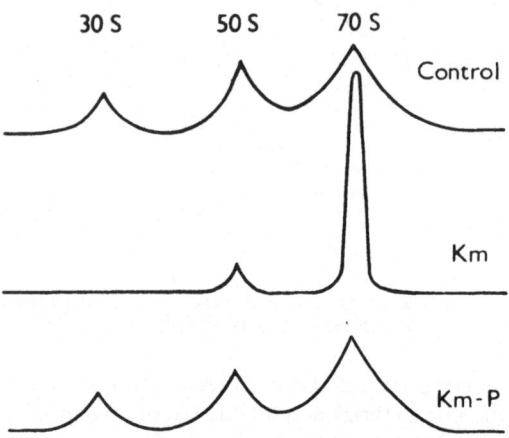

Fig 4. Sticking effect of streptomycin, streptomycin-3"-phosphate and streptomycin dinitrophenyl hydrazone on the ribosomes of *E. coli* (Mg^{2+} 1 mM).

It is of interest to note that streptomycin dinitrophenyl hydrazone which was effective as an inhibitor of protein synthesis in the cell-free system also inhibited the ribosome dissociation. As it is evident from Fig. 4, unlike phosphorylated streptomycin (400 µg/ml), this derivative in a concentration of 100 µg/ml had a sticking effect which was not less pronounced than the effect of streptomycin itself (200 µg/ml).

In conclusion it should be noted that testing of enzymatically modified aminoglycosides with respect to the functional activity of the ribosomes in various systems and conversion of the ribosomes during the ribosomal cycle may provide some data for differentiation of the specific and nonspecific effects of the aminoglycoside antibiotics at the ribosomal level.

REFERENCES

COUSIN, M. A., D. LANDO, T. OJASOO, and I. – P. RAYNAUD (1977): Biochemie, 59, 59–63.
DICKIE, P., L. E. BRYAN, and M. A. PICKARD (1978): Antimicrob. Agents and Chemother. 14, 569–580.
GANELIN, V. L., I. O. RAZLOGOVA, R. M. PETYUSHENKO, A. I. CHERNYSHEV, S. E. ESIPOV, YU. O. SAZYKIN, and S. M. NAVASHIN (1978): Biokhimiya 43, 2154–2162.
YAMAGUCHI, M., F. KOBAYASHI, and S. MITSUHASHI (1973): Comparative studies on the activities of lividomycin 5"-phosphate and lividomycin on polypeptide synthesis directed by poly-U in *E. coli* cell-free extracts. J. Antibiot. 26, 602–605.

*S.M.N., National Research
Institute of Antibiotics,
Nagatinskaya 3a, Moscow M-105, U.S.S.R.*

STRUCTURAL AND FUNCTIONAL RELATIONSHIPS BETWEEN AMINOGLYCOSIDE-MODIFYING ENZYMES FROM STREPTOCOCCI AND STAPHYLOCOCCI

P. COURVALIN, C. CARLIER and E. COLLATZ

Laboratory of Biochemistry, L. A. CNRS, 271, Institute of Pasteur,
Paris, France

Plasmid-mediated resistance towards aminoglycoside antibiotics in Gram-negative bacteria has been extensively studied during the past decade (for a recent review see Davies and Smith, 1978). Probably because aminoglycosides are less commonly used with staphylococcal infections and because streptococci are "naturally" resistant to these drugs, enzyme-mediated resistance towards aminoglycosides in Gram-positive cocci has been reported only recently (Dowding 1977, Courvalin and Davies, 1977, Le Goffic et al., 1977 and Courvalin et al., 1978a, Krogstad et al., 1978, Slocombe, 1978). The study of aminoglycoside resistance in Gram-positive cocci revealed certain mechanisms of modification unknown in Gram-negative bacteria (Table I.).

Gram-positive and Gram-negative bacteria are both human pathogens which have common eco-systems (e.g. the human gut), and it has been demonstrated that antibiotic

TABLE I

Classes of aminoglycoside-modifying enzymes found in Gram-positive
and Gram-negative bacteria*

Enzyme	Gram positive		Gram negative
	Staphylococcus	Streptococcus	
Phosphotransferases			
APH (6)	—	—	+
APH (3′)	+	+	+
APH (2″)	+	+	—
APH (3″)	+	—	+
APH (5″)	—	—	+
Nucleotidyltransferases			
AAD (6)	+	?	—
AAD (4′)	+	—	—
AAD (2″)	?	—	+
AAD (3″) (9)	+	?	+
Acetyltransferases			
AAC (3)	?	—	+
AAC (2′)	—	—	+
AAC (6′)	+	+	+

* modified, from Davies and Smith, 1978
+, Presence of enzyme;
—, absence of enzyme;
?, preliminary evidence that an enzyme of this type is present, but without complete characterization.

TABLE II.

Properties of the bacterial strains used

S. aureus strain	Phenotype (a)			Genotype (b)	Origin or obtention
BM 4600	Gen Tc	Kan Em$_c$	Str Pc	APH(2″) AAC(6′) APH(3′) AAD(3″ × 9)	Wild strain
RN450-Ag^R	Gen	Kan		APH(2″) AAC(6′)	Transduction
BM 4600-1	 Tc	Kan Em$_c$	Str Pc	APH(3′) AAD(3″ × 9)	Curing (Eb)
PALM	Gen	Kan		APH(2″) AAC(6′)	Wild strain (c)
S. faecalis BM 4100	Gen Tc Hly	Kan Em$_c$ Bcn	 Cm	APH(2″) AAC(6′)	Wild strain (d)

(a) Bcn, Bacteriocin production and resistance; Cm, chloramphenicol; Em$_c$, erythromycin constitutive; Hly, hemolysin; Pc, penicillin; Tc, tetracycline.
(b) Eb, ethidium bromide.
(c) Le Goffic et al., 1977a.
(d) Courvalin et al., in preparation.

resistance determinants originating from Gram-positive bacteria can be expressed in and confer drug resistance to *Escherichia coli* (Chang and Cohen, 1974, Courvalin et al., 1976, Courvalin and Fiandt, in preparation). Transfer of genetic information between Gram-positive and Gram-negative bacteria may then occur and should influence the effectiveness of aminoglycoside therapy.

We have studied the aminoglycoside-modifying enzymes from independently obtained Gram-positive clinical isolates and compared them with similar enzymes from Gram-negative organisms. The sources and properties of the strains used are listed in Table II.

RESISTANT PHENOTYPES

The minimal inhibitory concentrations (MICs) of nineteen antibiotics for staphylococcal strains are listed in Table III. The plasmid-free strain RN450 (Novick and Brodsky, 1972), susceptible to all antibiotics, was used as a control. Strain BM4600 represents a typical clinical isolate, it is resistant to all commercially available aminoglycosides and to penicillin, tetrycyclines, macrolides, lincosamides and group B streptogramins. This multi-resistant phenotype is due to the synthesis of four aminoglycoside-modifying activities (Fig. 1, Table II.). By transduction and curing experiments we were able to separate the APH(2″) and AAC(6′) activities from the APH(3′)-III and AAD(3″)(9) activities. Strain BM4600 harbours a single plasmid (pIP850) with a molecular weight of 4 kb (as determined by agarose gel electrophoresis-data not shown). No plasmid DNA could be detected in strain RN450-Ag^R. It is therefore possible that the genes coding for the enzymatic activities in this strain are integrated into the host chromosome.

TABLE III.

Minimal inhibitory concentrations (MIC) of various antibiotics for S. aureus strains

S. aureus	MIC of antibiotics (pg/ml)									
	Gen	Sis	Net	Kan A	Kan B	Ami	Tob	Neo A	Neo B	Par
RN 450	≤0.125	≤0.125	≤0.125	0.5	≤0.125	≤0.125	≤0.125	≤0.125	≤0.125	0.25
BM 4600	16	8	1	256	128	2	16	512	32	128
RN 450-Ag^R	8	16	0.5	64	8	0.5	16	32	0.25	0.5
BM 4600-1	0.25	≤0.125	≤0.125	64	32	0.5	2	128	64	128

S. aureus	MIC of antibiotics (pg/ml)								
	Liv A	But	Rib	Apr	Str	Spc	Tet	Min	Ero
RN 450	0.5	1	1	0.5	2	32	0.5	≤0.125	≤0.0625
BM 4600	512	16	4096	1	64	4096	128	1	8192
RN 450-Ag^R	1	1	16	1	2	16	0.5	≤0.125	≤0.0625
BM 4600-1	512	16	4096	1	128	>8192	128	1	8192

MICs were determined by the method of Steers et al., (1959) using Mueller-Hinton agar medium. Tet, tetracycline; Min, minocycline; Ero, erythromycin.

CHARACTERIZATION OF THE ENZYMES

I n an attempt to compare the acetyl- and phosphotransferases from strains Palm BM4600 and BM4100, four criteria were used: the substrate profiles of the enzymes, their pH optima, their apparent molecular weight and their isoelectric point.

In vitro substrate profiles. The substrate profiles of the enzymes extracted from strain BM4600 and its derivatives are shown in Fig. 1. The substrate specificities of aminoglycoside-modifying enzymes from staphylococci and streptococci are very similar but differ from those of similar enzymes extracted from Gram-negative organisms. The 6′-acetyltransferases from *S. aureus* (Fig. 1 and Le Goffic et al., 1977a, Huang and Davies, 1978) and from *S. faecalis* (Courvalin et al., in preparation) modify Neo A much more efficiently than the structurally related compound Neo B. They also acetylate Gen C_{1a} more efficiently than Sis and much more so than Net. By contrast, these three compounds are equally good substrates for the corresponding enzymes from *E. coli* or *P. aeruginosa*. (Le Goffic et al. 1977a, Huang and Davies 1978). Furthermore, the 3′-phosphotransferases from Gram-positive cocci (Fig. 1 and Courvalin and Davies, 1977, Krogstad, 1978) including *S. pneumoniae* (Courvalin et al., in preparation) modify Ami, whereas the corresponding enzymes of Gram-negative do not.

pH optima for enzymatic activity. The pH optima for the activity of AAC(6′) and APH(2″) from strains BM4600 and Palm have been determined (Fig. 2). They are manifest for the AAC(6′) enzymes at approximately pH 7, whereas there is no distinct optimum, between pH 5 and 8, for the APH(2″) enzymes. This is in contrast to the results obtained for the corresponding enzymes of *S. faecalis*, which have discrete optima at pH 5.5 and 8, respectively (Courvalin et al., in preparation).

Molecular weights. The apparent molecular weight of the enzymes was estimated after filtration through Sephadex (Fig. 3). The AAC(6′) and the APH(2″) from any of strains

Palm, BM4600 and BM4100 could not be separated from each other and all appeared to be monotonously similar. Their molecular weight was 31,000. The APH(3′)-III from strain BM4600 was separated from the two former enzymes and its apparent molecular weight was 22,000. Similar enzymes have been described in Gram-positive cocci earlier. Dowding (1977) separated a (larger) APH(2″) from a (smaller) APH(3′)-III by Sephadex G 100 chromatography. The APH(2″) was not separated by this techniques from the

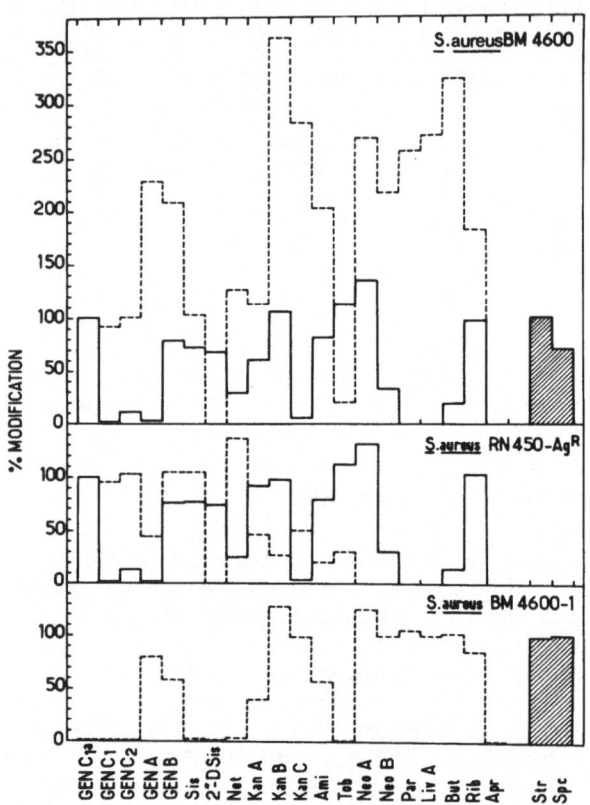

Fig. 1. Substrate profiles of enzymes extracted from *S. aureus* strain BM4600 and its derivatives. Adenylylation (hatched area) is expressed relative to streptomycin as 100%. Acetylation (solid lines) and phosphorylation (dashed lines) are expressed relative to gentamicin C1a as 100% for strains BM4600 and RN450–AgR. Phosphorylation is expressed relative to neomycin B as 100% for strain BM4600–1. The enzyme assays were performed as described by Haas and Dowding (1975). BM4600–1 contains adenylylating and phosphorylating activities. The adenylylating enzyme is equally effective against streptomycin and spectinomycin [AAD(3″) (9)]. Since kanamycin A is modified but tobramycin is not, it is inferred that the 3′–hydroxyl group is the site of phosphorylation; the fact that both butirosin and lividomycin A, are modified indicates that the enzyme is of type III [APH(3′)–III]. Strain RN450–Ag– contains both a phosphotransferase and an acetyltransferase. That sisomicin is phosporylated and that 2″–deoxy–sisomicin is not indicates that the 2″–hydroxyl group is the site of phosphorylation [APH(2″)]. That kanamycin B is a substrate for acetylatio nand kanamycin C is notindicates that the 6′–amino group is the site of acetylation [AAC(6′)]. The phosphorylation profile of strain BM4600 is consistent with the presence of the APH(2″) and the APH(3′)–III detected in strains RN450–AgR and BM4600–1, respectively. No adenylylation of gentamicin C1a, tobramycin, amikacin or kanamycin A was detected in strain BM4600.

AAC(6'). Le Goffic et al. (1977b) obtained similar results with the *S. aureus* strain Palm. Using an acrylamide-agarose column, they found an identical molecular weight, 28,000, for both the AAC(6') and the APH(2″). An APH(3')-III was isolated from *S. aureus* strain RN450/pSH2 (Smith, 1978). Its molecular weight was 29,000, determined after polyacrylamide gel electrophoresis containing sodiumdodecylsulfate.

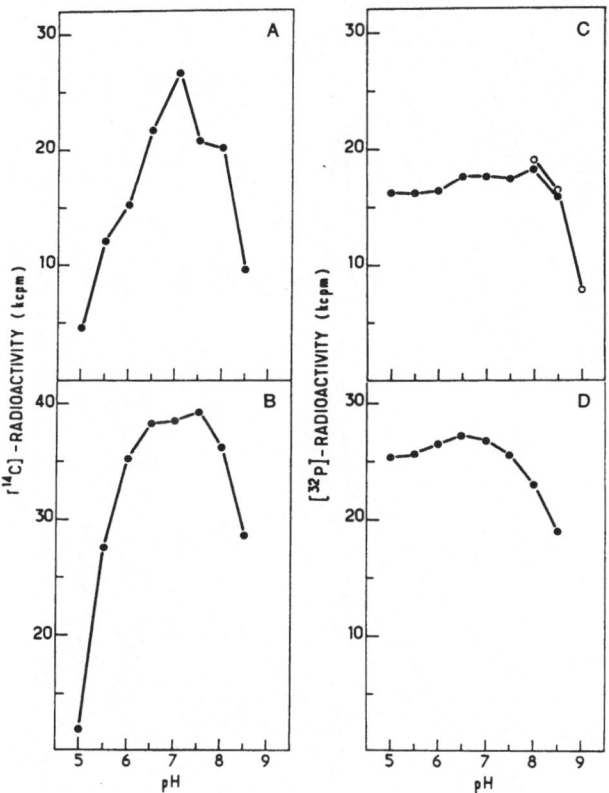

Fig. 2. pH optima for aminoglycoside-phosphorylating or-acetylating activities of *S. aureus* strains BM4600 (A, C) and Palm (B, D). The reaction was carried out in the presence of gentamicin Cla and [γ³²–P] ATP or [¹⁴C] CoA SAc. The buffers used were: ●——●, 60 mM Tris, 45 mM MgCl₂, 450 mM NH⁵Cl, 1.5 mM Dithiothreitol, adjusted with maleic acid; ○——○, 150 mM glycine, 45 mM MgCl₂, 450 mM NH₄Cl, 1.5 mM Dithiothreitol, adjusted with NaOH.

Isoelectric points. An attempt was made to separate the AAC(6') and the APH(2″) from strains Palm, BM4600 and BM4100 by isoelectric focussing in polyacrylamide gels (Fig. 4). The two enzymes from any of the three strains could not be separated. Furthermore they all appeared indistinguishable with respect to their isoelectric point which was estimated to be 5.3 ± 0.1. Dowding (1977) separated an AAC(6') from anAPH(2″) by polyacrylamide gel electrophoresis. It seems likely that isoelectric focussing would have yielded a similar separation, and it can then be assumed that at least one of the enzymes studied by Dowding (1977) is different from the enzymes described in Fig. 4. Le Goffic et al. (1977a) used isoelectric focussing in liquid medium and found an isoelectric point of 5.7 for AAC(6′) and of 5.8 for APH(2″). They found both enzymes from strain Palm to be inseparable by several techniques (Le Goffic et al., 1977b).

The comparison of antibiotic modifying enzymes on the basis of their apparent molecular weight or their isoelectric point is tempting but almost certainly misleading, unless exactly the same experimental conditions are applied or unless the variations of these conditions are kept in mind. It seems clear, that the difference of 3,000 resulting

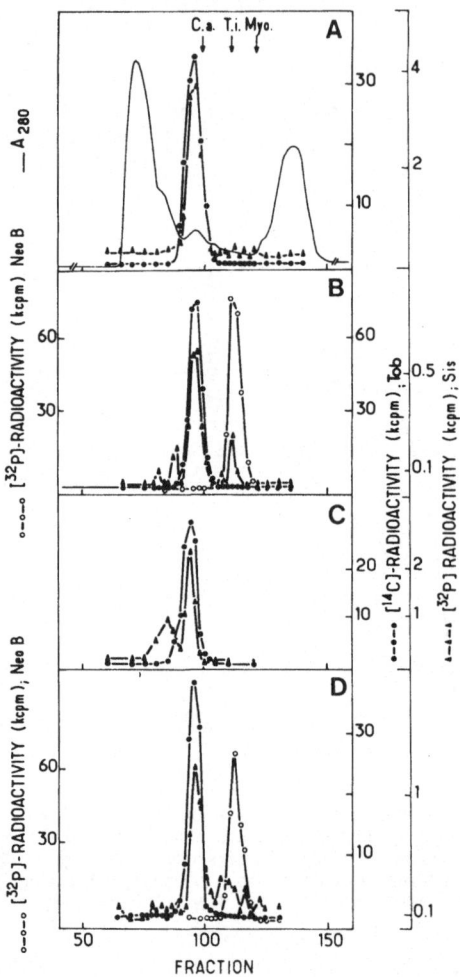

Fig.3. Gel filtration of crude enzyme preparations. Crude (S100) enzyme preparations (1 to 2 ml) were filtered through a column (1.5 × 190 cm) of Sephadex G100 SF, equilibrated with buffer: 50 mM Tris, 20 mM MgCl₂, 1mM NH₄Cl, 1%. β-mercaptoethanol, adjusted to ph 7.5 with acetic acid. The flow rate was 3 to 4 ml/h, and 1.5 ml fractions were collected. The determination of the enzyme activity was according to Haas and Dowding (1975). The molecular weight markers used were: Carbonic anhydrase (C.a.), 30.000; trypsin inhibitor (T.i.), 21.500; myoglobin (Myo), 17.200.

A., *S. aureus* Palm; B., *S. aureus* BM4600;
C., *S. faecalis* BM4100; D., BM4600 plus BM4100.

314

from the molecular weight determinations of the AAC(6′) and APH(2″) and of 0.4 to 0.5, resulting from the determinations of the isoelectric point of the same enzymes by Le Goffic et al., (1977a) and ourselves, reflects a methodological error rather than a true difference between the proteins. (The dilemma becomes obvious upon inspection of the two molecular weights, 29,000 (Smith, 1978) and 22,000 (our result) estimated for an

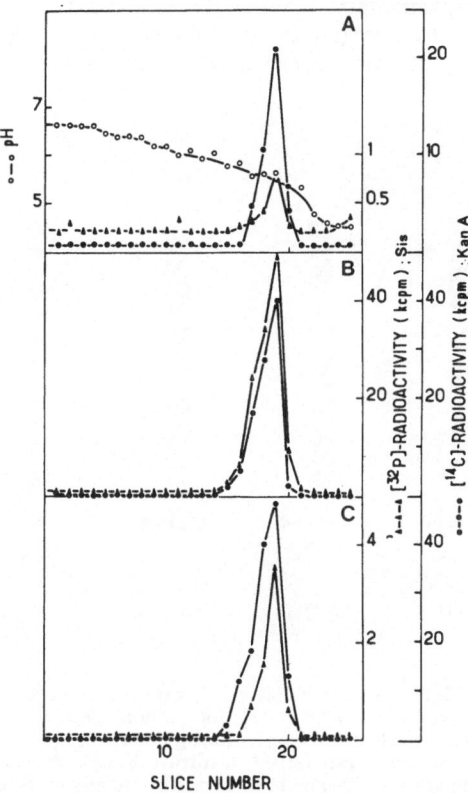

Fig. 4. Isoelectric focussing of crude enzyme preparations in polyacrylamide gels. Isoelectric focussing was of 30 μl of S100 preparations in polyacrylamide (5%) rods (0.3 × 13 cm) containing 2% ampholines (pH range 3.5 — 5 : 5,— 8 = 1 : 2 (as shown) or 2 : 1). The gels were sliced as indicated and every slice was cut in two. The determination in every slice was either of AAC(6′) and pH or AAC(6′) and (APH 2″). The enzyme activity was determined after incubation (3 h at 4°) of every half slice in 40 μl of buffer : 20 mM Tris, 20 mM MgCl₂, 300 mM NH₄Cl, 1%. β-mer-captoethanol, adjusted to pH 7.5 with acetic acid. A., *S. faecalis* BM4100; B, *S. aureus* 4600; C, *S. aureus* Palm.

APH(3′)-III from two different strains of *S. aureus*: 24% error? Two distinct proteins?).

Whatever the authentic values may be, it can be concluded that, in terms of size and charge, the AAC(6′) and the APH(2″) from the *S. aureus* strains Palm, BM4600 and from *S. faecalis* BM4100 are very similar. It remains open, whether the model proposed for the action of these two staphylococcal enzymes by Le Goffic et al. (1977b) holds true for their streptococcal counterparts.

The properties of the plasmids used are listed in Table IV., and the results are shown in Fig. 5. The gene encoding APH(3')-III from plasmid pSH2 has been shown to be representative of this type of enzyme in *S. aureus* strains (White et al., 1978, Courvalin and Fiandt, in preparation), and we used this probe in hybridization experiments. ^{32}P-labeled pSH2 complementary RNA (cRNA) hybridizes with *Eco*R1 restriction endonuclease generated DNA fragments of *S. faecalis* plasmids pJH1, pJH4 and pJH5. In addition, it has been demonstrated that the antiserum prepared against the 3'-phospho-

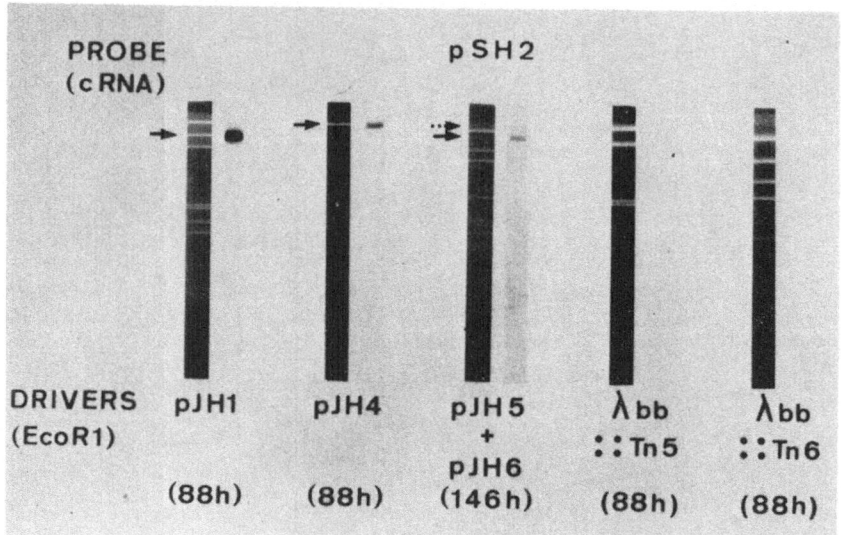

Fig. 5. Analysis of plasmid DNAs by hybridization. Two μg of the driver plasmid DNAs indicated on the bottom line were digested with restriction endonuclease *Eco*R1. The resulting DNA fragments were fractionated by electrophoresis through horizontal slab gels (0.8% agarose, 18 × × 13 × 0.4 cm) for 12 h at 3 V/cm, transferred to nitro-cellulose sheets as described by Southern (1975) and hybridized to *in vitro* − ^{32}P-labelled plasmid complementary RNA (cRNA) as indicated on the top line. The autoradiograms were exposed for the times (h) indicated.

transferase isolated from the *S. aureus* strain RN450/pSH2 cross-reacted with extracts of the *S. faecalis* strain JH2-7, harbouring the plasmids pJH4 or pJH5 plus pJH6 (White et al., 1978). By contrast, we have not found any homology, by hybridization, between the pSH2 gene coding for APH(3')-III and the resistance genes of the transposable elements Tn5 and Tn6, both from R-plasmids from Gram-negative bacteria. These two transposons encode phosphotransferases which are not homologous with each other (Courvalin et al., 1978b). The hybrid plasmid pAT5 codes for an APH(3')-III and an AAD(3″) (9) which originate from *S. aureus* (Courvalin and Fiandt, in preparation). In similar experiments we did not find any homology between ^{32}P-labeled pAT5 cRNA and DNAs from R-plasmids R6 and R100 which encode similar enzymes (data not shown). It would appear therefore that plasmid-mediated aminoglycoside-modifying enzymes with similar site specificities from Gram-positive cocci are at least partially related structurally and that they are not homologous to similar enzymes from Gram-negative bacteria.

TABLE IV.

Plasmids used

Plasmid	Name	Mol Wt (kb)	No	IR (kb)	Loop (kb)	EcoRI	APH(3″) type	AAD(3″) (9)	Plasmid	Host
	Vehicule		Passenger				Enzymes		Origin	
			Transposon							
Abb:: Tn5 Abb		40.5	Tn5	1.5	2.2		II		JR67	K. pneumoniae (1)
Abb:: Tn6 Abb		40.5	Tn6	None	4.1		I		JR72	E. coli (1)
R6		95	Tn601	I	I		I	+		E. coli (2)
R100		90	Tn8	ND	ND			+		S. flexneri (3)
pSH2		15					III			S. aureus (4)
pAT5	ColE1:: TnA (RSF2124)	11.1				10.2 2.8	III	+	pWA1 pWA2	S. aureus (5)
pJH1		75					III	+		S. faecalis (6)
pJH4		39					III	+		S. faecalis (6)
pJH5		114					III	+		S. faecalis (6)

TnA, transposable element directing resistance to ampicillin;
IR, inverted repetition; ND, not determined.

(1) (Berg et al., 1975), (2) (Lebek, 1962),
(3) (Nakaya et al., 1960), (4) (Courvalin a. Davies, 1977),
(5) (Courvalin a. Fiandt, in prep.) (6) (Courvalin et al., 1978)

MECHANISM(S) OF RESISTANCE

In Gram-negative bacteria, plasmid-mediated resistance towards aminoglycosides is not accompanied by gross inactivation of the antibiotics; it is, apparently, the modification of a rather small amount of the drug which prevents its further uptake (Davies and Benveniste, 1974). In this respect, the mechanism of resistance to aminoglycosides in Gram-positive cocci appears to be similar to that in Gram-negative bacteria (Table V.). When *S. aureus* strains BM4600 and RN450 — Ag^R were grown in the presence of sisomicin, which is a substrate for both APH(2″) and AAC(6′), no change in the antibiotic concentration in the medium was detected by microbiological methods.

Measurements of aminoglycoside-phosphorylating or — acetylating activities in strains grown in the presence or absence of antibiotics (Table VI.) revealed always substantial modifying activities. Thus, the aminoglycoside — modifying enzymes in *S. aureus* (as well as in streptococci — Courvalin et al., 1978a) are synthesized constitutively, like their counterparts in Gram-negative bacteria. In RN450 — Ag^R the levels of phosphorylation

TABLE V.

Induction of aminoglycoside-modifying enzymes

S. aureus	Induction	CPM	% Phosphorylation
BM 4600	0	6,295	100%
	+	35,186	559%
RN 450-Ag^R	0	6,015	100%
	+	7,115	118%
			% Acetylation
BM 4600	0	19,122	100%
	+	45,753	239%
RN 450-Ag^R	0	27,884	100%
	+	29,052	104%

Cells were grown to 8.5×10^8 colony forming units in broth in the presence or absence of sisomicin (20 µg/ml) and extracts for enzyme assays were prepared as described by Haas and Dowding (1975).

TABLE VI.

Inactivation of antibiotic in the culture medium

S. aureus	Inhibition zone diam (mm) ($\pm$SD) (range of values)
None (control)	(14 ($\pm$0.7)) (13.3—14.8)
BM 4600	13.6 ($\pm$0.6) (13—14.3)
RN 450-Ag^R	13.8 ($\pm$0.2) (13.5—14)

Cells were grown for twenty generations in broth containing 20 µg/ml of sisomicin and then harvested by centrifugation. Biologic activity of the antibiotic in the supernatant was determined using a bioassay, with *S. aureus* strain 209 P as the indicator organism. SD, Standard deviation.

and acetylation are similar with or without induction and, as mentioned above, the corresponding genes may be integrated into the chromosome. In BM4600 the levels of both activities are much higher after induction. This behaviour could be due to super-induction, gene amplification or increase of the copy number of the plasmid, although there is no direct evidence that the two enzyme activities are encoded by pIP850.

DISCUSSION

Staphylococci and streptococci are major human pathogens, especially in hospital environments where these two genera often exhibit broad antibiotic resistance spectra (Table III.). *A priori* the possibility of a transfer of plasmid-borne antibiotic resistance determinants between Gram-positive cocci and between Gram-positive and Gram-negative bacteria cannot be excluded, especially, since there is no barrier to the expression of certain resistance genes from Gram-positive bacteria in *E. coli* (Chang and Cohen,

318

1974, Courvalin et al., 1977, Courvalin and Fiandt, in preparation). In fact, there is recent evidence for structural homology between the genes encoding an APH(3')-II in a Gram-positive *Corynebacterium* and in Gram-negative bacteria (White et al., 1978).

We have therefore compared the aminoglycoside-modifying enzymes from Gram-positive and Gram-negative clinical isolates of antibiotic resistant and plasmid-containing bacteria. The majority of the enzymes described in staphylococci and in Gram-negative organisms differs with respect to their site specificity, and the set of enzymes found in streptococci is more closely related to that of stapylococci (Table I.). In addition, the *in vitro* substrate profiles of enzymes of the same type, e.g. APH(3'), AAC(6'), from staphylococci and streptococci, are very similar but differ from those of the corresponding enzymes in Gram-negative organisms. It remains to be established wheter subtle but distinct differences in substrate specificities between corresponding aminoglycoside-modifying enzymes from Gram-positive and Gram-negative bacteria represent a general phenomenon, i.e. wether such differences can also be found between the respective APH(3") and AAD(3") (9) enzymes. The (apparent) confinement of APH(2") activities to Gram-positive bacteria and the fact that the AAC(6') enzymes exist in a remarkable variety of types in Gram-negative organisms (Davies and Smith, 1978) but not in Gram-positive cocci, may be another reflection of the similarities the enzymes observed within the Gram-positive cocci and the distinction from those of the Gram-negative organisms. The similarity of the molecular weight and of the isoelectric point of the APH(2") and the AAC(6') from *S. aureus* and *S. faecalis* is striking. It could reflect a structural homology analogous to that, revealed by hybridization experiments, between the *S. aureus* and *S. faecalis* resistance genes encoding APH(3') enzymes (Fig. 5) which, in turn, are not homologous to the APH(3') genes of Gram-negative bacteria. No differences seem to exist, between the Gram-positive and -negative bacteria, in the general mechanism of resistance to aminoglycosides (Table VI.), and all the modifying enzymes are synthesized constitutively.

The fact that the aminoglycoside-modifying enzymes from staphylococci and streptococci are homologous could be due either to a recent divergence in evolution within the Gram-positive cocci or to a transfer of genetic information, also recent, between the two bacterial genera, even though there is no experimental evidence for a transfer of this kind. Whatever the explanation for the homologies may be, the mere fact that new aminoglycoside-modifying enzymes have been detected recently in Gram-positive cocci stresses their clinical and epidemiological relevance.

ABBREVIATIONS

ColE1, colicinogenic factor E1, λbb, λb515b519 cI857 Sam7, kb, kilobase, MIC, minimal inhibitory concentration, cRNA, complementary RNA, [γ-^{32}P] ATP, adenosine 5'-[γ-^{32}P] triphosphate, [^{14}C] CoASAc, [1-^{14}C]acetyl-coenzyme A, S100, supernatant (100,000xg) of a bacterial extract after sonication, phenotypes include Ag^R, resistant to aminoglycosides, Ami, amikacin, Apr, apramycin, But, butirosin, Gen, gentamicin, kan, kanamycin, Liv, lividomycin, Neo, neomycin, Net, netilmicin, Par, paromomycin, Rib, ribostamycin, Sis, sisomicin, 2"-D Sis, 2"-deoxy-sisomicin, Spc, spectinomycin, Str, streptomycin, Tob, tobramycin, genotypes include AAC(6'), aminoglycoside acetyltransferase 6', APH(3'), aminoglycoside phosphotransferase 3', APH(2"), aminoglycoside phosphotransferase 2", AAD(3") (9), aminoglycoside adenylyltransferase 3", 9.

319

REFERENCES

BERG, D. E., J. DAVIES, B. ALLET and J. D. ROCHAIX (1975): Transposition of R factor genes to bacteriophage λ, Proc. Natl. Acad. Sci. USA, 72, 3628–3632.

CHANG, A. C. Y. and S. N. COHEN (1974): Genome construction between bacterial species in vitro: Replication and expression of Staphylococcus plasmid genes in *Escherichia coli*, Proc. Natl. Acad. Sci. USA, 71, 1040–1034.

COURVALIN, P., B. WEISBLUM, and J. DAVIES (1977): Aminoglycoside-modifying enzyme of an antibiotic-producing bacterium acts as a determinant of antibiotic resistance in *Escherichia coli*, Proc. Natl. Acad. Sci. USA, 74, 999–1003.

COURVALIN, P. and J. DAVIES (1977): Plasmid-mediated aminoglycoside phosphotransferase of broad substrate range that phosphorylates amikacin, Antimicrob, Agents Chemother. 11, 619–624.

COURVALIN, P., W. V. SHAW and A. E. JACOB (1978a): Plasmid-mediated mechanisms of resistance to aminoglycoside-aminocyclitol antibiotics and to chloramphenicol in group D streptococci. Antimicrob. Agents Chemother. 13, 716–725.

COURVALIN, P., M. FIANDT and J. DAVIES (1978b): DNA relationships between genes coding for aminoglycoside-modifying enzymes from antibiotic-producing bacteria and R-plasmids. In Microbiology-1978, ed. D. Schlessinger. (American Society for Microbiology, Washington, D.C.). p. 262–266.

DAVIES, J. E. and R. E. BENVENISTE (1974): Enzymes that inactivate antibiotics in transit to their targets., Ann. N.Y. Acad. Sci. 235, 130–136.

DAVIES, J. and D. I. SMITH (1978): Plasmid-determined resistance to antimicrobial Agents, Ann. Rev. Microbiol. 32, 469–518.

DOWDING, J. E. (1977): Mechanisms of gentamicin resistance in *Staphylococcus aureus*. Antimicrob. Agents Chemother. 11, 47–50.

HAAS, M. and J. E. DOWDING (1975): Aminoglycoside-modifying enzymes, in Hash, J. H. (Ed.) Methods in Enzymology, Academic Press, New York, Vol. 43, pp. 611–628.

HUANG, T. S. R. and J. DAVIES (1978): Plasmid-determined aminoglycoside resistance in staphylococci. Proc. 2nd Tokyo Symp. Microb. Drug. Resistance. In press.

KROGSTAD, D. J., T. R. KORFHAGEN, R. C. MOELLERING, R. C., C. WENNERSTEN, M. N. SWARTZ, S. PERZYNSKI and J. DAVIES (1978): Aminoglycoside-inactivating enzymes in clinical isolates of *Streptococcus faecalis*. J. Clin. Invest. 62, 480–486.

LE GOFFIC, F., A. MARTEL, N. MOREAU, M. L. CAMPAU, C. J. SOUSSY and J. DUVAL (1977a): 2″-O-Phosphorylation of gentamicin components by a *Staphylococcus aureus* strain carrying a plasmid. Antimicrob. Agents Chemother. 12, 26–30.

LE GOFFIC, F., N. MOREAU and M. MASSON (1977b): Are some aminoglycoside antibiotics inactivating enzymes polyfunctional – Ann. Microbiol. (Inst. Pasteur), 128B, 465–469.

LEBEK, G. (1963): Über die Entstehung mehrfachresistenter Salmonellen. Ein experimenteller Beitrag. Zentralbl. Bakteriol. Parasitenk. 188, 494–505.

NAKAYA, R., A. NAKAMURA, A., and T. MURATA, (1970): Resistance transfer agents in *Shigella*. Biochem. Biophys. Res. Commun. 3, 654–659.

NOVICK, R. P., and R. BRODSKY (1972): Studies on plasmid replication, I. Plasmid incompatibility and establishment in *Staphylococcus aureus*. J. Mol. Biol. 68, 285–302.

SMITH, D. I. (1978): Purification and characterization of aminoglycoside-3′-phosphotransferases. Thesis, University of Wisconsin, Madison, Wisconsin, USA.

SLOCOMBE, B. (1978): Transmissible aminoglycoside resistance in strains of *Streptococcus faecalis*. In Current Chemotherapy. Proceedings of the 10th International Congress of Chemotherapy. Zürich, Switzerland, 2, 891–893.

SOUTHERN, E. M (1975): Detection of specific sequence among DNA fragments separated by gel electrophoresis, J. Mol. Biol. 98, 503–518.

STEERS, E., E. L. FOLTZ, B. S. GRAVES and J. RIDEN (1959): An inocula replicating apparatus for routine testing of bacterial susceptibility to antibiotics, Antibiot. Chemother. 9, 307–311.

WHITE, T. J., D. I. SMITH, S. ROSETHAL, and J. DAVIES (1978): Immunological comparison of aminoglycoside phosphotransferases, Progr. Abstr. Intersci. Conf. Antimicrob. Agents Chemother. 18th, Atlanta, Geo., Abstr. 287.

P. C., *Laboratoire de Biochimie,*
L.A. CNRS 271, Unite de Bacteriologie
Medicale, Inst. Pasteur, Paris, France

COULD A SINGLE ENZYME INACTIVATE AMINOGLYCOSIDE ANTIBIOTICS BY TWO DIFFERENT MECHANISMS?

F. LE GOFFIC, N. MOREAU, A. MARTEL and M. MASSON

C.N.R.S.—C.E.R.C.O.A., Thiais, France

INTRODUCTION

In a strain of *Staphylococcus aureus* from clinical origin resistant to all aminoglycoside antibiotics but neomycin and related compounds, we have found two mechanisms of inactivation: an acetyltransferase of the AAC(6') type and a new phosphotransferase. We have shown that the acetyltransferase was different from other known AAC(6'), and that the target of the phosphotransferase was the 2″hydroxylfunction of the aminoglycoside: 2″(o) phosphorylsisomicin was identified as inactivated compound (Le Goffic, Martel, Moreau, Capmau, Soussy and Duval 1977).

In attempts to separate both enzymatic activities, we performed affinity chromatography on agarose immobilized gentamicin and electrofocusing, we were unable to find any dissociation of the activities. We then made the hypothesis that the same enzymatic system could perform both reactions (Le Goffic, Moreau, Masson 1977).

We report here other tentatives of separation of both activities and some kinetic experiments leading us to the conclusion that our previous hypothesis could be reasonable.

MATERIAL AND METHODS

The strain is a *S. aureus* Chapman and coagulase positive from clinical origin. We call the corresponding plasmid R Palm. It confers the following resistances expressed as CMI in micrograms per milliliter: amikacin 16, gentamicin 64, kanamycin 128, sisomicin 128, tobramycin 64. The strain remains sensitive to neomycin 0.125, paromomycin 0.5, butirosin 2 and lividomycin 2.

Enzyme preparative and assay

The cells, grown in trypticase soy broth and washed twice were digested by lysostaphine in the buffer: Tris HCl 10 mM pH 7.4, MgCl$_2$ 10 mM, KCl 60 mM, β-mercaptoethanol 6 mM (5 g of bacteria, 5 mg of lysostaphine in 15 ml of buffer, 15 mn at 37°C). DNase was added and centrifugation was carried out for 20 mn at 10 000 g. The supernatant was centrifuged one night at 45 000 g and the upper 2/3 of the supernatant is taken.

The standard phosphocellulose paper (Whatman P81) binding assay (Haas and Dowding, 1975) is used to check enzymetic activity. ^{14}C acetyl CoA is the cofactor for acetylation and ^{32}P GTP the best cofactor for phosphorylation.

Purification of the enzyme

Crude extract is chromatographied on a column of indubiose A4-gentamicin as previously described (Le Goffic, Martel, Moreau, Capmau, Soussy and Duval, 1977). Active fractions are pooled, concentrated using polyethyleneglycol and chromatographied on ultrogel AcA 54. After concentration, the enzyme solution is kept in small portions in liquid nitrogen.

Electrofocusing

A LKB 81101 column is used according to the supplier's instructions, with 3.5–10 ampholine. Focusing is done at +4° during 3 days, energy applied to the column always inferior to 3 W.

Pseudo affinity chromatography

Agarose cibacron blue was prepared (Baird, Sherwood, Carr, Atkinson, 1976) or purchased from Pharmacia. Crude or purified extract may be used. The buffer is TrisHCl, pH 7.4 10 mM, $MgCl_2$ 10 mM, KCl 10 mM, β-mercaptoethanol 6 mM, glycerol 20 %.

Molecular weight determination

A 75×1.5 cm column filled with ultrogel ACA 54 is used. The buffer is TrisHCl pH 7.5 10 mM, $MgCl_2$ 10 mM, NH-Cl 60 mM, β-mercaptoethanol 6 mM, kanamycin 1 0.1 mM, glycerol 10% (Le Goffic, Martel, Moreau et al. 1977).

For sucrose gradient experiments, SW41 Beckman rotor is used. Tubes are filled with 11 ml of a 5–20% linear gradient in TrisHCl pH 7.4 10 mM $MgCl_2$ 10 mM, KCl 60 mM. Centrifugation is performed 23 h at 39 000 rpm.

Electrophoresis

The experiment is performed during 3 h at +4° according to Dowding (1977). After electrophoresis, gels are sliced in 1 mm fractions, each slice cut into two parts, each being assayed for acetyltransferase or phosphotransferase activity, after one night of incubation in the assay buffer, the cofactor having been omited.

Kinetic characteristics of both enzymatic activities

All experiments were done with purified enzyme. For thermal denaturation 500 µl of enzyme solution, alone or with a substrate or a cofactor are incubated at 48°C. 50 µl is assayed every 2 mn.

For determination of the nature of the inhibition by several substrates or cofactors, the amount of substrate modified (acetylated or phosphorylated) is measured in function of time at several concentrations of substrate, cofactor or inhibitor. The results are plotted in the form of Lineweaver Burk plot and Eadie plot.

RESULTS

The efficiency of different aminoglycosides as substrates for both activities has been given previously (Le Goffic, Martel, Moreau et al. 1977). Gentamicin C_1 is the best substrate for phosphorylation and is not a substrate for acetylation. Neamine is a good substrate for acetylation and is not a substrate for phosphorylation. Paromomycin is chosen as a non-substrate for both reactions.

Purification of the enzyme

The extent of purification after each step is given in Table I. The protein concentration is determined by the Lowry method. The purification is very efficient and both activities are always found in the same fractions (Fig. 1).

TABLE I.

TABLE I.

Purification of the enzyme

Purification step	Purification content (μg/ml)
Crude extract	40 000
Centrifugation 1 night at 45 000 g	20 000
Affinity chromatography on agarose gentamicin	80 to 100
Chromatography on ultrogel AcA 54	5 to 10

Protein content is given for similar activities and volumes.

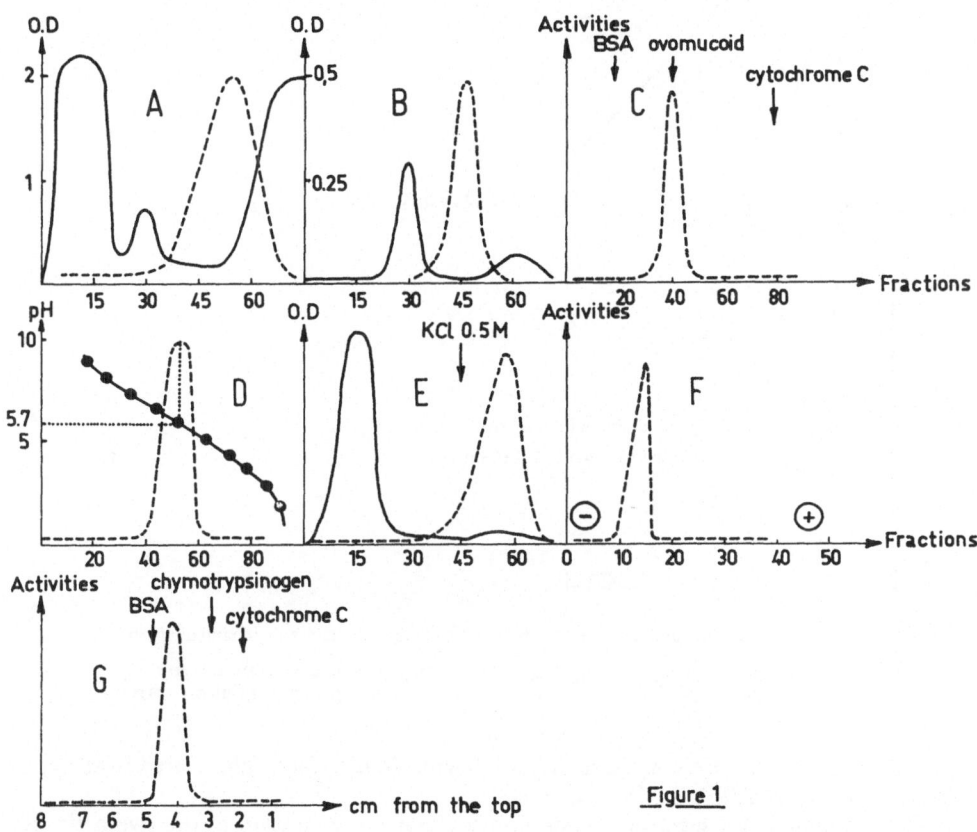

Fig. 1. A: affinity chromatography on agarose gentamicin – B: chromatography on AcAS-4 – C: MW determination on AcA S-4 – D: electrofocusing – E: pseudoaffinity chromatography on agarose – cibacron blue – F – polyacrylamide gel electrophoresis – G: MW determination by sucrose gradient sedimentation

——— OD
– – – both enzymatic activities, arbitrary units
○○○ pH

Electrofocusing, electrophoresis, pseudoaffinity, chromatography,
molecular weight determination

As seen from Fig. 1, both activities are never separated. The pI is 5.7, the MW is 28,000 as determined by gel exclusion, and 55,000 by centrifugation on sucrose gradient, showing that the enzyme seems to be dimeric.

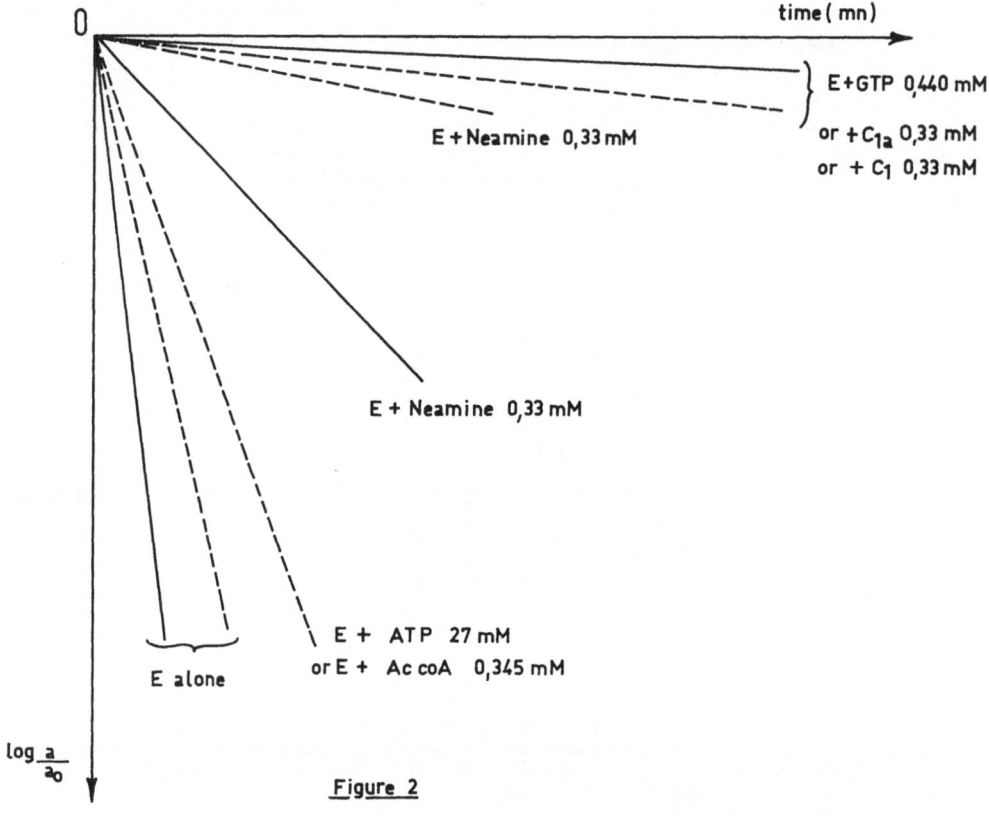

Figure 2

Fig.2. Protection by substrates or cofactors against thermic denaturation.

Log $\dfrac{\text{residual activity}}{\text{activity at time 0}}$ is plotted versus time

——— acetylation activity
– – phosphorylation activity

Both activities have the same affinity for cibacron blue, a molecule related in structure to cofactors such as ATP, ADP, AcCoA...

Polyacrylamide gel electrophoresis brings about no separation of the two activities. All these results are shown on Fig. 1.

Thermal denaturation

Fig. 2 shows that GTP, gentamicin C_{1A} (substrate of both activities), gentamicin C_1 (substrate of phosphorylation) and neamine (substrate of acetylation) confer good to excellent protection of both activities.

TABLE II.

TABLE II.

Competition studies

Constant S or C	Variable S or C	Inhibitor	Nature of the inhibition
	PHOSPHORYLATING ACTIVITY		
Genta C_1	GTP	AcCoA	Competitive
Genta C_1	ATP	AcCoA	Competitive
Genta C_1	ATP	GTP	Competitive
GTP	Genta C_1	Neamine	Competitive
GTP	Genta C_1	Paromomycin	Competitive
	ACETYLATING ACTIVITY		
AcCoA	Neamine	Genta C_1	Competitive
AcCoA	Neamine	Paromomycin	Competitive

Acetyl CoA and ATP at the concentration used in the experiment protect only slightly both activities.

Inhibition studies

Table II gives the results of some inhibition experiments. In the first column the substrate or cofactor is constant. In the second column the substrate or cofactor is variable and in the inhibitor of each the substrate or cofactor is variable.

The study is complete for the phosphorylating activity and shows that substrate or cofactor of acetylase are competitive inhibitors for phosphorylase. Paromomycin, a non substrate for both activities is also a competitive inhibitor of gentamicin C_1. ATP and GTP, two cofactors of phosphorylase are also mutual competitors.

For acetylating activity, we still lack the results for ATP and GTP as competitors. Gentamicin C_1 and paromomycin competitively inhibit neamine acetylation.

DISCUSSION

It was impossible to separate the two activities, 6' acetylate and 2" phosphorylase. They exhibit the same physicochemical properties (isoelectric point, molecular weight, electrophoresis migration, affinity for immobilized substrates and analogue of cofactors).

The molecular weight, 28,000 when determined by gel exclusion chromatography, was 55,000 as shown by sucrose gradient sedimentation. The enzyme might be dimeric, a fact which could allow for a single enzymatic system being responsible for two different activities.

The results of thermal inactivation and of inhibition studies are in strong agreement with this hypothesis: each activity is more or less protected or inhibited by the elements of the other reaction.

If the existence of a single enzyme seems to be proved by these results, one should confirm whether the binding site of the substrates on one hand and of the cofactors on the other hand are the same or different for both activities.

We will try to obtain more information about the mechanism of these enzymatic reactions in order to be able to determine the real (and not only the apparent) kinetic constants and to confirm the existence of a single site for cofactors on the one hand and for substrates on the other hand.

Comparison of our enzyme with the results given by other authors (Dowding, 1977) indicates that our enzymatic system is probably different from theirs, as suggested by Vogel et al. (1978), whose strain, in study in our laboratory, seems to be similar to our Palm strain.

REFERENCES

BAIRD, J. K., R. F. SHERWOOD, R. J. G. CARR and A. ATKINSON (1976): Enzyme purification by substrate elution chromatography from procion dye polysaccharide matrices. FEBS Letters, 70, 61–66.

DOWDING, J. E. (1977): Mechanisms of gentamicin resistance in Staphyloccus aureus. Antimicrob. Agents and Chemother. 11, 47–50.

HAAS, M. J. and J. E. DOWDING (1975): Aminoglycoside modifying enzymes. Methods Enzymol. 43, 611–628.

LE GOFFIC, F., A. MARTEL, N. MOREAU, M. L. CAPMAU, C. J. SOUSSY and J. DUVAL (1977): 2″o phosphorylation of gentamicin components by a *Staphylococcus aureus* strain carrying a plasmid. Antimicrob. Agents and Chemother. 12, 26–30.

LE GOFFIC, F., N. MOREAU and M. MASSON (1977): Are some aminoglycoside inactivating enzymes polyfunctional – Ann. Microbiol. (Inst. Pasteur 128B, 465–469.

VOGEL, L., C. NATHAN, H. M. SWEENEY, S. A. KABINS and S. COHEN (1978): Infections due to gentamicin-resistant staphyloccocus aureus strain in a nursery for neonatal infants. Antimicrob. Agents and Chemother. 13, 466–472.

F. Le G., C.N.R.S.–C.E.R.C.O.A.,
2 à 8, rue Henry Dunant, 94320 Thiais,
France

MULTIPLE ENZYMATIC MODIFICATION OF AMINOGLYCOSIDES BY *PSEUDOMONAS AERUGINOSA* AND *ENTEROBACTERICEAE* STRAINS RESISTANT TO GENTAMICIN, NETILMICIN, TOBRAMYCIN AND AMIKACIN

M. KETTNER, JANA NAVAROVÁ, IVETA MOLNÁROVÁ
and V. KRČMÉRY

*Institute of Experimental Pharmacology, Slovak Academy of Sciences
and Research Institute of Preventive Medicine,
Bratislava, Czechoslovakia*

The purpose of this paper is to present results of microbiological and biochemical studies examining the production of a number of aminoglycoside-modifying enzymes by a group of gram-negative bacteria. Many enterobacteria and *Pseudomonas aeruginosa* strains belong to causative agents which, especially in hospital environments, often exhibit broad antibiotic resistance spectra.

We have examined several strains of the aminoglycoside-resistant isolates collected from various clinical sources. The susceptibility of these resistant strains was determined against 14 selected aminoglycoside-aminocyclitol antibiotics including netilmicin and amikacin. The aminoglycoside-inactivating enzymes produced or assumed to be produced by the organisms were classified into genotypes according to Davies (1978) and Mitsuhashi (1975). All strains were obtained from Professor H. Knothe, Frankfurt.

MATERIAL AND METHODS

Determination of MIC. Minimal inhibitory concentrations (MIC) for a series of aminoglycoside-aminocyclitol antibiotics were determined by the agar dilution method (Ericsson and Sherris, 1971) using Mueller-Hinton agar containing twofold dilutions of the respective antibiotic solutions ranging in concentration from 0,25 µg/ml to 1024 µg/ml. Bacterial inocula were prepared by growing individual strains overnight in Mueller-Hinton broth and were adjusted to contain approximately 1×10^5 CFU/ml. The plates were incubated at 37°C for 18 h. The MIC was defined as the lowest concentration of antibiotic inhibiting growth by the presence of 5 discrete colonies or less at the site of inoculation.

Antibiotics and chemicals. Gentamicins C_1, C_{1a} and C(complex), sisomicin and netilmicin were provided by Schering Corporation, Bloomfield, N. J., kanamycins A and B, butirosin (complex) and amikacin were supplied by Bristol Laboratories, Syracuse, N.Y., lividomycin (complex) was kindly provided by Kowa Co., Tokyo Research Laboratories, tobramycin was supplied by Eli Lilly, Indianapolis, spectinomycin by Upjohn Company, Kalamazoo, Mich. and DKB was a gift of Dr. H. Kawabe, Gunma University, Maebashi.

Radio-labelled chemicals were bought from the Radiochemical Centre (Amersham, England) or from the Institute for Isotopes of the Hungarian Academy of Sciences, Budapest. Dithiothreitol, ATP and acetyl-CoA were obtained from Calbiochem (Lucerne, Switzerland). All other chemicals were of analytical grade.

TABLE I

Minimal inhibitory concentrations (MIC) of various aminoglycoside antibiotics for Ps. aeruginosa and anterobacteria strains

Bacterial	MIC of antibiotics (μg/ml)											
	*Gm C_1 C_{1a}	Sis	Net	Tob	Ak	KmA	KmB	Liv	But	Nm	Sm	Spc
E. coli 7	16	16	1	32	4	>1024	>1024	>1024	4	512	128	256
E. coli 9	8	4	0.5	16	4	1024	512	>128	2	256	64	256
K. pneumoniae 1	32	16	0.5	32	1	>1024	>1024	>1024	2	1024	2	4
Ps. aeruginosa 70	>128	>128	64	16	16	64	64	64	64	16	>1024	>1024
Ps. aeruginosa 93	4	2	16	16	8	>1024	>1024	>1024	64	512	128	64
Ps. aeruginosa 97	16	4	16	8	16	>1024	>256	64	>128	>1024	64	256
Ps. aeruginosa 113	>1024	>1024	>1024	256	32	>1024	>1024	512	>1024	>1024	>1024	>1024

MICs were determined according to Ericsson and Sherris (1971) using Mueller-Hinton agar medium
* Values for Gm C_{1a} are identical

Preparation of cell-free extracts. Bacterial strains were grown to the late log phase in Nutrient broth (Difco). Cells were then pelleted by centrifugation and washed twice with a solution of 10 mM tris buffer according to Smith et al. (1975). The suspension was maintained on ice and was sonicated with a Schoellerschall sonifier (Frankfurt/M, FRG) at 250 W for 5 min in 30-s bursts. Particulates were removed by centrifugation of the mixture for 2 h at 105.000xg. The supernatant fraction was then frozen and stored at —185°C in liquid nitrogen until use.

Phosphorylation and adenylylation of aminoglycoside antibiotics was assayed by the phosphocellulose paper binding assay as described by Devaud and Kayser (1977).

Acetylation of aminoglycoside antibiotics was also assayed by the phosphocellulose paper binding assay by the modified method of Korfhagen et al. (1976).

Residual antibiotic activity was assayed according to Kabins et al. (1974) using *Bacillus subtilis* ATCC 6633.

RESULTS

In vitro inhibitory activities of aminoglycoside antibiotics and their derivatives against gram-negative resistant microorganisms are presented in Table I. Practically all of the strains studied were resistant against gentamicin, sisomicin, netilmicin, tobramycin and amikacin. In *Pseudomonas aeruginosa* strains 70 and 113, remarkably high levels of resistance agaist gentamicin were observed.

All strains were examined for the presence of enzymes that acetylate, phosphorylate, or adenylylate aminoglycosides. The results are summarized in Tables II., III. and IV. A rate of 100 was assigned for acetylation and adenylylation of gentamicin, or for phos-

TABLE II.

Substrate specificity of aminoglycoside acetyltransferases

Drug	Acetylation relative to Gentamicin C_{1a} (%)					
	E. coli 7*	K. pneumoniae 1*	Ps. aeruginosa 70	Ps. aeruginosa 93	Ps. aeruginosa 97	Ps. aeruginosa 113
Gm C_{1a}	66	83	100	100	100	100
Gm C_1	100	100	75	61	79	0
Sis	139	117	81	51	118	31
Net	55	19	110	72	157	19
Tob	143	126	40	34	42	196
DKB	159	125	47	9	24	49
Km A	0	57	94	65	107	0
Km B	118	107	47	61	41	0
Ak	78	38	48	21	43	20
Liv	125	98	31	27	32	0
But	0	0	0	24	0	0
Nm B	0	0	4	50	0	27

* Acetylation relative to Gentamicin C_1

TABLE III.

Substrate specificity of aminoglycoside adenylyltransferases

Drug	Adenylylation relative to Gentamicin C_1 (%)						
	E. coli 7	E. coli 9	K. pneumoniae 1*	Ps. aeruginosa 70	Ps. aeruginosa 93*	Ps. aeruginosa 97	Ps. aeruginosa 113
Gm C_1	100	100	33	100	51	100	100
Gm C_{1a}	111	52	100	79	100	78	70
Sis	86	85	60	62	68	116	67
Net	61	42	0	73	75	73	74
Tob	75	48	67	47	65	159	79
DKB	42	75	50	43	65	140	47
Km A	130	48	50	68	57	180	74
Km B	86	81	49	36	81	102	79
Ak	0	0	0	0	0	0	0
Sm	55	130	15	54	51	63	66
Spc	61	100	22	33	53	136	114

* Adenylylation relative to Gentamicin C_{1a}

phorylation of neomycin. All other rates are expressed as relative rates. The reactions were performed at respective optimal pH, examined in the microbiological assays.

Assumed occurrence of various aminoglycoside-modifying enzymes classified according to Davies (1978) and Mitsuhashi (1975) is presented in Table 5.

DISCUSSION

Using radioenzymatic methods, in all bacterial isolates tested, the presence of several modifying enzymes has been proved (Table V.). In *E. coli 9* no acetylating activity was observed. The mentioned strain was sensitive to netilmicin (Table I.) and this observation

TABLE IV.

Substrate specificity of aminoglycoside phosphotransferases

Drug	Phosphorylation relative to Neomycin B (%)				
	E. coli 9	*K. pneumoniae* 1	*Ps. aeruginosa* 70	*Ps. aeruginosa* 97	*Ps. aeruginosa* 113
Km A	54	119	116	110	124
Km B	101	134	116	198	115
Tob	0	0	0	0	0
Liv	37	104	93	111	115
But	0	0	67	132	125
Nm B	100	100	100	100	100
Sm	0	0	0	0	107

TABLE V.

Classes of aminoglycoside-modifying enzymes found in the bacterial strains studied

Bacterial strain	Acetylating enzymes	Adenylylating enzymes	Phosphorylating enzymes
E. coli 7	AAC(3)-III AAC(6')-IV	AAD(2″) AAD(3″) (9)	—
E. coli 9	—	AAD(2″) AAD(3″) (9)	APH(3')-I
K. pneumoniae 1	AAC(3)-III	AAD(2″)	APH(3')-I
Ps. aeruginosa 70	AAC(3)-III AAC(6')-IV	AAD(2″) AAD(3″) (9)	APH(3')-I+II
Ps. aeruginosa 93	AAC(3)-III AAC(6')-IV	AAD(2″) AAD(3″) (9)	—
Ps. aeruginosa 97	AAC(3)-III AAC(6')-IV	AAD(2″) AAD(3″) (9)	APH(3')-I+II
Ps. aeruginosa 113	AAC(6')-IV	AAD(2″) AAD(3″) (9)	APH(3')-I+II APH(3″)

is in agreement with results of Habwe et al. (1979) obtained with a series of resistant *E. coli* strains, sensitive to netilmicin, which in many cases were resistant to AAD(2″).

All *Pseudomonas aeruginosa* strains in our studies were resistant to lividomycin and butirosin. In radioenzymatic assays both antibiotics were good substrates for APH(3'). We therefore so far assume the presence of aminoglycoside-phosphotransferases I plus II. Similarly, in all *Ps. aeruginosa* strains the presence of 2 adenylylating enzymes, i.e. AAD(2″) and AAD(3″) (9) was noted. Comparing the substrate specificity of gentamicin-adenylyltransferases in resistant *Ps. aeruginosa* strains, as described by Kabins et al. (1974), there exists a good agreement with our results.

In addition to adenylylating enzymes, in our *Ps. aeruginosa* strains also the presence of acetylating activity has been shown. When comparing the substrate specificities of acetyltransferases in *Ps. aeruginosa* strains described by Holmes et al. (1974) with our results, certain similarities may be noted with the exception that in our *Pseudomonas* strains very probably the simultaneous presence of two acetyltransferases may be assumed, i.e. AAC(3) and AAC(6′), based on the fact that Gm C_1 and Ak are good substrates at the same time. Davies and O'Connor (1978) could not exclude the possibility of existence of more than one acetyltransferase in multiresistant gram-negative strains. In cases of amikacin-resistant *Ps. aeruginosa* strains, all the described strains (Jacoby, 1974, Kawabe et al. 1975) were sensitive to Gm C_1 which is no substrate for AAC(6′). Especially in *Ps. aeruginosa* strains 70 and 113 high levels of resistance to Gm C_1 were observed. In these strains extremely high levels of resistance to netilmicin, tobramycin and also amikacin were found. This might be due to synergistic action of adenylylating and acetylating enzymes in these strains (Kontomichalou and Papachristou, 1976).

The obtained data testify to a wide range of occurrence of aminoglycoside-modifying enzymes in all bacterial isolates tested. Further attempts have been made to characterize the mechanisms of enzymatic inactivation more in detail. The results of these studies will be reported elsewhere.

REFERENCES

DAVIES, J. and S. O'CONNOR (1975): Enzymatic modification of aminoglycoside antibiotics: 3-N-acetyltransferase with broad specificity that determines resistance to the novel aminoglycoside apramycin. Antimicrob. Ag. Chemother. 14, 69–72.

DAVIES, J. and D. I. SMITH (1978): Plasmid-determined resistance to antimicrobial agents. Ann. Rev. Microbiol. 32, 469–518.

DEVAUD, M., F. H. KAYSER and U. HUBER (1977): Resistance of bacteria to the newer aminoglycoside antibiotics: an epidemiological and enzymatic study. J. Antibiotics 30, 655–664.

ERICSSON, H. M. and J. C. SHERRIS (1971): Antibiotic sensitivity testing. Report of an international collaborative study. Acta Pathol. et Microbiol. Scand., Sect B, Suppl. 217, p. 65–73.

HABWE, V. and S. SHADOMY (1979): Comparative in vitro studies with netilmicin, amikacin, gentamicin, sisomicin and tobramycin. J. Antimicrob. Chemother. 5, 73–79.

HOLMES, R. K., B. MINSHEW, K. GOULD and J. P. SANFORD (1974): Resistance of *Pseudomonas aeruginosa* to gentamicin and related aminoglycoside antibiotics. Antimicrob. Ag. Chemother. 6, 253–262.

JACOBY, G. (1974): Properties of an R plasmid in *Ps. aeruginosa* producing amikacin, kanamycin, tobramycin and sisomicin resistance. Antimicrob. Ag. Chemother. 6, 807–810.

KABINS, S., C. NATHAN and S. COHEN (1974): Gentamicin-adenylyltransferase activity as a cause of gentamicin resistance in clinical isolates of *Pseudomonas aeruginosa*. Antimicr. Ag. Chemother. 5, 565–570.

KAWABE, H., S. KONDO, H. UMEZAWA and S. MITSUHASHI (1975): R-factor mediated aminoglycoside antibiotic resistance in *Ps. aeruginosa*: a new aminoglycoside 6′-N-acetyltransferase. Antimicrob. Ag. Chemother. 7. 494–499.

KONTOMICHALOU, P. and E. PAPACHRISTOU (1975): Multiresistance plasmids in *Ps. aeruginosa* highly resistant to gentamicin. In: Microbial Drug Resistance (S. Mitsuhashi and H. Hashimoto, eds.), University Park Press, Tokyo, pp. 329–349.

KORFHAGEN, T. R., J. A. FERREL, C. L. MENEFEE and J. C. LORER (1976): Resistance plasmids of *Pseudomonas aeruginosa*: change from conjugative to nonconjugative in a hospital population. Antimicr. Ag. Chemother. 9, 810–816.

MITSUHASHI, S. (1975): Proposal for a rational nomenclature for phenotype, genotype and aminoglycoside-aminocyclitol modifying enzymes. In: Drug-Inactivating Enzymes and Antibiotic Resistance, S. Mitsuhashi, L. Rosival and V. Krčméry (eds), Avicenum, Prague, pp. 115–119.

SMITH, D. I., R. GOMEZ-LUS, M. C. RUBIO-CALVO, N. DATTA, A. E. JACOB and R. W. HEDGES (1975): Third type of plasmid conferring gentamicin resistance in *Pseudomonas aeruginosa*. Antimicrob. Ag. Chemother. 8, 227–230.

M.K., Institute of experimental Pharmacology, Slovak Academy of Sciences, Dúbravská cesta, P.O.B. 1041, 881 05 Bratislava, Czechoslovakia

BIOCHEMICAL ASPECTS OF BACTERIAL RESISTANCE TO NEW β-LACTAM DRUGS NON-HYDROLYSABLE BY β-LACTAMASES

T. YOKOTA and E. AZUMA

Department of Bacteriology, School of Medicine,
Juntendo University, Tokyo, Japan

INTRODUCTION

It has been well established that the resistance to β-lactam drugs in clinical isolates of bacteria is mainly due to the production of β-lactamases. Cross-resistance levels to various penicillins (PC), and cephalosporins (CES), however, are rather complicated because of the variety of drugs and β-lactamases. The resistance level of a β-lactamase-producing bacterial strain can be assumed from the index value of *Vmax* devided by *Km* of the enzyme and drug (Pollock, 1965). Although the physiological efficiency of a β-lactamase thus calculated was found to be in agreement with gram-positive bacteria in which the enzyme is produced extracellularly, discrepancies are often observed between values of *Vmax/Km* and resistance levels in gram-negative bacteria that produce β-lactamase as an epienzyme. At the 3rd Symposium on Antibiotic Resistance, Castle of Smolenice, one of the authors presented a paper showing that strains of *Escherichia coli* become resistant by inheriting R plasmids that encode TEM-type β-lactamase higher to carbenicillin (CBPC) and sulbenicillin (SBPC), which are hardly hydrolyzed by the enzyme, than to ampicillin (ABPC) easily hydrolyzed (Yokota and Yamamoto, 1977). A hypothesis was postulated suggesting that an unknown barrier directed by β-lactamase, that hydrolyzes the drugs but binds to them hardly is a possible mechanism of the resistance in gram-negative bacteria preventing the passage of drugs through the periplasmic space to the PC-sensitive site (Yamamoto and Yokota, 1977). This paper deals with the generality of the concept.

NEW β-LACTAM DRUGS NON-HYDROLYSABLE BY β-LACTAMASES

Many new derivatives of CES that are almost non-hydrolysable by various types of β-lactamases have been devised in the world including Japan. They are classified in two groups, i.e. 7-α-methoxy derivatives such as cefoxitin (CFX), cefmetazole (CMZ) and 6059-S, and 7-Z(2)-methoxyimino derivatives such as cefuroxime (CXM), cefotaxime (HR756), FK749 and SCE-1365, as shown in Fig. 1. Neither penicillinase (PCase)-type nor cephalosporinase (CESase)-type β-lactamases can hydrolyze those new drugs with little exceptions, i.e. CXM is hydrolyzed in rather high extent by the CESase produced by *Proteus vulgaris*, and HR756 is hydrolyzed to a little extent by the oxacillin-hydrolyzing enzyme specified by some R plasmids. Other new derivatives of CES, cefamandole (CMD), cefotiam (CTM) and T-1551, were excluded from the present experiment because of their susceptibilities to the TEM-type β-lactamase encoded by the major part of R plasmids.

Ent. cloacae has been well known as one of the opportunistic pathogens that can be a source of microbial substitution after the CES-treatment. To clarify the mechanism of resistance to β-lactam drugs in the microbe, the clinical isolates resistant to CFX were selected and their susceptibilities to CMZ, 6059-S and HR756 were examined by the plate dilution method together with that to cefazolin (CEZ), as shown in Fig. 2. Some

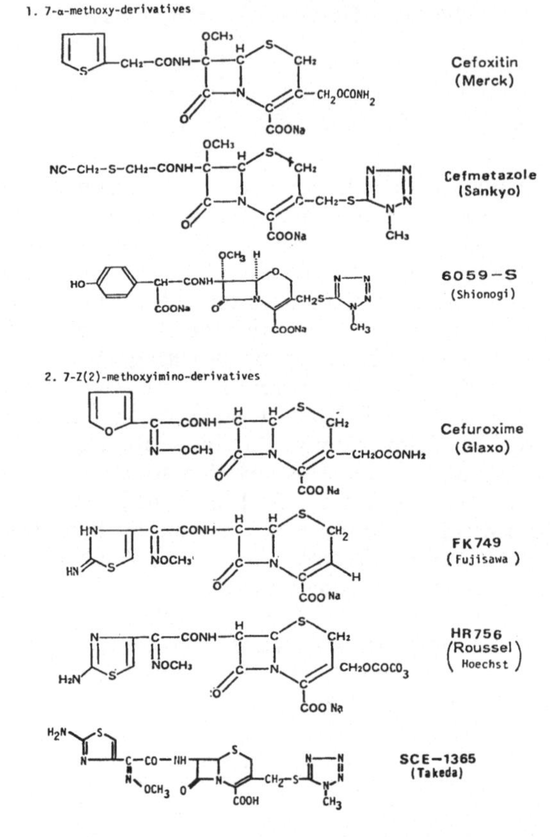

Fig. 1. New β-lactamase-tolerant derivatives of cephalosporin.

strains of them were found to be resistant not only to CFX but also to CMZ, 6059-S and HR756. MICs of the strains of *Ent. cloacae* were 25 to 3,600 μg/ml, 6.25 to 1,600 μg/ml, 0.2 to 100 μg/ml and 0.39 to 50 μg/ml to CFX, CMZ, 6059-S and HR756, respectively. None of the drugs were hydrolyzed by the CESase prepared from any strain. This evidence indicates that the reason why some strains of *Ent. cloacae* are still highly resistant to the new drugs should be explained either by the poor penetrability of the outer membrane, or by the lowered susceptibility of CES-sensitive sites (murein-trans-peptidases: PC-binding proteins), or by unknown barriers directed by β-lactamase preventing the passage of drugs through the periplasmic space. Since cripticities of strains (Richmond and Curtis, 1974) highly resistant to CFX and the other drugs were not different from those of ones moderately resistant to CFX but sensitive to the others when measured with CEZ as the substrate of CESase, the first possibility seemed to be untenable. The second possibility was also untenable, because the resistant bacteria due to the alternation of susceptibility of PC-binding proteins are usually resistant to all of PC and CES without relations to β-lactamases, (Lacey, 1975, Appelbaum, Hallett, Bhamjee, Scragg, Bowen, and Cooper, 1978). The third possibility was most

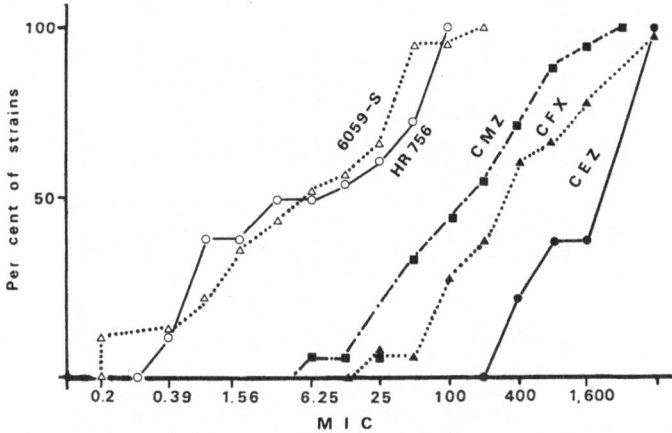

Fig. 2. Cumulative sensitivity curve of 19 clinical isolates of *Enterobacter cloacae* resistant to cefoxitin.

likely since the resistance levels of *Ent. cloacae* were almost parallel with the amounts of produced β-lactamase, even though the enzyme did not hydrolyze the drugs. Furthermore, it was confirmed that the MICs of *Ent. cloacae* strains were decreased by one half to one eight to CFX when a β-lactamase irreversible inhibitor was added with concentrations not manifesting the antibacterial activity. The results obtained indicate that the resistance of *Ent. cloacae* to the new CES derivatives is based upon an unknown barrier of periplasmic space directed by β-lactamase, and can be reduced by irreversible inactivation of the enzyme.

SHIFT OF RESISTANCE LEVELS OF *E. COLI* STRAINS BY INHERITING R PLASMIDS AND AFFINITIES OF THE NEW DRUGS TO TEM-TYPE β-LACTAMASE

The TEM-type and oxacillin-hydrolyzing β-lactamases could not hydrolyze CFX, CMZ, 6059-S and HR756 with little exceptions, although the resistance levels of 53 substrains of *E. coli* 20SO were increased to the drugs except HR756 by inheriting R plasmids specifying the enzymes, as shown in Fig. 3, i.e. 10 times higher in the average to 6059-S, 4 times to CFX and twice to CMZ. An increase of resistance levels to CEZ that is hydrolyzed by the enzymes was, of course, more significant.

The affinities of new drugs to various types of β-lactamases were measured by the acidimetric method with phenol red (Rubin and Smith, 1973) as the *Ki* values. As shown on Table I., *Ki* values of 6059-S, CFX, CMZ and HR756 were found to be 158 μM, 3,160 μM, 960 μM and 15,500 μM, respectively. It may be note worthy that the resistance to 6059-S possessing the smallest *Ki* value (highest affinity to the enzyme) was increased

335

most significantly and that to HR756 possessing the highest Ki value (lowest affinity) it was not changed by inheriting the R plasmids.

The table indicates also the Ki values of new drugs to the CESase of *Ent. cloacae*. It can be understood that the reason why some strains of *Ent. cloacae* are still resistant to the new CES derivatives is because the microbe produces the CESase possessing high binding affinity to them even though the enzyme can not hydrolyze the drugs.

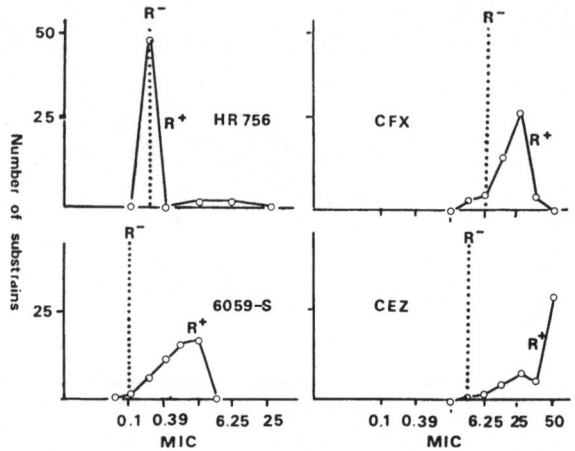

Fig. 3. Shift of sensitivity distribution of *Escherichia coli* 20SO carrying 53 various R plasmids to new β-lactamase-tolerant cephalosporins. All R plasmids were obtained from clinical isolates of *E. coli* and *Klebsiella spp.* Among them 51 plasmids and 2 plasmids encode TEM-type and oxacillin-hydrolyzing β-lactamases, respectively.

TABLE I.

Affinities of new β-lactamase-tolerant derivatives of cephalosporin to various types of β-lactamases

Enzyme		Specific acitivity*	Km**	Ki			
Types	Source			CFX	CMZ	6059-S	HR756
Ia	*Ent. cloacae* Nek 39	21.4	323.	0.002	0.04	0.0036	0.26
Ic	*P. vulgaris* GN76	4.1	111.	5.4	12.0	65.7	3,300.
IIb	*P. mirabilis* GN79	158.1	148.	6,230.	2,960.	503.	11,200.
III	*E. coli* CSH2 (RK1)	1,200.	54.	3,160.	960.	153.	15,500.
IV	*Klebsiella* GN69	84.7	61.	5,100.	1,200.	36.3	10,100.
V	*E. coli* KL 1410 (RGN328)	40.4	35.	191.	80.	5.4	120.

* Specific activities are expressed as μ moles of substrates hydrolyzed for 60 min. at 30 C per mg protein. CER and ABPC are employed as substrates for Ia and Ic, and for IIb, III, IV and V β-lactamases, respectively.

** Km values are measured with the substrates as described above.

DISCUSSION

The authors postulated again a hypothesis of unknown barrier of periplasmic space directed by β-lactamases as one mechanism of β-lactam-resistance in gram-negative bacteria. Although the resistance level of gram-positive bacteria to β-lactam drugs can be imagined from the amount and index value of $Vmax/Km$ of β-lactamase produced, it is not sufficient for the explanation of the resistance in gram-negative bacteria. Since the latter microbes produce β-lactamase as an epienzyme localizing in the periplasmic space between the outer and cytoplasmic membranes and the site of action of β-lactams exists on the inner membrane, a binding affinity of enzyme to the drugs may influence the passage of drugs through the space. Conformational changes of β-lactamases by irreversible inhibitors, such as clavulanic acid (Reading and Cole, 1977) and CP-45,899 (English, Retsema, Girard, Lynch, and Barth, 1978) will give the answer because these inhibitors can reduce the resistance level due to the binding. Details of interrelationships between the resistance and inhibitors for β-lactamase are now under investigation.

REFERENCES

APPELBAUM, P. C., A. F. HALLET, A. BHAMJEE, J. N. SCRAGG, A. J. BOWEN, and R. C. COOPER (1978): Penicillin- and chloramphenicol-resistant *Streptococcus pneumoniae* strains from clinical specimes. *In* Current Chemotherapy, Vol. 1, p. 462–464, Amer. Soc. Microbiol., Washington, D.C.

ENGLISH, A. R., J. A. RETSEMA, A. E. GIRARD, J. E. LYNCH, and W. E. BARTH (1978): CP-45,899, a beta-lactamase inhibitor that extends the antibacterial spectrum of beta-lactams: Initial bacteriological characterization. Antimicrob. Agents Chemother. 14, 414–419.

LACEY, R. W. (1975): Anbiotic resistance plasmids of *Staphylococcus aureus* and their clinical importance. Bacteriol. Rev. 39, 1–32.

POLLOCK, M. R. (1965): Purification and properties of penicillinases from two strains of *Bacillus lichenformis*: a chemical, physicochemical and physiological comparison. Biochem. J. 94, 666–675.

REDING, C. and M. COLE (1977): A beta-lactamase inhibiting beta-lactam from *Streptomyces clavuligerus*. Antimicrob. Agents Chemother. 11, 852–857.

RICHMOND, M. H. and A. C. CURTIS (1974): The interplay of β-lactamases and intrinsic factor in the resistance of gram-negative bacteria to penicillins and cephalosporins. Ann. New York Acad. Sci. 235, 553–567.

RUBIN, F. A. and D. H. SMITH (1973): Characterization of R factor β-lactamases by the acidimetric method. Antimicrob. Agents Chemother. 3, 68–73.

YOKOTA, T. and T. YAMAMOTO (1977): β-Lactamase-directed barrier of *Escherichia coli* carrying R plasmids for penicillins and cephalosporins. *In* Plasmids, Medical and Theoretical Aspects. Avicenum, Prague and Springer-Verlag, New York, p. 345–351.

YAMAMOTO, T. and T. YOKOTA (1977): Beta-lactamase-directed barrier for penicillins of *Escherichia coli* carrying R. plasmids. Antimicrob. Agents Chemther. 11, 936–940.

T.Y., *Dept. of Bacteriology,*
School of Medicine,
Juntendo University, Tokyo,
Japan

II. MEDICAL PART

D) ECOLOGY AND EPIDEMIOLOGY OF RESISTANT STRAINS

Chairman: L. ROSIVAL

STABILITY OF THE DRUG RESISTANCE OF *STAPHYLOCOCCUS AUREUS* UNDER DIFFERENT CONDITIONS

Spontaneous loss of some plasmids in experimental conditions *in vivo*

V. ZUEVA, O. DMITRENKO and YU. LINEVICH

The Gamaleya Institute of Epidemiology and Microbiology
Academy Medical Sciences USSR
Moscow, U.S.S.R.

INTRODUCTION

The use of antibiotics will undoubtedly be one of the most important weapons in the struggle against the drug resistance of microorganisms. Such a strategy must include rational use of antibiotics not only for single patients but in the whole hospital, city or even country. One of the elements of this use under modern conditions of spreading multiple drug-resistant staphylococci can be restriction of the use of some antibiotics and even a complete ban of some preparations in medical practice. Data on genetic control of the drug-resistance of staphylococci, obtained lately, enables us to provide a theoretic basis for this sensible policy.

This kind of microorganism has genes controlling the resistance to separate antibiotics which are located, as a rule, on isolated plasmids. Such a resistance may be not stable, because the plasmid genes can be lost by the microbial cell without loss of its vital functions. At the same time, information on the stability of the drug-resistance of staphylococci, even *in vitro*, is very scarce, incomplete and contradictory. In the literature there are only isolated reports on the stability of the drug-resistance of staphylococci in the organisms of experimental animals. (Borowski, 1964, Schenderov, 1974) and in the organism of man (Lacey, 1975; Annear and Baron-Hay, 1976).

MATERIAL AND METHODS

Bacterials strains: the *Staphylococcus aureus* strains and plasmids used in this study, together with phenotype of staphylococci and origins, are given in Table I.

Cultural media: nutrient broth. Hottinger agar and aseptic exudate.

Methods: Loss of plasmids *in vitro*: the cultures were grown for 18 hours at 37°C in broth. Then 0.2 ml of these cultures were transfered in ten test tubes with 1.8 ml of broth and ten tubes with 1.8 ml of aseptic exudate. All the tubes were incubated for ten days at 20–25°C.

Loss of plasmids *in vivo* (in the organism of experimental animals): the cultures were grown for 18 hours at 37°C in broth. Then suspensions of staphylococci were washed three times by centrifugation and concentrated five-fold. The suspensions obtained were injected into animals.

0.1 ml of each suspension was introduced into the tail veins of mice (18–20 g). The mice were sacrificed after ten days. Kidneys were obtained from each animal and were homogenized separately in 2 ml saline with 10% horse serum.

TABLE I.
Bacterial cells and plasmids

Strain No.	Pheno-type	Chromo-somal Markers	Plasmid	Molecular Weight ($\times 10^6$ dalton)	Number of Copies	Refe-rences
8325-I	Sm	Str	—			Novick, 1967
8325φII de	Em	—	P1258 recombined with phage φII	24*	2–5*	Novick, 1967
8325-IEM	SmEm	Str	P1258 recombined with phage φII	24*	2–5*	Novick, 1967
MS353 rms 5	Cm	—	rms 5	2.9	5–7	Inoue et al., 1976
MS353 rms 6	Tc	—	rms 6	3.0	22–25	Inoue et et al., 1976
796/75	PnCm	—	pen 796	—	—	Witte, 1977
			Chl 796	—	—	Witte, 1977

* Molecular weight and the number of copies of the initial penicillinase plasmid.

Rats were infected subcutaneously in the dorsal area with 20 ml of air and 1 ml of 10% suspension of mustard in persicorum oil. In 3–4 days, when the aseptic exudate gathered in the district of the introduced mustard, 1.8 ml exudate was obtained by individual syringe from each animal and was placed in a separate tube. These samples were used for experiments *"in vitro"*.

Then 0.5 ml of staphylococcal suspension was introduced into the inflammatory foci. The exudate was obtained in ten days and was placed in individual tubes for each animal.

The determination of the ratio of drug-sensitive cells: all samples (broth, exudate and homogeneous) were inocrelated in dilution on media without antibiotics and some isolated colonies were obtained. The plates were incubated for 18 hours at 37°C. The colonies were replicated into agar containing antibiotics: 100 mkg of erythromycin per ml (for *Staphylococcus aureus* 8325φ II de and *Staphylococcus aureus* 8325-IEm), 12.5 mkg of tetracycline per ml (for *Staphylococcus aureus* MS 353 rms6); 12.5 mkg chloramphenicol per ml (for *Staphylococcus aureus* MS 353 rms5 and *Staphylococcus aureus* 796(75); 0.1 mkg penicillin per ml (for *Staphylococcus aureus* 796(75).

The ratio of drug-sensitive cells was counted by the standard method.

The reversion of drug-resistance: the sensitive variants were stored for six months at 4°C and every month they were grown on the same medium without any antibiotics. Then their sensitivity to antibiotics was determined.

The data obtained were processed statistically to find the average error ($\pm$m) and the criteria "t" and "P". The difference was reliable when P $\leq$ 0.05.

RESULTS

I. Stability of drug resistance staphylococcal plasmids *in vitro* and *in vivo*.

Two cultural media were used to study the stability of drug-resistance *in vitro* (broth and aseptic exudate) and two experimental models were used to study the stability of drug-resistance *in vivo* (subcutaneous inflammatory focus and inflammatory process in kidneys). The experiments were carried out with *S. aureus* strain 8325ψ II de. The data obtained are shown on Figure 1.

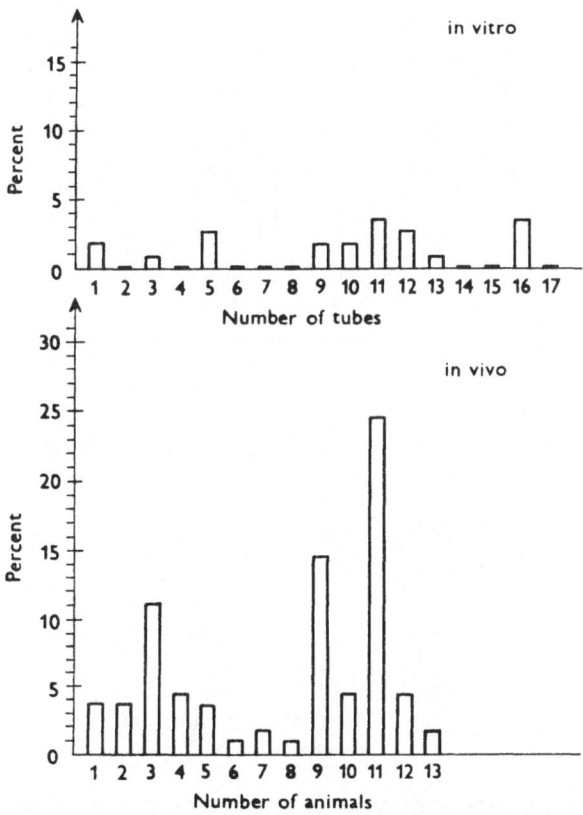

Fig.1. Frequency of appearance of spontaneous sensitive variants in the population of S. aureus 8325 μ II de persisting in exudate in tubes and in the substaneous inflammatory focus in rats. Conventional signs: poles show the number of erythromycin-sensitive cells in percentage related to all the cells in the population obtained from separate tubes or animals.

It was found that the spontaneous erythromycin-sensitive variants were found in only seven out of ten tubes with broth and their number fluctuated from 0.9 to 1.8%. Erythromycin-sensitive cells were found in nine out of sixteen tubes with exudate and their number obtained a value of 3.6% in some tubes.

Erythromycin-sensitive variants were found in the subcutaneous inflammatory lesions of all animals and their number fluctuated from 2 to 25%. The sensitive variants appeared more frequently in the inflammatory focus of kidneys: seven of sixteen mice had over 10% sensitive cells. The number of such cells in some animals varied from 1% to 49%.

The difference between the average number of sensitive cells in broth (0.9 ± 0.2) and in exudate (1.2 ± 0.3) *in vitro* was not reliable. However, sensitive variants were accumulated in exudate in vivo more intensively than in exudate *in vitro* with probability

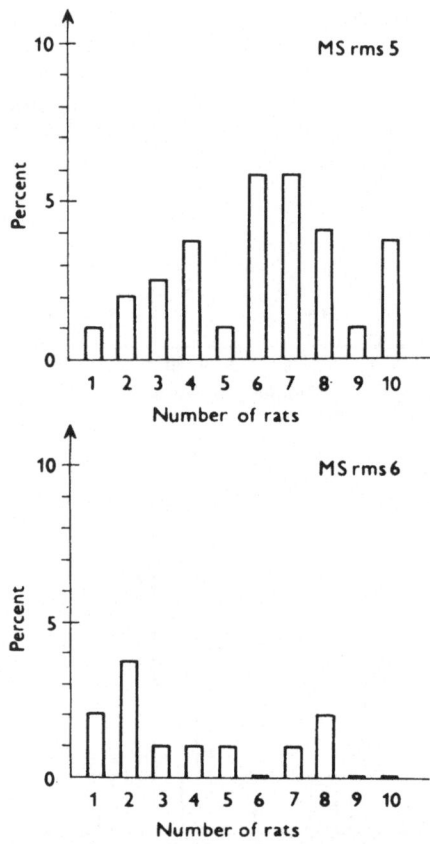

Fig. 2. Frequency of appearance of spontaneous variants sensitive to chloramphenicol or tetra-cycline in the populations of strains S. aureus MS 353 rms 5 and S. aureus MS 353 rms 6 per-sisting in the subcutaneous inflammatory focus in rats.
Conventional signs: poles show the number of sensitive cells in percentage related to all the cells in the population obtained from separate animals.

P = 0.01. The average number of sensitive cells found in kidney homogenotes (9.5 ± ± 2.6) was larger than that found in the exudate (6.5 ± 2.0), but the difference was not statistically significant.

II. The dependence of the stability of drug resistant plasmids from plasmid and from the *Staphylococcus aureus* strain.

A. The stability of two different plasmids in the same strain of *S. aureus* MS 353 was studied in the first series of experiments: one variant of the strain

344

had a plasmid which determined resistance to tetracycline, the other — to chloramphenicol. The data obtained are shown in Figure 2, where it is demonstrated that all the experimental animals had chloramphenicol-sensitive cells and their number fluctuated from 1 to 5.8%. Only 70% of the animals had tetracycline-sensitive cells and their number was less than 3.8%. The difference between the average number of chloramphenicol-sensitive and tetracycline-sensitive cells was statistically reliable. The determined chloramphenicol resistance of the plasmid was less stable than the determined penicillin resistance of plasmid in *S. aureus* 796/75. Both plasmids were present in this strain. All the animals

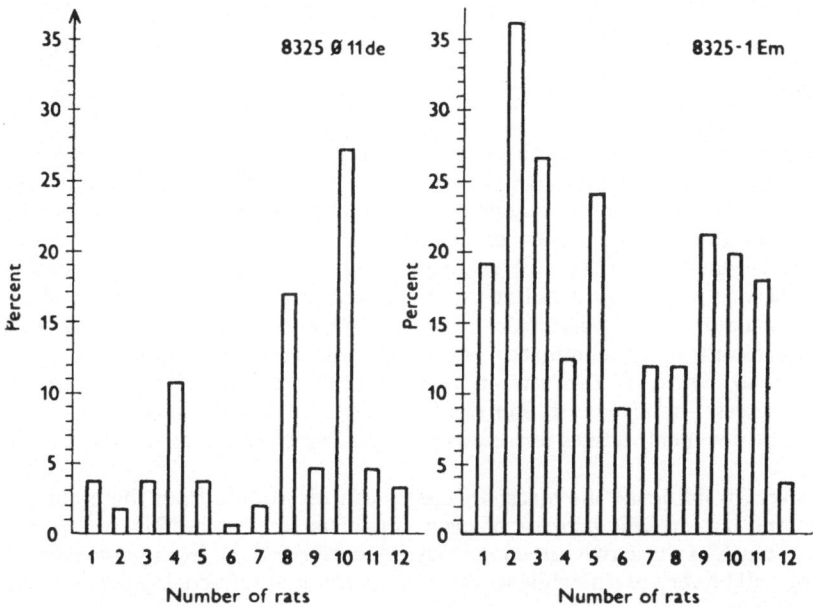

Fig.3. Frequency of appearance of spontaneous erythromycin-sensitive variants in the populations of strains S. aureus 8325 φ II de and S. aureus 8325-IEm persisting in the subcutaneous inflammatory focus in rats.
Conventional signs: poles show the number of sensitive cells in percentage related to all the cells in the population obtained from separate animals.

infected by a suspension of *S. aureus* 796/75, had chloramphenicol-sensitive cells and their number fluctuated from 1 to 10.8%. Only 40% of the animals had penicillin-sensitive cells and their number was less than 1%. The difference between the average number of chloramphenicol-sensitive and penicillin-sensitive cells was statistically significant (P = 0.001).

B. The stability of the same plasmid was compared in two strains in the second series of experiments: *S. aureus* 8325-ψII de and *S. aureus* 8325-IEm, which contained a defective penicillinase plasmid, were used to study this problem. The data obtained are shown in Figure 3. More than 10% of erythromycin-sensitive cells were found in the population of *S. aureus* 8325-IEm in the inflammatory focus of 80% of the animals. The same number of sensitive cells was found in the population of *S. aureus* 8325-φII de in the inflammatory focus of only 25% of animals. Thus the element φII de was more stable in *S. aureus* 8325 φII de (P = 0.001).

The reversion of resistance of six erythromycin-sensitive variants of *S. aureus* 8325-φ II de, of ten chloramphenicol-sensitive variants of *S. aureus* MS 353 rms 6, of two tetracycline-sensitive variants of *S. aureus* MS 353 rms 5 and of five penicillin sensitive variants of *S. aureus* 796/75 was studied. All the studied variants kept their sensitivity to the corresponding antibiotic.

Thus the experiments revealed a reliably much higher frequency of appearance of spontaneous plasmid-negative variants *in vivo* than *in vitro*. In addition, *in vivo* the stability of the staphylococcal resistance did not depend on the location of the inflammatory focus, but did depend on the plasmid, microbial cell and the organism of the animal.

DISCUSSION

At present it is difficult to make a more or less well-founded proposal for the mechanism of the greater instability of drug-resistance of *S. aureus in vivo*. Even without knowledge of the mechanism of this phenomenon, it is, however, safe to say that this fact has great practical significance. This process can be realized in two directions: 1) to influence the microbial cell and the resistant determinants, and 2) to influence the macroorganism. The former was studied in detail at the beginning of the sixties after the discovery of plasmid resistance of microorganisms. Among different groups of chemical compounds some were found which eliminated plasmids. However, the frequency of elimination was very small. Thus, for the last 5–8 years there has been a sceptical attitude towards these substances. In this connection the high intensity of spontaneous loss of plasmids *in vivo* opens new perspectives for seeking eliminating compounds directly on the model *in vivo*.

The latter (to influence the macroorganism) can be studied from the point of view of a specific and non-specific increase of macroorganism resistance. Now it is even difficult to suggest which of these two directions may be more effective. In any case, scientific work in this one will be very useful while working on methods of influencing the drug-resistance of staphylococci.

REFERENCES

ANNEAR, D. I. and G. S. BARON-HAY (1976): Med. J. Aust. I(12), 399–400.
BOROWSKI, J. (1964): Can. J. Microbiol. 10, 4, 595–603.
INOUE, M., H. OSHIMA, T. SAITO, T. OCUBO and S. MITSUHASHI (1976): Virology 73, 295–298.
LACEY, R. W. (1975): Bacteriol. Rev. 39, 1, 1–32.
NOVICK, R. P. (1967): 33, 155–156.
SCHENDEROV, B. A. (1974): Zh. Microb. Epid. Immunol. I, 148–150.
WITTE, W. (1977): Zbl. Bact. Ayg., 1 Abt. A 237, 147–159.

*V.S.Z. Gamaleya Epidemiology and
Microbiology Institute, Moscow, U.S.S.R.*

STUDIES ON THE ROLE OF COL V PLASMID IN THE PATHOGENICITY OF *E. COLI* STRAINS

H. MILCH and É. CZIRÓK

National Institute of Hygiene, Budapest, Hungary

INTRODUCTION

A severe outbreak of meningitis and enteritis was observed in a premature infants' ward in the autumn of 1974. It was shown to be associated with *E. coli* 0:78 (Czirók, Milch, Madár and Semjen, 1977). *E. coli* 0:78 strains were isolated from newborn infants' wards, nurseries, patients, healthy carriers and from diseased animals. The strains, different in origin and pathogenicity, were uniform in antigenic structure and phage pattern, but they differed in colicinogenicity (Milch, Czirók, Madár and Semjen, 1977). The epidemic strains and calf-pathogenic strains produced colicin V, the strains isolated from sporadic cases of infants and from healthy carriers were characterized by other types of colicin or were not colicinogenic. Mouse pathogenicity was determined for the strains carrying different Col and R plasmids, and the correlation between the plasmid carrier state and pathogenicity was examined by elimination.

MATERIAL AND METHODS

Strains examined: 27 strains isolated from diseased and healthy infants were examined for R and Col plasmid carriage.

Serological examinations: were performed as described previously (Czirók, Milch, Madár and Semjen, 1977).

Colicin sensitivity was tested with Frédéricq's colicin producing strains (Frédéricq, 1948).

Antibiotic sensitivity to chloramphenicol, ampicillin, streptomycin, tetracycline, neomycin, kanamycin, colistin, polymyxin B, nalidixic acid and gentamicin was examined with "RESISTEST" disks (Institute for Serobacteriological Production and Research Human, Budapest).

Conjugation was performed as described previously (Milch, Gyenes and Hérmán, 1974).

Recipient strain to detect R and Col plasmids:
E. coli HfrH lac⁻ nal^r and for Col plasmid also
E. coli PA J09 Sm^r L⁻ F⁻.

Test method for phage restriction and the phages used were described previously (Milch, Gyenes and Hérmán, 1975).

Culture medium. After conjugation the cultures were plated on eosin-methylene blue medium containing antibiotics at inhibitory concentrations: nalidixic acid (50 μg/ml), tetracycline (50 μg/ml), ampicillin (50 μg/ml), chloramphenicol 25 μg/ml), kanamycin (20 μg/ml) ans streptomycin (50 μg/ml).

Mouse pathogenicity was examined by intraperitoneal injection of CFLP mice weighing 16–18 g (Smith, 1974).

Elimination experiments. For plasmid elimination acriflavine (Hirota, 1960), acridine orange (Kahn and Helinski, 1964), sodium dodecyl sulphate (Tomoeda Umuzuka, Kubo and Nakamura, 1968) and ethidium bromide (Bouanchaud, Scavizzi and Chabbert, 1969) were used. Elimination by ethidium bromide was supplemented with UV irradiation (Watanabe and Fukasawa, 1961) and mitomycin C treatment (Hardy and Meynell, 1977). Eliminations were performed with agents of different concentrations in broth inoculated with *E. coli* 078 culture using 10^4 cells/ml. After overnight incubation at 37 °C samples were plated on agar for 24 hr on 3 to 5 occasions. Colonies were replicated to agar plates containing the adequate antibiotics for testing R plasmids, and to antibiotic free medium for testing colicin production.

RESULTS

Preliminary mouse-pathogenicity experiments. (Determining of LD_{50} values). The difference in mouse-lethality was examined between Col V$^+$ and Col V$^-$ strains. Each group of 5 mice was injected intraperitoneally with 14 strains of 078 Col V$^+$ and 13 of 078 Col V$^-$.

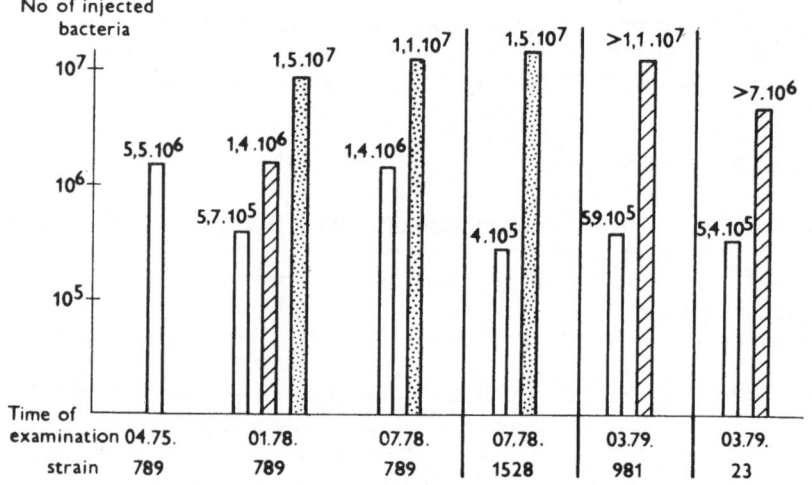

Of the 70 mice injected with Col$^+$ strains 50 died within 8 days, of the 65 injected with Col V$^-$ strains 15 died. The mouselethality of the Col V$^-$ strains was significantly lower (p < 0.1%).

Elimination experiments were carried out with four R$^+$ Col V$^+$ strains (designations: 1528, 789, 981, 23 [Rivie]) which showed high lethality in the preliminary mouse pathogenicity test. Elimination of plasmids failed with acridine orange, acriflavine, sodium dodecyl sulphate, it was efficient only with ethidium bromide supplemented with UV irradiation and mitomycin C treatment. Eliminants are shown on Table I. In the strain 789 – R$^+$, resistant to Cm, Tc, Ap, Col V$^+$ – Col plasmid and Cm determinant became eliminated. In the strain 981 – R$^+$, resistant to Cm, Tc, Ap, Col V$^+$ – Col V plasmid could not be eliminated, but of the examined 1128 colonies Cm and Tc resistance was lost in 375 colonies. The MIC value controlled by the retained Ap determinant was much

348

TABLE I.

E. coli 0:78 strains carrying Col V and R plasmids and their eliminants

Designation of untreated and eliminated strains	Antibiotic resistance	R plasmid	Col V plasmid	Treatment	No. of clones examined
789	Cm Tc Ap	R+	Col V+	—	1000
789/17	Tc Ap	R+	Col V+	Ethidium bromide 200 µg/ml - U.V.	
789/12	Tc Ap	R+	Col−	,,	
789/14	Tc Ap	R+	Col−	,,	1042
789/38	Tc Ap	R+	Col−	,,	
789/35 (freeze-dried)	Tc Ap	R+	Col−	Ethidium bromide 100 µg/ml	
1528	Cm Tc Ap	R+	Col V+	—	1000
1528/3	Cm Tc Ap	R+	Col−	Ethidium bromide 50 µg/ml	
1528/52	Cm Tc Ap	R+	Col−	,,	
1528/76	Cm Tc Ap	R+	Col−	,,	1728
1528/89	Cm Tc Ap	R+	Col−	,,	
981	Cm Tc Ap	R+	Col V+	—	800
981/46	Ap	R+	Col V+	Ethidium bromide 500 µg/ml	1128
23	Sm Tc	R+	Col V+	— .	100
23/1	Sm	R−	Col V+	Ethidium bromide 100 µg/ml	149

lower (MIC value: 30 µg/ml) than the level of the original ampicillin resistance (MIC value: 1600 ng/ml). The low level ampicillin resistance was also transferable. The possibility of the presence of two R plasmids was supported by phage restriction experiments too. Experiments are in progress to prove the presence of the two R plasmids.

In the strain 23 – R+ Col V+ – Col plasmid also did not become eliminated, but the original Sm, Tc resistant strain became only Sm resistant and this resistance was not transferable.

Mouse pathogenecity examinations following elimination.

Figure 1. shows the LD_{50} values of the strains examined. The LD_{50} values of the original strains varied between $4.10^5 - 2.3.10^6$. (LD_{50} of strain 789 was examined at different periods). During storage in agar stock culture for 3 years the LD_{50} value decreased. The LD_{50} level in the freeze dried condition was the same as three and a half years earlier. LD_{50} of the eliminant 789, which lost Cm resistance, increased from $5.7.10^5$ to $1.4.10^6$. Joint elimination of Cm resistance and Col V plasmid resulted in a greater increase in the LD_{50} value ($1.6.10^7$). These findings were reproducible for the 1528

TABLE II.

Differences of LD$_{50}$ values of the original and eliminated Strains

Designation of strain	Original strain		Eliminant strain		Significance
	Injected bacteria	No. of mice d/i	Injected bacteria	No. of mice d/i	
R789	$1.5 . 10^6$	11/20	$9 . 10^5$	0/20	p < 0,001
R1528	$2.8 . 10^6$	14/20	$2.1 . 10^6$	10/80	p < 0.001
981	$8.3 . 10^4$	10/20	$1.3 . 10^5$	0/20	p < 0.001
23	$1.6 . 10^5$	12/20	$1.3 . 10^5$	1/20	p < 0.001

d = died
i = infected

strain, having lost Col V plasmid. In case of eliminants with the same properties, the averages of LD$_{50}$ values of the different inocula are given. The LD$_{50}$ value of strain 981 probably carrying two types of R plasmids and Col plasmid increased from $5 . 9 . 10^5$ to $1 . 1 . 10^7$, as it has lost one of its R plasmids (resistance to Cm, Tc and high level resistance to Ap).

☐ R$^+$ (Cm Tc Ap) Col V$^+$

▨ R$^+$ (Tc Ap) Col V$^+$

▩ R$^+$ (Tc Ap) Col V$^-$

☐ R$^+$ (Cm Tc Ap) Col V$^+$

▩ R$^+$ (Cm Tc Ap) Col V$^-$

☐ R$^+$ (Cm Tc Ap) Col V$^+$

▨ R$^+$ (Ap) Col V$^+$

☐ R$^+$ (Sm Tc) Col V$^+$

▨ R$^-$ (Sm) Col V$^+$

In the course of elimination, strain 23 retained the Col plasmid and lost the R plasmid which mediated resistance to tetrycycline, so that the LD$_{50}$ values increased from $5 . 4 . 10^5$ to $7 . 1 . 10^6$. Further experiments were aimed at determining whether the observed differences were significant.

Groups of 20 mice were injected with a dose corresponding to the LD$_{50}$ value of each of the untreated strains $(1 . 5 . 10^6, 2 . 8 . 10^6, 8 . 3 . 10^4, 1 . 6 . 10^5)$. The data were analysed using the 2×2 contingency table and examined for significance by the X^2 test. The lower lethal effect observed in the group injected with the eliminant strain was significant (Table II.).

DISCUSSION

A significant difference was observed in mouse lethality when these were injected with Col V$^+$ and Col V$^-$ strains isolated from patients and healthy carriers. It was our aim to confirm the connection between the pathogenicity and Col plasmid by elimination experiments. In two strains Col plasmids were eliminated, in two strains R plasmids, and in all cases the LD$_{50}$ values of eliminants increased significantly.

Our findings on the pathogenic role of Col plasmids are in partial agreement with those of Smith and Huggins (1978), who demonstrated increased pathogenicity in chicken by curing the Col V plasmid with lauryl sulphate in the Col V producing invasive *E. coli* strains.

Those results, however, which indicate an LD$_{50}$ value increase of Col V plasmid-retaining and resistance determinants lost eliminants complicate the problem and suggest the presence of a pathogenicity influencing plasmid that has not yet been demonstrated by genetic studies.

REFERENCES

BOUANCHAUD, D. H., M. R. SCAVIZZI, Y. A. CHABBERT (1969): Elimination by ethidium bromide of antibiotic resistance in *Enterobacteruaceae* and *Staphylococcus aureus*. J. gen. Microbiol. 54, 417–425.

CZIRÓK, É., H. MILCH, J. MADÁR and G. SEMJÉN (1977): Characterization of *Escherichia coli* serogroups causing meningitis, sepsis and enteritis. Acta microbiol. Acad. Sci. hung. 24, 115–126.

FRÉDÉRICQ, P. (1948): Actions antibiotiques réciproques chez les *Enterobacteriaceae*. Rev. belge Path. méd. exp. 19. Suppl. 4. 1–5.

HARDY, K. G., G. G. MEYNELL (1972): Induction of colicin factor E2-P9 by mitomycin C. J. gen. Microbiol. 73, 547–549.

HIROTA, J. (1960): The effect of acridine dyes on mating type in *Escherichia coli*. Proc. Natl. Acad. Sci. U.S. 46, 57–64 (1960).

KAHN, Ph., D. R. HELINSKI (1964): Relationship between colicinogenic factors El and V and F factor in *Escherichia coli*. J. Bacteriol. 88, 1573–1579.

MILCH, H. É. CZIRÓK, J. MADÁR and G. SEMJÉN (1977): Characterization of *Escherichia coli* serogroups causing meningitis, sepsis and enteritis. Acta microbiol. Acad. Sci. hung. 24, 127, 137.

MILCH, H., M. GYENES and G. HÉRMÁN (1975): Characterization on the basis of phage restriction of R plasmids from *Escherichia coli* strains. In: Drug inactivating enzymes and antibiotic resistance (Eds. S. Mitsuhashi, L. Rosival and V. Krčméry, pp. 391–396.

SMITH, H. W. (1974): A search for transmissible pathogenic characters in invasive strains of *Escherichia coli*: the discovery of a plasmid-controlled toxin and a plasmid-controlled lethal character closely associated, or identical, with colicine V. J. gen. Microbiol. 83, 95–111.

SMITH, H. W., M. B. HUGGINS (1978): The effect of plasmid-determined and other characteristics on the survival of *Escherichia coli* in the alimentary tract of two human beings. J. gen. Microbiol. 109, 375–379.

TOMOEDA, M., M. INUZUKA, N. KUBO and S. NAKAMURA (1968): Effective elimination of drug resistance and sex factors in *Escherichia coli* by sodiumdodecylsulfate. J. Bacteriol. 95, 1078–1089.

WATANABE, T., T. FUKUSAWA (1961): Episome mediated transfer of drug resistance in the *Enterobacteriaceae*. II. Elimination of resistance factors with acridine dyes. J. Bacteriol. 81, 679–683.

H.M., OKI Institute
Gyáli út. 2–6,
Budapest, Hungary

STUDIES ON R AND COL PLASMIDS
IN SHIGELLA SONNEI STRAINS FREQUENTLY ISOLATED
FROM OUTBREAKS

A. HAJNAL, V. LÁSZLÓ, I. FINANCSEK, and I. STRAUB

*Pest County Public Health Station, Budapest, National Institute of Hygiene,
Hungary and F.J.C. National Researce Institute for Radiobiology and Radiohygiene,
Budapest, Hungary*

INTRODUCTION

In many parts of the world, *S. sonnei* has become the predominant species of *Shigella* (Center for Disease Control, 1975, Kostrzewsky and Stiputkowska-Misiurewicz, 1968).

This species has been predominant in Hungary since 1975, before that time *S. sonnei* prevalence was observed here first in 1968 (László, Milch and Madár, 1970). The phage type 2 was dominant in Hungary in 1975 (it occurred in 30%, of 2773 strains examined). In 1978 phage type 2 was found in 49,5% of the examined strains (2596). Simultaneously, an increase was observed in the occurrence of colicine type 12, streptomycin and sulphonamid resistant strains (23.1% of all typed strains). This type was also prevalent among the strains causing epidemics mainly during seasonal periods. In this study, the plasmids of these epidemiologically important strains are examined.

MATERIAL AND METHODS

Strains

Donors: 42 *S. sonnei* strains isolated from 5 different epidemics in 1978 (phage type 2, colicine type 12, streptomycin and sulphonamid resistant).

Recipients: *E. coli* J 5-3 rifampicin resistant, *E. coli* K12 HfrH nalidixic acid resistant.

Indicators: from dr. Kallings (*S. sonnei* 4, 15, 22, 24, 28, 31, 33, 41, 53, 55) *E. coli* K12 ROW Colicin standards: from dr. Fredericq (Ca 31, Ca 18, K 53, Ca 42, 185 M4 (E3)b/23, Ca 46, Ca 58, Ca 53, MR 2, K 49, P 15, 32T 19/VT 5,1-7a, 18511/S7a, Ca 62, P 9, Ca 38).

Phages

λ, ∅ 2, T_1–T_7, *E. coli* phage-set (Milch and Gyenes, 1972), MS2, f_2 male specific phages.

Culture media

Eosin methylene blue agar containing streptomycin (30 μg/ml) and nalidixic acid (50 μg/ml), nutrient broth agar complemented with rifampicin 250 μg/ml and streptomycin (30 μg/ml) were used as selective media.

Crosses

The 3 hr broth cultures of the donor and recipient strains were centrifuged, subsequently the sediments were mixed in a ratio of 10:1 and plated on agar plates. After incubation at 37°C for 3 hr the growth was suspended in broth (2 ml), and 0,1 ml of the undiluted mixture was plated on selective media and incubated at 37°C for 18 hr or 42 hr. The colonies were subcultured three times on agar plates supplemented with adequate antibiotics.

Determination of fi character

Each R plasmid was classified according to inhibition of pili synthesis after transfer to *E. coli* K 12 HfrH, tested by visible lysis of phages MS2 and f_2 in HfrH R$^+$ culture.

Phage restriction

Changes of lysis pattern of R$^+$ and R$^+$Col$^+$ transconjugant *E. coli* were examined compared to recipient R$^-$Col$^-$ strain using the phages mentioned above.

Colicin sensitivity

The strains producing known colicins were dropped on agar plates and incubated at 37°C for 18 hr. Cultures were exposed to chloroform vapour for 30 min and then over-layered by two-hour broth cultures of Col$^+$ transconjugant strains in soft agar (0.5 ml: 4.5 ml). The inhibition zones were determined after overnight incubation at 37°C.

Lysogeny

Eleven indicator strains were used for testing lysogeny. The test strains were treated with 0.1 µg/ml mitomycin C at 37°C for 30 min, followed by centrifugation for 20 min at 4000 rpm (Janetzki T 23). The sediments were suspended in 5 ml nutrient broth incubated at 37°C for 1 hr and centrifuged (20 min, 4000 rpm). The supernatants were plated on indicator strains and incubated at 37°C for 18 hr.

Compatibility test

This test was carried out according to Datta and Hedges (1971).

TABLE I.

Lysis pattern of transconjugant E. coli with E. coli phage-set* and with T1-T7, λ, φ2 phages

Outbreaks	No. of strains		Phages																		
	R$^+$	R$^+$Col$^+$	λ	φ2	T1-T7	2	3	4	4a	4b	6	7	12	14	15	16	17	18	23	24
Hajdúszo-boszló	17	14	+*	+	+	+	+	+	+	+	+	+	−	+	+	+	+	+	+	+
Szolnok	14	10	+	+	+	+	+	+	+	+	+	+	+	+	+	+	+	+	+
Budapest IX.	4	2	+	+	+	+	+	+	+	+	+	+	+	+	+	+	+	+	+
Budapest II.	3	3	+	+	+	+	+	+	+	+	+	+	+	+	+	+	+	+	+
Karanc-salja	4	4	+	+	+	+	+	+	+	+	+	+	+	+	+	+	+	+	+
Recipient strain	−	−	+	+	+	+	+	+	+	+	+	+	+	+	+	+	+	+	+

+ lysis

* only the phages indicated which caused lysis

Agarose gel electrophoresis

The electrophoresis was performed according to Meyers, Sanchez, Elwell and Falkow (1976), modified by Financsek (1979).

RESULTS AND DISCUSSION

The 42 *S. sonnei* strains selected from 5 different outbreaks carried fi⁻ R plasmids on the basis of the lysis by MS2, f_2 male-specific phages. The Col plasmids could be transferred from 33 strains together with R plasmid. Phage restriction mediated by R and Col plasmids was analysed in transconjugants. The R plasmids, R and Col plasmids, respectively, failed to cause phage restriction with phages λ, $\varnothing_2$ T_1–T_7 and *E. coli* type phages.

TABLE II.

Sensitivity of Col⁺ transconjugant E. coli strains to known colicins

Outbreaks which are the sources of Col plasmids	A	B	E1	E2	E3	G	H	Ia	Ib	K	M	S	V	X	J+Ia	S3+Ib	E+I
Hajdúszoboszló	+	+	+	+	+	+	+	±	+	+	+	+	+	+	+	+	+
Szolnok	+	+	+	+	+	+	+	±	+	+	+	+	+	+	+	+	+
Budapest IX.	+	+	+	+	+	+	+	±	+	+	+	+	+.	+	+	+	+
Budapest II.	+	+	+	+	+	+	+	±	+	+	+	+	+	+	+	+	+
Karancsalja	+	+	+	+	+	+	+	±	+	+	+	+	+	+	+	+	+

+ inhibition
± weak inhibition

TABLE III.

Lysis spectra of temperate phages liberated from lysogenic S. sonnei strains

Outbreaks, No. of strains	Indicator strains										
	S. sonnei										E. coli
	4	15	22	24	28	31	33	41	53	55	K12
Hajdúszoboszló 17	—	+	+	—	+	+	—	+	—	—	+
Szolnok 14	—	+	+	—	+	+	—	+	—	—	+
Budapest IX. 4	—	+	+	—	+	+	—	+	—	—	+
Budapest II. 3	—	+	+	—	+	+	—	+	—	—	+
Karancsalja 4	—	+	+	—	+	+	—	+	—	—	+

+ lysis
— no lysis

355

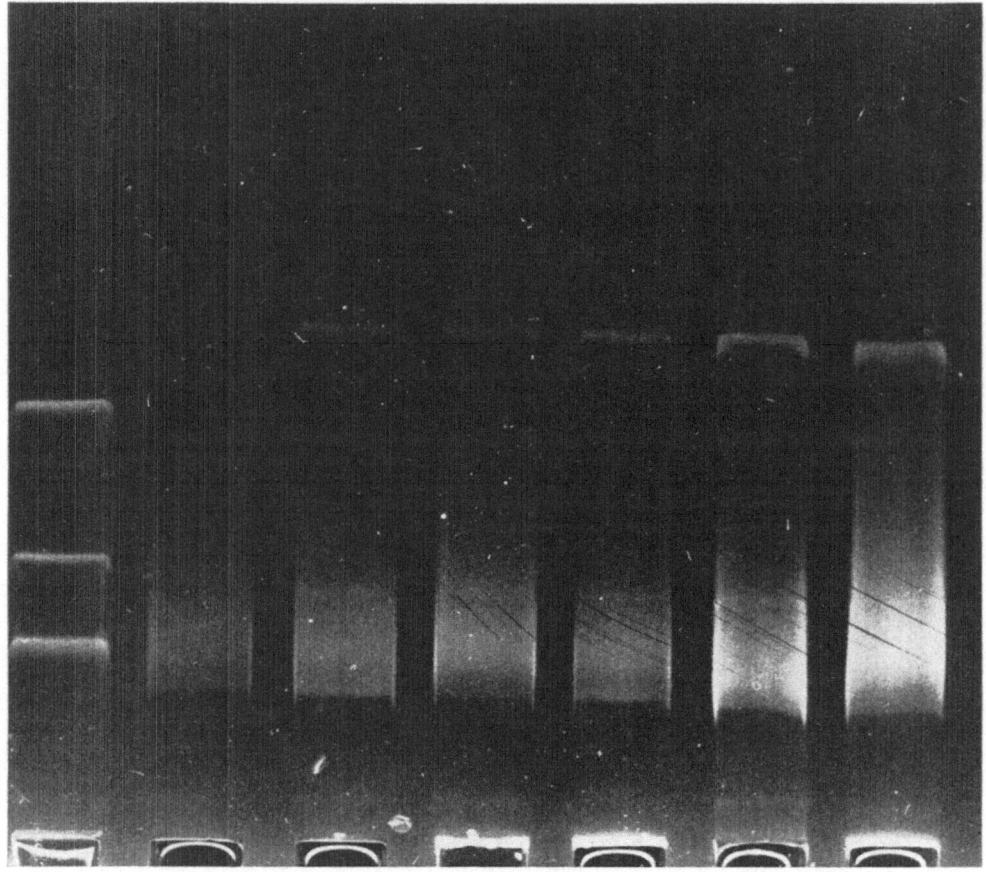

Fig. 1   A-E Plasmids from five strains of five different outbreaks.
           F   Recipient strain without plasmids.
           G   pCRI plasmid m.w. 8.7×10^6.

The colicin sensitivity of col$^+$ transconjugant strains was tested using colicins A B E$_1$ E$_2$ E$_3$ G H Ia Ib K M S V X J$+$ Ia S3 $+$Ib E $+$ I.

The colicin type 12 of the Abbott-Shannon system corresponds to the production of colicin E3 (Horák, 1975). In contrast the strains examined in this study were sensitive to all of the colicins. The sensitivity was moderate to colicin Ia. As regards the sensitivity for the colicins, the strains behaved identically. The strains examined were lysogenic on the basis of phage liberation experiments by mitomycin C. The liberated phages (temperated phages) were checked by 11 indicator strains. The results showed the homogeneity of the phages according to their host range.

Using agarose gel electrophoresis we could demonstrate two plasmids in the strains. The smaller one is probably the col plasmid itself of about 7.0–7, 7 $\times$ 10^6 Dalton compared to pCRI standard plasmid (8.7 $\times$ 10^6 Dalton). The bigger plasmid probably is identical with the R plasmid. The plasmids from different strains seem to be identical on the basis of agarose gel electrophoresis.

According to phage liberations, colicin production and antibiotic resistance examinations there are at least three plasmids in the strains and they are identical in different strains.

Compatibility properties were tested in one strain from each of the five outbreaks. The plasmids were related to I, N and H compatibility groups.

Accordingly, the strains of different origin are uniform even by compatibility of their plasmids.

We have concluded that randomly selected strains which are homogenous in phage type, colicin type and antibiotic resistance are uniform in the harboured plasmids as well. The results indicate that all strains of phage type 2, colicin type 12, resistant to streptomycin and sulphonamid, may be genetically homogenous.

REFERENCES

CENTER FOR DISEASE CONTROL (1975): – *Shigella* surveillance report no. 36, 1st and 2nd quarters. Center for Disease Control, Atlanta.
DATTA, W. and R. W. HEDGES (1971): Compatibility groups among fi⁻R factors. Nature (London) 234, 222–223.
FINANCSEK, I. (1979): (Personal communication.)
HORÁK, V. (1975): Typing of Shigella sonnei colicins by means of specific indicators. Zbl. Bakt. Hyg., I. Abt. Orig. A 233, 58–66.
KOSTRZEWSKI, J. and H. STIPUŁKOWSKA-MISIUREWICZ (1968): Changes in the epidemiology of dysentery in Poland and the situation in Europe. Arch. Immunol. Ther. Exp. 16, 429–451.
LÁSZLÓ, V., H. MILCH and J. MADÁR (1970): Some characteristics of the dysentery in Hungary, 1968, revealed by phage typing. Egészségtudomány 14, 274–284.
MEYERS, J. A., D. SANCHEZ, L. P. ELEWELL and S. FALKOW (1976): Simple agarose gel electrophoretic method for the identification and characterization of plasmid DNA. J. Bacteriol. 127, 1529–1538.
MILCH, H. and M. GYENES (1972): Subdivision and correlation studies of serologically grouped *Escherichia coli* strains by phage typing. Acta Microbiol. Acad. Sci. Hung. 19, 213–244.

A. H., Gyáli út. 2–6,
Budapest, Hungary

R PLASMID ECOLOGY IN SALMONELLA IN HUMAN GUT

J. C. PALOMARES and E. J. PEREA

Department of Microbiology and Medicine Preventive University Hospital,
School of Medicine, University of Sevilla, Sevilla, Spain

Our paper presents a study of antibiotic resistance and R Plasmids in the intestinal aerobic microflora of 19 patients with Salmonellosis. In each patient we isolated the *Salmonella* and the *Enterobacteriaceae* present in faeces.

In each isolate we studied the resistance pattern, conjugative ability with *E. coli* and the plasmid's molecular weight by agarose gel electrophoresis.

Seven (37%) *Salmonella* isolates were drug resistant. Six of them had transferable plasmids. Three had plasmids coding for Sm^rTc^r, with a molecular weight of 30 Mdal (2) and 65 Mdal (1). Two were Km^rTc^r, with a molecular weight of 46 and 66 Mdal, respectively, and one was $Sm^rAm^rTc^r$ with a molecular weight of 36 Mdal. One had a non-transferable plasmid with a molecular weight of 48 Mdal.

The *Enterobacteriaceae* isolated were *E. coli* (21), *K. pneumoniae* (7) and *P. vulgaris* (2). 20 (66%) had transferable drug resistance.

The same transferable resistance pattern (Sm^rTc^r) was found simultaneously in *Salmonella* and *E. coli* in only one patient. One plasmid of 30 Mdal was found both in the *Salmonella* and E. coli strains.

We found 5 plasmids with $Sm^rAm^rTc^r$ and two with Sm^rTc^r with similar resistance patterns and molecular weight to the *Salmonella* plasmids but in different patients.

The remaining plasmids had from 1 to 6 drug resistances (Sm^r and $Km^rSm^rCm^rAm^r$ Tc^rSu^r) and a molecular weight of 30 Mdal to 80 Mdal.

Thus the resistance pattern of plasmids is broader in non-pathogenic bacteria than in *Salmonella* and plasmids occur more frequently in non-pathogenic bacteria. Plasmid transfer is not difficult *"in vitro"* between *Salmonella* and *E. coli* but there must be other factors acting in its transfer *"in vivo"*.

Introduction

Drug resistance and R plasmids have been found to be widespread among pathogenic bacteria, e.g. *Salmonella* and *Shigella*, and among other species which are primarily non-pathogenic for man. (S. Falkow, 1975, Mitsuhashi, 1977).

Because of the clinical and public health implications of R-factors, data on their prevalence and ecology are of great importance. Widespread epidemics caused by *Salmonella* have become relatively less common due to improvements in sanitation and water-quality control, but there is still a high incidence of some forms of Salmonellosis, notably gastroenteritis, and the spread of strains with acquired R factors is now a problem (Falkow, 1975). This possibility makes the study of the origin and ecological relationships between *Salmonella* R factors and those present in the bacterial intestinal flora of man of prime importance.

To date there does not seem to have been any survey of R plasmids in human intestinal flora on a community scale. The present study was carried out to investigate the presence of R plasmids in the intestinal flora of patients with Salmonellosis.

We have tested for the presence of R plasmids both in the *Enterobacteria* and in the *Salmonella* isolated from the faeces of each of 19 patients with Salmonellosis. The bacteria isolated were tested for resistance to antibiotics and the existing R plasmids were characterized by molecular weight (M.W.), fi character and enzyme restriction analysis in order to determine the relationships between plasmids of pathogenic and of non-pathogenic intestinal flora in these patients.

MATERIAL AND METHODS

Bacteria. — 19 *Salmonella* (13 *S. typhimurium*, 2 *S. typhi*, 2 *S. paratyphi* B and 2 *S. enteriditis*), 21 *E. coli*, 7 *Klebsiella pneumoniae* and 2 *Proteus vulgaris*, all isolated from 19 salmonellosis patients in the Bacteriology Department of the University Hospital of Seville.

- *E. coli* K12 E711 *lac⁻, his, pro, trip.* nalidixic acid resistant (Perea and Daza, 1974).
- *E. coli*: A rifampicin-resistant mutant of HfrC *met* (Perea and Daza, 1974).
- *E. coli* K12 C⁻ Thy⁻ provided by N. Datta.

Antibiotic Sensitivity Tests. — The antibiotic sensitivity of all the strains was determined as described previously (Perea and Daza, 1974).

Transfer of R factors by conjugation. — The method used was described previously (Perea and Daza, 1974) and the mating procedure was always carried out at both 25°C and 37°C.

fi character. — The method of Coetzee et al., 1972, was used to determine this character.

Molecular weight (M.W.) determination. — The method described by T. Eckhard, 1978, was used.

Isolation of supercoiled DNA. — We used a method described previously (Perea and Daza, 1974).

Restriction Enzyme Digestion. — Restriction enzyme, *Eco*R1, was obtained from Miles Biochemicals Ltd, and the digestion buffers were those recommended by Miles. The procedure was that of N. Datta (private communication). The enzyme reaction was carried out at 37°C in a total volume of 30–50 ml. The enzyme reaction wad determined by heating to 65°C for 10 min. Glycerol and bromophenol blue were added to final concentrations of 5% and 0,5%, respectively, ready for application to slab gels.

RESULTS

We isolated 19 *Salmonella* strains. Eight (40%) of them were resistant to one or more antibiotics. The resistance pattern was completely transferred to *E. coli* from six (30%) strains (Table I.). The most frequent pattern in this group was SmTc and there was no differences between the number of "fi⁻" and "fi⁻" R plasmids. The molecular weight (M.W.) of these plasmids ranged from 30 to 65 Mdal.

In the *Salmonella* strains that were sensitive to all the antibiotics tested, we did not find any plasmid by agarose gel electrophoresis.

We also isolated 30 primarily non-pathogenic enterobacteria from 19 patients. Twenty (66%) of them showed transfer in all or in part of their antibiotic resistance pattern. The highest percentage of transferable drug resistant strains was found in *E. coli* (17 strains, 81% of the total *E. coli* isolates). Only three *Klebsiella pneumoniae* of a total of seven isolates transferred their resistance and the two *Proteus vulgaris* did not transfer their resistance but exhibited plasmids in gel electrophoresis. (Table II.) The resistance patterns ranged from one to six resistance.

TABLE I.

R factors from 19 Salmonella strains

Salmonella species	Resistance pattern	Resistance transferred	M.W.	fi type
S. typhimurium	Sm Tc	Sm Tc	30	fi⁻
S. typhimurium	Sm Tc	Sm Tc	30	fi⁻
S. typhimurium	Sm Tc	Sm Tc	65	fi⁺
S. typhimurium	Km Tc	Km Tc	46	fi⁺
S. typhimurium	Km Tc	Km Tc	66	fi⁺
S. enteritidis	Sm Tc Am	Sm Tc Am	40	fi⁻
S. paratyphi B	Km	—	48	—
S. paratyphi B	Sm	—	—	—

The remaining eleven strains were completely sensitive.

TABLE II.

R factors from non pathogenic Enterobacteria

Species	Resistance pattern	Resistance transferred	M.W.	fi type
E. coli	Sm Tc Km	Sm	68	fi⁻
	Sm Am Km	Am Km	42	fi⁻
	Sm Tc Am Su	Am Tc	36	fi⁻
	Sm Am Km Cm	Sm Am	65	fi⁻
	Sm Tc	Sm Tc	30	fi⁻
	Sm Tc Am Cm	Sm Tc	30	fi⁻
	Sm Tc Su	Sm Tc Su	43	fi⁺
	Sm Tc Am Km	Sm Tc Am	40	fi⁻
	Sm Tc Am Km Su	Sm Tc Am	40	fi⁻
E. coli	Sm Tc Cm	Sm Tc Cm	44	fi⁻
	Sm Tc Am Cm	Sm Tc Am Cm	60	fi⁻
	Sm Tc Am Cm	Sm Tc Am Cm	60	fi⁻
	Sm Tc Am Cm Su	Sm Tc Am Cm Su	75	fi⁻
	Sm Tc Am Cm Su	Sm Tc Am Cm Su	75	fi⁻
	Sm Tc Am Cm Su[1]	Sm Tc Am Cm Su	75	fi⁻
	Sm Tc Am Cm Su Km	Sm Tc Am Cm Su Km	57	fi⁻
	Sm Tc Am Cm Su Km	Sm Tc Am Cm Su Km	57	fi⁻
K. pneumoniae	Sm Tc Am Cm Su Km	Sm Am Km	80	fi⁻
	Sm Tc Am	Sm Tc Am	40	fi⁻
	Sm Tc Am Km	Sm Tc Am	40	fi⁻
P. vulgaris	Sm Am Km	—	42	—
	Sm Tc Am Cm	—	60	—

There were 4 E. coli and 4 K. pneumoniae that did not have any plasmid.

The "fi⁻" plasmids were in the majority in this group and the M.W. of these plasmids ranged from 30 to 80 Mdal.

Table III. lists the overall results of the 19 Salmonellosis cases studied. The same transferable resistance pattern (SmTc) was found both in *S. typhimurium* and *E. coli* in only one patient. One plasmid of 30 Mdal was found both in the *Salmonella* and *E. coli* strains.

TABLE III.

Overall pattern of R factors in 19 salmonellosis patients

Patient	Microbial flora	R. pattern	Plasmid*	R. transferred	M.W.	fi type
1	*S. typhimurium*	—	—	—	—	—
	E. coli	Sm Tc Cm Am Su Km	pSE 1	Sm Tc Cm Am Su Km	57	fi⁻
2	*S. typhimurium*	Sm Tc	pSE 2	Sm Tc	30	fi⁻
	E. coli	Sm Tc	pSE 3	Sm Tc	30	fi⁻
3	*S. typhimurium*	—	—	—	—	—
	E. coli	Sm Tc Cm Am Su Km	pSE 4	Sm Tc Cm Am Su Km	57	fi⁻
	P. vulgaris	Sm Tc Cm Am	—	—	60	—
4	*S. paratyphi B*	Sm	—	—	—	—
	E. coli	Sm Tc Cm Am Su	pSE 5	Sm Tc Cm Am Su	75	fi⁻
5	*S. typhimurium*	Km Tc	pSE 6	Km Tc	46	fi⁺
	E. coli	Sm Tc Cm	pSE 7	Sm Tc Cm	44	fi⁻
6	*S. enteritidis*	—	—	—	—	—
	E. coli	Sm Tc Cm Am Su	pSE 8	Sm Tc Cm Am Su	75	fi⁻
7	*S. typhimurium*	—	—	—	—	—
	K. pneumoniae	Sm Tc Cm Am Su Km	pSE 9	Sm Am Km	80	fi⁻
	K. pneumoniae	Sm Tc Am Su	—	—	—	—
8	*S. typhimurium*	—	—	—	—	—
	E. coli	Sm Tc Cm Am	pSE 10	Sm Tc	30	fi⁻
	E. coli	Sm Tc Su	pSE 11	Sm Tc Su	43	fi⁺
9	*S. typhimurium*	Sm Tc	pSE 12	Sm Tc	30	fi⁻
	E. coli	—	—	—	—	—
	K. pneumoniae	—	—	—	—	—
10	*S. typhimurium*	Sm Tc Km	pSE 13	Km Tc	66	fi⁺
	E. coli	—	—	—	—	—
11	*S. paratyphi B*	Km	—	—	48	—
	E. coli	Sm Tc Cm	—	—	—	—
	K. pneumoniae	Sm Tc Am	pSE 14	Sm Tc Am	40	fi⁻
12	*S. typhimurium*	—	—	—	—	—
	E. coli	Sm Tc Am Su	pSE 15	Am Tc	36	fi⁻
13	*S. typhimurium*	Sm Tc	pSE 16	Sm Tc	65	fi⁺
	E. coli	Sm Am Km	pSE 17	Am Km	42	fi⁻
	P. vulgaris	Sm Am Km	—	—	42	—
14	*S. typhimurium*	—	—	—	—	—
	E. coli	Am Tc Am Su Km	pSE 18	Sm Tc Am	40	fi⁻
	E. coli	—	—	—	—	—
15	*S. typhi*	—	—	—	—	—
	E. coli	Sm Tc Am Km	pSE 19	Sm Tc Am	40	fi⁻
	K. pneumoniae	Sm Tc Am Km	pSE 20	Sm Tc Am	40	fi⁻
16	*S. typhimurium*	—	—	—	—	—
	E. coli	Sm Tc Cm Am Su	pSE 21	Sm Tc Cm Am Su	75	fi⁻
	K. pneumoniae	Tc	—	—	—	—
17	*S. typhimurium*	—	—	—	—	—
	E. coli	Sm Cm Am Km	pSE 22	Sm Am	65	fi⁻
	K. pneumoniae	Su	—	—	—	—

TABLE III. (*cont.*)

Patient	Microbial flora	R. pattern	Plasmid*	R. transferred	M.W.	fi type
18	*S. typhi*	—	—	—	—	—
	E. coli	Sm Tc Km	pSE 23	Sm	68	fi⁻
19	*S. enteritidis*	Sm Tc Am	pSE 24	Sm Tc Am	40	fi⁻
	E. coli (*lac⁻*)	Sm Tc Cm Am	pSE 25	Sm Tc Cm Am	60	fi⁻
	E. coli	Sm Tc Cm Am	pSE 26	Sm Tc Cm Am	60	fi⁻

* The nomenclature is that recommended by Novick et al. (Bact. Revs., 40(1) 168–189) 1976.

Plasmids with similar resistance patterns (Sm Tc Am), M.W. and fi character were observed in *Salmonella* and in non-pathogenic bacteria, but never simultaneously in the same patient.

In each of three patients (nos. 13, 15, 19) a possibly identical R plasmid was found in all the non-pathogenic bacteria examined, but not in the *Salmonella* strains (the R plasmid being different in each patient).

In 13 patients the non-pathogenic bacteria examined had transferable drug resistance, but not the *Salmonella* strains, only two *Salmonella* showing transferable resistance with sensitive non-pathogenic accompanying bacteria (patients 9 a 10).

Digestion with *Eco*R1 did not break plasmids pSE 2 and pSE 3.

DISCUSSION

The frequency of transferable drug resistance among the Salmonella was 30%, a figure comparable to that reported by other workers (Mitsuhashi 1977, Rodriguez et al. 1977).

For non-pathogenic bacteria this frequency was 66%, similar to that found previously here (Perea and Daza, 1974) and elsewhere (Nakaya et al., 1975, Mitsuhashi, 1977, Falkow, 1975).

Among multiple resistance patterns, that of Sm Tc was the most predominant in *Salmonella* (50%), similar to findings published in some reports (Nakaya et al., 1975, Marsik et al., 1975, Falkow, 1975) but not in others (Timoney, 1978, Panse and Wadhwa, 1976) that reported broader resistance patterns.

These facts lead us to suggest that the antibiotic selective pressure is not as high in our country as elsewhere, because this pressure would probably select a broader resistance pattern.

Sm Tc is the basic unit in plasmid resistance patterns. We observed, in general, that when a new drug resistance was added, the M.W. was incremented by some Mdal.

S. typhimurium was the most common *Salmonella* isolated and had the highest incidence of R plasmids. Similar finding were reported by Jonson et al., 1972, Nakaya et al., 1975, Falkow, 1975, Anderson, 1977.

The resistance patterns in non pathogenic bacteria were similar to those previously reported here (Perea and Daz, 1974) and showed more frequency and variety in size than those of Salmonella (Mitsuhashi, 1977, Falkow, 1975).

Plasmids with the same resistance pattern (Sm Tc Am) M.W. and fi character were observed either in *Salmonella* or in non-pathogenic bacteria but never concurrently in both in the same patient.

Only one patient had an identical plasmid both in *Salmonella* and in the accompanying *E. coli*.

These factors would suggest that the acquisition of R factors by *Salmonella* occurs outside the patient's intestine, possibly due to factors inimical to the transmission "*in vivo*" of R plasmids (Lacey, 1975, Anderson et al., 1973, Williams, 1969).

More weight could be given to this hypothesis by the findings that, in each of three patients (13, 15 and 19), the same R plasmid was found in the non-pathogenic bacteria examined but not in the *Salmonella*.

Moreover, there were also 13 patients where only the non-pathogenic bacteria were drug resistant and not the *Salmonella*, while there were only 2 cases with the opposite finding. The easy transferability of R plasmids from *Salmonella* to *E. coli* "*in vitro*" corroborates our hypothesis.

Thus we may conclude that the R factors present in the Salmonella strains isolated here do not originate in the patient's normal intestinal flora which is present at the beginning of the disease. Of course, the possibility of acquiring new plasmids during convalescence due to antibiotic selective pressure remains.

REFERENCES

ANDERSON, J. D., INGRAM, L. C., RICHMOND, M. H., and WIEDEMAN, B. (1973): Studies on the nature of plasmids arising from conjugation in the human gastro-intestinal tract. J. Med. Microbiol. 6:475–486.

COETZEE, J. N., DATTA, N., and HEDGES, R. W. (1972): R factors from *Proteus retggeri*. J. Gen. Microbiol. 72:543–552.

ECKHARD, T. (1978): A Rapid Method for the Identification of Plasmid Desoxyribonucleic Acid in Bacteria. Plasmid. 1:584–588.

FALKOW, S. (1975): The prevalence and ecology of R-factors in Infectious Multiple Drug Resistance. Pion Limited, London.

JONSSON, M., RUTBERG, L., and TUNEVALL, G. (1972): Transferable resistance to antibiotics in gram-negative bacteria isolated in a hospital for infectious diseases. Scand. J. Infect. Dis. 4:209–219.

MARSICK, F. J., PARISI, J. T., and BLENDEN, D. C. (1975): Transmissible Drug Resistance of *Escherichia coli* and *Salmonella* from Humans, Animals, and their Rural Enviroments. J. Infect. Dis. 123:296–302.

MITSUHASHI, S. (1977): Epidemiology of R Factors. *In* R Factor. Drug Resistance Plasmid. S. Mitsuhashi Ed. University Park Press. Tokyo.

NAKAYA, R., YOSHIDA, Y., and TERAWAKI, Y. (1975): Antibiotic resistance and R Plasmids in *Salmonella* isolated from Humans in Japan (1966–1972). *In* Microbial Drug Resistance. S. Mitsuhashi and H. Hashimoto Eds. University Park Press. Tokyo.

NOVICK, R. P., CLOWES, R. C., COHEN, S. N., CURTISS III, R., DATTA, N., and FALKOW, S. (1976): Uniform Nomenclature for Bacterial Plasmids: A Proposal. Bacteriol. Rev. 40:168–189.

PANSE, M. V., and WADHWA, K. M. (1976): Transmissible Drug Resistance Among *Enterobacteriaceae* II. Incidence of R Factor among Enteric Organisms Isolated from Diarrhoea Cases and Normal Healthy Individuals. Indian, J. Med. Res. 64:399–404.

PEREA, E. J., and DAZA, R. M. (1974): R Factors to aminoglycoside antibiotics. *In* Antibiotic Resistance. Drug-Inactivating Enzymes and Antibiotic Resistance. S. Mitsuhashi, L. Rosival, V. Krčméry Eds. Avicenum Prague and Springer, Heidelberg.

TIMONEY, J. F. (1978): The Epidemiology and Genetics of Antibiotic Resistance of *Salmonella typhimurium* Isolated from Diseased Animals in New York. J. Infect. Dis. 137:67–73.

WILLIAMS, H. (1969): Transfer of antibiotic reistance from animal and human strains of *Escherichia coli* to resident *E. coli* in the alimentary tract of man. Lancet. June 14, 1174–1176.

J. P., Hospital Universitario DPTO,
Mikrobiologia,
Auda Dr Fedriani Sa, Sevilla, Spain

DRUG RESISTANCE IN MAIN UDDER PATHOGENS

F. FEDERIČ, O. J. VRTIAK and V. JORDÁNOVÁ

*Laboratory of Bovine Mastitis, Institute of Experimental Veterinary Medicine,
Košice, Czechoslovakia*

INTRODUCTION

In the early period of the first stage of intensive mastitis control in dairy cows a single examination of a representative number of animals (altogether 17816 dairy cows) was carried out in Slovakia in 1974 in order to determine the objective situation. It was found that the mammary gland was in a good state of health in only 53.29% of the animals; bacterial agents were isolated from the milk in 23.5% of the animals. From the bacterial agents of mastitis *Streptococcus agalactiae* and *Staphylococcus aureus* were isolated (70.6% and 14.3%, respectively). At present the latter are the most important pathogens of the mammary gland.

While the epidemiological importance of *Staphylococcus aureus* has been rather well known for a longer time, that of *Streptococcus agalactiae* has been constantly increasing from the sanitary point of view. In USA and in many European countries a markedly increasing tendency has been observed in the occurrence of neonatal infections by the above agent. In infants pneumoniae, meningitis and sepsis with a high frequency of fatal cases have been observed (Horn, Meyer, Wyrick and Zimmerman, 1974). In adults infections of the upper respiratory tract and infections of the urogenital tract (McDonald, 1977) are rather frequent.

In nature, *Staphylococcus aureus*, from the viewpoint of acquiring antibiotic resistance, belongs to the most plastic microorganisms. Resistance is frequently multiple and determined by plasmids (Lacey, 1975). *Streptococcus agalactiae* lacks such marked features, although in 1976 a plasmid was described for the latter, carrying resistance markers to chloramphenicol, erythromycin, linkomycin and pristinamycin as well as a plasmid carrying marker resistance to tetracycline (Horodniceanu, Bouanchaud, Bieth and Chabbert, 1976).

MATERIAL AND METHODS

In order to determine the frequency rate of antibiotic resistance, 400 *S. aureus* strains isolated from mastitic cows were tested each year throughout 1976, 1977 and 1978 one the whole territory of Czechoslovakia. Using the previously described semi-quantitative method (Vrtiak, Federič and Rabiu, 1977) susceptibility to eight antibiotics, was examined i.e. penicillin, cloxacillin, ampicillin, streptomycin, neomycin, tetracycline, chloramphenicol and erythromycin.

In 100 strains of *Streptococcus agalactiae* isolated from the milk of dairy cows in three vast areas of Czechoslovakia, the minimum inhibitory concentration (MIC) for the following 11 antibiotics was stated by the agar dilution method: penicillin, cloxacillin, ampicillin, streptomycin, neomycin, gentamicin, tobramycin, oleandomycin, tetracycline, chloramphenicol and erythromycin. An overnight broth culture of each strain was spotted

TABLE I.

Antibiotic resistance in S. aureus isolated from mastitic cows
in 1976–1978 (400 strains yearly)

Antibiotic	% of resistant strains		
	1976	1977	1978
Penicillin	16.0	13.5	13.5
Cloxacillin	1.5	1.5	2.0
Ampicillin	6.0	4.5	5.0
Streptomycin	15.0	9.7	15.0
Neomycin	3.0	2.5	4.0
Tetracycline	10.0	16.0	14.0
Chloramphenicol	1.0	9.5	3.5
Erythromycin	3.0	3.5	2.0

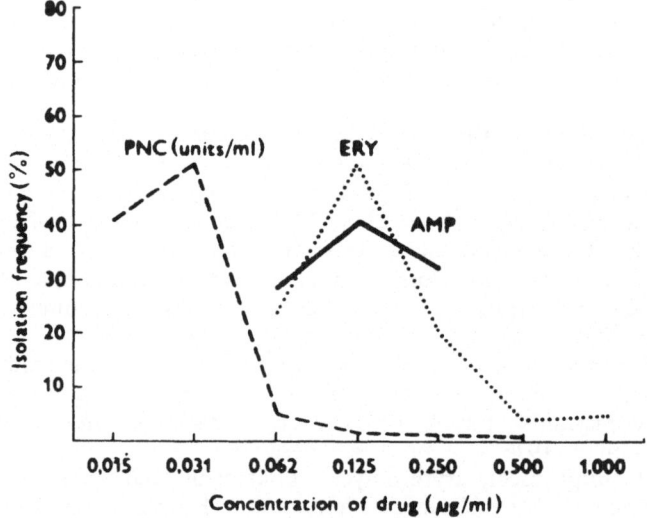

Fig. 1.

on DST agar containing serial twofold dilutions of each drug. The susceptibility was determined after incubation at 37°C for 24 hours.

RESULTS

The examination results considering S. aureus strains are reported in Table I. During the 3 year-period of investigation the highest frequency rate of resistance was the following: penicillin 14.3%, tetracycline 13.3% and streptomycin 13.2%, less frequently to ampicillin 5.2% and chloramphenicol 4.7%, seldom to neomycin 3.2%, erythromycin 2.8% and to cloxacyllin 1.7%.

From the data in the table it is evident that the following could be observed during the examination period:

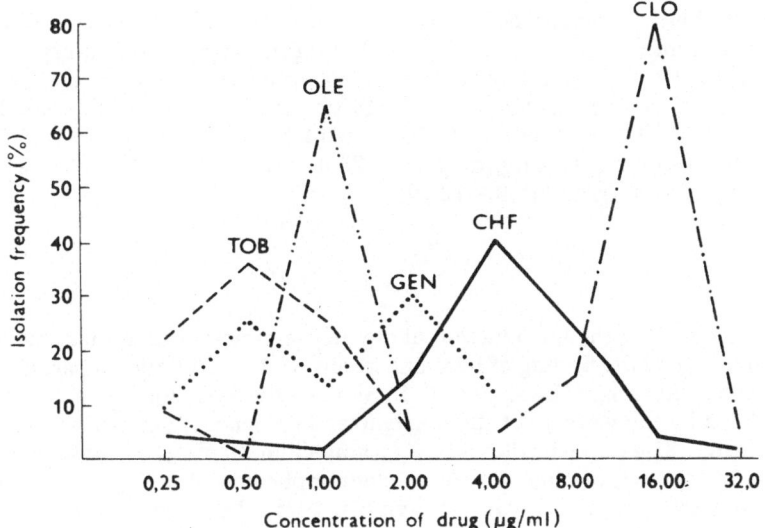

Fig. 2.

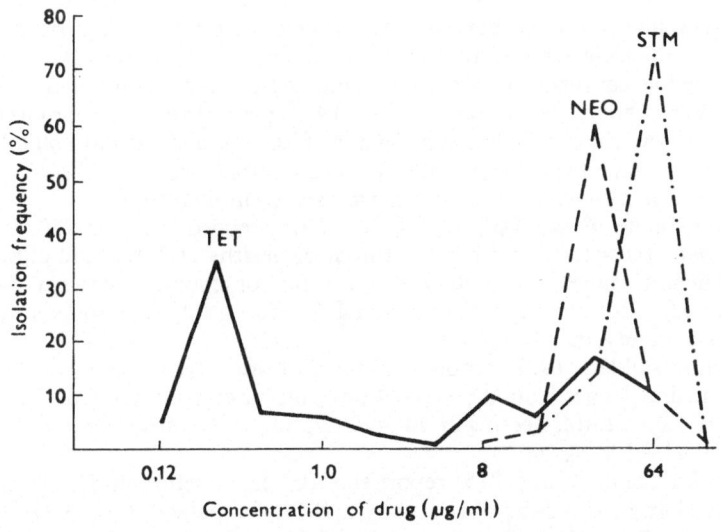

Fig. 3.

– a slight decrease in the occurrence of strains resistant to penicillin, ampicillin and erythromycin,

– an increase in the occurrence of strains resistant to tetracycline and chloramphenicol and

– a relatively constant, low occurrence of strains resistant to cloxacillin, neomycin and erythromycin.

The MIC of drugs among 100 Streptococcus agalactiae strains is given in Figs. 1, 2 and 3.

The lowest value of average MIC was observed in penicillin – 0.032 unit/ml (range 0.015 – 0.5 unit/ml), then in ampicillin – 0.149 µg/ml (0.062 – 0.25), erythromycin – 0.180 µg/ml (0.062 – 1.0), tobramycin – 0.596 µg/ml (0.25 – 2.0), oleandomycin – 0.942 µg/ml (0.25 – 1.0), gentamicin – 1.548 µg/ml (0.25 – 4.0), chloramphenicol – 4.970 µg/ml (0.25 – 32.0), cloxacillin – 15.24 µg/ml (4.0 – 32.0), tetracycline – 15.224 µg/ml (0.125 – 64.0), neomycin 42.4 µg/ml (8.0 – 128.0) and the highest value was found in streptomycin – 58.889 µg/ml (16.0 – 128.0).

DISCUSSION

It is well known that the introduction of any new antibiotic into human medicine was generally followed by the finding of resistant strains to this antibiotic, in particular among the strains of hospital origin (Lacey, 1975). In Slovakia the following resistance in *S. aureus* strains, isolated by the workers of the hygienic and epidemiological service in 1973, was found, penicillin 82.41%, oxacillin 11.54%, ampicillin 40.65%, streptomycin 27.57%, neomycin 13.23%, tetracycline 35.56%, chloramphenicol 13.69% and erythromycin 25.65% (Krčméry, Grunt, Rosival and Výmola, 1975). In strains isolated from animal sources generally antibiotic resistant strains appear less frequently in dependence on the range and amount of antibiotics applied. In Czechoslovakia the use of intramammary preparations for mastitis treatment is increasing at the present time; these preparations contain tetracycline, chloramphenicol and cloxacillin, on the other hand, the use of preparations containing streptomycin, neomycin and Zn bacitracin is decreasing. The results obtained as compared to the data resulting from the testing of *S. aureus* strains, isolated in 1975 (Vrtiak, Federič and Rabiu, 1977) prove that this microorganism is the most important one among the bacterial agents of mastitis and the most suitable indicator of changes in the tendencia of antibiotic resistance initiation.

Data on MIC in *Streptococcus agalactiae* are scarce in literature.

Mitsuhashi, Inoue, Fuse, Kaneko and Oba (1974) report that with 908 tested *Streptococcus pyogenes* strains isolated from the throat of healthy children and clinical samples, MIC for penicillin, ranges from 0.0005 – 0.2 unit/ml, for chloramphenicol 0.4 – 100 µg/ml, tetracyccline 0.2 – 100 µg/ml, streptomycin 1.6 – 50 µg/ml, erythromycin 0.012 – 6.25 µg/ml and oleandomycin 0.2 – 25 µg/ml.

Matsen and Coghlan (1972) report that for 55 tested strains of group B streptococci MIC for penicillin ranges from 0.003 – 0.4 unit/ml, for erythromycin 0.003 – 0.02 µg/ml, oxacillin 0.01 – 0.8 µg/ml, neomycin 50 – 200 µg/ml, streptomycin 100 – 200 µg/ml and tetracycline 0.02 – 100 µg/ml.

Anthony and Concepcion (1975) report that the examination of 511 group B streptococci strains showed in susceptible strains a MIC for penicillin of 0.05 – 0.4 unit/ml, ampicillin 0.1 – 0.2 µg/ml, erythromycin 0.05 – 0.1 µg/ml, chloramphenicol 3.1 – 6.2 µg/ml and tetracycline 0.4 – 1.6 µg/ml and in resistant strains a MIC for kanamycin of 50 and more, erythromycin 0.8 and more, chloramphenicol 25 and tetracycline more than 50 µg/ml.

Baker, Webb and Barrett (1976) testing the MIC of 244 B group streptococci isolated from patients, obtained the following average values: penicillin 0.083 unit/ml, ampicillin 0.2 µg/ml, tetracycline 50 µg/ml, chloramphenicol 0.8 µg/ml and gentamicin 25 µg/ml.

The usual MIC of *Streptococcus agalactiae* is for penicillin 0.005 – 0.01 unit/ml, ampicillin 0.02 and cloxacillin 0.06 µg/ml (Neu, 1977).

From the above data and from the results obtained the necessity follows to revalue the effectiveness of mastitis therapy using preparations containing streptomycin and neomycin with regard to their relatively high MIC for *Streptococcus agalactiae* strains.

REFERENCES

ANTHONY, B. F. and N. F. CONCEPCION (1975): Group B *Streptococcus* in a general hospital. J. Infect. Dis. 132, 561–567.

BAKER, C. J., B. J. WEBB and F. F. BARRETT (1976): Antimicrobial susceptibility of group B streptococci isolated from a variety of clinical sources. Antimicrob. Agents Chemother. 10, 128–131.

HORN, K. A., W. T. MEYER, B. C. WYRICK and R. A. ZIMMERMAN (1974): Group B streptococcal neonatal infection. Journal of American Medical Association 230, 1165–1167.

HORODNICEANU, T., D. H. BOUANCHAUD, C. BIETH and Y. A. CHABBERT (1976): R plasmids in *Streptococcus agalactiae* (group B). Antimicrob. Agents Chemother. 10, 795–801.

KRČMÉRY, V., J. GRUNT, L. ROSIVAL and F. VÝMOLA (1975): Nationwide survey of antibiotic resistance by means of a computer. Zbl. Bakt. I. Orig. 231, 250–258.

LACEY, R. W. (1975): Antibiotic resistance plasmids of *Staphylococcus aureus* and their clinical importance. Bact. Rev. 29, 1–32.

MATSEN, J. M. and C. R. COGHLAN (1972): Antibiotic testing and susceptibility patterns of streptococci. In: L. W. Wannamaker, J. M. Matsen (ed), *Streptococci* and Streptococcal Disceases-197, Academic Press, New York, 189–204.

McDONALD, J. S. (1977): Streptococcal and staphylococcal mastitis. Journal of American Veterinary Medical Association 170, 1157–1159.

MITSUHASHI S., M. INOUE, A. FUSE, Y. KANEKO and T. OBA (1974): Drug resistance in *Streptococcus pyogenes*. Japan. J. Microbiol. 18, 98–99.

NEU, H. C. (1977): The penicillins I. Overview of microbiology. New York State J. Med. 77, 768–771.

VRTIAK, O. J, F. FEDERIČ and M. A. RABIU (1977): Drug resistance in *Staphylococcus aureus* strains isolated from raw milk. Plasmids Medial and Theoretical Aspects (ed. Mitsuhashi, S., Rosival L. and Krčméry V.) Avicenum, Springer Verlag, 53–57.

F. F., Institute of Experimental
Veterinary Medicine, Komenského, 73,
041 81 Košice, Czechoslovakia

SAMONELLA HEIDELBERG PENICILLINASE PLASMIDS FOUND DURING A NOSOCOMIAL EPIDEMIC

K. KOLTSIDA, P. PARASKEVOPOULOU and P. KONTOMICHALOU

Department of Clinical Therapeutics, School of Medicine,
University of Athens, Greece

INTRODUCTION

The frequency of *S. heidelberg* has increased in Scotland, United States, Finland and Israel (World Health Organisation, 1978). Foodborn outbreaks associated with that serotype were also noted. The frequency of this serotype has also increased in Greece. According to the reports of the Hellenic National Salmonella Center, *S. heidelberg* was in the last three years the second or the third serotype in frequency (WHO, 1978). In our hospital, which includes Maternity and Internal Medicine Departments, a nosocomial outbreak from *Salmonella* of this serotype started in 1975 and it caused most of the hospital acquired salmonellosis until 1978. We report here the results of our study on the epidemic in the Internal Medicine Department, the causative organisms and the resistance plasmids detected.

THE SALMONELLA HEIDELBERG NOSOCOMIAL EPIDEMIC

In October 1975 we had four febrile enteritis cases from where the same serotype of *S. heidelberg* was isolated. A total of hundred samples were obtained from the surveillance of kitchen environment and food handling personnel.* Twenty of those samples were positive for *Salmonella* of serotype: *S. tennessee*, *S. heidelberg*, *S. albany* and *S. thompson*. The *S. heidelberg* isolates were only two: one from fish and the other from the chicken cutting benches. Nurses and Medical staff were also examined for carriers. Positive cases of the *S. heidelberg* serotype were found predominantly in the personnel. *S. heidelberg* isolated from all sources (patients, environment and nurses) was always resistant only to ampicillin. On the contrary all other serotypes were sensitive to all antibiotics. The number of nosocomial and community acquired salmonellosis from October 1975, when the epidemic started, until the end of 1977 is shown in Table I. There were 74 nosocomial and four community cases. In 1978 we had only one case of *S. heidelberg* in May.

The distribution of infections from *S. heidelberg* comparing to 12 salmonellosis of other serotypes is given in Table I. In all years all *S. heidelberg* were found to be resistant only to ampicillin. The distribution of the epidemic among the wards in our Clinic had the following features: the Department is divided to A, B, C of the third floor and D on the fourth floor. At the beginning of the epidemic (1975) only one case was found in ward B, on the third floor, comparing to 16 cases in ward A, 10 cases in ward D and

* We are grateful to Professor J. Papadakis, Hellenic National Salmonella Center, for his help in performing this survey.

This work has been supported by grants from the Ministry of Hygiene of Greece and the Hellenic National Research Foundation (E.I.E.)

TABLE I.

TABLE I.

Infections with Salmonellae

Year	1975	1976	1976	1977
Months	10th–12th	2d–8th	9th–12th	1st–12th
Cases	33	29	11	5
S. heidelberg	31	27	3	5
Other serotypes	2	2	8	—
Nosocomial infections	32	28	9	5
Community infections	1	1	2	—

5 cases in ward C. In January 1976 a post-partum patient with a severe infection of *S. heidelberg* was admitted in that ward. The patient was transferred from the Maternity unit of our hospital. Since then the number of *S. heidelberg* nosocomial cases in ward B, where this patient was transfered, was significantly increased and we thought that the severe case in ward B could cause the reappearance of the *S. heidelberg* epidemic. Until the end of the epidemic there were 18 cases in ward B, 12 in ward A, 4 in ward C and 7 in ward D.

DETECTION OF PLASMIDS AND β-LACTAMASES

A total of thirty ampicillin resistant *S. heidelberg* strains were tested for conjugal transferability of ampicillin resistance to an *E. coli* K_{12} host- strain. In all cases the ampicillin resistance was transmissible. Eleven plasmids from *S. heidelberg* donors isolated from environment, patients and carriers were further studied. The sources of strains and their distribution during the two years period are shown in Table II.

Table III. comprises the resistance phenotype to ampicillin and carbenicillin confered to an *E. coli* K_{12} by the eleven plasmids, the results of specific activities and the substrates profiles of the plasmid-mediated β-lactamases. The resistance to carbenicillin was four

TABLE II.

Plasmids studied

Culture E. coli RC85 + plasmid	Source	Culture speciments	Active disease	Carrier	Ward	Date
pPK90	nurse	stools	—	+	—	10–10–1975
pPK91	nurse	stools	—	+	—	10–10–1975
pPK92	patient	stools	+	—	D	10–10–1975
pPK93	patient	stools	+	—	A	11–10–1975
pPK94	kitchen personnel	stools	—	+	—	29–10–1975
pPK95	environment	Kitchen's swabin	—	+	—	25–11–1975
pPK96	patient	blood	+	—	B	22–1–1976
pPK97	patient	stools	+	—	C	6–3–1976
pPK98	patient	stools	+	—	C	7–4–1976
pPK100	patient	stools	+	—	B	22–4–1977
pPK101	patient	blood	+	—	A	4–5–1977

TABLE III.

Minimal inhibitory concentration of β-lactam antibiotics, specific activity and substrate profile of penicillinases mediated by the plasmids

Cultures E. coli RC85 +plasmid	Minimal inhibitory concentrations (μg/ml)			Specific activity*	Relative rate of hydrolysis						Type of penicillinase
	Amp.	Carb.	Ceph.		PenG	Amp	Carb	Clox	CR	Cth	
pPK90	250	4000	<1	408	100	135	24	4	90	20	TEM
pPK91	125	2000	<1	314	100	142	28	3	80	5	TEM
pPK92	250	4000	<1	394	100	135	14	2	79	10	TEM
pPK93	125	2000	<1	206	100	106	10	1	67	19	TEM
pPK94	125	1000	<1	197	100	116	16	5	128	33	TEM
pPK95	500	2000	<1	242	100	122	18	5	68	9	TEM
pPK96	8000	32000	4	425	100	140	34	6	123	17	TEM
pPK97	125	2000	<1	172	100	118	24	3	55	11	TEM
pPK98	125	1000	<1	206	100	110	29	3	56	6	TEM
pPK100	500	4000	<1	350	100	137	15	2	66	6	TEM
pPK101	500	2000	<1	185	100	133	16	2	63	7	TEM

* Specific activity is expressed in units of penicillinase per hour per mgr of protein in 5mM PenG as substrate, in crude extracts.

to sixteen times higher than to ampicillin. The specific activity of the plasmid-coded β-lactamases was high, ranging in crude preparations of the *E. coli* $K_{12}R^+$ cultures from 172–425 units per hour per mgr of protein (Iodometric method, Perret, 1954, units by Pollock and Torriani, 1953). Their substrates profiles indicate that the penicillinases are very similar to each other and belong to TEM-type (Kontomichalou P., Papachristou E. and Levis G. 1974). Experiments are in progress to identify the subtypes of the penicillinases

COMPATIBILITY REACTIONS AND MOLECULAR WEIGHTS OF THE PLASMIDS

Subsequently we investigated the compatibility taxonomy of the plasmids. All eleven plasmids were found to be *fi⁻*. In experiments of sensitivity to male specific phages by testing both visible lysis and increase of phage titre, we found that *E. coli* K_{12} host harboring our plasmids was not sensitive to I-, N- and P-specific phages. Therefore we conclude that the plasmids did not belong to these three compatibility groups.

Classical compatibility experiments were further performed with reference plasmids of known compatibility groups W, C, N, O and M (Datta, 1974).

As shown in Table IV, the plasmid PK96 which was found in the *S. heidelberg* strain, isolated from the post-partum patient, was the only one that gave typical compatibility reactions with group W reference plasmid. It was characterised as belonging to group W. Plasmids from all the other sources showed similar behaviour to each other, consisting of exclusion and recombination with the group M reference plasmid. Thus we classified these plasmids to compatibility group M.

The plasmids were further studied by agarose gel electrophoresis on single cell lysates, using reference plasmids of known molecular weights (Meyers et al. 1976 and Eckhardt T. 1978). In the culture carrying the plasmid of compatibility group W (PK96) only one band was detected, corresponding to a plasmid of 58 Mdaltons. One of the group M compatibility plasmids (PK91) gave also one band corresponding to 63 Mdaltons.

TABLE IV.

Incompatibility group determination and agarose gel electrophoresis of plasmids on study

Culture E. coli RC85 + plasmid	Exclusion	Compatibility	Recombination	Conclusion found	Number of plasmids	Molecular weight (Md)
pPK96	(+) W (−) C, N	(−) W (+) C, N		group W	1	58
pPK91	(+) M (−) W, O, C		(+) M	group M	1	63
pPK90 ⎫ pPK92 ⎬	(−) W, O, C (−) W, O, C		(+) M (+) M	group M group M	3	a) 63 b) 35 c) 5.5
pPK93 ⎫ pPK94 ⎪ pPK95 ⎪ pPK97 ⎬ pPK98 ⎪ pPK100 ⎪ pPK101 ⎭	(−) W, O, C (−) W, O, C (−) W, O, C (−) W, O, C (−) W, O, C (−) W, O, C (−) W, O, C		(+) M (+) M (+) M (+) M (+) M (+) M (+) M	group M group M group M group M group M group M group M	2	a) 63 b) 5.5

Seven of the group M compatibility plasmids gave two bands: one large of 63 Mdaltons and one small of 5.5 Mdaltons. The remaining two group M plasmids gave three bands of 63, 35, 5.5 Mdaltons. Thus all ten group M plasmids have a common plasmid band of 63 Mdaltons and nine out of ten have also a common small molecule of 5.5 Mdaltons. Two plasmids have an additional band of 35 Mdaltons as well.

CONCLUSION

In conclusion the plasmids from *S. heidelberg* nosocomial epidemic belong to compatibility group M. They are very similar to each other, they code for TEM-type β-lactamase and have the same molecular weights of 63 and 5.5 Mdaltons. Only one plasmid from the *S. heidelberg* epidemic was found to belong to compatibility group W and also had a different size (58 Md). The hypothesis that the strain carrying this plasmid was responsible for the continuation of the epidemic is no more valid, since strains from all following cases have plasmids with different properties.

In a nosocomial *Salmonella* outbreak the strains are associated because they belong to the same serotype and usually have the same antibiotic pattern. Our data demonstrate that an additional useful epidemiological tool is the identification of the plasmids carried by the pathogenic strains.

REFERENCES

DATTA N. (1975): Epidemiology and classification of plasmids. Microbiology 1974 ed. by D. Schlessinger, American Society for Microbiology, Washington D. C. p. 9–15.

ECKHARDT T. (1978): A rapid method for the identification of plasmids deoxyribonucleic acid in Bacteria. Plasmid, I p. 584–588.

KONTOMICHALOU P., PAPACHRISTOU E. and LEVIS, G. (1974): R-mediated β-lactamases and episomal resistance to the β-lactam drugs in different bacteria hosts. Antimicrob. Ag. and Chemother. p. 60–72.

MEYERS et al (1976): Simple agarose gel electrophoretic method for the identification and characterisation of plasmid deoxyribonucleic acid. J. of Bacteriology, 127: 1529–1537.

PERRET J. C. (1954): Iodometric assay of penicillinase. Nature (London) 174: 1042–1043.

POLLOCK M. R. and A. M. TORRIANI (1953): Purification et characteristiques physicochimiques de la penicillinase de Bacillus cereus. C. R. Academie Sci. 237: 276–278.

WORLD HEALTH ORGANISATION (WHO) 1978: Wkly Epidem. Rec. No 8, p. 53–56.

K. K.
Alexandra Hospital, Vass. Sofias
K. Lourou Str., Athens, Greece

II. MEDICAL PART
E) COMPUTER SURVEILLANCE
OF ANTIBIOTIC RESISTANCE
Chairman: B. WIEDEMANN

COMPUTERIZATION OF A CLINICAL MICROBIOLOGY LABORATORY

F. H. KAYSER, J. WÜST and J. MUNZINGER

Institute of Medical Microbiology, University of Zuerich, Switzerland

INTRODUCTION

The laboratory of clinical microbiology of the Institute of Medical Microbiology in Zuerich is a large, busy laboratory that serves the university hospital and further general and private hospitals of the Kanton of Zuerich. In addition, this laboratory does the work of a microbiology public health laboratory and serves also as a reference centre for diagnostic microbiology in the area. Approximately 120 000 specimens of all types are processed annually. More than 220 000 bacteriological, virological and mycological tests are performed with these samples.

In the past, reporting of patient's test results was accomplished by type-written reports on special forms. Permanent records were kept in the laboratory mostly by manual entry of data into log books. For most of the tests performed, hospitals or physicians have to be charged. The numerous, individual type-written bills made a rather large invoicing department necessary.

In order to reduce the workload of this department and in order to collect and conveniently store the many microbiological data for later retrieval, we decided to computerize our laboratories. Although several "package" systems for microbiology were available in 1974, none seemed ideally suited to our operation for economical and organisational reasons. A special system was therefore developped and introduced. This report briefly describes our system and discusses the benefits and the drawbacks we observed in running the system for more than 5 years.

Description of the system

Figures 1 and 2 show a simplified flow chart for the system. Physicians send specimens and request forms to the institute. In the reception area, the samples are unpacked and distributed to the different special laboratories. There, a serial number is given to each specimen and the corresponding request form. In the electronic data processing bureau of the administrative department, the patient's data and the physician's or hospital ward's code number are transferred from the request form into the IBM 3741 recording system and stored on magnetic discettes. Every afternoon, the patient data and adresses of physicians, recalled by the code number from the "address discette", are printed onto special report forms. In addition, daybook listings are printed, containing, in alphabetical order, the name of the patient, as well as the serial number of the specimen and the name and adress of the medical customer. Identical alphabetical listings are performed at the end of the week in order to facilitate the tracking of a patient's sample in case of telephone requests. Daybook listings, the preprinted report forms and the request forms are then sent back to the laboratories.

When the laboratory work has been completed sufficiently for a report to be sent, the results are marked or written manually on the report forms. Antibiotic resistance is reported by markings in a special block of the report form. In addition, further test results and the fees are entered onto the report as code numbers in special sections of the form.

A carbon copy of the finished report is immediately sent to the physician, the original form is withhold for further treatment. Every week, the discettes and the report forms are sent to a commercial bureau, where the data of the discettes are converted to the tape

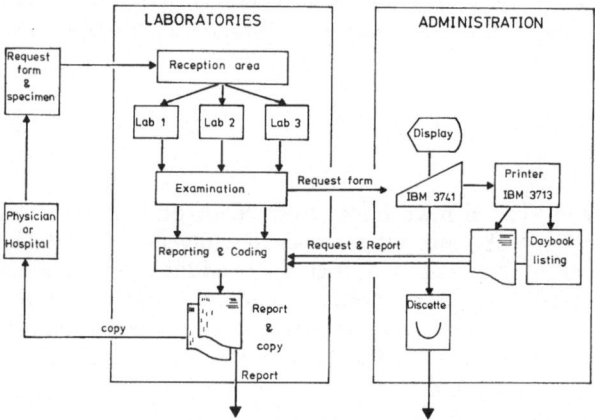

Fig. 1. Flow chart of the system showing the internal handling of data and material.

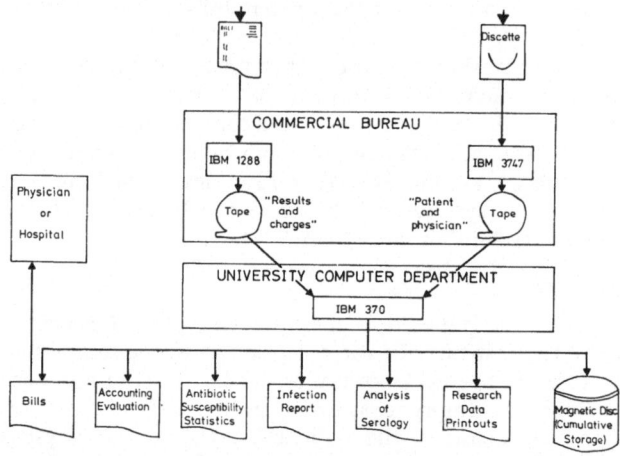

Fig. 2. Flow chart of the system showing the external handling of data.

"patient and physician", and the data of the reports are read by an optic reader and converted to the tape "results and charges". Both tapes are sent to the computer department of the university and stored in a cumulative microbiology data bank.

Special lists, report and statistics

From the data bank, data can be retrieved at convenient time intervals with special programs. Hospitals or physicians receive weekly cumulative bills. Invoices are controlled by periodically transferring the appropriate information to the computer's accounting file. Summaries of total test data support the laboratory's request for money, additional space, personnel and/or equipment.

TABLE I.

Percentages of susceptibility to sulphamethoxazole/trimethoprim in staphylococci
and Enterobacteriaceae, isolated from outpatients and hospitalized patients

Bacteria	Percentage (%) of susceptible strains				
	1974	1975	1976	1977	1978
Staphylococcus aureus	81	81	82	74	82
Staphylococcus epidermidis	83	77	70	65	63
Escherichia coli	94	93	92	90	88
Citrobacter freundii	86	82	80	80	79
Klebsiella spp.	78	71	70	70	69
Enterobacter spp.	90	88	86	91	86
Serratia spp.	66	59	33	30	35
Proteus mirabilis	89	88	86	84	86
Proteus spp. (indole-positive)	79	78	84	78	76
Total number of strains	10 117	16 185	16 111	17 557	17 445
% susceptible	86	84	82	78	80

Standardized disc test (FDA). Susceptibility = inhibition zone of 16 mm or more around
25 μg Su/Tp (19:1) disc.

The accumulated laboratory test results provide an immense data source, which reflects both primary and nosocomial infectious disease events. Reports can be generated from this source reflecting the distribution of a selected species according to the specimen source, the distribution of a selected species according to a hospital's medical service, or the distribution of a selected species according to patient location inside a hospital. Such reports can alert the epidemiologist and the infection control committee and pinpoint the source of nosocomial infection. Statistics can be produced indicating the overall antimicrobial susceptibility percentage of bacteria to various drugs in a hospital, the susceptibility percentage according to the specimen source, the susceptibility percentage according to the site of isolation or the susceptibility percentage according to medical service. Antimicrobial resistance patterns can be compared between the hospital population and the extrahospital population. Patterns can be grouped to indicate those strains resistant to multiple antimicrobics. This would represent a statistical incidence of R plasmids in the bacterial flora of an institution or in bacteria isolated from outpatients. Clearly, whether or not R plasmids exist would have to be proved by the ability or representative strains to transfer the resistance factor and/or by the demonstration of the physical presence of such factors. As an example of the numerous reports which can be generated by the system, Table I. summarizes the frequency of susceptibility to sulphamethoxazole/trimethoprim (Su/Tp) in nearly 80 000 strains isolated in 1974–1978. Despite the fact that Su/Tp is the most used of all antimicrobials in our area, the decrease in susceptibility is only small. These data stimulated investigations on the genetics of Su/Tp resistance. It was found that resistance to Tp in most gramnegative strains was determined by chromosomal genes. When resistance was plasmid determined, many of the plasmids were either nonconjugative or contained only the markers for Su^R and Tp^R and no further R-determinants. This situation would explain the slow decrease in susceptibility to Su/Tp among gramnegative organisms isolated in our area. Table II., as another example, presents comparative data on the activity of the newer cephalosporins and ureidopenicillins to enterobacterial strains isolated from hospitalized patients.

TABLE II.

Comparative susceptibility of Enterobacteriaceae to β-lactam antibiotics.
(Strains from hospitalized patients. January – April 1979, Zürich)

Bacteria	Percent of strains susceptible to					
	Ampicillin	Cephalothin	Cefamandol	Cefuroxim	Cefoxitin	Mezlocillin
E. coli	71	79	92	96	96	76
Klebsiella spp.	3	79	85	91	94	44
Enterobacter spp.	31	13	86	80	27	89
Serratia spp.	13	0	18	18	84	42
Proteus mirabilis	90	93	95	95	97	92
Indole-positive						
Proteus spp.	13	7	38	23	93	94
Total number of strains	2670	2670	2670	2670	2670	2670
% susceptible	55	73	87	89	92	72

Standardized disc test (FDA).

Further statistics can evaluate serology data. Comparison of antibody titres to virus antigens with titres obtained in preceding months, for instance, enables monitoring of ongoing virus or bacterial diseases in the population.

CONCLUSIONS

The system described in this report is an electronic data processing system for clinical microbiology, tailored for the specific situation and the needs of our institution. A major disadvantage of the system is that it is off-line. Thus, reporting by computer and direct access to the file of a patient cannot be obtained. When we introduced the system 6 years ago, the buying of our own computer was not possible, for economical reasons. At that time we were also not able to use a computer power for daily reporting under conditions necessary to maintain a tight time table for work processing. Using our system reporting, on the other hand, is not afflicted by a possible computer breakdown. The data collecting devices of our system (IBM 3741) proved to be very reliable. Immediate service is guaranteed should a breakdown occur. As a matter of fact, we have not had to postpone reporting because of problems with data collection in nearly 6 years. Approximate estimates of the cost of the system revealed that costs on balance have remained similar or increased only slightly, mainly because of a considerable reduction in the clerical workload. Not included in the estimate, however, are costs for computer power, because as a university institution, we are not billed by the computer department of the university.

The described system collects and stores all microbiological as well as administrative data in our laboratories. Special programs process all data necessary for invoicing and for the control of the invoicing system. Bacteriology, mycology and virology data are summarized as special lists, reports or statistics. These provide valuable information to enhance patient care directly or indirectly.

F. K., University of Zürich,
CH-8028 Zürich, Switzerland

IDENTIFICATION OF
SPECIFIC ANTIBIOTIC RESISTANCE MECHANISMS
FROM ROUTINE SUSCEPTIBILITY TESTING RESULTS

T. F. O'BRIEN, M. A. GUZMAN, J. J. FARRELL, R. L. KENT,
A. A. MEDEIROS, and J. F. ACAR

From the Department of Medicine, Peter Bent Brigham Hospital and Harvard Medical School, Boston, Massachusetts, U.S.A. From the Miriam Hospital and Brown University School of Medicine, Providence, Rhode Island, U.S.A. From Hopital St. Joseph, Paris, France

INTRODUCTION

Laboratories everywhere measure the resistance of clinical bacterial isolates to various antibacterial agents and report the resistance to each agent. The ratios between the levels of resistance of an isolate to different agents may provide additional information. Its comparative resistance to chemically related agents may indicate which of several alternative mechanisms of resistance to those agents the isolate possesses, and by inference, which resistance gene. Its comparative resistance to unrelated agents may express linkage of different resistance genes. We examine here some of these relationships for various classes of antibacterial agents.

MATERIAL AND METHOD

The data presented here are derived mostly from routine clinical laboratory test results. Disk susceptibility test results are performed on Mueller-Hinton agar with standardized disks and inocula, and the data computer-filed and analyzed as described previously (O'Brien, Acar, Medeiros, Goldstein and Kent, 1978). Minimal inhibitory concentration determinations were performed in doubling dilutions of Mueller-Hinton broth inoculated with 10^5 organisms. Beta lactamase enzyme identification was performed by isoelectric focusing developed with a chromogenic cephalosporin (Matthew and Hedges, 1976).

RESULTS

Figure 1 consists of computer-generated plots of the diameters of the zones of inhibition around the gentamicin and tobramycin disks of a large number of consecutive clinical isolates of 4 species of gram negative bacilli tested routinely over a period of 8 months in one hospital laboratory. Disks for each isolate were tested on the same plate. For isolates of *Escherichia coli* the diameters of gentamicin zones matched very closely those of tobramycin. This is also true for those isolates of *Klebsiella pneumoniae* which have larger zone diameters, but for those with smaller diameters tobramycin zones can be seen to be appreciably smaller than those of gentamicin. A sample of this subgroup was found to all have the 2″ adenylating aminoglycoside-inactivating enzyme.

Virtually all of the isolates of *Pseudomonas aeruginosa* have smaller zones for gentamicin than for tobramycin, and for the small number with small (resistant) zones the difference is greater. The reverse is true for isolates of *Serratia marcescens*. Since virtually all of the other variables affecting reproducability of zone size such as inoculum, media composition etc are controlled for when both disks are tested on one plate, consistent differences as seen here can only represent basic differences in the susceptibility of the isolates to each of the agents.

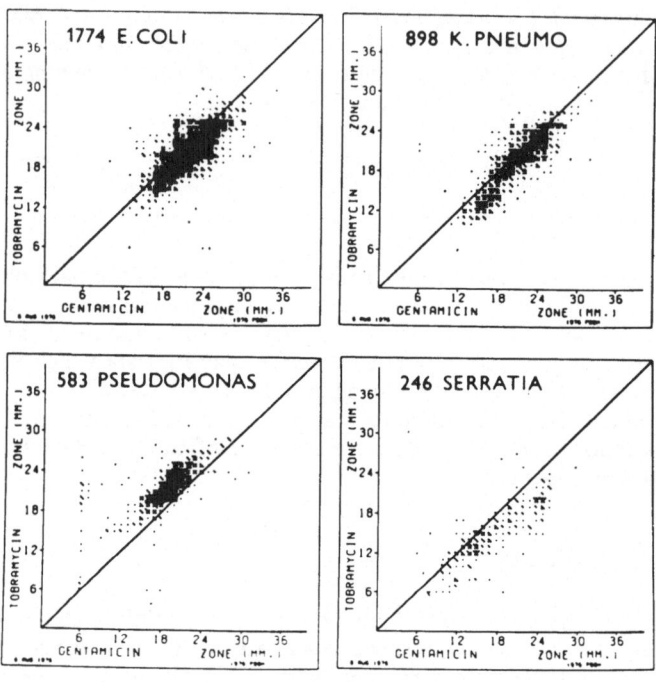

Fig. 1.

Figure 2 exemplifies the same principle for a different antibiotic family using tube dilution minimal inhibitory concentration dilutions (MIC) rather than the disk diffusion method. MIC values to two members of the tetracycline family of antibiotics, tetracycline and minocycline were determined for each of a series of clinical isolates of enterobacteriaceae. One large grouping of isolates have MIC values less than 32 μg/ml to each of the agents and are unlikely to have resistance mechanisms. A second grouping consisting of *Proteus mirabilis* and *Serratia marcescens* isolates have MICs of tetracycline of 62 or 125 μg/ml and of minocycline between 32 and 500. Their resistance was not transferred to *E. coli* K-12 in mating experiments.

Another grouping consisting of *E. coli* isolates had MICs of tetracycline mostly of 250 or 500 μg/ml, but with MICs of minocycline ranging from 16 to 125 μg/ml. Six of these *E. coli*, as well as the *Klebsiella pneumoniae* and the *Serratia marcescens* isolates which had MICs of 500 for tetracycline and of 125 μg/ml for minocycline, were tested for transfer of tetracycline resistance to *E. coli* K-12, and all transferred it. Precise mechanisms of tetracycline resistance are not yet fully elucidated (Davis and Smith, 1978) but comparison

GRAM NEGATIVE BACILLI

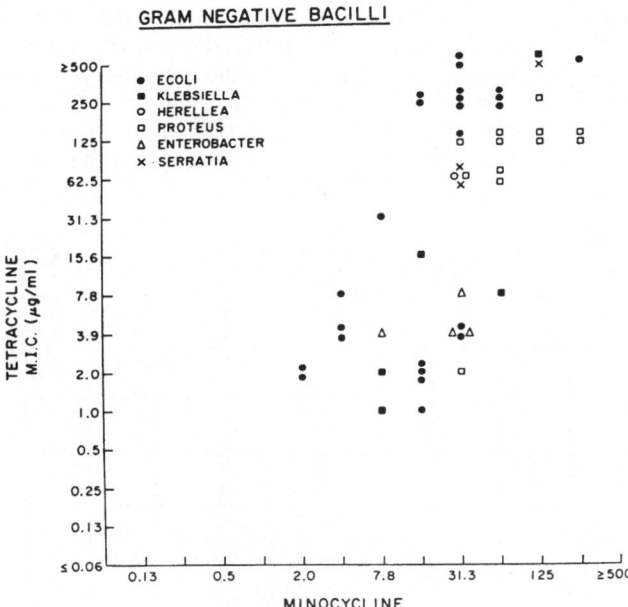

Fig. 2.

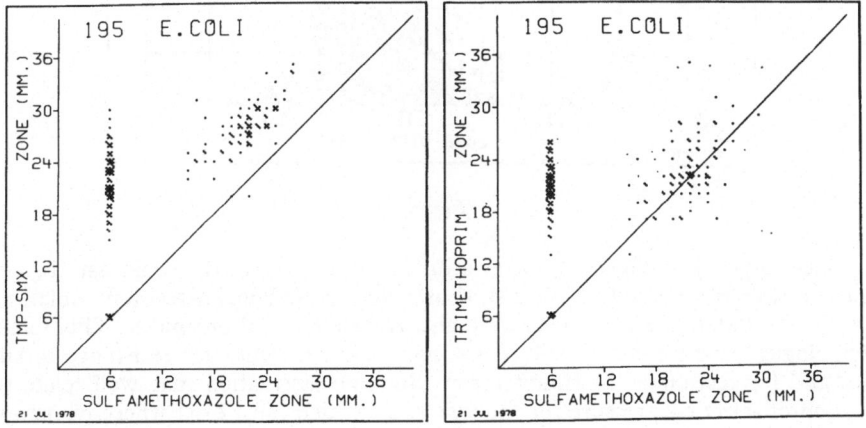

Fig. 3.

of resistance levels to two different tetracyclines was capable here of discriminating between a plasmid-mediated mechanism and one which appeared non-transferable.

Figure 3 shows a different type of information derived from similar two-dimensional plots of simultaneous disk diameter values for 195 clinical isolates of *E. coli* using the sulfamethoxazole disk and the combination trimethoprim-sulfamethoxazole disk in the first panel, and substituting the trimethoprim disk for the combination disk in the second panel. In the first panel it can be seen that those isolates with zone diameters greater than

16 mm for sulfamethoxazole have a median zone diameter value of about 28 mm for the trimethoprim-sulfamethoxazole disk. In contrast, the isolates which are resistant to sulfamethoxazole (zone diameter = 6) have a median zone diameter to the combination disk of approximately 20 mm. This presumably reflects, again revealed in routine clinical data, that the synergy between the sulfonamide and trimethoprim (Bushby and Hitchings, 1968) which would account for the increment, is lost or diminished in isolates which are sulfonamide resistant.

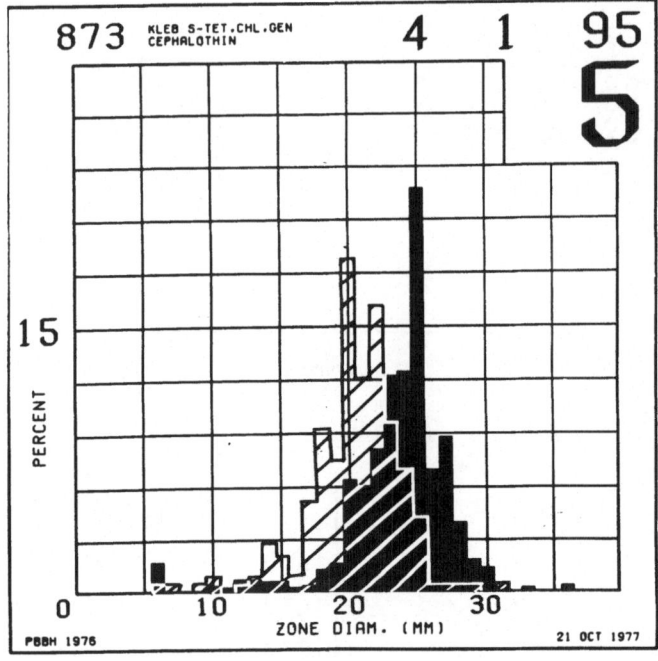

Fig. 4.

The second panel of Figure 3 makes a different point from the same data base. The separate testing of a trimethoprim disk shows that trimethoprim-resistant isolates were relatively rare but that all of them were also resistant to sulfonamides. This makes an epidemiological point about the close association of the genes for resistance to sulfonamides and for resistance to trimethoprim. Such an association may well relate to the association of selection pressure for the two genes which must exist where trimethoprim is used therapeutically only in combination with a sulfonamide. However a trimethoprim resistance gene is known to be carried on a transposon (Richards, Sojka, Datta and Wray, 1978) and so pressumably could distribute separately. The type of analysis used here could serve as a mechanism of surveillance using routine laboratory data to detect early dissemination of a trimethoprim resistance gene separately from sulfonamide resistance, as might be anticipated in areas where trimethoprim is used alone as a therapeutic agent.

Figure 4 is another computer-generated plot of the distribution of zone diameter values around the 30 μg cephalothin disk. The black bars represent the zone diameter values for 873 consecutive clinical isolates of *Klebsiella pneumoniae* specified as susceptible to tetracycline, chloramphenicol and gentamicin. The reason for elimination of isolates with

386

resistance to any of these agents is that previous studies had shown that it would eliminate the majority of isolates from which antibiotic resistance transfer could be demonstrated. Since the TEM β-lactamase is usually plasmid mediated the elimination was seen as a way of getting a comparison group relatively, but not absolutely, free of TEM β-lactamase containing strains.

The shaded bars represent 259 isolates of *Klebsiella pneumoniae* of a particular biotype (API 5205773) known from previous studies to contain a plasmid carrying the gene for the TEM-β-lactamase, and confirmed by isoelectric focusing electrophoresis of a sample of these isolates. The distribution of zone diameters of this group is shifted to the left with a peak approximately 5 mm smaller than those of the control group. We have previously shown that in *E. coli* clinical isolates the presence of several different β-lactamases can be predicted from routine laboratory zone diameter values to ampicillin, cephalothin and carbenicillin (Medeiros, Kent and O'Brien, 1974). In *Klebsiella* isolates the analysis is complicated by the existence of species associated resistance to ampicillin and carbenicillin, but Figure 4 suggests that cephalothin zone diameter values could be statistically discriminating.

DISCUSSION

The data presented here illustrate several principles. One is that the ratios of the diameters of zones of inhibition around disks impregnated with related antibiotics placed on the same monoinoculated plates are more reproducable from plate-to-plate and time-to-time than are the diameters, presumably because inoculum and media differences between plates affect the zones of related disks similarly. A second is that different mechanisms producing resistance to a family of related antibacterial agents will tend to produce different patterns of levels of resistance to the various members of that family. Given the reproducability of these levels when tested on the same plate, as mentioned above, it should be possible to derive presumptive identification of the resistance mechanisms from analysis of the relative levels of resistance as reflected in zone diameter values or, alternatively, in MIC values.

The examples above illustrate these principles using, at most, two agents and two-dimensional plots. However, the growing number of related agents being used clinically is resulting in a correspondingly increased number being used in routine susceptibility testing. Thus in many laboratories 3 or 4 β-lactam and aminoglycoside antibiotics are now being tested routinely and others may be added in the future. This should increase the discriminatory power of the procedure but it would no longer be analyzable in two-dimensional plots. However, more complex analytical methods are available for handling this type of data such as Stepwise Discriminant Analysis (Jennrich and Sampson, 1977) and their use will be explored.

The value of this type of analysis is that it may permit surveillance of antibiotic resistance at the ultimate level of understanding — the epidemiology of the resistance genes. We know now that most clinically important antibiotic resistance is due to a relatively small number of very specific resistance mechanisms, many of them specific antibiotic-inactivating enzymes coded for by specific genes. The resistance genes are mobile between plasmids on transposons, between bacterial strains on plasmids, and between people on bacterial strains. If it proves possible to discriminate mechanisms and, by inference, genes in the enormous volume of clinical laboratory data that is generated daily in all parts of the world it should be possible to develop an overall view of the actual working of this complex and interrelated global system (O'Brien, Norton, Kent and Medeiros, 1977).

REFERENCES

O'BRIEN, T. F., ACAR, J. F., MEDEIROS, A. A., NORTON, R. A., GOLDSTEIN, F. and KENT, R. L. (1978): International comparison of prevalence of resistance to antibiotics. J. Amer. Med. Assoc. 239, 1518–1523.

MATTHEW, M. and HEDGES, R. W. (1976): Analytical isoelectric focusing of R factor-determined β-lactamases; correlation with plasmid compatibility. J. Bacteriol. 125, 713.

DAVIES, J. and SMITH, D. I. (1978): Plasmid-determined resistance to antimicrobial agents. Am. Rev. Microbiol. 32, 469–518.

BUSHBY, S. R. M. and HITCHINGS, G. H. (1968): Trimethoprim, a sulfonamide potentiator. Brit. J. Pharmacol. Chemother. 33, 72.

RICHARDS, H., SOJKA, W. J., DATTA, N. and WRAY, C. (1978): Trimethoprim-resistance plasmids and transposons in *Salmonella*. Lancet ii, 1194–1195.

MEDEIROS, A. A., KENT, R. L. and O'BRIEN, T. F. (1974): Characterization and prevalence of the different mechanisms of resistance to betalactam antibiotics in clinical isolates of *Escherichia coli*. Antimicrob. Ag. and Chemother. 6, 791–801.

JENNRICH, R. and SAMPSON, P. (1977): Stepwise Discriminant Analysis. in BMDP Biomedical Computer Programs P-series. ed. by. W. J. Dixon and M. B. Brown. Univ. Calif. Press. 711–734.

O'BRIEN, T. F., NORTON, R. A., KENT, R. L. and MEDEIROS, A. A. (1977): International surveillance of prevalence of antibiotic resistance. J. Antimicrob. Chemother, 59–66.

T. F. O'B., Peter Bent Brigham Hospital,
Boston, Massachusetts 02115, U.S.A.

CLINICAL USE OF ANTIBIOTICS —
A NATIONWIDE COMPUTER STUDY

J. GRUNT, V. KRČMÉRY, L. ROSIVAL
Research Institute of Preventive Medicine, Bratislava, Czechoslovakia

INTRODUCTION

Rational use of antibiotics is dependent on the agreement of bacteriological and clinical diagnoses and the adequacy of the clinical indication and use of antibiotics.

In 1975 we published results from monitoring of antibiotic resistance in more than 180 000 bacterial strains belonging to 8 species of selected important (problem) bacteria which are interesting as well as significant from the view point of acquiring multiple drug resistance. Since then, this computer-assisted surveillance of antibiotic resistance in Slovakia has been performed annually (Grunt et al., 1975, Grunt et al., 1978).

During processing of the obtained data, it became obvious that studies on antibiotic resistance should be complemented with those concerning the true clinical effectiveness of selected reserve antibiotics. These are represented by newer antibiotics, that are mostly endangered from the viewpoint of resistance, but whose effectiveness should be preserved as much as possible.

In this work the results of our second project are given: The clinical use and effectiveness of reserve antibiotics. We included the following topics:

– prophylactic and therapeutic use of reserve antibiotics in individual hospital wards,
– their subsequent use or use in combination with reserve antibiotics as well as with other antibacterials,
– comparison of the clinical effectiveness of reserve antibiotics with bacteriological results,
– rationality of the use of reserve antibiotics according to the adequacy of obtaining bacteriological results and the proper choice of the drug(s),
– adequacy of dosage and duration of application of reserve antibiotics.

METHODS

In a project for computer processing of data on the clinical use of antibiotics, we designed a record sheet where data were filled in by the ward doctor in each case when reserve antibiotic(s) were administered during hospitalization. The record is divided into four parts:

1. Basic data on the patient, diagnosis and ward,
2. Data from bacteriological examination, on which the application of reserve antibiotics was based,
3. Data on antibiotics administered, i.e. reserve as well as all other antibiotics (termed non-recorded drugs),
4. Explanation of the coding procedure, abbreviation of antibiotics, bacteria, pathological material, etc.

Detailed instructions with a key code have been distributed to all doctors in collaborating hospitals.

A pilot study has been performed, in which over a one month period, the project was tested and finally formed.

Ten hospitals collaborated in this study for 18 months. In this period of time, 8,411 applications of 8 reserve antibiotics were recorded and submitted to a central review group of specialists who checked the formal coherence of the records. Before submitting, the record was reviewed, and discussed if necessary, by a local chemotherapeutical supervisor appointed by the director of each cooperating hospital.

Following antibiotics have been monitored: linomycin (LIN), oxacillin (OXA), kanamycin (KAN), gentamicin (GEN), colistin (COL), carbenicillin (CAR), cephalosporins (CEF) and cotrimoxazole (COT). They are predominantly used in hospitals and not in general practice in this country.

RESULTS

Of a total of 8,411 patients who received reserve antibiotics in the hospitals monitored, in 943 or 11.2%, the reserve antibiotic was administered prophylactically. In this group, 120 patients (12.7%) received two reserve antibiotics prophylactically. In 9 patients, three such antibiotics were administered together or subsequently.

Concerning the mode of application, patients also received other antibiotics (non-recorded ones) together with reserve drugs. Of patients receiving one reserve antibiotic, 30% received additional non-recorded drugs. In patients receiving two reserved drugs, 25% also received non-recorded antibiotic(s).

Concerning the drugs administered, Table I. presents the number of applications of individual reserve antibiotics and the mode of their administration, i.e. singly or in combinations with other reserve drugs, In combination, gentamicin was administered most frequently. When three antibiotics were administered simultaneously, gentamicin was always included (9 cases). 7 of 8 reserve antibiotics were present in these three-drug combinations.

Prophylactically least frequently were administered lincomycin (LIN) (3%) and most freuqently oxacillin (OXA) (24%), kanamycin (KAN) (23%) and gentamicin (GEN)

TABLE I.

The mode of prophylactical administration of reserve antibiotics

Mode of applications	Number (in %) of applications of									
	OXA	KAN	GEN	CAR	SEP	COL	CEF	LIN	Total	
									ABS	%
Alone	27.0	24.0	18.4	10.4	7.6	4.7	4.8	3.1	815	75.5
With one additional reserve drug	17.6	22.3	22.7	19.7	6.3	8.4	1.7	1.3	238	22.0
With two additional reserve drugs	11.1	14.9	33.3	11.1	11.1	11.1	7.4	—	27	2.5
Total	24.5	23.4	19.8	12.5	7.4	5.6	4.2	2.6	1080	100

TABLE II.

Reserve antibiotics which were used prophylactically on individual wards

Reserve drug	Number (in %) of application of reserve antibiotics								
	Pediatric	Newborn	Prema-ture born	Gynaecology +obstetrics	Surgery +orthop.	Urology	Others	Total	
								ABS	%
OXA	19.2	26.8	25.4	3.2	43.3	1.0	29.5	265	24.5
KAN	19.2	36.3	29.8	9.7	2.4	4.8	9.1	253	23.4
GEN	23.1	15.2	13.6	22.6	23.6	41.8	22.7	213	19.8
CAR	9.6	18.1	19.3	—	0.8	1.0	9.1	135	12.5
SEP	3.9	0.3	—	16.1	10.2	46.6	6.8	80	7.4
COL	15.4	1.9	10.1	9.1	5.5	4.8	5.7	61	5.6
CFR	9.6	0.7	1.8	40.3	2.4	—	5.7	45	4.2
LIN	—	0.7	—	—	11.8	—	11.4	28	2.6
Total	4.8	39.0	21.1	5.7	11.8	9.5	8.1	1080	100

(20%). Differentiation of reserve antibiotics administered prophylactically according to individual hospital ward is given in Table II.

Of non-recorded antibacterials, penicillin (PEN) was administered most frequently for prophylaxis (48%), followed by ampicillin (19%), streptomycin (STR) (13%) and chloramphenicol (CMP) (10%).

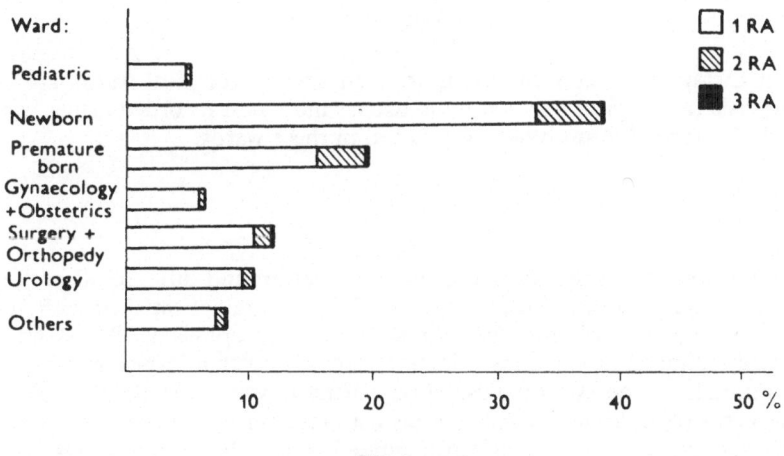

Fig. 1.

Concerning individual wards, reserved antibiotics were administered prophylactically most frequently in newborn wards and in prematurely-born children's wards (Fig. 1). These two wards, together with other pediatric wards, used 65% of all prophylactically administered reserve antibiotics. In Table II. the participation of individual drugs in these as well as in other wards is listed. It can be seen that, in pediatric wards, gentamicin was used most while, in newborn and premature-born units, the most frequently used prophylactical antibiotic was kanamycin. The high frequency of prophylactic usage of gentamicin was also found in urogical wards, where contrimoxazole (COT) was also rather popular for this purpose.

Therapeutical use of reserve antibiotics was studied and analyzed from numerous standpoints. Only the following introductory data are given here.

Fig. 2 shows that pediatric wards use reserve antibiotics by far most frequently (44% of all therapeutical applications). Together with newborn and prematurely-born units,

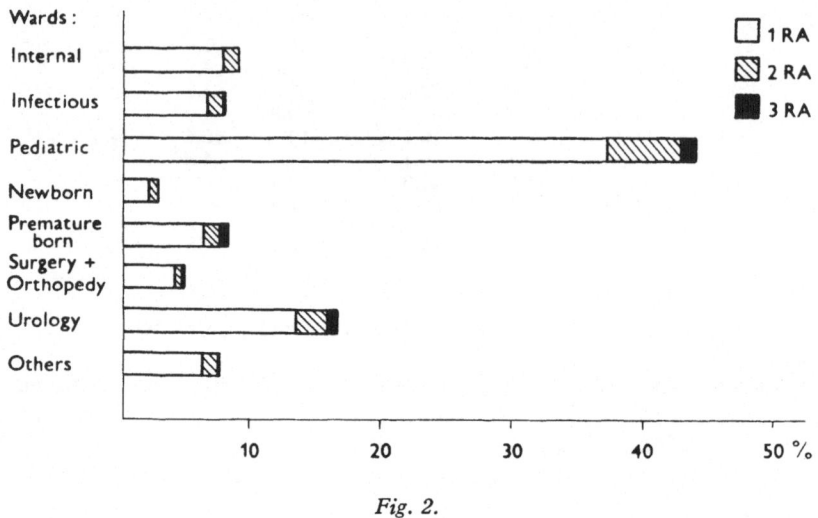

Fig. 2.

55% of all therapeutical applications have been used. Urological wards are in second place. Surprisingly, surgical wards seem to use far more non-recorded drugs, as we found that only 4.6% of reserve antibiotics were used in these wards.

DISCUSSION

Although reports from the literature indicate that around 30% of all antibiotics are administered prophylactically (Study Group, 1977), we are not satisfied with our finding that 11.2% of reserve antibiotics are administered for prophylaxis. We assume that, in a decision for antibiotic prophylaxis, clinicians should and could consider the use of more common antibiotics. Too extensive use of new drugs in wards was reported to contribute to the emergence of multiple-resistant strains with resistance to newer, reserve antibiotics. Moreover, our finding as to which wards tended to use the reserve drugs is somewhat disturbing. While surgery is blamed in the literature as being the biggest consumer of antibiotics for prophylaxis (Study group, 1977) we found that the pediatric sector is more to blame. Thus babies are stigmatized from an early age with drugs which should be considered as antibiotic reserve. In a previous paper (Grunt, Krčméry and Rosival, 1978) we showed mutual selection of resistant strains by some pairs of reserve antibiotics. The use of gentamicin and/or colistin, as well as oxacillin and/or lincomýcin, for example, selected bi- and multiple-drug resistance to a remarkable extent. This development might further contribute to the deterioration in the position of important antibacterials. Thus, the use of reserve drugs should be rational and supervised as much as possible.

In conclusions, we assume that the overall indications for the use of reserve as well as non-restricted antibacterials should be restricted. Reserve antibiotics should be protected much more from broader and/or indiscriminate use, and, if possible, used as prophylactic agents only exceptionally.

REFERENCES

GRUNT, J., KRČMÉRY, V., VÝMOLA, F.: Nationwide survery of antibiotic resistance by means of a computer. Zbl. Bakteriol. I. Abt. Orig. A, 224, 1973, 380–390.

KRČMÉRY, V., GRUNT, J., ROSIVAL, L.: Nationwide survey of antibiotic resistance by means of a computer: analysis of 200 000 strains of problem bacteria isolated in 1973. Zbl. Bakt. I. Orig. A, 231, 1975, 250–258.

GRUNT, J., KRČMÉRY, V., ROSIVAL, L.: Monitoring bacterial resistance to restricted antibiotics. Amer. J. Hosp. Pharm. 35, 1978, 1387–9.

STUDY GROUP: Prophylatic antimicrobial drug therapy at five London teaching hospitals. Lancet 1, 1977, 1351–3.

J. G., Research Institute of Preventive Medicine, Limbová 14, Bratislava-Kramáře, Czechoslovakia

PROBLEMS CONCERNING MULTICENTER STUDIES ON DRUG RESISTANCE

B. WIEDEMANN

Institute of Microbiology and Immunology, University of Bonn, F.R.G.

INTRODUCTION

Drug resistance is spreading all over the world like pandemic infections. In order to cope with this problem epidemiological studies on drug resistance are necessary. Unlike infectious diseases the underlying infective principle is more complex and therefore an epidemiological study is much more difficult. One has to study genes for drug resistance in bacteria, which are situated in chromosomes, plasmids, transposons, and perhaps in genomes of bacteriophages. One phenotypic character can be expressed by different mechanisms. For example resistance to penicillins by production of various β-lactamases, by change of the binding proteins or by a penetration barrier.

Multicenter studies like that of the Paul-Ehrlich-Gesellschaft (1976) can help to provide scientists with more detailed information on the spread of drug resistance characters. A real epidemiology however cannot be provided by these tools. The problems that can easily be handled by this sort of study are the following:

1. Alteration in the prevalence of bacterial species resistant to the different antimicrobials.
2. Development of resistance to newly developed drugs.
3. Development of resistance to drugs used only in husbandary, to follow a possible flow of drug resistance or resistant bacteria from animals to man.
4. Differences in the drug resistance in geographical distinct aeras.
5. Detection of hospital epidemics with resistant bacteria.
6. Planning of chemotherapy without sensitivity test.
7. More sophisticated evaluation of the data can even lead to information about the spread of specific resistance mechanisms.

I. Methodological prerequisites for multicenter resistance studies

If experimental data shall be realiable the methods for all experiments have to be exactly identical in all laboratories. Standardisation of sensitivity tests is concerned with many problems as described elsewhere (Ericson and Sherris, 1971, Paul-Ehrlich-Gesellschaft, 1976). The media, the inoculum of bacteria, and the drug content of the discs are the most important experimental elements which can influence the result. In the study of the Paul-Ehrlich-Gesellschaft we use the same batch of Müller Hinton Agar over a period of several years in order to avoid differences in the media. Each laboratory follows a detailed prescription of methods.

Furthermore, a set of control strains has to be checked every day. By these measures the laboratory staff can look for mistakes if the zone sizes do not match with the given standards and on the other hand data from those laboratories which have controls out of certain limits can be eliminated. The control of the disc content during the Paul-Ehrlich-study has shown that there is a variety to more than 100% in the drug content on the discs in some instances. We even found cartridges correctly labelled without any drug. (Wiedemann and Klaus, paper in preparation).

But not only the sensitivity test has to be standardized but also the differeetiation of bacterial strains. A wrong classification and especially a different classification in the laboratories can lead to a totally misleading amount of resistant bacteria in one species,

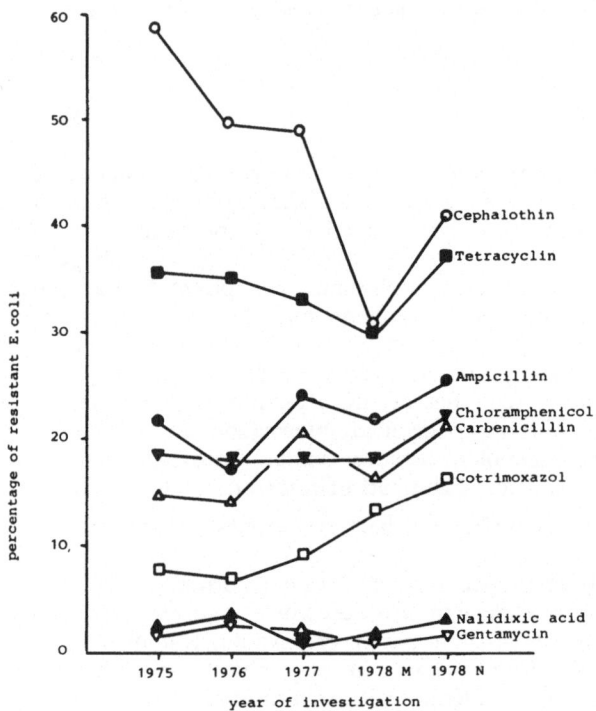

Fig. 1. Development of drug resistance in *E.coli* from 1974 to 1978. The deviation of the percentage of cephalothin resistant strains is probably due to the technique of determination (see text).

due to different naturally occurring resistance markers in some genera or species. In our study all laboratories use the same kit for the identification. In the case of gram negative rods we use Enterotube and Fermotube (Roche). In this respect it is more important to use a system which gives reproducible results in each laboratory than to have a great accuracy.

In a computer program all the available data especially zone sizes, biochemical reactions, source of strains must be stored, so that in a retrospective evaluation with as many data as possible can be used.

II. Results of a multicenter study

Figure 1 represents data from the study of the Paul-Ehrlich-Gesellschaft showing the resistance of *E. coli* to selected drugs. The graph demonstrates, that over a wide area (F. R. Germany, Austria and Zwitzerland) there is a steady state in the percentage of resistant strains, with only few exceptions. Comparing these data with those of a single laboratory one finds drastic changes in one hospital, due to the special ecological situation in this area. The number of *E. coli* strains resistant to contrimoxazole, however, seems to rise steadily from about eight percent in 1975 to sixteen percent in 1978. The reason could be the distribution of plasmids with trimethoprim resistance (Richards, in this volume)

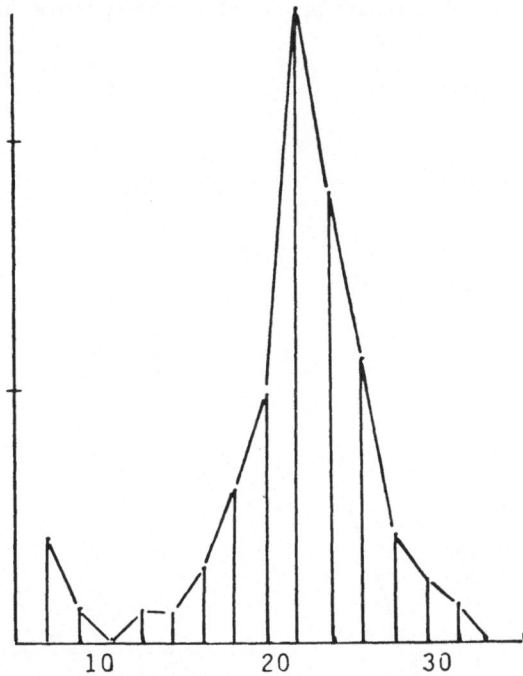

Fig.2. A plot of zone diameter against the number of strains gives a clear cut Gauß'ien distribution of the natural sensitive population with a peak at 22 mm.

Table I.

Comparison of the recommended breakpoints with those zone sizes which give the end of sensitive population of bacteria (*E. coli*)

Drug	DIN	NCCLS	End of sensitive population (see text)
Ampicillin	13	11	11
Carbenicillin	13	17	17
Cephalothin	21	14	11
Tetracyclin	16	14	16
Chloramphenicol	20	12	16
Gentamicin	14	12	19
Kanamycin	18	13	17
Sulfamethoxazol + Trimethoprim	10	10	14
Nalidixic acid	13	13	18
Nitrofurantoin	10	14	10
Streptomycin	10	—	8

(Saroglon, Paraskevopoulou and Kontomichalou, in this volume). The other exception is the level of resistance to cephalothin, ranging from 32 to 60%. There is evidence, that this variation is no reflexion of real shift in the resistance of the bacterial population, but a matter of the method used. The German DIN commission recommended a breakpoint of 21 mm considering the pharmacokinetic of the drug in man. This breakpoint however divides the wild type sensitive population od *E. coli* just into two halves (Fig. 2). A slight change in the method therefore results in a big shift of the population to either side of the

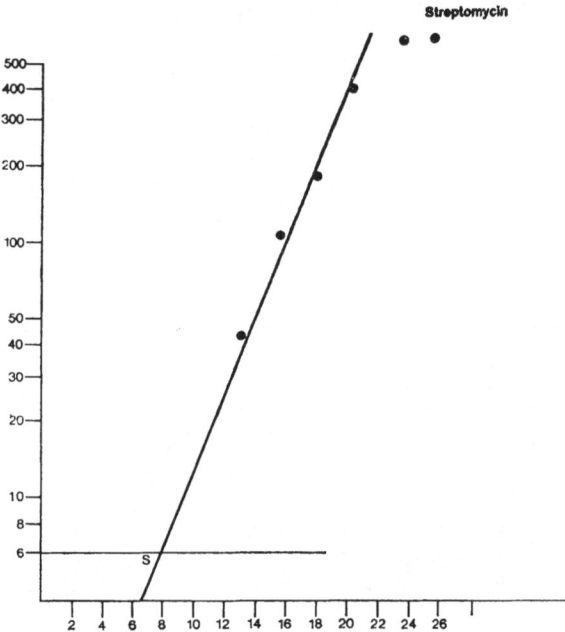

Fig. 3. The log of the cumulative number of strains is plotted against the zone diameter of the sensitive population. The resulting regression line gives the theoretical end of the Gauß'ien distribution. By elimination of 1% of the population we find a point of insertion which gives a practical breakpoint between the sensitive and the resistant portion of the *E. coli* population.

breakpoint. We believe that a breakpoint should never divide a bacterial population but should mark the end of the sensitive population if this is not contradictory to the clinical experience and to the pharmacokinetic of the drug. At least the population distribution should be considered (Alestik et al., 1978).

To find the end of the natural Gauß'ien distribution of the sensitive wild type population we plot the frequency of strains from the sensitive population against the zone diameter (Fig. 3) and use the point of intersection of the regression line with the horizontal line that eliminates one percent of the population as a "natural" breakpoint. In Table I. these breakpoints for *E. coli* are compared with those recommended by the NCCLS- and the DIN commission. These data have to be completed by other species, grampositive and gramnegative, but they probably give a good guide line for finding a reasonable breakpoint that can be used for most species.

In epidemiological terms the prevalence of special resistance patterns is much more interesting than the percentage of single resistance markers. From these data the com-

Table II.

The resistance pattern of *E.coli* strains give a good information on the frequency of multiresistant strains and the occurence of combinations of resistance genes. 38% of all resistant strains e.g. show a combined Sm Su resistance.

No.	Resistance pattern											Frequency Total	%
1	wild type											863	37.95
2		Su										204	8.97
3						Te						176	7.74
4		Su				Te						92	4.04
5	—▷	Su		Sm		Te						59	2.59
6	—▷	Su		Sm								52	2.28
7		Su	TS									42	1.85
8	—▷	Su		Sm	Ka	Te	Cm	Am	Ca			41	1.80
9		Su				Te	Cm	Am	Ca			34	1.50
10		Su				Te	Cm					28	1.23
11	—▷	Su		Sm		Te		Am	Ca			24	1.05
12	—▷	Su	TS	Sm		Te						24	1.05
13	—▷	Su	TS	Sm	Ka	Te	Cm	Am	Ca			19	0.84
14	—▷	Su			Ka	Te	Cm	Am	Ca			19	0.84
15					Ka							18	0.79
16										Ce		17	0.74
17		Su				Te		Am	Ca			15	0.66
18								Am	Ca			14	0.62
19								Am		Ce		13	0.57
20							Cm					13	0.57
21	—▷	Su		Sm		Te	Cm	Am	Ca			12	0.53
22		Su	TS			Te						12	0.53
23		Su					Cm					11	0.48
24				Sm		Te						11	0.48
25	—▷	Su		Sm		Te	Cm					11	0.48
26		Su						Am	Ca			10	0.44
27				Sm								10	0.44
28	—▷	Su		Sm	Ka	Te	Cm					10	0.44
29						Te		Am	Ca			9	0.40
30	—▷	Su		Sm	Ka		Cm	Am	Ca			9	0.40
31	—▷	Su	TS	Sm	Ka		Cm	Am	Ca			9	0.40
32	—▷	Su	TS	Sm		Te	Cm	Am	Ca			9	0.40
33	—▷	Su	TS	Sm	Ka	Te	Cm	Am	Ca	Ce		9	0.40
34	—▷	Su	TS	Sm								8	0.35
35					Ka	Te						7	0.31
36		Su			Ka	Te						7	0.31
37	—▷	Su		Sm				Am	Ca			7	0.31
38		Su									Ni	6	0.26
39		Su			Ka							6	0.26
40		Su								Ce		6	0.26
41								Am				6	0.26
42		Su					Cm	Am	Ca			6	0.26
43	—▷	Su		Sm	Ka	Te	Cm	Am	Ca	Ce		6	0.26
44		Su	TS			Te	Cm	Am	Ca			6	0.26
45									Ca			5	0.22
46						Te					Ni	5	0.22
47						Te	Cm					5	0.22
48	—▷	Su		Sm		Te					Ni	5	0.22
49	—▷	Su		Sm	Ka	Te						5	0.22
50	—▷	Su		Sm			Cm	Am	Ca			5	0.22
51	—▷	Su		Sm	Ka	Te	Cm	Am		Ce	Ni	5	0.22
52		Su	TS			Te		Am	Ca			5	0.22
53	—▷	Su	TS	Sm	Ka	Te		Am	Ca			5	0.22

bination of markers is obvious, for example the SmSu resistance which is present in 38% of all resistant *E. coli* strains. Table II. gives the data from Mannheim, Rehm and Hamza 1979, which were taken from the multicenter study of the Paul-Ehrlich-Gesellschaft.

More detailed information can be drawn if one plots the zone diameter of 2 β-lactam antibiotics on one graph, as shown by O'Brien in this volume. This sort of data however

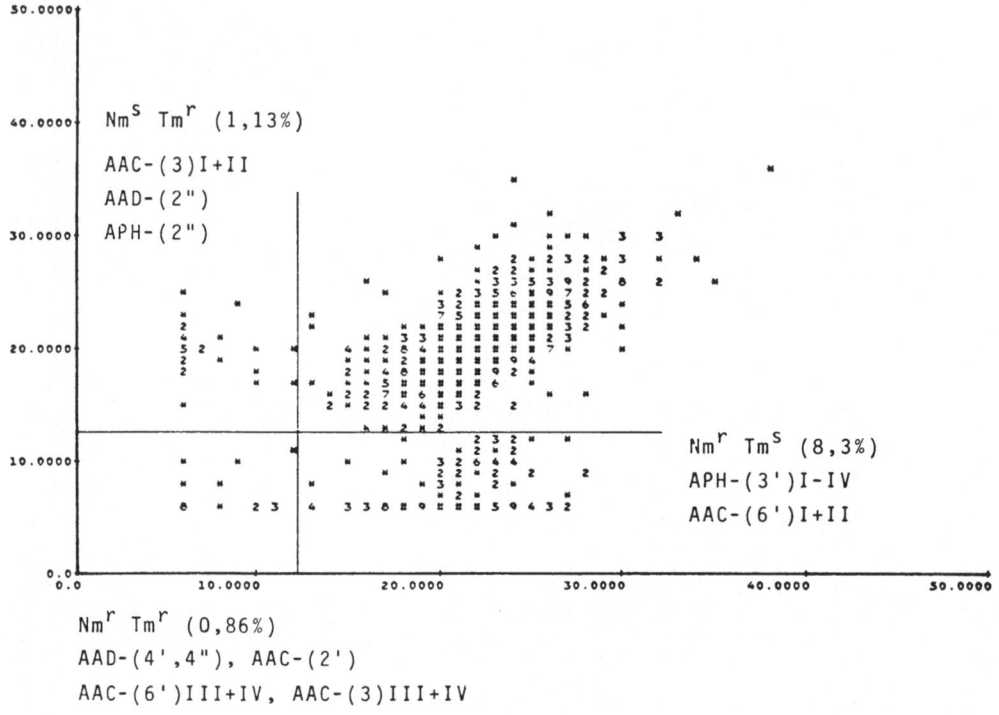

Fig. 4. Computer plot of zone sizes to tobramycin against zone sizes to neomycin. The major group is sensitive to both drugs. * represents one strain, a number gives the number of strains with the respective data and stands for more than 9 strains. Bacteria resistant to both drugs, lower left field (0,86% of total strains tested), produce either AC–(2') or AAC–(6')II, AAC–(6')III or AAC–(6')IV. Strains sensitive to tobramycin and resistant to neomycin, lower right field (8,31%) produce one of the phosphotransferases APH–(3')I–IV. Strains resistant to tobramycin and sensitive to neomycin, upper left field (1,13%) produce acetyltransferase AAC–(3) or adenyl-transferase AAD–(4', 4'').

can only be deduced from strains of a small area as here special R-factors with a β-lacta-mase with a characteristic substrate profile spread, in a way, that all strains bearing this R-factor give a characteristic cloud in this plot. If one uses strains from many different sources, the diversity of strains does not allow a clear cut conclusion.

We tried to do a similar evaluation of our data using the zone sizes of aminoglycoside antibiotics. In Fig. 4 computer graph is shown. It gives the date of 2.500 *E. coli* strains showing that there are some distinct classes of bacteria which probably produce the aminoglycoside inactivating enzymes indicated in the graph. From the plots with combinations of other aminoglycoside antibiotics the picture is not as easy to interpret as that one shown. It seems easier to draw conclusions concerning the epidemiology of

400

Table III.

The prevalence of various aminoglycoside inactivating enzymes in gram negative bacteria shows at least for the AAD–(2″) a species specificity for *E. coli*, Klebsiella and Enterobacter

Enzyme	Prevalence of enzymes in			*Pseudomonas*	*Proteus*	Total
	E. coli	*Klebsiella*	*Enterobacter*			
APH–(3′) I		3	2		5	10
APH–(3′) II				6		6
APH–(3′) I+II/III				3		3
APH–(3″)	1	2		6		9
AAC–(6′) I		2				2
AAC–(6′) IV				3		3
AAC–(3) I					2	2
AAC–(3) II		1			1	2
AAD–(3″)	7	2	3	2		14
AAD–(2″)	1	2	2			5
STR ——740++						

Table IV.

Frequency of Aminoglycoside inactivating enzymes in correlation to the resistance pattern of gram negative bacteria.

Frequency in %	Resistance pattern								Enzyme
37	Sm	Sp							AAD–(3″)
16	Sm								APH–(3″)
32			Km				Nm	Pm	APH–(3′) I–IV
7		Gm	Km	SiS	Tm	Dm			AAD–(2″)
3		Gm	Km	SiS	Tm	Dm	Ami	Nm	AAC–(6′) IV
~2		Gm		SiS					AAC–(3) I
~2		Gm	Km	SiS	Tm				AAC–(3) II
0,8			Km						AAC–(6′) I

aminoglycoside inactivating enzymes from the resistance pattern. Therefore we tested 130 gramnegative unselected strains from clinical sources and tried to correlate the resistance pattern with the occurrence of the inactivating enzymes. Table III. shows the prevalence of the enzymes in those 43 resistant strains. From the experience with these strains in correlating the enzyme present in the strains with the resistance pattern, we deduced the data shown in Table IV. For this correlation it is necessary to test the sensitivity to at least Kanamycin, Neomycin, Gentamycin, Sisomycin, Tobramycin, Dibekacin, Amikacin, Ribostomycin, Lividomycin, Butirosin, Streptomycin and Spectinomycin. However, errors in this calculation cannot be avoided, especially in multiple resistant strains. This example should demonstrate that a sophisticated evaluation of the data of a multicenter study can give much more information on the epidemiology of resistance than just the percentage of resistant strains.

CONCLUSIONS

The data from a multicenter study give a variety of a valuable information to the medical science. But one has to consider the prerequisites necessary in respect to the methods employed. A critical evaluation is necessary in order to avoid mistakes of interpretation.

REFERENCES

ALESTIK, K., K. DORNBUSCH, C. ERICSON, L. O. KALLINGS, C. KAMME, F. NORDBRING, R. NORRBY, and G. WALLMARK (1978): Resistensbestämming av bakterier: Ny indeling i kanslighetsgrupper S. I och R. Läkartidningen 75 – 4346–4348.
DEUTSCHES INSTITUT FÜR NORMUNG e. V. (1976): Methoden zur Empfindlichkeitsprüfung von bakteriellen Krankheitserregern (ausser Mycobakterien) gegen Chemotherapeutika. Entwurf November 1976, DIN 58940, Teil 3. Beuth Verlag Berlin.
ERICSON, H. and J. C. SHERRIS (1971): Antibiotic sensitivity testing. Report of an international collaborative study. Acta Path. Scand. Sect B Suppl. 217, 3–90.
MANNHEIM, W. and W. F. REHM (1979): Wildtypische und erworbene Resistenzmuster klinischer Isolate von Escherichia coli. Infection in press.
NATIONAL COMMITTEE FOR CLINICAL LABORATORY STANDARDS (1975): Performance standards for antimicrobial disc susceptibility tests. Villanova, U.S.A.
O'BRIEN, T. (1979): Global deployment of antibiotic resistance genes. This volume.
RICHARDS, H. (1979): The spread of transposon 7 among gramnegative bacteria. This volume.
SAROGLOU, G., P. PARASKEVOPOULOU, and P. KONTOMICHALOU (1979): Tp resistant plasmids from Enterobacteriaceae isolated in Greece. This volume.

B. W., Inst. für Medizinische Mikrobiologie
und Immunologie der Universität,
Endenisch, An der Immenburg 4 AVZ II
5300 Bonn 1, B.R.D.

SUBJECT INDEX

403

AUTHOR INDEX *

*) **Numbers in bold types refer to the first page of the authors communication.**

410